Intermediate Algebra

Michael N. Payne

College of Alameda

WEST PUBLISHING COMPANY
St. Paul   New York   Los Angeles   San Francisco

To my parents,
Mary and Otis Payne

Production: Greg Hubit Bookworks
Text design: Marvin R. Warshaw
Technical illustrations: John Foster
Typesetting: Interactive Composition Corporation
Cover photo: The Image Bank
Cover design: The Quarasan Group, Inc.

Library of Congress Cataloging in Publication Data

Payne, Michael (Michael Noel)
  Intermediate algebra.

  Includes index.
  1. Algebra.   I. Title.
QA154.2.P38   1985   512.9   84-22081
ISBN 0-314-85285-9
1st Reprint—1985

1992

LOGARITHM PROPERTIES

1. $\log_b MN = \log_b M + \log_b N$
2. $\log_b \dfrac{M}{N} = \log_b M - \log_b N$
3. $\log_b M^P = p \log_b M$

VARIATION

1. Direct variation: $y = kx$, $k \neq 0$
2. Inverse variation: $y = \dfrac{k}{x}$, $k \neq 0$
3. Joint variation: $w = kxy$, $k \neq 0$

COMPLEX NUMBERS

1. $\sqrt{-1} = i$ and $i^2 = -1$
2. Complex number: $a + bi$, where a, b are real numbers
3. $a + bi = c + di$ if and only if $a = c$ and $b = d$
4. $(a + bi) + (c + di) = (a + c) + (b + d)i$
5. $(a + bi) - (c + di) = (a - c) + (b - d)i$
6. $(a + bi)(c + di) = (ac - bd) + (ad + bc)i$

CONIC SECTIONS

1. Circle: $(x - h)^2 + (y - k)^2 = r^2$
2. Parabola: $y^2 = 4px$ or $x^2 = 4py$
3. Ellipse: $\dfrac{x^2}{a^2} + \dfrac{y^2}{b^2} = 1$
4. Hyperbola: $\dfrac{x^2}{a^2} - \dfrac{y^2}{b^2} = 1$

   or $\dfrac{y^2}{a^2} - \dfrac{x^2}{b^2} = 1$

FUNCTION NOTATION

1. $f(x)$ Value of f at x.
2. $f^{-1}(x)$ Value of inverse at x.

TYPES OF FUNCTIONS

1. Linear function f: $f(x) = ax + b$
2. Quadratic function f:

   $f(x) = ax^2 + bx + c$

3. Exponential fun

   $f(x) = b^x$, $b >$

4. Logarithmic fun
   where $y = \log_b$

GEOMETRIC FORMULAS

Let A = area, C = circumference, P = perimeter, S = surface area, V = volume, r = radius, h = altitude, l = length, w = width, b (or a) = length of base, and s = length of a side.

1. Circle

   $A = \pi r^2$; $C = 2\pi r$

2. Cone (right circular)

   $S = \pi r \sqrt{r^2 + h^2}$; $V = \frac{1}{3}\pi r^2 h$

3. Cube

   $S = 6s^2$; $V = s^3$

4. Cylinder (right circular)

   $S = 2\pi rh$; $V = \pi r^2 h$

5. Parallelogram

   $A = bh$

6. Rectangle

   $A = lw$; $P = 2l + 2w$

7. Rectangular Box

   $S = 2(lw + wh + lh)$; $V = lwh$

8. Sphere

   $S = 4\pi r^2$; $V = \frac{4}{3}\pi r^3$

9. Square

   $A = s^2$; $P = 4s$

10. Trapezoid

    $A = \frac{1}{2}(a + b)h$

11. Triangle

    $A = \frac{1}{2}bh$

Intermediate Algebra

Contents

v

Preface

No textbook can replace the presence of a skillful, motivating, and imaginative teacher. If a teacher had unlimited time, the only written materials needed would be an enormous number of exercises. In reality, of course, time is limited, and a textbook therefore becomes a necessary adjunct to an instructor in the classroom.

In writing this book, I have given utmost importance to the following objectives:

- Readability
- Mathematical correctness
- Motivation through examples
- Interesting and varied applications
- Extensive exercises

Readability

One of my main guiding principles has been clarity of presentation. The style of the text is straightforward, readable, and understandable without being dull. In addition, graphic devices, including the functional use of color, draw attention to important items. Definition boxes highlight important terms and concepts, caution boxes give warnings against common mistakes, and procedure boxes provide quick reference to rules and procedures. Each chapter concludes with a summary presented in outline form for review and easy reference.

Mathematical Correctness

In addition to the active pursuit of readability, great care has been taken to ensure the mathematical correctness and accuracy of all material. Fundamental principles have been explained and proved as completely and rigorously as feasible in a textbook of this level.

Motivation Through Examples

Many detailed and fully explained examples are presented to help motivate the reader. In fact, examples are often presented *before* a general principle is introduced. In this way, the student has a model to follow while general principles are developed; these principles are then reinforced by additional examples.

Interesting and Varied Applications

A wide range of applications—practical, computational, and theoretical—is given throughout the book. These serve to maintain interest and to emphasize the relevance and power of algebra in solving problems encountered in the real world.

Extensive Exercise Sets

A large number of exercises ranging from the routine to the more challenging (marked with an asterisk) are offered to provide sufficient practice and to reinforce and expand fundamental concepts.

ACKNOWLEDGMENTS

I wish to thank the following individuals, who reviewed the manuscript and offered many helpful suggestions:

Phillip W. Bean
Mercer University, Georgia

Ben P. Bockstege
Broward Community College, Florida

Mary Jane Causey
University of Mississippi

Edgar M. Chandler
Phoenix College, Arizona

Barbara Cohen
West Los Angeles College

Richard R. Conley
Ashland Community College, Kentucky

Mike Farrell
Carl Sandburg College, Illinois

Dan Kemp
South Dakota State University

Anne F. Landry
Dutchess Community College, New York

Myrna L. Mitchell
Pima Community College, Arizona

Jeff Mock
Diablo Valley College, California

Gilbert B. Perez
North Harris County College, Texas

Mark Phillips
Cypress College, California

John J. Saccoman
Seton Hall University, New Jersey

Rudy Svoboda
Indiana University, Purdue University of Fort Wayne

Robert B. Thompson
Lane Community College, Oregon

Jack B. Twitchell
Mesa Community College, Arizona

I would like to express my special appreciation to John Spellmann of Southwest Texas State University for proofreading and checking the exercises and examples in the book. I am very grateful to Ruth Suzuki for expert typing, and to Greg Hubit for his excellent work during the production of the book. I am also grateful to the staff of West Publishing Company for their outstanding work in the production of this book. In particular, I wish to thank Peter Marshall, Executive Editor, and Mark Jacobsen, Assistant Production Editor.

Finally, a very special thanks to Paul Johnson for his help, encouragement, and friendship.

To the Student

This textbook was written for you. Learning algebra is essentially a matter of obtaining an understanding of certain basic concepts and developing manipulative skills. As you work your way through the book, you will gain an understanding of the basic concepts and their applications and achieve proficiency in manipulative skills. To this end, many examples and illustrations stressing geometric and physical intuition are given. It is hoped that this treatment of combining problem solving, manipulative skills, and theory will open up to you the exciting world of mathematics.

1 The Set of Real Numbers

1.1 SETS

The days of the week, the even integers, a bunch of grapes, and the names listed in a telephone book have something in common; they are all examples of objects that have been grouped together and considered as a single entity. This idea of grouping objects together gives rise to the mathematical notion of a **set**, a concept that was developed by the German mathematician George Cantor (1845–1918).

A **set** is any collection of objects. The objects that constitute the set are called the **elements** or **members** of the set.

EXAMPLE 1   (a) The set $A = \{2, 4, 6\}$ has the elements 2, 4, and 6.

(b) The set $V = \{a, e, i, o, u\}$ has the elements a, e, i, o, and u.

In talking about a given set, we use the symbol $\in$ to indicate whether a certain object is an element of the set.

$x \in A$ means "x is an element of A."

$x \notin A$ means "x is not an element of A."

EXAMPLE 2   (a) If $A = \{2, 4, 6\}$, then $4 \in A$ and $5 \notin A$.

(b) If $V = \{a, e, i, o, u\}$, then $u \in V$ and $m \notin V$.

A set having no elements is called the empty set or null set and is represented by the symbols $\emptyset$ or $\{\ \}$.

EXAMPLE 3   (a) The set of all men who weigh a ton is the null set.

(b) The set of odd numbers divisible by 2 is the empty set.

CAUTION

The set $\{0\}$ is not the same as the empty set because it is a set that has one element, the number zero, 0.

There are two common ways of describing a set: the roster or listing method, and rule or set-builder notation.

The Roster Method

To describe a set using the roster (listing) method, list the elements between braces, $\{\ \}$.

EXAMPLE 4   The set S of positive integers smaller than 10 can be written $S = \{1, 2, 3, 4, 5, 6, 7, 8, 9\}$.

EXAMPLE 5   The set C of the first five letters in the English alphabet can be written $C = \{a, b, c, d, e\}$.

EXAMPLE 6   The set of negative integers can be written

$$\{\ldots, -3, -2, -1\}$$

The three dots, read "and so on," indicate that certain elements of the set have not been listed. Nevertheless, it is clear what the missing members are.

Set-Builder Notation

Sometimes it is inconvenient or impossible to list the members of a set. In such cases, the set can often be described by stating a common characteristic of its elements. A convenient way of doing this is to use set-builder notation.

EXAMPLE 7   Suppose P is the set of all states in the United States. Using set-builder notation, we write

$$P = \{x \mid x \text{ is a state of the United States}\}$$

Read: "the set" "of all x" "such that" "x is a state of the United States" where the symbol x is called a variable. A variable is a symbol used to represent any element of a given set that contains at least two elements.

In general, set-builder notation takes the form

$$A = \{x \mid x \text{ has property } P\}$$

which is read, "A is the set of all elements x such that x has the property P."

EXAMPLE 8   The set $S = \{1, 2, 3, 4, 5, 6, 7, 8, 9\}$ can be written

$$S = \{x \mid x \text{ is a positive integer less than 10}\}$$

EXAMPLE 9   The set $E = \{2, 4, 6, 8, \ldots\}$ can be written

$$E = \{x \mid x \text{ is a positive even integer}\}$$

A set whose elements can be counted (and this counting process eventually stops) is called a finite set. A set that is neither finite nor empty is called an infinite set.

EXAMPLE 10   The set $B = \{1, 3, 5, 7, 9\}$ is finite.

EXAMPLE 11   Let $G = \{x \mid x \text{ is a grain of sand on Earth}\}$. Although it may be difficult to count the number of grains of sand on Earth, G is a finite set.

EXAMPLE 12   The set $E = \{2, 4, 6, 8, \ldots\}$ is an infinite set.

DEFINITION

Two sets, A and B, are said to be equal if they contain the same elements, in which case we write

$$A = B$$

EXAMPLE 13 $\{3, 7, 9\} = \{9, 7, 3\}$. Both sets have exactly the same elements, even though the elements are listed in different orders.

EXAMPLE 14 $\{a, b, c\} = \{a, b, c, c, c\}$. Both sets have exactly the same elements, even though the element c is repeated.

When considering whether two sets are equal, order and repetition of the elements are not important.

EXAMPLE 15 Let $G = \{x \mid x$ is an even integer greater than 1 and less than 10$\}$, $H = \{2, 4, 6, 8\}$, and $J = \{2, 4, 4, 6, 8, 8\}$. Then $G = H = J$.

Consider the sets $A = \{1, 2, 3\}$ and $B = \{2, 1, 3, 5\}$. Note that each element of A is also an element of B. However, each element of B is not an element of A. For example, $5 \in B$ but $5 \notin A$. Thus, the set A is not equal to the set B, and we write $A \neq B$. While the sets A and B are not equal, they do illustrate the set relationship called **subset** .

DEFINITION

Let A and B be two sets and assume that each element of A is also an element of B. Then A is said to be a **subset** of B and we write $A \subseteq B$.

EXAMPLE 16 Let $U = \{1, 2, 3, 4, 5, 6, 7, 8, 9\}$, $A = \{4, 5, 6\}$, and $B = \{3, 4, 5, 6, 8\}$. Then $A \subseteq U$ and $B \subseteq U$. Also, $A \subseteq B$. These relationships are illustrated in Figure 1.1, where the region inside and on the rectangle represents U, the region inside and on the larger circle represents B, and the region inside and on the smaller circle represents A.

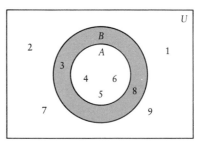

Figure 1.1

Diagrams that picture set relationships are called **Venn diagrams** in honor of the English mathematician John Venn (1834–1923).

EXAMPLE 17 Let $C = \{x \mid x$ is a playing card$\}$ and $A = \{x \mid x$ is an ace$\}$. Then $A \subseteq C$.

EXAMPLE 18 Let $R = \{4, -1, 7\}$, $S = \{7, -1, 8, 4\}$, and $T = \{-1, 4, 7, 6\}$. Then $R \subseteq S$ and $R \subseteq T$. However, $S \not\subseteq T$ because while $8 \in S$, $8 \notin T$.

EXAMPLE 19 The set $\{a, b, c\}$ has the following subsets: $\{a, b, c\}$, $\{a, b\}$, $\{a, c\}$, $\{b, c\}$, $\{a\}$, $\{b\}$, $\{c\}$, and $\emptyset$.

As Example 19 illustrates,

---

1. Every set is a subset of itself; $A \subseteq A$.
2. The empty set is a subset of every set; $\emptyset \subseteq A$. This is true because there is no element of $\emptyset$ that is not also an element of A.

We would now like to emphasize the difference between the notations $\in$ and $\subseteq$. The symbol $\in$ is used when talking about elements of a set. The symbol $\subseteq$ is used when talking about subsets. For example, the element a is not the same as the set containing the element a. That is, $a \neq \{a\}$. Thus, $a \in \{a, b, c\}$, but $a \not\subseteq \{a, b, c\}$. Also, $\{a\} \subseteq \{a, b, c\}$ but $\{a\} \not\in \{a, b, c\}$.

Just as the operations of addition, subtraction, multiplication, and division of real numbers can be used to solve an assortment of problems, there are also several operations on sets that can be used to solve many problems. These operations are the complement, intersection, and union of sets.

DEFINITION

A universal set, U, is a set that consists of all elements being considered in a given problem.

EXAMPLE 20 When talking about human population, the universal set consists of all the people in the world.

EXAMPLE 21 When discussing the set of letters in the English alphabet, the universal set $U = \{a, b, c, \ldots, x, y, z\}$.

If U is a universal set and A is a subset of U, then the complement of A in U, denoted by A', is the set of elements of U that are not elements of A. This concept is illustrated in Figure 1.2.

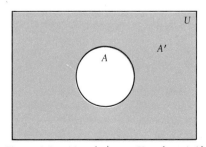

Figure 1.2 $A' = \{x \mid x \in U \text{ and } x \not\in A\}$

EXAMPLE 22 If $U = \{2, 4, 6, 8, 10\}$ and $A = \{4, 8, 10\}$, then $A' = \{2, 6\}$.

EXAMPLE 23 If $U = \{x \mid x$ is a letter in the English alphabet$\}$ and $A = \{x \mid x$ is a consonant$\}$, then $A' = \{x \mid x$ is a vowel$\}$.

EXAMPLE 24 Let $A = \{1, 3, 6, 8\}$ and $B = \{2, 6, 8, 9\}$. Consider the set consisting of all elements in A that are also in B. This set can be written $\{6, 8\}$. Now consider the set formed by combining all the elements of A and all the elements of B. This set can be written $\{1, 2, 3, 6, 8, 9\}$.

The two sets formed in Example 24 are called the intersection $(A \cap B)$ and union $(A \cup B)$, respectively, of A and B.

DEFINITIONS

Intersection: $A \cap B = \{x \mid x \in A \text{ and } x \in B\}$

Union: $A \cup B = \{x \mid x \in A \text{ or } x \in B\}$

The intersection and union of the two sets A and B are seen graphically in Figure 1.3.

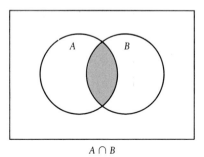

$A \cap B$ $A \cup B$

Figure 1.3

EXAMPLE 25 Let $A = \{a, b, c, d, e\}$, $B = \{a, c, e, g\}$, and $C = \{b, e, f, g\}$.

(a) Find $(A \cup B) \cup C$.   (b) Find $A \cup (B \cup C)$.

Solution (a) We first find $(A \cup B) = \{a, b, c, d, e, g\}$. Then the union of $(A \cup B)$ and C is

$$(A \cup B) \cup C = \{a, b, c, d, e, f, g\}$$

(b) We first find $(B \cup C) = \{a, b, c, e, f, g\}$. Then the union of A and $(B \cup C)$ is

$$A \cup (B \cup C) = \{a, b, c, d, e, f, g\}$$

Notice that $(A \cup B) \cup C = A \cup (B \cup C)$. ∎

EXAMPLE 26 Let $A = \{4, 5, 9\}$ and $B = \{1, 3, 8\}$. Then $A \cap B = \emptyset$. The sets A and B are said to be disjoint.

DEFINITION
Disjoint sets: $A \cap B = \emptyset$

PROBLEM 1 In an experiment with hybrid roses, the rose plants were classified into sets as follows:

$$W = \text{all-white petals} \qquad R = \text{all-red petals}$$

$$S = \text{fast-growing} \qquad T = \text{long stem}$$

Describe the characteristics of the plants in the following sets:

(a) $S \cap R$
(b) $T \cup W$
(c) $(T \cap S) \cup (T \cap R)$
(d) $S \cup (W \cap R)$

Answer (a) The flowers in $S \cap R$ are in both S and R. Thus $S \cap R$ consists of fast-growing, all-red roses.

(b) The flowers in $T \cup W$ are either in T or W. Thus $T \cup W$ consists of flowers that are either long-stemmed, or all-white, or both.

(c) The flowers in $(T \cap S)$ are long-stemmed and fast-growing. The flowers in $(T \cap R)$ are long-stemmed and all-red. Thus, $(T \cap S) \cup (T \cap R)$ consists of flowers that are either long-stemmed and fast-growing or long-stemmed and all-red.

(d) The set $(W \cap R)$ is empty since the roses cannot be both all-white and all-red. Thus,

$$S \cup (W \cap R) = S \cup \emptyset = S$$

and $S \cup (W \cap R)$ is the set of fast-growing roses. ∎

EXERCISES 1.1

In Exercises 1–10, fill in the blanks with the correct symbol, $\in$ or $\notin$.

1. 4 __ $\{3, -1, 0, 6, 4, 5\}$.
2. -7 __ $\{-3, 8, a, 7, -5\}$.
3. x __ $\{a, b, c, w, x, y, z\}$.
4. s __ $\{a, e, i, o, u\}$.
5. -3 __ $\{0, 1, 2, \ldots\}$.
6. 3 __ $\{x \mid x \text{ is an odd integer}\}$.

7. $-8 \underline{} \{x \mid x$ is a nonnegative integer$\}$.
8. $0 \underline{} \{x \mid x$ is a multiple of 5$\}$.
9. May $\underline{} \{x \mid x$ is a month of the year$\}$.
10. Canada $\underline{} \{x \mid x$ is a European country$\}$.

In Exercises 11–20, let $A = \{-4, m, a, 0, 7, \{5\}, e, 1, -2\}$. Answer the following as true or false.

11. $4 \in A$
12. $m \in A$
13. $3 \notin A$
14. $-2 \in A$
15. $5 \notin A$
16. $-4 \in A$
17. $e \notin A$
18. $\emptyset \in A$
19. $\{5\} \in A$
20. $\{a\} \in A$

In Exercises 21–28, use the brace notation $\{\ \}$ to list or give a partial listing of the given sets:

21. The set of all the days in the week whose names begin with S.
22. The set of all the months in the year that have exactly 30 days.
23. The set of positive integers less than 8.
24. The set of even numbers between 10 and 20.
25. The set of all negative integers less than -5.
26. The set of all positive integers greater than 12.
27. The set of even numbers between 10 and 30 that are divisible by 7.
28. The set of even numbers between 10 and 30 that are not divisible by 7.

In Exercises 29–34, use the set-builder notation to rewrite each of the following sets.

29. $E = \{2, 4, 6, 8, 10, 12, 14\}$
30. $O = \{1, 3, 5, 7, 9, 11, 13\}$
31. $T = \{$Alabama, Alaska, Arizona, Arkansas$\}$
32. $W = \{$Washington, Truman, Eisenhower, Kennedy, . . . $\}$
33. $P = \{. . . -8, -7, -6, -5\}$
34. $Q = \{. . . , -12, -9, -6, -3, 0, 3, 6, 9, 12, . . .\}$

In Exercises 35–40, identify each of the following sets as finite or infinite.

35. The names of all the students in your mathematics class
36. The set of all counting numbers, $\{1, 2, 3, . . .\}$
37. The odd numbers between 7 and 7 million
38. The set of points on a given line
39. The set of all multiples of 5
40. The set of all divisors of 24

In Exercises 41–49, let $A = \{1, 3, 5, 7\}$, $B = \{1, 3, 5\}$, and $C = \{1, 3, 6\}$. Indicate which statements are true and which are false.

41. $A \subseteq B$
42. $B \subseteq A$
43. $A \subseteq C$
44. $C \subseteq A$
45. $A \subseteq A$
46. $B \subseteq B$
47. $\emptyset \subseteq C$
48. $B = C$
49. $A \subseteq \emptyset$

In Exercises 50–55, each set in column I is a subset of one or more of the sets in column II. Determine which.

I	II
50. $C = $ the set of all citizens of California	$N = $ the set of all integers
51. $E = $ the set of all even integers	$U = $ the set of all citizens of the United States
52. $D = $ the set of all divisors of 6	$R = $ the set of all rectangles
53. $F = $ the set of all divisors of 15	$A = $ the set of all divisors of 60
54. $\emptyset = $ empty set	$Q = $ the set of all quadrilaterals
55. $S = $ the set of all squares	$B = $ the set of all divisors of 12

In Exercises 56–67, let $P = \{0, 1, 2\}$ and indicate which statements are true and which are false.

56. $1 \subseteq P$
57. $1 \in P$
58. $\{0\} \subseteq P$
59. $\{0\} \in P$
60. $2 \in \{2\}$
61. $\emptyset \in P$
62. $\emptyset \subseteq P$
63. $0 \subseteq P$
64. $0 \in P$
65. $0 = \emptyset$
66. $0 \in \emptyset$
67. $\{2\} = 2$

In Exercises 68–75, which statements are true?

68. $\emptyset \subseteq \emptyset$
69. $\emptyset \in \emptyset$
70. $\emptyset = \{\emptyset\}$
71. $\emptyset \in \{\emptyset\}$
72. $\emptyset = \{\ \}$
73. $\emptyset = \{0\}$
74. $\emptyset \subseteq \{\emptyset\}$
75. $\emptyset \in \{\{\emptyset\}\}$

In Exercises 76–88, let $U = \{l, m, n, p, r\}$, $S = \{l, m, n, r\}$, and $T = \{l, n, p\}$. List the following sets:

76. S'
77. T'
78. $S' \cup T'$
79. $S' \cap T'$
80. $S \cup T$
81. $S \cap T$
82. $(S \cup T)'$
83. $(S \cap T)'$
84. $S \cup S'$
85. $S \cap S'$
86. $T \cup T'$
87. $T \cap T'$
88. What can be said about Exercises 78 and 83? Exercises 79 and 82?

In Exercises 89–97, let $A = \{1, 2, 3, 4\}$, $B = \{3, 4, 5\}$, $C = \{3, 4, 5, 6, 7\}$, and $D = \{6, 7, 8, 9\}$. Find:

89. $A \cup B$
90. $A \cap B$
91. $A \cup D$
92. $A \cap D$
93. $A \cup C \cup D$
94. $A \cap C \cap D$
95. $A \cup (C \cap D)$
96. $A \cap (C \cup D)$

97. $(A \cup C) \cap (A \cup D)$

In Exercises 98–109, refer to Figure 1.4. Let C = the set of points inside the circle, T = the set of points inside the triangle, and R = the set of points inside the rectangle. Shade the following sets:

98. $R \cap T$
99. $R \cap C$
100. $T \cap C$
101. $R \cap T \cap C$
102. $R \cup T$
103. $R \cup C$
104. $T \cup C$
105. $R \cup T \cup C$
106. $C \cup (R \cap T)$
107. $C \cap (R \cup T)$
108. $(C \cap R) \cup T$
109. $(C \cup R) \cap (C \cup T)$

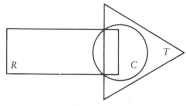

Figure 1.4

For Exercises 110–117, consider the following facts. A person's blood type can be classified according to the antigens present or not present in his or her blood. Every person is doubly classified. If a person has the Rh antigen, that person is said to be Rh positive. If a person does not have the antigen, he or she is Rh negative. If the blood contains the A antigen but not the B antigen, it is type A. If the blood contains the B antigen but not the A antigen, it is type B. If the blood contains both the A and B antigens, it is type AB. If the blood contains neither antigen, it is type O. Let

$A = \{x \mid x$ is a person whose blood has the A antigen$\}$

$B = \{x \mid x$ is a person whose blood has the B antigen$\}$

$R = \{x \mid x$ is a person whose blood has the Rh antigen$\}$

Determine the blood types of the following people:

110. $R \cap A \cap B$
111. $R' \cap A \cap B$
112. $R' \cap A \cap B'$
113. $R \cap A' \cap B$
114. $R \cap (A \cap B')$
115. $R' \cap (A' \cap B)$
116. $R \cap (A' \cap B')$
117. $R' \cap A' \cap B'$

In Exercises 118–126, assume that A is any subset of a universal set U. Identify each of the following sets as equal to A, A', U, or $\emptyset$.

118. $A \cup A'$
119. $A \cap A'$
120. $A \cap A$
121. $A \cup A$
122. $A \cup \emptyset$
123. $A \cap \emptyset$
124. $\emptyset \cap U$
125. $\emptyset \cup U$
126. $A \cup U$

In Exercises 127–138, state the conditions that must be imposed on C or D to make the statement true. Recall that U is a universal set.

127. $C \cup D = \emptyset$
128. $C \cap D = \emptyset$
129. $C \cap D = U$
*130. $C \cup D = U$
131. $C \cup D = D$
132. $C \cap D = D$
133. $C \cap U = C$
134. $C \cup U = C$
135. $C \cup \emptyset = \emptyset$
136. $C \cap \emptyset = C$
137. $C \cap \emptyset = \emptyset$
138. $C \cup U = U$

1.2 THE SET OF REAL NUMBERS

Much of our work in algebra deals with sets of numbers, and the language of sets is useful in describing these sets. The first set of numbers we encounter are the numbers used in counting $\{1, 2, 3, \ldots\}$. This set is called the set of **natural numbers** or **counting numbers**. The letter N is used to designate the set of natural numbers. Thus,

$$N = \{1, 2, 3, \ldots\}$$

This set is also referred to as the **positive integers**. Note that each successive natural number (except the first) is obtained by adding 1 to its predecessor: 1,

$1 + 1 = 2$, $2 + 1 = 3$, $3 + 1 = 4$, and so on. Furthermore, we observe that the addition of two natural numbers always gives a natural number. However, if we subtract two natural numbers, the result may not be a natural number. For example, $4 - 4$ is not a natural number. To give such a result meaning, we introduce a new number called zero, which is denoted symbolically by 0. The union of the set N and the set $\{0\}$ is called the set of whole numbers and is denoted by the letter W. That is,

$$W = \{0, 1, 2, 3, \ldots\}$$

The subtraction of two whole numbers may not give a whole number. Consider such expressions as $3 - 6$ and $5 - 9$, which are not whole numbers. In order to give these expressions meaning, we introduce the set of negative integers $\{\ldots, -3, -2, -1\}$, and the union of this set with the whole numbers (W) extends our system of natural numbers (N) to the set of integers, denoted by the letter J.

$$J = \{\ldots, -3, -2, -1, 0, 1, 2, 3, \ldots\}$$

Thus, the set of integers consists of the natural numbers, the number zero, and the negative natural numbers.

While the multiplication of two integers results in an integer, the division of two integers does not necessarily result in an integer. For example, expressions such as $\frac{3}{5}$, $\frac{1}{2}$, and $\frac{7}{-4}$ are meaningless in the set of integers. Thus, we extend our set of integers to include numbers that are the ratio of two integers.

DEFINITION

A rational number is any number that can be written in the form a/b, where $a, b \in J$, and $b \neq 0$. That is, a rational number is a number that can be expressed as a quotient (ratio) of two integers.

The set of rational numbers is denoted by Q; that is,

$$Q = \left\{ \frac{a}{b} \;\middle|\; a, b \in J \quad \text{and} \quad b \neq 0 \right\}$$

EXAMPLE 1   The numbers $-\frac{4}{3}$, $\frac{1}{5}$, $\frac{17}{9}$, and $\frac{0}{8}$ are rational numbers.

Note that the set of integers J is a subset of the set of rational numbers Q since every integer can be expressed as the ratio of two integers.

EXAMPLE 2 $3 = \frac{6}{2}$, $8 = \frac{8}{1}$, and $0 = \frac{0}{7}$.

All rational numbers can be written in a decimal form. For example, $\frac{6}{5} = 1.20$ and $\frac{9}{11} = 0.818181\ldots$. The decimal 1.20 is called a terminating decimal, while the decimal $0.818181\ldots$ is called a repeating decimal because the two digits 81 repeat themselves indefinitely. All rational numbers can be represented by terminating or repeating decimals.

There are also numbers used in algebra that are not rational numbers; that is, numbers that cannot be expressed as the quotient of two integers. These numbers form a set called the irrational numbers.

EXAMPLE 3   The numbers

$$\pi = 3.1415926535\ldots$$
$$\sqrt{2} = 1.1414213\ldots$$
$$e = 2.7182818283\ldots$$

and $0.030030003\ldots$

are irrational numbers.

The decimal representations of irrational numbers are nonterminating and nonrepeating. For example, while the decimal $0.030030003\ldots$ has a pattern, the pattern is not repeating.

The rational numbers and the irrational numbers together form a set called the real numbers, denoted by R. The set of real numbers and the subset relations existing between the sets of numbers introduced in this section are shown in Figure 1.5.

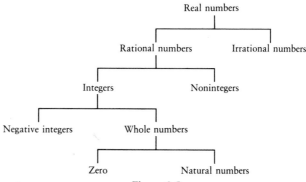

Figure 1.5

In this book we shall deal with the set of real numbers and the set of complex numbers. Complex numbers will be introduced in Chapter 7 so that we may solve certain equations that cannot be solved using only the real numbers.

EXERCISES 1.2

In Exercises 1–10, choose the correct set(s) from the following: (a) natural numbers, (b) integers, (c) rational numbers, (d) irrational numbers, (e) real numbers.

1. The number 5 is   2. The number -9 is
3. The number 0.7 is
4. The number $-\frac{4}{5}$ is
5. The numbers $-4, -3, -2, -1$ are
6. The numbers $0, 1, 2, 3$ are
7. The numbers $-6\frac{2}{3}, -1, 0, 3, \frac{7}{9}$ are
8. The numbers $-\sqrt{5}, \sqrt{3}, e, \pi$ are
9. The numbers $-0.6, -0.2, -0.1$ are
10. The numbers $-\sqrt{2}, -1, 0, \pi, 4$ are

In Exercises 11–20, determine whether the given statements are true.

11. 27 is a whole number.
12. -6 is a natural number.
13. $\sqrt{4}$ is a rational number.
14. $-\sqrt{25}$ is a natural number.
15. -0.345345 is an irrational number.
16. $\dfrac{2\pi}{5}$ is a real number.
17. 4.1367 is a rational number.
18. $8 + \sqrt{3}$ is an irrational number.
19. 0.01001001. . . is a rational number.
20. 0.3333. . . is a rational number.

In Exercises 21–32, express each rational number as a terminating decimal or repeating decimal.

21. $\frac{5}{9}$ 22. $\frac{5}{8}$ 23. $\frac{5}{16}$ 24. $\frac{3}{11}$
25. $-6\frac{1}{2}$ 26. $-\frac{7}{12}$ 27. $\frac{7}{9}$ 28. $3\frac{2}{3}$
29. $-\frac{35}{99}$ 30. $7\frac{13}{33}$ 31. $\frac{11}{6}$ 32. $5\frac{1}{7}$

In Exercises 33–47, let $U = R$, where

$R = \{x \mid x \text{ is a real number}\}$, and

$J = \{x \mid x \text{ is an integer}\}$,

$N = \{x \mid x \text{ is a natural number}\}$,

$\overline{N} = \{x \mid x \text{ is a negative integer}\}$,

$Q = \{x \mid x \text{ is a rational number}\}$,

$\overline{Q} = \{x \mid x \text{ is an irrational number}\}$,

$Z = \{0\}$.

Describe the following sets.

33. $\overline{N} \cup Z$ 34. $J \cup R$
35. $\overline{Q} \cap \overline{N}$ 36. $N \cap J'$
37. $\overline{N} \cap Z$ 38. $(R \cup Q)'$
39. $N' \cap J$ 40. $J' \cap Q$
41. $(J \cap Q)'$

Let $P = \{-2, \frac{4}{5}, -1, 0, \sqrt{5}, -\sqrt{2}, -6, \frac{7}{8}, 1\}$. List the elements of the following sets:

42. $P \cap R$ 43. $P \cap J$ 44. $P \cap N$
45. $P \cap \overline{N}$ 46. $P \cap Q$ 47. $P \cap \overline{Q}$

In Exercises 48–57, fill in "All," "Some," or "No."

48. _____ counting numbers are natural numbers.
49. _____ natural numbers are real numbers.
50. _____ integers are whole numbers.
51. _____ rational numbers are natural numbers.
52. _____ natural numbers are integers.
53. _____ real numbers are irrational numbers.
54. _____ negative integers are real numbers.
55. _____ irrational numbers are rational numbers.
56. _____ whole numbers are negative integers.
57. _____ irrational numbers are counting numbers.
58. Give an example that shows that the sum of two irrational numbers is not always an irrational number. Do the same for the product of two irrational numbers.
*59. Prove that the sum of two rational numbers is always a rational number.
*60. Prove that the product of two rational numbers is always a rational number.

1.3 SOME PROPERTIES OF THE REAL NUMBERS

In this section we discuss several basic properties of real numbers. These properties can be expressed in terms of the operations of addition and multiplication. The real numbers are said to be **closed** with respect to the operations of addition (denoted by +) and multiplication (denoted by ·). That is, if a and b are real numbers, there is a unique real number $a + b$, called the **sum** of a and b, and a unique number $a \cdot b$, called the **product** of a and b. Addition and multiplication have the following properties for all real numbers a, b, and c.

Commutative Properties

For addition: $a + b = b + a$

For multiplication: $a \cdot b = b \cdot a$

EXAMPLE 1   Addition: $5 + 9 = 9 + 5$ because $5 + 9 = 14$ and $9 + 5 = 14$

$\frac{1}{3} + \frac{4}{5} = \frac{4}{5} + \frac{1}{3}$ because $\frac{1}{3} + \frac{4}{5} = \frac{17}{15}$ and $\frac{4}{5} + \frac{1}{3} = \frac{17}{15}$

Multiplication: $6 \cdot 7 = 7 \cdot 6$ because $6 \cdot 7 = 42$ and $7 \cdot 6 = 42$

$\frac{5}{7} \cdot \frac{3}{4} = \frac{3}{4} \cdot \frac{5}{7}$ because $\frac{5}{7} \cdot \frac{3}{4} = \frac{15}{28}$ and $\frac{3}{4} \cdot \frac{5}{7} = \frac{15}{28}$

Associative Properties

For addition: $a + (b + c) = (a + b) + c$

For multiplication: $(a \cdot b) \cdot c = a \cdot (b \cdot c)$

EXAMPLE 2   Addition: $(3 + 1) + 6 = 3 + (1 + 6)$ because $(3 + 1) + 6 = 4 + 6 = 10$ and $3 + (1 + 6) = 3 + 7 = 10$

Multiplication: $7 \cdot (3 \cdot 5) = (7 \cdot 3) \cdot 5$ because $7 \cdot (3 \cdot 5) = 7 \cdot 15 = 105$ and $(7 \cdot 3) \cdot 5 = 21 \cdot 5 = 105$

Identity Properties

For addition: There exists a unique real number called the additive identity, denoted by 0 (*zero*), such that $a + 0 = 0 + a = a$.

For multiplication: There exists a unique real number called the multiplicative identity, denoted by 1 (*one*), such that $a \cdot 1 = 1 \cdot a = a$.

EXAMPLE 3 $4 + 0 = 0 + 4 = 4$ and $9 \cdot 1 = 1 \cdot 9 = 9$

Inverse Properties

For addition: For each real number, a, there is a unique real number, $-a$, called the additive inverse or negative of a, such that

$$a + (-a) = (-a) + a = 0$$

For multiplication: For each real number, $a \neq 0$, there is a unique real number, $1/a$, called the multiplicative inverse or reciprocal of a, such that

$$a \cdot \frac{1}{a} = \frac{1}{a} \cdot a = 1$$

EXAMPLE 4 Addition: $8 + (-8) = (-8) + 8 = 0$ and $\frac{3}{7} + (-\frac{3}{7}) = (-\frac{3}{7}) + \frac{3}{7} = 0$

Multiplication: $(-4) \cdot (\frac{1}{-4}) = (\frac{1}{-4}) \cdot (-4) = 1$, and the reciprocal of $\frac{2}{5}$ is $1/(2/5) = \frac{5}{2}$, since $\frac{2}{5} \cdot \frac{5}{2} = \frac{5}{2} \cdot \frac{2}{5} = 1$

The properties discussed above dealt separately with operations of addition and multiplication. Next, we present a property that combines both addition and multiplication. Let's begin by considering the following example.

EXAMPLE 5 A saleswoman travels two days a week, on Monday and Tuesday. Her car gets 20 miles per gallon. If she uses 13 gallons of gasoline on Monday and 17 gallons on Tuesday, how many miles did she travel?

Solution We first note that the term *per* can be expressed using fractions. Thus, 20 miles per gallon can be written 20 miles/gallon. Since the saleswoman used a total of (13 + 17) gallons, her total mileage is given by 20 miles/gallon · (13 + 17) gallons = 20 miles/gallon · 30 gallons = 600 miles. Her total miles are *also* the sum of her miles travelled on Monday and her miles travelled on Tuesday. That is,

$$\left(20 \, \frac{\text{miles}}{\text{gallon}} \cdot 13 \text{ gallons}\right) + \left(20 \, \frac{\text{miles}}{\text{gallon}} \cdot 17 \text{ gallons}\right)$$
$$= \qquad 260 \text{ miles} \qquad + \qquad 340 \text{ miles}$$
$$= \qquad 600 \text{ miles}$$

Either way, we compute her total miles and find that they are the same; that is, using only the numbers, we have

$$20(13 + 17) = 20 \cdot 13 + 20 \cdot 17 \quad \blacksquare$$

This example illustrates a property that we will use a great deal in working with real numbers. It is called the distributive property.

Distributive Property

$a \cdot (b + c) = a \cdot b + a \cdot c$ and $(b + c) \cdot a = b \cdot a + c \cdot a$

EXAMPLE 6
$$4(3 + 5) = (4 \cdot 3) + (4 \cdot 5)$$

because $4(3 + 5) = 4(8) = 32$ and $(4 \cdot 3) + (4 \cdot 5) = 12 + 20 = 32$. Also,

$$(2 + 6) \cdot 7 = (2 \cdot 7) + (6 \cdot 7)$$

because $(2 + 6) \cdot 7 = 8 \cdot 7 = 56$ and $(2 \cdot 7) + (6 \cdot 7) = 14 + 42 = 56$

PROBLEM 1   State the properties that justify each of the following equalities:

(a) $\frac{3}{5} + (-\frac{3}{5}) = 0$ (b) $(-4) + (6) = (6) + (-4)$

(c) $6 \cdot (\frac{7}{12} + 3) = \frac{7}{2} + 18$ (d) $13 \cdot \frac{1}{13} = 1$

(e) $(5 \cdot 7) \cdot 9 = 5 \cdot (7 \cdot 9)$ (f) $(4 + 3) \cdot (-5) = (4)(-5) + (3)(-5)$

(g) $(\frac{-3}{7})(\frac{2}{5}) = (\frac{2}{5})(\frac{-3}{7})$ (h) $6 + (3 + 2) = (6 + 3) + 2$

Answer   (a) additive inverse

(b) commutative property for addition

(c) distributive property

(d) multiplicative inverse

(e) associative property for multiplication

(f) distributive property

(g) commutative property for multiplication

(h) associative property for addition   ■

   We now use these properties to derive other very important properties of real numbers. As above, we assume that a, b, and c are real numbers.

Cancellation Properties

For addition: If $a + c = b + c$, then $a = b$.
For multiplication: If $ac = bc$ and $c \neq 0$, then $a = b$.

EXAMPLE 7   (a) If $x + 4 = y + 4$, then $x = y$.

(b) If $2x = 2y$, then $x = y$.

Double Negative Property

For any real number a,

$$-(-a) = a$$

EXAMPLE 8 $-(-13) = 13$ and $-[-(-7)] = -7$

Zero-Factor Properties

1. $a \cdot 0 = 0 \cdot a = 0$
2. If $a \cdot b = 0$, then either $a = 0$ or $b = 0$.

EXAMPLE 9   If the expressions $(x - 3)$ and $(x + 4)$ represent real numbers and $(x - 3)(x + 4) = 0$, then the zero-factor property says

$$(x - 3) = 0 \quad \text{or} \quad (x + 4) = 0$$

EXERCISES 1.3

In Exercises 1–20, state the axiom of real numbers that is illustrated.

1. $3 + \sqrt{5} = \sqrt{5} + 3$
2. $-7 + 0 = -7$
3. $8 + (-5) = (-5) + 8$
4. $(13 + 6) + 24 = 13 + (6 + 24)$
5. $-\frac{1}{3} + \frac{1}{3} = 0$ **6.** $(0.475)(1) = 0.475$
7. $\sqrt{3} + (\sqrt{3})(0) = \sqrt{3}$
8. $\frac{1}{5}(7 + 9) = (\frac{1}{5})7 + (\frac{1}{5})9$
9. $-4[5 + (-9 + 4)]$
   $= (-4)(5) + (-4)(-9 + 4)$
10. $(\frac{1}{9})9 = 9(\frac{1}{9})$ **11.** $(5 + 6) \cdot 1 = 5 + 6$
12. $\left(\frac{1}{6}\right)\left(\frac{1}{1/6}\right) = 1$ **13.** $(2 \cdot 4)7 = 2(4 \cdot 7)$
14. $y \cdot 0 = 0 \cdot y$ **15.** $z(x + y) = (x + y)z$
16. $(-b) + b = 0$ **17.** $x(y + z) = x(z + y)$
18. $y + (-y) = 0$ **19.** $y[x + (-x)] = y \cdot 0$
20. $b + a(c + d) = b + ac + ad$

In Exercises 21–25, answer as true or false.

21. Each integer has an additive inverse.

22. Each integer has a multiplicative inverse.
23. The reciprocal of a real number between 0 and 1 is a number larger than 1.
24. Subtraction is commutative.
25. Division is associative.
26. Name the additive inverse and multiplicative inverse for each of the following:
   (a) 1   (b) -2 (c) $\sqrt{3}$
   (d) π (e) $-\frac{7}{13}$
27. State the property for each step.
   (i) $a(-1) + a = a(-1) + a(1)$
   (ii) $= a[-1 + 1]$
   (iii) $= a[0]$
   (iv) $= 0$
28. State the property for each step.
   (i) $b + [a + (-b)] = b + [(-b) + a]$
   (ii) $= [b + (-b)] + a$
   (iii) $= [0] + a$
   (iv) $= a$

29. State the property for each step.

(i) $(x + y) + [(-x) + (-y)]$
$$= (y + x) + [(-x) + (-y)]$$

(ii) $\qquad = [(y + x) + (-x)] + (-y)$

(iii) $\qquad = [y + \{x + (-x)\}] + (-y)$
(iv) $\qquad = [y + 0] + (-y)$
(v) $\qquad = y + (-y)$
(vi) $\qquad = 0$

1.4   PROPERTIES OF ORDER

The idea of one number being greater than or less than another number plays a vital role in mathematics. We now give a precise definition of these concepts.

DEFINITION

If a and b are real numbers, we say that $a > b$, read "*a is greater than b*," if $a - b$ is a positive number. Thus,

$$a > b \quad \text{means} \quad a - b > 0$$

Similarly, we say that $a < b$, read "*a is less than b*," if $b - a$ is a positive number. Thus,

$$a < b \quad \text{means} \quad b - a > 0$$

EXAMPLE 1   We say $2 < 5$, read "2 is less than 5," because $5 - 2 = 3$ is a positive number. The statement $2 < 5$ can also be written as $5 > 2$, read "5 is greater than 2."

Note. The symbols $>$ and $<$ point toward the smaller number, and are called **inequality signs**. Expressions such as $a > b$ or $a < b$ are called **inequalities**.

EXAMPLE 2   (a) $7 > 4$ since $7 - 4 = 3 > 0$

(b) $\frac{2}{3} > \frac{1}{6}$ since $\frac{2}{3} - \frac{1}{6} = \frac{4}{6} - \frac{1}{6} = \frac{3}{6} = \frac{1}{2} > 0$

(c) $-3 < 1$ since $1 - (-3) = 1 + 3 = 4 > 0$

(d) $-2 > -4$ since $-2 - (-4) = -2 + 4 = 2 > 0$

EXAMPLE 3   The statement $2 < 2$, read "2 is *not* less than 2," is true since $2 - 2 = 0$ is not positive. Similarly, $2 > 2$, since $2 - 2$ is not positive.

It would be convenient if we also defined the ideas of "greater than or equal to" and "less than or equal to."

DEFINITION

If a and b are real numbers, we say $a \geq b$, read "*a is greater than or equal to b*," if $a - b$ is a nonnegative real number; that is, if $a - b$ is a positive real number or zero. Thus,

$$a \geq b \quad \text{means} \quad a - b \geq 0$$

Similarly, we say that $a \leq b$, read "*a is less than or equal to b*," if $b - a$ is a nonnegative real number; that is, if $b - a$ is a positive real number or zero. Thus,

$$a \leq b \quad \text{means} \quad b - a \geq 0$$

EXAMPLE 4   (a) $17 \leq 21$ since $21 - 17 \geq 0$

(b) $2 \geq 2$ since $2 - 2 \geq 0$

(c) $-4 \leq -4$ since $-4 - (-4) = -4 + 4 = 0 \geq 0$

We make the following assumptions about order in the set of real numbers.

Trichotomy Property

If a and b are real numbers, then one and only one of the following statements is true:

$$a > b, \quad a = b. \quad a < b$$

The trichotomy property states that if a and b are any real numbers, then the first number is less than the second, or the two numbers are equal, or the first is greater than the second.

Transitive Property of Order

If a, b, and c are real numbers, and if $a < b$ and $b < c$, then $a < c$.

EXAMPLE 5   (a) Since $-2 < 0$ and $0 < 4$, then $-2 < 4$.

(b) Since $x < y$ and $y < z$, then $x < z$.

(c) Since $4 < 8$ and $8 < x$, then $4 < x$.

> **Additive Property of Order**
>
> If a, b, and c are real numbers, and if $a < b$, then $a + c < b + c$. Also, if $a > b$, then $a + c > b + c$.

The additive property of order states that when the same number is added to both sides of an inequality, the order of the inequality remains unchanged.

EXAMPLE 6   (a) Since $11 < 20$, then $11 + 3 < 20 + 3$, or $14 < 23$.

(b) Since $15 > 8$, then $15 + 4 > 8 + 4$, or $19 > 12$.

(c) Since $x > y$, then $x + 2 > y + 2$.

> **Multiplicative Property of Order**
>
> Let a, b, and c be real numbers.
> 1. If $a < b$ and c is positive, then $ac < bc$.
> 2. If $a > b$ and c is positive, then $ac > bc$.
> 3. If $a < b$ and c is negative, then $ac > bc$.
> 4. If $a > b$ and c is negative, then $ac < bc$.

The multiplicative property of order states that when both sides of an inequality are multiplied by a positive number, the order of the inequality remains unchanged; when both sides of an inequality are multiplied by a negative number, the order of the inequality is reversed.

EXAMPLE 7   (a) Since $4 < 7$, then $4 \cdot 2 < 7 \cdot 2$, or $8 < 14$.

(b) Since $10 > 3$, then $10 \cdot (-6) < 3 \cdot (-6)$, or $-60 < -18$.

The Real Number Line

The real numbers can be represented geometrically by points on a straight line in such a way that for each real number there corresponds one and only one point, and conversely, to each point on the line there corresponds exactly one real number. Such an association between the set of real numbers and the set of points on a straight line is called a **one-to-one correspondence**. This number line can be constructed by starting with a line that extends indefinitely in opposite directions. We then choose a point on the line to associate with the number 0 and call this point the **origin**. We define the numbers represented by the points of the line to the right of 0 to be **positive** and those to the left of 0 to be **negative**. The number 0 is neither positive nor negative. Points associated

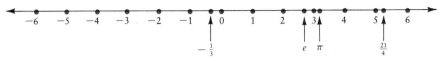

Figure 1.6

with the integers are then determined by laying off successive units to the right and left of the origin, as illustrated in Figure 1.6.

We can obtain rational numbers such as $-\frac{1}{3}$ and $\frac{21}{4}$ by subdividing the successive units. For example, the rational $-\frac{1}{3}$ can be located by subdividing the segment of the number line from 0 to -1 into three equal parts. The endpoint of the first such subdivision is then associated with the number $-\frac{1}{3}$. Similarly, the point associated with $\frac{21}{4} = 5\frac{1}{4}$ is twenty-one fourths units to the right of the origin, which is one-fourth of the distance from the unit point 5 to the unit point 6. Certain irrational numbers such as $\sqrt{2}$ can be found by geometric construction. However, other irrational numbers such as π and e cannot be located by a geometric construction. Such irrationals can, however, be approximated by decimals to any degree of accuracy. For example, $\pi = 3.1415$ and $e = 2.7182$. See Figure 1.7.

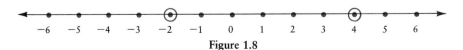

Figure 1.7

The point associated with a number on a number line is called the **graph** of that number, and the number is called the **coordinate** of that point. The number line that results from associating each of its points with a real number is called a **real number line.**

From the way in which the real number line was constructed, we can illustrate the idea of order of the real numbers. If A and B are points with coordinates a and b, respectively, then $a > b$ if and only if A lies to the right of B. For example, $4 > -2$, since the point that corresponds to 4 lies to the right of the point that corresponds to -2. See Figure 1.8.

Figure 1.8

EXERCISES 1.4

In Exercises 1–10, insert the proper inequality sign, $<$ or $>$, between the following pairs of numbers:

1. $-9 \_\_ -2$
2. $-1 \_\_ -7$
3. $7 - 2 \_\_ 4 - 1$
4. $13 - 8 \_\_ 5 + 1$
5. $\frac{1}{3} \_\_ .7$
6. $-\frac{1}{4} \_\_ -\frac{1}{5}$
7. $-\frac{22}{7} \_\_ -\pi$
8. $\sqrt{2} \_\_ 1.4$
9. $\frac{1}{9} \_\_ .10$
10. $e \_\_ 2.75$

In Exercises 11–16, express the given statement in terms of inequalities.

11. $\sqrt{3}$ is less than e
12. z is negative
13. y is nonnegative
14. a is greater than or equal to 5
15. x is less than or equal to -3
16. y is not greater than 6
17. Arrange the following numbers in order of increasing value.
    (a) $3, -\frac{2}{3}, -\sqrt{6}, 0, \pi, \frac{22}{7}, 1.62, e$
    (b) $0, -2, \sqrt{11}, 5, -\frac{\pi}{3}, -3.13$

In Exercises 18–25, fill in the blank with one of the symbols <, >, ≤, or ≥ so that the resulting statement will be true.

18. Since $-7 \le 3$, then $-7 + 4$ ___ $3 + 4$.

19. Since $10 > 2$, then $10 - 8$ ___ $2 - 8$.

20. Since $4 \le 14$, then $(-2)(4)$ ___ $(-2)(14)$.

21. Since $36 > 18$, then $\frac{36}{3}$ ___ $\frac{18}{3}$.

22. If $3 \le x$, then -3 ___ $-x$.

23. If $y < 5$, then $\frac{y}{2}$ ___ $\frac{5}{2}$.

24. If $x > -6$ and $-6 > y$, then x ___ y.

25. If $a > b$ and $c < b$, then a ___ c.

*26. If $a < b$, $c < d$, and $e < f$, prove that $a + c + e < b + d + f$.

*27. If $x < 0$, show that $0 < -x$. Also, if $y > 0$, show that $0 > -y$.

*28. Prove that the product ab is positive if and only if a and b are both positive or both negative.

*29. Prove that the product ab is negative if and only if a and b have opposite signs.

*30. Prove that, if a and b are positive and $a > b$, then $a^2 > b^2$.

*31. If $a > b$, show that $a > \dfrac{a + b}{2} > b$.

*32. If the number x is represented by a point to the left of zero on the real number line, determine which of the following is true:

(a) $2x > x$ (b) $2x = x$ (c) $2x < x$

*33. (a) If $2x < x$, where must the point corresponding to x lie on the real number line?

(b) If $2x = x$?

(c) If $2x > x$?

*34. Explain why we cannot tell whether x or $2x$ is larger.

*35. Describe the position on the real number line of the points corresponding to the numbers that are:

(a) equal to their squares

(b) equal to their square roots

(c) greater than their square roots

(d) less than their squares

(e) less than their square roots

1.5   ADDITION AND SUBTRACTION OF SIGNED NUMBERS

The operations of addition, subtraction, multiplication, and division are defined for the set of real numbers. We now review the rules for operating with signed numbers. But before stating these rules, we need to define the idea of the absolute value of a real number. The absolute value of a real number a, denoted $|a|$, is the distance between the point on the real number line with coordinate a and the origin. For example, Figure 1.9 shows that $|2| = 2$, since 2 is two units away from the origin. Since -2 is also two units away from the origin, we also have $|-2| = 2$. Note that if a is a nonnegative number, then $|a| = a$. If a is a negative number, then $|a| = -a$. Thus we are able to write a formal definition for absolute value.

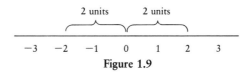

Figure 1.9

DEFINITION

If a is a real number, then the absolute value, $|a|$, of a is defined by

$$|a| = \begin{cases} a & \text{if } a \ge 0 \\ -a & \text{if } a < 0 \end{cases}$$

EXAMPLE 1   Find $|4|, |-4|, |0|, |-\sqrt{2}|, |-\pi|$.

Solution   Since 4 and 0 are nonnegative, we have $|4| = 4$ and $|0| = 0$. Since $-4, -\sqrt{2}$, and $-\pi$ are negative, we use the formula $|a| = -a$ to obtain $|-4| = -(-4) = 4, |-\sqrt{2}| = -(-\sqrt{2}) = \sqrt{2}$, and $|-\pi| = -(-\pi) = \pi$. ∎

Given the definition of absolute value, we now state the rules for the addition, subtraction, multiplication, and division of signed numbers.

Addition of Signed Numbers

Rule for Addition of Signed Numbers

To add two numbers with

like signs: Add the absolute values of the numbers and use the common sign.

unlike signs: Subtract the absolute values (the smaller from the larger) and use the sign of the number with the larger absolute value.

EXAMPLE 2   Find the sum $(+6) + (+5)$.

Solution   Since the signs are alike, add the absolute values:

$$|+6| + |+5| = 6 + 5 = 11$$

Use the common sign $+$ in the sum. Thus,

$$(+6) + (+5) = +11 \quad ∎$$

EXAMPLE 3   Find the sum $(-6) + (-5)$.

Solution   Since the signs are alike, add the absolute values:

$$|-6| + |-5| = 6 + 5 = 11$$

Now, use the common sign $-$. Thus,

$$(-6) + (-5) = -11 \quad ∎$$

EXAMPLE 4   Find the sum $(-13) + (+7)$.

Solution   Since the signs are unlike, subtract the smaller absolute value from the larger absolute value:

$$|-13| - |+7| = 13 - 7 = 6$$

Use the $-$ sign on the result 6 because -13 has a larger absolute value than 7. Thus,

$$(-13) + (+7) = -6 \quad \blacksquare$$

EXAMPLE 5   Find the sum $(+13) + (-7)$.

Solution   Since the signs are unlike, subtract the smaller absolute value from the larger absolute value:

$$|+13| - |-7| = 13 - 7 = 6$$

Use the $+$ sign on the result 6 because $+13$ has a larger absolute value than -7. Thus,

$$(+13) + (-7) = +6 \quad \blacksquare$$

Often signed numbers are used to represent gains and losses, $(+)$ numbers denoting gains, $(-)$ numbers denoting losses. Also, signed numbers may represent such ideas as future or past time, or the level above or below some preassigned zero in temperature. See Figure 1.10.

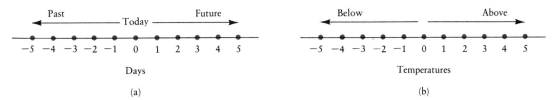

Figure 1.10

EXAMPLE 6   A student has \$37 in assets but receives a bill for \$76. What are his new assets?

Solution   This problem can be represented by the addition of signed numbers. The \$37 can be represented by $+37$; the \$76 bill by -76. His new assets are given by

$$(+37) + (-76) = -39$$

that is, the student has a debt of \$39. $\blacksquare$

When three or more signed numbers are to be added, we use the commutative and associative laws for addition to group all the positive numbers together and all the negative numbers together and add the result of these two sums.

EXAMPLE 7   Add $(+3) + (-5) + (+4) + (-1) + (-2)$.

Solution   The commutative law allows us to rearrange the order of the numbers; the associative law allows us to group them.

$$(+3) + (-5) + (+4) + (-1) + (-2) =$$
$$\underbrace{(+3) + (+4)}_{(+7)} + \underbrace{(-5) + (-1) + (-2)}_{(-8)} = -1 \quad \blacksquare$$

EXAMPLE 8   Mrs. Jane Brown's normal temperature is 98° Fahrenheit. During a recent illness she kept a record of the number of degrees her temperature rose or dropped every day for a week.

Monday:	rose 4°F	Friday:	dropped 2°F
Tuesday:	rose 1°F	Saturday:	dropped 2°F
Wednesday:	dropped 3°F	Sunday:	rose 1°F
Thursday:	rose 3°F		

(a) How many degrees did Mrs. Brown's temperature rise during the week?

(b) How many degrees did Mrs. Brown's temperature drop during the week?

(c) What was the total change in her temperature during the week?

(d) What was Mrs. Brown's temperature at the end of the week?

Solution   Let + indicate a rise in temperature and − a drop in temperature. Then

(a) $(+4)°F + (+1)°F + (+3)°F + (+1)°F = (+9)°F$

(b) $(-3)°F + (-2)°F + (-2)°F = (-7)°F$

(c) $(+9)°F + (-7)°F = (+2)°F$ net change

(d) $(+98)°F + (+2)°F = (+100)°F$ $\blacksquare$

Subtraction of Signed Numbers

The operation of subtraction is the inverse operation of addition. That is,

$$a - b = c \quad \text{means} \quad a = c + b$$

For example, $7 - 3 = 4$ because $4 + 3 = 7$. In general, if $a = c + b$, then adding $(-b)$ to both sides gives

$$a + (-b) = (c + b) + (-b) = c + [b + (-b)] = c + 0 = c$$

Thus,

$$a - b = a + (-b)$$

Rule for Subtraction of Signed Numbers
Change the sign of the number to be subtracted and use the rule for addition of signed numbers.

EXAMPLE 9 (a) $(+10) - (+3) = (+10) + (-3) = +7$

(b) $(-21) - (-6) = (-21) + (+6) = -15$

(c) $(+30) - (-12) = (+30) + (+12) = +42$

(d) $(-15) - (-18) = (-15) + (+18) = +3$

EXAMPLE 10 Compute $(-74) - (-21) + (+30) - (+10)$.

Solution $(-74) - (-21) + (+30) - (+10) = (-74) + (+21) + (+30) + (-10)$

$$= (-74) + (-10) + (+21) + (+30)$$

$$= (-84) + (+51)$$

$$= -33 \quad \blacksquare$$

Often we omit the $+$ sign for positive numbers when adding and subtracting signed numbers.

EXAMPLE 11 $120 - 47 + 83 - 28 + 5$

Solution $120 - 47 + 83 - 28 + 5 = (120) + (-47) + (83) + (-28) + (5)$

$$= 133 \quad \blacksquare$$

EXERCISES 1.5

In Exercises 1–10, state the absolute value of each number.

1. 0   2. -1.6 3. -5 4. 7

5. $-\frac{1}{3}$ 6. $\frac{1}{5}$ 7. $-e$ 8. π

9. -1 10. -0.76

In Exercises 11–16, evaluate each expression.

11. $6 + |6|$ 12. $6 + |-6|$

13. $-6 + |-6|$ 14. $-|6| + |-6|$

15. $-(|-2| + |5|)$ 16. $-(-|-4| + |-4|)$

In Exercises 17–32, state whether the statement is true or false.

17. $|-30| = |30|$ 18. $|-4| = -|4|$

19. $|0| = -|0|$ 20. $|-10| = -10$

21. $|5| + |-5| = 0$ 22. $|8| - |-8| = 0$

23. $|2\frac{1}{2}| + |-2\frac{1}{2}| = 5$ 24. $-|-6| = -6$

25. $-|-3| = 3$

26. $-[|7| - |-7|] = 0$

27. The absolute value of every real number is greater than zero.

28. For all real numbers a, $|a| = |-a|$.

29. Some real numbers do not have absolute values.

30. For all real numbers a and b, $|a + b| = |a| + |b|$.

145,983

31. For all real numbers a and b, $|a - b| = |a| - |b|$.

32. For all real numbers a and b, $|a - b| = |b - a|$.

In Exercises 33–52, evaluate the given expression.

33. $6 - 2$ **34.** $25 - 10$

35. $1 - 9$ **36.** $-5 - (-12)$

37. $-6 - (-9)$ **38.** $4 - (-3)$

39. $2 - 8 - 3 + 6$ **40.** $-5 + 9 - 1 + 4$

41. $13 - (-7) - 4$ **42.** $7 - (10 + 2 - 6)$

43. $9 - (2 - 7 + 1)$ **44.** $4 - (-5) - 11$

45. $14 + (-6) - (-2)$

46. $-8 + (-3) - (-7)$

47. $(25 - 17) + (30 - 16)$

48. $(6 - 9) - (2 - 4)$

49. $7 + 16 - 33 + 10 - 5 + 3$

50. $-6 - 4 + 3 - 5 + 3 - 1$

51. $-34 + 72 - 8 + 3 - 5 - 11 + 7 - 10$

52. $20 - 30 - 3 + 14 - 21 + 27 - 5 - 7$

53. Meggin owns a stock that is traded on the New York Stock Exchange. On Monday it closed at $52 per share; it rose $7 on Tuesday; it fell $13 on Wednesday; rose $2 on Thursday; and fell $5 on Friday. What was the closing price of the stock on Friday?

54. A man who weighs 315 pounds went on the Melt-It-Away Diet. He kept a daily record of the number of pounds he gained and lost for an entire week. His record shows that on Monday he gained 3 pounds; Tuesday he lost 2 pounds; Wednesday he lost 1 pound; Thursday he gained 2 pounds (party!); and Friday lost 4 pounds.

(a) How many pounds did the man gain for the week?

(b) How many pounds did he lose for the week?

(c) What was the total change in weight for the week?

(d) How much did the man weigh at the end of the week?

55. During a heat wave, the temperature rose 4 degrees in an hour. The next hour it rose 3 degrees more, and the third hour it rose another 5 degrees, but the fourth hour it fell 6 degrees. Determine the total change in temperature.

1.6 MULTIPLICATION AND DIVISION OF SIGNED NUMBERS

Multiplication of Signed Numbers

Having discussed addition and subtraction of signed numbers, we now turn to multiplication and division of signed numbers. Recall that multiplication is a form of addition; that is, the product $(+3)(+4)$ means the sum of three 4's.

$$(+3)(+4) = (+4) + (+4) + (+4) = +12$$

Thus, it seems reasonable to define the product of two positive real numbers to be a positive real number.

Similarly,

$$(+3)(-4) = (-4) + (-4) + (-4) = -12$$

and since $(+3)(-4) = (-4)(+3)$, by the commutative law of multiplication, we have $(-4)(+3) = -12$. Thus, it would appear that the product of a positive real number and a negative real number should be a negative real number.

To determine the product of two negative real numbers, consider the following series of statements:

1. $(-3)(0) = 0$ — The product of any real number and zero is zero.
2. $[(+4) + (-4)] = 0$ — The sum of a real number and its additive inverse is zero.
3. Then $(-3)[(+4) + (-4)] = 0$ — Substituting $[(+4) + (-4)]$ for 0 in statement 1
4. $(-3)(+4) + (-3)(-4) = 0$ — Distributive law for multiplication
5. $(-12) + (-3)(-4) = 0$ — Since $(-3)(+4) = -12$, as shown above
6. $(+12) + (-12) + (-3)(-4) = 0 + (+12)$ — Addition property of equality
7. $0 + (-3)(-4) = 0 + (+12)$ — Additive inverse, $(+12) + (-12) = 0$
8. $(-3)(-4) = (+12)$ — Additive identity

Thus, the product of two negative real numbers is a positive real number. We now state a rule for the multiplication of signed numbers.

Rule for Multiplication of Signed Numbers

1. If two numbers have like signs, their product is positive.
2. If two numbers have unlike signs, their product is negative.

EXAMPLE 1   (a) $(-1)(-1) = 1$

(b) $(-\frac{1}{3})(+\frac{1}{5}) = (-\frac{1}{15})$

(c) $(+4)(-\frac{1}{2}) = -2$

(d) $(+6)(+2) = 12$

The product of more than two signed numbers can be determined by successive applications of the rule for multiplication of signed numbers.

EXAMPLE 2   Multiply $(-2)(+5)(-3)(-2)$.

Solution   Multiplying the first two factors gives

$$(-2)(+5)(-3)(-2) = (-10)(-3)(-2), \quad \text{repeating yields}$$
$$= (+30)(-2)$$
$$= -60 \quad \blacksquare$$

EXAMPLE 3   Multiply $(-3)(-3)(-3)(-3)$.

Solution   Multiplying the first two factors and the last two factors, we have

$$(-3)(-3)(-3)(-3) = (+9)(+9) = 81 \quad \blacksquare$$

The preceding examples illustrate the fact that we can find the sign of the product by counting the number of negative factors.

General Rule of Signs for Multiplication

1. If an even number of factors are negative, the product is positive.
2. If an odd number of factors are negative, the product is negative.

EXAMPLE 4   Multiply $(-7)(-3)(+1)(-1)$.

Solution   Multiplying the numbers without considering signs, we obtain 21. Since there are three negative factors, the product is negative. Thus,

$$(-7)(-3)(+1)(-1) = -21 \quad \blacksquare$$

EXAMPLE 5   Multiply $(-\frac{1}{3})(+9)(-4)(-\frac{1}{2})(+6)(-2)$.

Solution   We have an even number of negative factors, so the product is positive. Thus,

$$(-\tfrac{1}{3})(+9)(-4)(-\tfrac{1}{2})(+6)(-2) = 72 \quad \blacksquare$$

EXAMPLE 6   Ric Beach has a savings account balance of $462. He makes two deposits of $37 each. Then he withdraws $50 and on three occasions withdraws $25 each. What is the new balance of Ric's savings account?

Solution   Let the deposits be represented by a plus sign $(+)$ and withdrawals by a minus sign $(-)$. Then Ric made a total deposit of

$$(2)(37 \text{ dollars}) = 74 \text{ dollars}$$

and a total withdrawal of

$$(-1)(50 \text{ dollars}) + (-3)(25 \text{ dollars}) = -125 \text{ dollars}$$

Thus, Ric has a new balance of

$$(462 \text{ dollars}) + (74 \text{ dollars}) + (-125 \text{ dollars}) = 411 \text{ dollars} \quad \blacksquare$$

Division of Signed Numbers

The rules developed for multiplying signed numbers can be used for dividing signed numbers. This follows from the relationship between multiplication and division.

Relationship Between Multiplication and Division

For any two real numbers a and b, $b \neq 0$,

$$\frac{a}{b} = a \cdot \frac{1}{b}$$

That is, dividing a by b is equivalent to multiplying a by the multiplicative inverse of b, $1/b$. The number a is called the numerator (dividend) and the number b is called the denominator (divisor).

Note. $\frac{a}{b}$ is also written $a \div b$.

EXAMPLE 7   (a) $\frac{4}{7} = 4 \cdot \frac{1}{7}$ (b) $\frac{-3}{5} = -3 \cdot \frac{1}{5}$

EXAMPLE 8   Find the quotient $\frac{-8}{4}$.

Solution   We are looking for a real number that when multiplied by 4 becomes -8. Using our rules for the multiplication of signed numbers, we have

$$\frac{-8}{4} = -2 \quad \text{because} \quad (4)(-2) = -8 \quad \blacksquare$$

EXAMPLE 9   Find the quotient $\frac{-10}{-2}$.

Solution $\frac{-10}{-2} = 5 \quad \text{because} \quad (-2)(5) = -10. \quad \blacksquare$

Rule for Division of Signed Numbers

1. If the numerator and the denominator have like signs, the quotient is positive.
2. If the numerator and the denominator have unlike signs, the quotient is negative.

EXAMPLE 10 Simplify $\dfrac{(+7)(-8)}{(-5)(-2)}$.

Solution $\dfrac{(+7)(-8)}{(-5)(-2)} = \dfrac{-56}{10} = \dfrac{-28}{5}.$ ■

It should be noted that $\frac{-28}{5} = \frac{28}{-5} = -\frac{28}{5} = -\frac{-28}{-5}$. In general,

$$\frac{-a}{b} = \frac{a}{-b} = -\frac{a}{b} = -\frac{-a}{-b}$$

What about division involving zero?

(a) $\frac{0}{4} = ?$ (b) $\frac{4}{0} = ?$ (c) $\frac{0}{0} = ?$

In case (a), the quotient is 0 because $(0)(4) = 0$. In case (b), we see that there exists *no* real number c such that $(0)(c) = 4$; thus, the quotient $\frac{4}{0}$ is not defined. In case (c), we see that for *any* real number c $(0)(c) = 0$, and the quotient $\frac{0}{0}$ is not uniquely determined. Hence,

> division by zero is not defined for real numbers

Order of Operations

What is the value of an expression such as $6 - 2 - 7 + 9 \cdot 2 \div 3$, which has no grouping symbols to indicate the order in which operations are to be performed?

Order of Operations

1. Do all operations inside parentheses.
2. Simplify any number expression containing exponents.
3. Do multiplication and division as they occur from left to right.
4. Do addition and subtraction as they occur from left to right.

EXAMPLE 11 Simplify $6 - 2 - 7 + 9 \cdot 2 \div 3$.

Solution $6 - 2 - 7 + 9 \cdot 2 \div 3 = 6 - 2 - 7 + 18 \div 3$

$$= 6 - 2 - 2 + 6$$

$$= 4 - 2 + 6$$

$$= 2 + 6$$
$$= 8 \quad \blacksquare$$

EXAMPLE 12 Simplify $(-\frac{1}{3})[-2 + 3(4 - 7)]$.

Solution $(-\frac{1}{3})[-2 + 3(4 - 7)] = (-\frac{1}{3})[-2 + 3(-3)]$
$$= (-\frac{1}{3})[-2 + (-9)]$$
$$= (-\frac{1}{3})(-11)$$
$$= \frac{11}{3} \quad \blacksquare$$

EXERCISES 1.6

In Exercises 1–30, multiply.

1. $(9)(-7)$
2. $(-3)(-5)$
3. $(-8)(0)$
4. $(5)(3)$
5. $(-6)(6)$
6. $(10)(12)$
7. $(-1)(-1)$
8. $(-3)(-3)$
9. $(\frac{1}{2})(\frac{-1}{3})$
10. $(-4)(\frac{1}{-4})$
11. $(7)(\frac{-1}{7})$
12. $(\frac{2}{5})(\frac{3}{7})$
13. $(1.5)(.5)$
14. $(-2)(3.5)$
15. $(-1.1)(-1.1)$
16. $(3.3)(-1.7)$
17. $(-1)(-2)(-3)$
18. $(-3)(4)(-5)$
19. $(6)(-2)(1)$
20. $(-4)(-4)(-4)$
21. $(3)(3)(-3)$
22. $(-1)(-1)(-1)$
23. $(-2)(4)(3)(-2)$
24. $(-2)(-3)(-4)(-5)$
25. $(-3)(-2)(-1)(6)$
26. $(7)(1)(-1)(4)$
27. $|-10| \cdot |2| \cdot (-3)$
28. $|-8| \cdot (-2) \cdot |-2|$
29. $(9)(-7)(3)(0)$
30. $(\frac{1}{2})(9)(-8)(\frac{1}{3})$

In Exercises 31–40, simplify.

31. $3 + 2[3(-5) + 4]$
32. $(-3)(5) + (-4)(-2)$
33. $-2(-3) + 5(-6)$ 34. $5 + 2[3(-4) + 1]$
35. $3(-2)(-2) + 5(-2) - 6$
36. $2(-3)(-3) - 5(-3) + 4$
37. $[(7) - (-3)] \cdot [-12 - (-5)]$
38. $[(-2) - (-4)] \cdot [(-7) + (-3)]$
39. $[(22 - 18) - 2(-3 + 7)] \cdot [-5 - 8(3 - 4)]$
40. $[2(-5 - 2) + 3(-2 + 5)] \cdot$
     $[-3(1 - 4 - 3) - 15]$

In Exercises 41–60, find the quotient, if possible.

41. $\dfrac{12}{-3}$
42. $\dfrac{-12}{3}$
43. $\dfrac{12}{3}$
44. $\dfrac{-12}{-3}$
45. $\dfrac{-6}{-3}$
46. $\dfrac{-14}{7}$
47. $\dfrac{27}{3}$
48. $\dfrac{64}{-8}$
49. $\dfrac{-42}{-7}$
50. $\dfrac{87}{-3}$
51. $\dfrac{121}{-11}$
52. $\dfrac{-84}{4}$
53. $\dfrac{0}{-2}$
54. $\dfrac{0}{3}$
55. $\dfrac{4}{0}$
56. $\dfrac{-5}{0}$
57. $\dfrac{-240}{-60}$
58. $\dfrac{600}{-30}$
59. $\dfrac{500}{25}$
60. $\dfrac{-183}{3}$

In Exercises 61–80, evaluate each expression.

61. $\dfrac{(-5)(3)}{15}$
62. $\dfrac{(-9)(-6)}{-27}$
63. $\dfrac{-48}{(-4)(-4)}$
64. $\dfrac{(-10)(-6)}{(5)(4)}$
65. $\dfrac{(-14)(5)}{(7)(-5)}$
66. $\dfrac{(-12)(6)}{(-8)(-9)}$
67. $-\left(\dfrac{-36}{-9}\right)$
68. $-\left(\dfrac{42}{-6}\right)$
69. $-\left[-\left(\dfrac{35}{-5}\right)\right]$
70. $-\left[-\left(\dfrac{-63}{-9}\right)\right]$
71. $\dfrac{3 + (-12)}{7 - 16}$
72. $\dfrac{-3 - 2}{10 - 15}$
73. $\dfrac{-(-9) + 21}{-(-3) - 8}$
74. $\dfrac{-10 + (-8)}{-4 + 6}$
75. $\dfrac{10(-7)}{-15 - (-8)} - \dfrac{-14 - 6}{(-2)(-2)}$

76. $\dfrac{16}{(-2)(4)} + \dfrac{12 + 8}{-10}$

77. $\dfrac{27 - (-13)}{(-4)(2)} \cdot \dfrac{-34 - (-4)}{(-5)(-2)}$

78. $\dfrac{(-3)(-4)(-2)}{3 - 2 - 4} \cdot \dfrac{(-6)(-4)}{-2 - (-10)}$

79. $\dfrac{(-3)(-2) - 4}{(2 - 3)(-5)} \div \dfrac{6[(-1)(-5) + 7]}{[-1 + (-2)]\cdot[5 - 2]}$

80. $\dfrac{1 - (-2)(-3)}{(-1)(5) - 0} \div \dfrac{-4 - (-1)(-6)}{(-3)(4) + (2)(7)}$

81. Pamela Hill has assets of $2000. After withdrawing $1100, she then divides her assets equally among her three children. How much does each child get?

82. During a certain week, IBM stock rose 4 points on Monday, rose 7 points on Tuesday, fell 5 points on Wednesday, rose 2 on Thursday, and fell 10 points on Friday. Find the average change in IBM stock for the week.

83. During one week in the winter, the highest daily temperature readings were 4°C, −6°C, 8°C, 10°C, −4°C, −10°C, and −5°C. Find the average temperature reading.

84. A woman who weighs 152 pounds goes on a diet. She keeps a daily record of the number of pounds she loses and gains. On Monday she loses 2 pounds, Tuesday she loses 2 pounds, Wednesday she gains 1 pound, Thursday she gains 3 pounds, and Friday she loses 5 pounds.
(a) How many pounds did she gain during the week?
(b) How many pounds did she lose during the week?
(c) What was the total result of the diet for the week?
(d) How much did she weigh at the end of the week?

85. Kelly bought three rugs for $2.50 each, two pillows for $1.75 each, and five candles for $1.00 each. She later sold the rugs for $3.00 each, the pillows for $2.00 each, and the candles for 50 cents each. What was the net financial result of the transactions?

Chapter 1 SUMMARY

SETS

Symbol	How Read	Meaning
1. $A = \{a, b, c, \ldots\}$	The set A whose elements are $a, b, c, \ldots$	Roster (list) notation for sets
2. $A = \{x \mid x \text{ has property } p\}$	The set A whose elements x satisfy property p	Set-builder notation for sets
3. $x \in A$	x "is an element of" the set A.	
4. $x \notin A$	x " is *not* an element of" the set A.	
5. $\emptyset$ or $\{ \}$	The empty set.	The set that contains no elements
6. $A = B$	Set A equals set B.	A and B contain the same elements.
7. $A \subseteq B$	A is a subset of B.	Every member of set A is also a member of set B.

8. $A \subset B$	A is a **proper subset** of B.	A is a subset of B but A is not equal to B.
9. U	The **universal set**	The set that consists of all elements being considered in a particular problem
10. A'	**complement** of set A	The elements that belong to the universal set but not to set A
11. $A \cap B$	A **intersect** B	The elements that belong to both set A *and* set B
12. $A \cup B$	A **union** B	The elements that belong to A or B or both A and B

REAL NUMBER SYSTEM

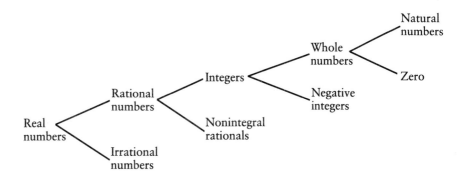

A **rational number** is any number that can be written in the form $\dfrac{a}{b}$, where a and b are integers and $b \neq 0$.

PROPERTIES OF THE REAL NUMBERS

1. *Commutative Property for Addition*

$$a + b = b + a$$

2. *Associative Property for Addition*

$$(a + b) + c = a + (b + c)$$

3. *Additive Identity: Zero*

$$a + 0 = 0 + a = a$$

4. *Additive Inverse: Negative*

$$a + (-a) = (-a) + a = 0$$

1. *Commutatve Property for Multiplication*

$$a \cdot b = b \cdot a$$

2. *Associative Property for Multiplication*

$$(a \cdot b) \cdot c = a \cdot (b \cdot c)$$

3. *Multiplicative Identity: One*

$$a \cdot 1 = 1 \cdot a = a$$

4. *Multiplicative Inverse: Reciprocal*

$$a \cdot \frac{1}{a} = \frac{1}{a} \cdot a = 1, \text{ if } a \neq 0$$

5. *Distributive Properties*

$$a \cdot (b + c) = a \cdot b + a \cdot c \quad \text{and} \quad (a + b) \cdot c = a \cdot c + b \cdot c$$

6. *Cancellation Property for Addition*
   If $a + c = b + c$, then $a = b$.
7. *Double Negative Property*

$$-(-a) = a$$

6. *Cancellation Property for Multiplication*
   If $ac = bc$ and $c \neq 0$, then $a = b$.

8. *Zero-Factor Property*

$$a \cdot 0 = 0 \cdot a = 0, \text{ and if } a \cdot b = 0, \text{ then } a = 0 \text{ or } b = 0.$$

9. *Trichotomy Property*
   Exactly one of the following is true: $a > b$, $a = b$, or $a < b$.
10. *Transitive Property of Order*
   If $a > b$ and $b < c$, then $a < c$.
11. *Additive Property of Order*
   If $a < b$, then $a + c < b + c$.
   If $a > b$, then $a + c > b + c$.

11. *Multiplicative Property of Order*
   (i) If $a < b$ and $c > 0$, then $ac < bc$.
   (ii) If $a > b$ and $c > 0$, then $ac > bc$.
   (iii) If $a < b$ and $c < 0$, then $ac > bc$.
   (iv) If $a > b$ and $c < 0$, then $ac < bc$.

OPERATIONS OF SIGNED NUMBERS

1. Absolute value $|a| = \begin{cases} a, \text{ if } a \geq 0 \\ -a, \text{ if } a < 0 \end{cases}$

2. Addition of signed numbers:
   like signs: Add absolute values and use common sign.
   unlike signs: Subtract the absolute values (the smaller from the larger) and use the sign of the number with the larger absolute value.

3. Subtraction of signed numbers: $a - b = a + (-b)$

4. Multiplication of signed numbers:

(a) If two numbers have like signs, their product is positive.

(b) If two numbers have unlike signs, their product is negative.

5. General rule for signs of multiplication:
   (a) If an even number of factors are negative, the product is positive.
   (b) If an odd number of factors are negative, the product is negative.

6. Relation between multiplication and division:
   $$\frac{a}{b} = a \cdot \frac{1}{b}$$

7. Division of signed numbers:
   (a) If the numerator and denominator have like signs, then the quotient is positive.

(b) If the numerator and denominator have unlike signs, then the quotient is negative.

8. $-\dfrac{a}{b} = \dfrac{-a}{b} = \dfrac{a}{-b} = -\dfrac{-a}{-b}$

9. Division by zero is not defined for real numbers.

ORDER OF OPERATIONS

1. Do all operations inside parentheses.
2. Simplify any number expression containing exponents.
3. Do multiplication and division as they occur from left to right.
4. Do addition and subtraction as they occur from left to right.

Chapter 1 EXERCISES

In Exercises 1–5, tell which statements are true and which are false.

1. $7 \in \{0, -2, 3, -7\}$
2. $9 \notin \{-9, 7, 4, 9\}$
3. $\emptyset \notin \{4, -3, 7, 6, 11\}$
4. $\{7\} \in \{5, 17, -8, 7, 10\}$
5. $0 \in \{x \mid x \text{ is a multiple of } 3\}$

In Exercises 6–10, list the members of each set.

6. $\{x \mid x \text{ is one of the five Great Lakes of the United States}\}$
7. $\{x \mid x \text{ is a U.S. Senator from your state}\}$
8. $\{x \mid x \text{ is the planet(s) on which you have lived}\}$
9. $\{x \mid x \text{ is a suit in a deck of playing cards}\}$
10. $\{x \mid x \text{ is a person living in the U.S.A. and whose age is over 300 years old}\}$

In Exercises 11–15, use set-builder notation to write the set.

11. $A = \{\ldots, -8, -6, -4, -2\}$
12. $B = \{41, 42, 43, 44, 45, 46, 47, 48, 49\}$
13. $C = \{\text{French, Spanish, Russian, German, Swahili, } \ldots\}$
14. $D = \{\text{oxygen, sodium, hydrogen, chlorine, } \ldots\}$
15. $E = \{Hamlet, Taming of the Shrew, Macbeth, \ldots\}$

In Exercises 16–20, identify each set as finite or infinite.

16. The set of numbers that are multiples of 6.
17. $\{x \mid x \text{ is an odd integer}\}$
18. $\{1, 2, 3, 4, \ldots\}$
19. The number of subsets that can be formed from a set containing n elements.
20. The set of all points exactly 3 meters away from a given point.

In Exercises 21–25, let $A = \{1, 2, 3, 4, 5\}$ and determine which of the following sets are equal to A.

21. $\{5, 3, 2, 1, 4\}$ 22. $\{1, 2, 3, 4\}$
23. $\{4, 5, 5, 1, 3, 2, 2\}$
24. $\{x \mid x \text{ is a positive integer less than } 5\}$
25. $\{x \mid x \text{ is a positive integer less than or equal to } 5\}$
26. Prove that $C = \{4, 5, 6, 8\}$ is not a subset of $D = \{x \mid x \text{ is a positive even integer}\}$.
27. Let $M = \{0, 1, 3\}$. List all subsets of M.
28. Let $N = \{0, \{1, 3\}\}$. List all subsets of N.
29. Let $P = \{\emptyset\}$. List all subsets of P.
30. List all subsets of $\emptyset$.

In Exercises 31–40, determine which statements are true.

31. $\{4, 6, 9\} \subseteq \{7, 5, 4, 0, 3, 6, 9\}$
32. $\{-7, 3, -15\} \subseteq \{-15, -7, 3\}$
33. $\emptyset \subseteq \{-3, -4, 5, 1, 0\}$
34. $\{a, b, c, d\} \subseteq \{b, a, e, d, c\}$
35. $\{x, y, z\} \nsubseteq \{z, y, x\}$ 36. $\{0\} \subseteq \emptyset$
37. $\{\{4\}\} \subseteq \{4\}$ 38. $\emptyset \subseteq \{\{0\}\}$
39. $\{x \mid x \text{ is a prime number}\} \subseteq \{x \mid x \text{ is an odd number}\}$
40. $\{x \mid x \text{ is an even prime number}\} \subseteq \{x \mid x \text{ is a divisor of } 12\}$

In Exercises 41–50, consider the following sets: $\emptyset$,

$M = \{2\}$, $N = \{2, 5\}$, $P = \{2, 6, 9\}$, $Q = \{2, 3, 5, 6, 8\}$, $R = \{2, 5, 6, 8, 9\}$, and $U = \{1, 2, 3, 4, 5, 6, 7, 8, 9\}$. *Insert the correct symbol* $\subseteq$ *or* $\nsubseteq$ *between each pair of sets.*

41. $M \_\_ M$ 42. $\emptyset \_\_ M$ 43. $M \_\_ N$
44. $N \_\_ P$ 45. $N \_\_ R$ 46. $P \_\_ Q$
47. $P \_\_ R$ 48. $Q \_\_ R$ 49. $Q \_\_ U$
50. $U \_\_ U$
51. Use Venn diagrams to illustrate that if $A \subseteq B$ and $B \subseteq C$, then $A \subseteq C$.
52. Let $S = \{\{a, b\}, \{m\}, \{p, q, r\}\}$. How many subsets does S have? List all the subsets of S.

In Exercises 53–63, let $U = \{1, 2, 3, 4, 5, 6, 7, 8, 9, 10\}$, $P = \{2, 4, 6, 8, 10\}$, $Q = \{6, 7, 8, 9, 10\}$, $R = \{1, 3, 5, 7, 9\}$, and $S = \{6, 8, 5, 1\}$. *List the elements in the given sets.*

53. $P \cap Q$ 54. $R \cup S$
55. $P' \cap R$ 56. $P' \cap R'$
57. $(P \cap Q)'$ 58. $P \cup (R \cap S)$
59. $P \cap (Q \cup S)$ 60. $P \cap S'$
61. $(P' \cup Q)'$ 62. $Q \cap (R \cup S)'$
63. $(P \cap Q)' \cup R'$

In Exercises 64–67, let $U = \{x \mid x \text{ is a college student}\}$, $S = \{x \mid x \text{ is a senior}\}$, and $M = \{x \mid x \text{ is a mathematics major}\}$.

64. The set of seniors who are mathematics majors is represented by:
(a) $S \cup M$ (b) $S' \cup M$
(c) $S' \cap M$ (d) $S \cap M'$
(e) $S \cap M$
65. The set of students that are not mathematics majors and are not seniors is represented by:
(a) $S \cup M'$ (b) $S' \cap M'$
(c) $S' \cap M$ (d) $(S \cap M)'$
(e) $S' \cup M'$
66. The set of seniors that are not mathematics majors is represented by:
(a) $S \cap M'$ (b) $S' \cup M$
(c) $S' \cap M$ (d) $S' \cap M'$
(e) $S' \cup M'$
67. The set of students that are either seniors or mathematics majors is represented by:
(a) $S \cap M$ (b) $S' \cap M'$
(c) $S \cup M$ (d) $S' \cup M'$
(e) $S' \cup M$

In Exercises 68–70, use Venn diagrams to verify the given statement.

68. $(A \cup B') \cap (A \cup B) = A$

69. $(A' \cup B') \cap (A' \cup B) = A'$
70. $(A' \cup B') \cap (A' \cup B) \cap (A \cup B) = A' \cap B$

In Exercises 71–74, complete the statement:

71. If $A \subseteq B$, then $A \cup B = ?$
72. If $A \subseteq B$, then $A \cap B = ?$
73. If $A \cup B = B$, then $? \subseteq ?$
74. If $A \cap B = A$, then $? \subseteq ?$
75. Give an example to show that we can have $A \cap B = A \cap C$ without $B = C$.

In Exercises 76–87, Let $R = $ real numbers, $Q = $ rational numbers, $Q' = $ irrational numbers, $J = $ integers, $N = $ natural numbers, and $N' = $ negative numbers. State whether each of the following is true.

76. $-4 \in N$ 77. $-7 \in Q$
78. $\sqrt{16} \in N$ 79. $\sqrt{7} \in Q'$
80. $0 \in J$ 81. $\frac{2}{3} \in Q'$
82. $\pi \in Q'$ 83. $0 \in N$
84. $J \subseteq Q$ 85. $N' \subseteq R$
86. $N \subseteq Q$ 87. $R = Q \cup Q'$

In Exercises 88–108, state the property of the real numbers that is illustrated by each example.

88. $7 + (-3) = (-3) + 7$
89. $(-14)(5) = (5)(-14)$
90. $(x + y) + 3 = x + (y + 3)$
91. $(4x)y = 4(xy)$
92. $(-7)[4 + y] = (-7)(4) + (-7)(y)$
93. $(-3) + 0 = -3$ 94. $25 \cdot 1 = 1 \cdot 25$
95. $4 + (-4) = 0$ 96. $16 \cdot \frac{1}{16} = \frac{1}{16} \cdot 16$
97. $-(-\frac{2}{3}) = \frac{2}{3}$
98. If $u + 3 = v + 3$, then $u = v$.
99. $14 \cdot 0 = 0 \cdot 14$ 100. $-[-(-6)] = -6$
101. If $a - 7 = b - 7$, then $a = b$.
102. If $3x = 3y$, then $x = y$.
103. If $x < 5$ and $5 < 12$, then $x < 12$.
104. $(7 + 2) \cdot 3 = 7 \cdot 3 + 2 \cdot 3$
105. If $\frac{2}{3}a = \frac{2}{3}b$, then $a = b$.
106. If $x < y$, then $4x < 4y$.
107. If $x < y$, then $-x > -y$.
108. If $x(x - 1) = 0$, then either $x = 0$ or $x - 1 = 0$ or both $x = 0$ and $x - 1 = 0$.

In Exercises 109–114, evaluate:

109. $|37| + |-37|$ 110. $|-42| - |42|$
111. $|0|$ 112. $2 \cdot |7| + |-7|$
113. $-3|5| + |-15|$
114. $-[3 \cdot |-5| + (-4)]$

In Exercises 115–128, perform the indicated operations.

115. $(-1) + (-5)$ 116. $7 - (-5)$

117. $(3) + (-6) - (-9)$
118. $(-3) - (-8) + (-1)$
119. $(10)(-3)(-2)$ 120. $(14)(-1)(-1)(1)$
121. $(-5)(-4)(-3)$ 122. $-(-6)(7)(-4)$
123. $(-20) \div 4$ 124. $-(-36) \div 6$
125. $\frac{-8}{4} - \frac{12}{-3}$
126. $[2 - (-10)] \div [2 - (-4)]$
127. $\frac{(-1)(-12)}{11 - (+5)} + \frac{-3 + (-21)}{(-2)(-6)}$
128. $\frac{(-5)(2)}{(-2) - (-12)} \cdot \frac{(-7) + (-2)}{1 - (-2)}$

In Exercises 129–133, complete the chart:

129.

If $n =$	-1	3	0	5	-2
then $n + (-2) =$					

130.

If $n =$	-7	0	4	-3	5
then $(-3) - n =$					

131.

If $n =$	2	$-\frac{1}{3}$	-1	0	-3
then $6(-n) =$					

132.

If $n =$	4	-1	2	5	-3
then $(-4) \div n =$					

133.

If $n =$	10	0	-14	5	-20
then $(-n) \div (-2) =$					

134. A ship sails directly west from Port Rand for 103 miles and then sails 154 miles directly east. Where is the ship then located relative to Port Rand?

135. Gilbert Bradberry owns 10 shares of the Widget Company of America. On Monday it closed at \$42 per share; on Tuesday it fell \$5; onWednesday it rose \$4; on Thursday it rose \$13; and on Friday it fell \$20. Using signed numbers, determine the worth of Gilbert's stock in the Widget Co. of America at the end of the week.

136. A student obtains exam grades of 67, 72, 94, and 69 on his mathematics exams. Without computing his average exam grade, determine whether his average is over or under 73; over or under 79; over or under 77.

137. At a meeting of the Alameda Jazzercise Club, the members reported the following changes in weight for the previous week: three members gained 2 pounds, two members lost 4 pounds, four members lost 1 pound, five members' weights remained the same, twelve members gained 1 pound, and five members lost 5 pounds. What was the total pounds lost or gained by the club?

138. Three explorers climb out of a cave (located at the base of a mountain) whose lowest point is 300 feet below sea level. They proceed to climb to the top of the mountain whose peak is 5700 feet above sea level. If the temperature at the lowest point of the cave was 120°F and drops 20°F for every 600 feet as they climb to the peak of the mountain, what is the temperature of the peak?

*139. Let a, b, and c be real numbers. Determine which of the following statements are false. If false, give an example that shows it to be false (this is called a counterexample).
   (a) $|a + b| = |a| + |b|$
   (b) $|a - b| = |a| - |b|$
   (c) $|ab| = |a|\,|b|$
   (d) $\left|\dfrac{a}{b}\right| = \dfrac{|a|}{|b|}$

*140. Show that the given expression is true by stating the reason that justifies each step of the proof.

Statement: $a[b + (-c)] = ab + (-ac)$

Proof:
$$
\begin{aligned}
a[b + (-c)] &= ab + a(-c) & \text{Step 1} \\
&= ab + a[(-1)c] & \text{Step 2} \\
&= ab + a[c(-1)] & \text{Step 3} \\
&= ab + (ac)(-1) & \text{Step 4} \\
&= ab + (-ac) & \text{Step 5}
\end{aligned}
$$

*141. If $a = -b$, show that $-a = b$.

*142. Prove that $-(ab) = (-a)b$.

*143. Prove that $(-a)(-b) = ab$.

*144. Prove that if $a \neq 0$ and $b \neq 0$, then $(1/a)(1/b) = 1/ab$.

*145. Show that the set of prime numbers is not closed under addition or multiplication.

2 Polynomials

Arithmetic is the branch of mathematics that deals with the addition, subtraction, multiplication, and division of numbers. In algebra, however, these operations are performed not only on numbers, but also on symbols that represent arbitrary elements of any set. Recall that in Chapter 1 we used letters such as x and y to represent any element of a given set and that such letters are called **variables**. In this chapter we discuss how the fundamental operations of addition, subtraction, multiplication, and division can be applied to expressions involving variables and/or numbers. In particular, we discuss a very important type of algebraic expression called a **polynomial**.

2.1 DEFINITIONS

If two or more numbers (or variables) are to be multiplied, each of the numbers (or variables), as well as the product of any of them, is called a **factor** of the product.

EXAMPLE 1 In the product $5xy$, the factors are 1, 5, x, y, $5x$, $5y$, xy, and $5xy$.

If the factors in a product are identical numbers (or variables), we can use the notation of exponents to indicate how many times the number (or variable) is a factor.

EXAMPLE 2 (a) $4 \cdot 4 \cdot 4 \cdot 4 \cdot 4$ can be written as 4^5. The expression 4^5 is read "4 to the fifth power" or the "fifth power of 4."

(b) $x \cdot x \cdot x \cdot x \cdot x$ can be written x^5 and is read "x to the fifth power" or the "fifth power of x."

DEFINITION

In general, if a is any real number and n is a positive integer, then

$$a^n = \underbrace{a \cdot a \cdot a \cdot \ldots \cdot a}_{n \text{ factors}}$$

The number a is called the **base** and n is called the **exponent**, and we say that a^n is the nth power of a. When $n = 1$, we simply write a rather than a^1.

EXAMPLE 3 (a) $y^6 = y \cdot y \cdot y \cdot y \cdot y \cdot y$. The base is y and the exponent is 6.

(b) $(\frac{1}{3})^4 = \frac{1}{3} \cdot \frac{1}{3} \cdot \frac{1}{3} \cdot \frac{1}{3}$. The base is $\frac{1}{3}$ and the exponent is 4.

Note. An exponent refers only to the number (or variable) that is directly to the left of the exponent.

EXAMPLE 4 (a) $a^3 b^2 = a \cdot a \cdot a \cdot b \cdot b$

(b) $xy^2 = x \cdot y \cdot y$

(c) $7z^3 = 7 \cdot z \cdot z \cdot z$

(d) $(7z)^3 = (7z)(7z)(7z) = 7 \cdot 7 \cdot 7 \cdot z \cdot z \cdot z$

PROBLEM 1 Is $(-3)^4 = -3^4$?

Answer $(-3)^4 = (-3)(-3)(-3)(-3) = 81$
However, $-3^4 = -(3 \cdot 3 \cdot 3 \cdot 3) = -81$.
Since $81 \neq -81$, we have $(-3)^4 \neq -3^4$. ■

PROBLEM 2 Given that the area of a rectangle equals length times width, find the area of the rectangle whose length and width are both x inches.

Answer

$$A = x^2 \text{ in.}^2 \qquad x \text{ inches}$$

x inches

Area $= (x \text{ inches}) \cdot (x \text{ inches}) = x^2$ square inches

Recall that a rectangle whose dimensions (length and width) are equal is called a **square**. ■

EXERCISES 2.1

In Exercises 1–12, name the base(s) and exponent(s) of the expression.

1. 3^5 **2.** $(-5)^2$

3. r^4 **4.** z

5. 10^7 **6.** $(2t)^3$

7. $(a + b)^6$ **8.** $4x^2$

9. $(4x)^2$ **10.** $-4t^5$

11. $9x^2y^3$ **12.** $-7m^4n^5$

In Exercises 13–24, write the given expression using exponents.

13. $2 \cdot 2 \cdot 2$ **14.** $10 \cdot 10 \cdot 10 \cdot 10 \cdot 10$

15. $z \cdot z \cdot z \cdot z$ **16.** $3 \cdot a \cdot a \cdot a \cdot a$

17. $(-4y)(-4y)(-4y)$

18. $-3 \cdot a \cdot a \cdot b \cdot b \cdot b$ **19.** $-5 \cdot \dfrac{1}{y} \cdot \dfrac{1}{y} \cdot \dfrac{1}{y}$

20. $(x + y)(x + y)$ **21.** $\pi \cdot r \cdot r$

22. $\dfrac{4}{3} \cdot \pi \cdot r \cdot r \cdot r$

23. $a \cdot a \cdot b \cdot b - a \cdot b \cdot b \cdot b \cdot c \cdot c$

24. $x \cdot x \cdot x - y \cdot y \cdot y$

In Exercises 25–40, write the given expression without exponents.

25. x^5 **26.** y^7

27. a^3b^4 **28.** c^2d^3

29. $(-2x)^4$ **30.** $(5a)^3$

31. $-2x^4$ **32.** $5a^3$

33. $(a - b)^2$ **34.** $(x + y)^3$

35. $(-a)^3(-b)^4$ **36.** $-a^2(-b)^2$

37. $x^3y - 4xy^2$ **38.** $a^2b^4 - ab^3$

39. $(2a)^2b^2 - c^2$ **40.** $5p^3q + 2p^2q^4$

In Exercises 41–50, write each expression using exponents.

41. The second power of nine (often called the square of nine).

42. The third power of negative five (often called the cube of negative five).

43. One-sixth the square of h.

44. One-fourth the cube of t.

45. The square of $8M$.

46. The cube of $-5i$.

47. The ninth power of xy.

48. The fifth power of $(a - 3)$.

49. The cube of the sum of x plus 15.

50. The fourth power of x squared minus y cubed.

51. The volume of a cube is length times width times height. Find the volume of a cube with length equal to x inches.

2.2 ALGEBRAIC EXPRESSIONS AND POLYNOMIALS

An **algebraic expression** is an expression formed from any combination of numbers and variables by using the operations of addition, subtraction, multiplication, division, or taking roots.

EXAMPLE 1 The following are algebraic expressions:

(a) -3 (e) $\dfrac{5x^2 - 3y}{6}$

(b) $5x^3y^2$ (f) $\dfrac{x^2 - 3x + 7}{4x^2 - 5y}$

(c) $4x^2 - 2x + 1$ (g) $\dfrac{1}{x}$

(d) $-13xy + 3y - 2x + 6$ (h) $\sqrt{3x}$

When specific numbers are substituted for the variables in an algebraic expression, the resulting number is called the **value** of the expression for these values of the variables.

EXAMPLE 2 Find the value of $\dfrac{x^2 - 3x + 7}{4x^2 - 5y}$ when $x = -1$ and $y = 2$.

Solution Substituting $x = -1$ and $y = 2$, we obtain

$$\frac{(-1)^2 - 3(-1) + 7}{4(-1)^2 - 5(2)} = \frac{1 + 3 + 7}{4 - 10} = \frac{11}{-6} \quad \blacksquare$$

By an algebraic expression *in* certain variables, we mean an algebraic expression that contains only those variables. A **constant** is an algebraic expression that contains no variables. When an algebraic expression is written as the sum of other algebraic expressions, each of the expressions is called a **term** of the given algebraic expression.

EXAMPLE 3 The algebraic expression (c) in Example 1 can be written as the sum $(4x^2) + (-2x) + 1$; hence $4x^2$, $-2x$, and 1 are terms of $4x^2 - 2x + 1$.

Any factor or group of factors in a term is said to be the **coefficient** of the remaining factors in the term. For example, in the term $-13xy$, the $-13x$ is the coefficient of y; y is the coefficient of $-13x$; -13 is the numerical coefficient of xy.

Throughout the rest of the text, unless otherwise indicated, the word *"coefficient"* will refer to the numerical coefficient.

EXAMPLE 4 The coefficient of $5xy$ is 5, the coefficient of $-9x^2$ is -9, and the coefficient of x^3 is understood to be 1.

PROBLEM 1 Determine the terms and their coefficients for the algebraic expression $4x^3 - 5x^2 - x + 7$.

Answer Algebraic expressions: $4x^3 - 5x^2 - x + 7$
Terms: $4x^3$, $-5x^2$, $-x$, 7
Coefficients: 4, -5, -1, 7 $\blacksquare$

> **DEFINITION**
> An algebraic expression that involves *only* addition, subtraction, multiplication, division, and in which all variables occur only as nonnegative integral powers, is called a **rational expression**.

EXAMPLE 5 The following are rational expressions:

(a) $\dfrac{x + y}{3}$ (b) $z + \dfrac{1}{z}$ (c) $\dfrac{5x^2 - 3y}{6}$

(d) -10 (e) $\dfrac{x^2 - 3x + 7}{4x^2 - 5}$ (f) $4x^2 - 2x + 1$

DEFINITION

Any rational expression in which *no* variable occurs in a denominator is called a **polynomial**.

EXAMPLE 6 In Example 5, the rational expressions (a) $\dfrac{x + y}{3}$, (c) $\dfrac{5x^2 - 3y}{6}$, (d) -10, and (f) $4x^2 - 2x + 1$ are polynomials. However (b) $z + \dfrac{1}{z}$ and (e) $\dfrac{x^2 - 3x + 7}{4x^2 - 5}$ are *not* polynomials.

PROBLEM 2 Determine which of the following are polynomials and which are not polynomials.

(a) $4x^{1/3} + x^3y^2 + 12y$ (b) $-6y^4z + 9$

(c) $y^4 + y^{-2} + 3$ (d) $3x^2 - \dfrac{1}{x}$

(e) -10

Answer Since every polynomial is a rational expression, every exponent in a polynomial must be a nonnegative integer number, and no variable may occur in a denominator. Thus,

(a) $4x^{1/3} + x^3y^2 + 12y$ is *not* a polynomial since it contains a fractional exponent, $x^{1/3}$.

(b) $-6y^4z + 9$ is a polynomial.

(c) $y^4 + y^{-2} + 3$ is *not* a polynomial since it contains a negative exponent, y^{-2}.

(d) $3x^2 - \dfrac{1}{x}$ is *not* a polynomial since it contains a variable in a denominator.

(e) $-10 = -10x^0$ and zero is a nonnegative number. Thus, -10 is a polynomial. ∎

Figure 2.1 shows the set relationship between algebraic expressions, rational expressions, and polynomials.

It is sometimes convenient to describe a polynomial by the number of terms it contains. A polynomial consisting of exactly one term is called a

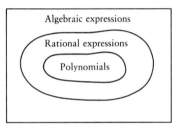

Figure 2.1 {polynomials} ⊆ {rational expressions} ⊆ {algebraic expressions}

monomial. For example, 7, $4xy$, and $-3a^2bc^2d$ are all monomials. A polynomial consisting of exactly two terms is called a **binomial**. For example, $x - y, -2y^2 + 3y$, and $\frac{1}{3}x + 5$ are all binomials. A polynomial consisting of exactly three terms is called a **trinomial**. For example, $x + y + z$, $4x^3 + 2x - 7$, and $ax^2 + bx + c$ are all trinomials.

Degree of a Polynomial

The **degree of a term** of a polynomial is given by adding the exponents of the variables in that term.

EXAMPLE 7   The terms of $-3x^4 + 5xy^2 + 7x^2y^2 - xy + 2$ have the following degrees:

$$-3x^4 \qquad \text{is of degree 4}$$
$$5xy^2 = 5x^1y^2 \qquad \text{is of degree 3}$$
$$7x^2y^2 \qquad \text{is of degree 4}$$
$$-xy = -x^1y^1 \qquad \text{is of degree 2}$$
$$2 = 2x^0 \qquad \text{is of degree 0}$$

Note. 2 is called the **constant term** and constant terms have degree zero.

Since a polynomial consists of terms, the degree of a polynomial can be defined by computing the degrees of its terms. The **degree of a polynomial** is the degree of the term(s) with the highest degree.

EXAMPLE 8   In the polynomial $-3x^4 + 5xy^2 + 7x^2y^2 - xy + 2$, the highest degree terms are of degree 4. Thus, the polynomial $-3x^4 + 5xy^2 + 7x^2y^2 - xy + 2$ is of degree 4.

EXAMPLE 9   Evaluate each of the following polynomials.

(a) $5x^2 + 3y - 12x + 1$ for $x = -2$ and $y = 1$

(b) $x^2 - 3x - 4$ for $x = 4$

Solution (a) Letting $x = -2$ and $y = 1$, we obtain $5(-2)^2 + 3(1) - 12(-2) + 1 = 5(4) + 3(1) - 12(-2) + 1 = 20 + 3 + 24 + 1 = 48$.

(b) Letting $x = 4$, we get $(4)^2 - 3(4) - 4 = 16 - 12 - 4 = 0$. ■

Many of the polynomials we shall encounter contain only one variable. We now give the general form of a polynomial in a single variable.

DEFINITION

A polynomial in a single variable x is an algebraic expression of the form

$$a_n x^n + a_{n-1} x^{n-1} + \cdots + a_1 x + a_0$$

where n is a nonnegative integer and $a_n, a_{n-1}, \ldots, a_1, a_0$ are numerical coefficients.

The coefficient a_n of the highest power of x is called the **leading coefficient** of the polynomial, and if $a_n \neq 0$, then the polynomial is said to be of **degree** n. If *all* the coefficients $a_n, a_{n-1}, \cdots, a_1, a_0$ are zero, the polynomial is called the **zero polynomial** and is denoted by 0. We usually do not assign a degree to the zero polynomial.

EXAMPLE 10 The polynomial $-6x^3 + 5x^2 - 2x + 4$ is a polynomial in the variable x of degree 3 and leading coefficient -6.

PROBLEM 3 A shopper bought x sheets, costing \$15 per sheet, and y pillowcases, costing \$9 per pillowcase. What does the polynomial $15x + 9y$ represent?

Answer The first term

$$15 \, \frac{\text{dollars}}{\text{sheet}} \cdot x \, \text{sheets} = 15x \, \text{dollars}$$

gives the total cost, in dollars, of the sheets. The second term

$$9 \, \frac{\text{dollars}}{\text{pillowcase}} \cdot y \, \text{pillowcases} = 9y \, \text{dollars}$$

gives the total cost, in dollars, of the pillowcases. Thus, the polynomial $15x + 9y$ represents the total cost of the sheets and pillowcases. ■

EXERCISES 2.2

In Exercises 1–10, determine whether each expression is or is not a polynomial.

1. $-5x^2 + 3x - 3$
2. $-7x^2 z$
3. $4x^{1/2} + 7x + 6$
4. $-x^{-2} + x + 1$

5. $\dfrac{3x^2 - 2x + 7}{52}$

6. $\dfrac{1}{3}x^2 - \dfrac{1}{x} + 8$

7. 147

8. $\dfrac{1}{4}x^3 + \dfrac{1}{3}xy - \dfrac{1}{5}$

9. $x^{4/3}y - 3x + 1$

10. $x^2 + y^2 + 2xy$

In Exercises 11–20, give the terms and coefficients of the terms for each given polynomial.

11. $15x^2y + 16y$
12. $3x^2 - 5xyz + 2$
13. $x^2 - 7xy + 3y^2$
14. $2x^4 - \sqrt{3}xy + 13$
15. $\dfrac{2}{3}x^3z - \dfrac{1}{2}xz + \dfrac{4}{7}z^2$
16. $\dfrac{1}{2}y^5 + \dfrac{1}{3}y^4x - 6x^2 + 7x^3 + 4$
17. $3.1xy^3 - 10x^2y^2 + 1.5x^3$
18. $4x^5y^2 - x^3y + 2xy^2 - 1.7y$
19. $\sqrt{3}x^2 + \pi x + 15$
20. $\pi y^3 + \sqrt{2}y^2 - 0$

In Exercises 21–35, find the degree of each polynomial.

21. z^2
22. $5x^3$
23. x
24. $2s^2t$
25. $1 - x - x^2$
26. $\sqrt{3}t$
27. $x^2 - x$
28. $y^4 - wxyz$
29. -5
30. $xy^2 + x^2y^2 - xy$
31. $rstuv$
32. $4x^4y^3 + x^2y^2 - 2xy^3 + 1$
33. $\dfrac{3}{5}x^6 - 2x^4 - 1$
34. $\sqrt{3}x^{10} + 14x^6y^5 + y^{10}$
35. $5x^2y^2 - 2x^4 + 3xy^3 - 7x^3y + y^4$

In Exercises 36–50, evaluate the given expression when $x = 1$, $y = 2$, $z = 3$, and $w = 6$.

36. $2xw^2 + y^4$
37. $xyzw$
38. $yw - xz$
39. $(x + y)(w - 1)$
40. $x^2 + y^2 + z^2$
41. $xyz - w$
42. $y^2 - w^3$
43. $2x + 3y - z$
44. $\dfrac{1}{2}(2w - y)$
45. $\dfrac{z + w}{x + y}$
46. $w\left(\dfrac{1}{y} - \dfrac{1}{z}\right)$
47. $\dfrac{z - y}{x} + \dfrac{w - z}{y - x}$
48. $(w - z - y)$
49. $x^2 + 2xy + y^2$
50. $x + y(z + w)$

*51. Consider the rectangle in Figure 2.2, whose sides have length l and w.

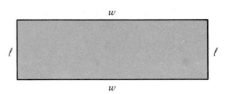

Figure 2.2

(a) Write the polynomial representing the perimeter of the rectangle.
(b) Write the polynomial representing the area of the rectangle.

*52. Consider the triangle in Figure 2.3, whose sides are x, x, and y, respectively.

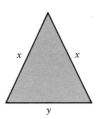

Figure 2.3

Write the polynomial representing the perimeter.

*53. Consider the geometric figure shown in Figure 2.4.

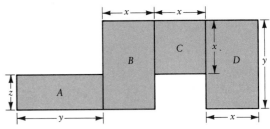

Figure 2.4

What does each of the following polynomials represent?
(a) $zy + 2xy + x^2$ (b) $2y + 2z$
(c) $(2x + 2y) + (2x + 2y)$ (d) $4x$
(e) $3y + 7x + 3z$

*54. If a car travels at a rate of 70 miles per hour for t hours, what does the polynomial $70t$ represent?

*55. Susan Newman had 75 cents to buy fruit. She bought x apples, which cost 11 cents each, and y oranges, which cost 13 cents each. Write a polynomial that represents the change Susan has after buying the fruit.

*56. If a ball is thrown upward at a speed of 64 feet per second, in the absence of air resistance, it reaches a height of h feet after t seconds according to the formula

$$h = 64t - 16t^2$$

What is the height of the ball after 3 seconds?

2.3 ADDITION AND SUBTRACTION OF POLYNOMIALS

In Chapter 1 we discussed the addition properties of real numbers. Since polynomials are expressions that represent real numbers, these properties also apply to polynomials. One of those properties that will be used in the addition and subtraction of polynomials is the distributive law. Recall that the distributive law states that

$$a(b + c) = ab + ac$$

However, using the symmetric law of equality and the commutative law of multiplication, the distributive law may be written in the form

$$ab + ac = a(b + c) = (b + c)a$$

We now show how these properties can be used to add or subtract polynomials.

Recall that a polynomial consists of terms. Those terms of a polynomial that differ only in their numerical coefficients are called **like terms**. For example,

$$3x \quad \text{and} \quad 7x$$
$$-5x^2 \quad \text{and} \quad 12x^2$$
$$6xy \quad \text{and} \quad 2xy$$
$$14x^2y^3 \quad \text{and} \quad -8x^2y^3$$

are like terms. However,

$$2x \quad \text{and} \quad 2x^2$$

are *not* like terms because they have different powers of the variable x. The term $2x$ has x to the first power, and the term $2x^2$ has x to the second power.

PROBLEM 1   Can we add the terms $7x$ and $3x$?

Answer   Yes, we can add the like terms $7x$ and $3x$. Using the distributive property, we have

$$7x + 3x = (7 + 3)x = 10x \quad \blacksquare$$

EXAMPLE 1   Add the like terms $14x^2y^3$ and $-8x^2y^3$.

Solution $14x^2y^3 + (-8x^2y^3) = [14 + (-8)]x^2y^3 = 6x^2y^3$ ▪

Adding Polynomials

To add polynomials, add the coefficients of like terms in the polynomials.

EXAMPLE 2 Find the sum of the polynomials $7xy + 10$ and $3xy + 5$.

Solution Grouping like terms and using the distributive law gives

$$(7xy + 10) + (3xy + 5) = (7xy + 3xy) + (10 + 5) \quad \text{[grouping like terms]}$$
$$= (7 + 3)xy + (10 + 5) \quad \text{[distributive law]}$$
$$= 10xy + 15 \quad ▪$$

EXAMPLE 3 Add $5x^2 - 3x$ and $7x^2 + 6x - 9$.

Solution $(5x^2 - 3x) + (7x^2 + 6x - 9)$
$$= (5x^2 + 7x^2) + (-3x + 6x) + (-9) \quad \text{[grouping like terms]}$$
$$= (5 + 7)x^2 + (-3 + 6)x + (-9) \quad \text{[distributive law]}$$
$$= 12x^2 + 3x - 9 \quad ▪$$

Sometimes it is convenient to use a vertical arrangement to perform addition.

EXAMPLE 4 Add $4x^3 - 3x^2 + 3x - 1$, $-5x^2 + 2$, and $5x^3 + x^2 - 7x$.

Solution
$$4x^3 - 3x^2 + 3x - 1$$
$$(+) \qquad - 5x^2 \qquad + 2$$
$$(+) \ \underline{5x^3 + \ x^2 - 7x}$$
$$9x^3 - 7x^2 - 4x + 1 \quad ▪$$

Many problems that we encounter in algebra contain parentheses, and, once again, the distributive law plays an important role in removing them.

Rule for Removal of Parentheses

1. When an expression within parentheses is preceded by a positive sign, $+$, the parentheses may be removed without making any change in the expression.
2. When an expression within parentheses is preceded by a negative sign, $-$, the parentheses may be removed if the sign of *every* term within the parentheses is changed.

EXAMPLE 5 Remove parentheses and combine terms: $(16x^2 - 2x + 7) - (3x^2 + x - 3)$.

Solution Since the first polynomial, $(16x^2 - 2x + 7)$, is assumed to be preceded by a positive sign, we can remove the parentheses without making any change in the expression. However, the second polynomial, $(3x^2 + x - 3)$, is preceded by a negative sign and the parentheses may be removed if the sign of every term in that polynomial is changed. Thus,

$$(16x^2 - 2x + 7) - (3x^2 + x - 3) = 16x^2 - 2x + 7 - 3x^2 - x + 3$$
$$= 16x^2 - 3x^2 - 2x - x + 7 + 3 \qquad \text{[rearranging terms]}$$
$$= 13x^2 - 3x + 10 \qquad \text{[combining like terms]} \quad \blacksquare$$

Note that this idea allows us to subtract polynomials.

EXAMPLE 6 Subtract the polynomial $(3x + 2y - 4)$ from the polynomial $(-3x + 4y)$.

Solution
$$\begin{aligned} (-3x + 4y) &\longrightarrow -3x + 4y \\ \underline{-(3x + 2y - 4)} &\longrightarrow \underline{-3x - 2y + 4} \\ &\qquad\quad -6x + 2y + 4 \end{aligned}$$

Thus, the answer is $-6x + 2y + 4$. $\blacksquare$

PROBLEM 2 Find the mistake in the following example:

$$(4x^2 - 3xy + 4) - (3x^2 - 2xy) = 4x^2 - 3xy + 4 - 3x^2 - 2xy$$
$$= 4x^2 - 3x^2 - 3xy - 2xy + 4$$
$$= x^2 - 5xy + 4$$

Answer The mistake is in the removal of the parentheses. The expression $(3x^2 - 2xy)$ is preceded by a negative sign, and the parentheses may be removed if the sign of **every term** within the parentheses is changed. In the above example only the sign of $3x^2$ was changed. The sign of $-2xy$ should have been changed to $+2xy$. Therefore, the example should read as follows:

$$(4x^2 - 3xy + 4) - (3x^2 - 2xy) = 4x^2 - 3xy + 4 - 3x^2 + 2xy$$
$$= 4x^2 - 3x^2 - 3xy + 2xy + 4$$
$$= x^2 - xy + 4 \quad \blacksquare$$

Rule for Removal of Brackets

When one set of parentheses is contained within a bracket, first remove the innermost parentheses and combine terms. Then continue this process until all parentheses, brackets, and so on, have been removed and like terms combined.

EXAMPLE 7 Simplify $12y + 4 - [(3 - y) - (5y + 6)]$.

Solution
$$12y + 4 - [(3 - y) - (5y + 6)] = 12y + 4 - [3 - y - 5y - 6]$$
$$= 12y + 4 - [-6y - 3]$$
$$= 12y + 4 + 6y + 3$$
$$= 18y + 7 \quad \blacksquare$$

PROBLEM 3 The triangle in Figure 2.5 has sides represented by $4x - 1$, $2x + 5$, and $9x - 5$. Find the perimeter of the triangle.

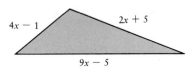

$4x - 1$ $2x + 5$

$9x - 5$

Figure 2.5

Answer The perimeter is obtained by adding the lengths of the three sides. That is,

$$(4x - 1) + (2x + 5) + (9x - 5) = 4x - 1 + 2x + 5 + 9x - 5$$
$$= 15x - 1 \quad \blacksquare$$

PROBLEM 4 A box is built in the form of Figure 2.6. The bottom costs 4 cents per square inch, the sides cost 5 cents per square inch, and the the top costs 7 cents per square inch. Give a polynomial that represents the total cost of the box.

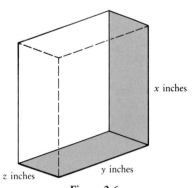

x inches

z inches y inches

Figure 2.6

Answer The area of the bottom is given by yz square inches, and its cost is

$$\frac{4 \text{ cents}}{\text{square inches}} \cdot yz \text{ square inches} = 4yz \text{ (cents)}$$

Similarly, the cost of the top is given by

$$\frac{7 \text{ cents}}{\text{square inches}} \cdot yz \text{ square inches} = 7yz \text{ (cents)}$$

Note that there are two pairs of sides: one pair having dimensions x and y; the second pair having the dimensions x and z. A side with dimensions x and y has an area given by xy square inches. Since there are two such sides, $2xy$ square inches represents the total area of the sides and

$$\frac{5 \text{ cents}}{\text{square inches}} \cdot 2xy \, \text{square inches} = 10xy \, (\text{cents})$$

represents the total cost of those two sides. Similarly, the sides with dimensions x and z have a total area of $2xz$ square inches and a total cost of $10xz$ (cents). Thus, the total cost of the box is

$$4yz\,(\text{cents}) + 7yz\,(\text{cents}) + 10xy\,(\text{cents}) + 10xz\,(\text{cents})$$

Combining like terms, we have

$$(11yz + 10xy + 10xz)\text{cents} \quad \blacksquare$$

EXERCISES   2.3

In Exercises 1–20, add the polynomial expressions.

1. $3x^2 + 4x^2$ 2. $7y^3 + 9y^3$
3. $2x^3 + 5x^3$ 4. $-2x^2 + 6x^2$
5. $3xy + 5xy + 2xy$
6. $7xyz + xyz + 4xyz$
7. $-8x^4 + 5x^4 + 3x^4$
8. $(-6y) + (-4y)$
9. $4x^2y + (-2x^2y) + x^2y$
10. $-4y^3 + (-3y^3) + (-2y^3)$
11. $(4x + 6y) + (9x + 3y)$
12. $(9x^2 + 5) + (2x^2 - 8)$
13. $(x^2 + 3x + 5) + (2x^2 - 4x - 1)$
14. $(-5x^2 + 2x + 4) + (5x^2 - 4x + 6)$
15. $(-4a^2b^2 + 2c^2d^2) + (-6a^2b^2 - 5c^2d^2)$
16. $(3xy^2 + 2xy + 5x^2y)$
    $+ (2xy^2 - 4xy + 2x^2y)$
17. $(4a^2 + 3 + 5a) + (6a - 2a^2 + 2)$
    $+ (2a^2 - 3a + 8)$
18. $(a^2 - ab + 2b) + (2ab + b^2)$
    $+ (-a^2 + ab) + (a^2 + 2b^2 - 3ab)$
19. $(5x^3y + 4x^2y^2 - xy^3 + a^2)$
    $+ (3xy^3 - x^2y^2 + 5a^2 - x^3y)$
    $+ (2a^2 - 9x^2y)$
20. $(17ax^2y + 6x + 6xy + 7)$
    $+ (36axy^2 + 4ax^2y - 5x + 2xy - y)$

In Exercises 21–40, simplify and combine like terms.

21. $(3x - 4) - (5x + 1)$
22. $(x - 5y) - (2x - 4y)$
23. $(4x - 2y) - (3x + 7y) + (8x - 3y)$
24. $(5x^2 + 6x - 7) - (x^2 - 2x + 4)$
25. $(y^2 + 9y + 2) + (-4y - y^2) + 7y^2$
26. $(4a^2 - 2a + 1) + (a^2 - 3a + 1)$
    $- (5a^2 - 2a - 3)$
27. $(6x - 3y + 1) - (2x - y - 1)$
    $+ (x - 3y + 5)$
28. $8 + [4 + (7 - y)]$
29. $14 - [-5 + (6x + 4)]$
30. $y^2 - [-3y + (7 - 4x)]$
31. $2x^2 - [6x - (3x - x^2) + 1]$
32. $9z - [15z^2 - (7 + 9z - 2z^2)] + z^2$
33. $(5x + 2y) - [(2x - 5y) - (3x + y)]$
34. $\{[(6x + 5y) - 2y] - (2x - 3y)\} - y$
35. $(2a - 3b) - [(a - b) - (4a - b)]$
36. $[(-a^2 + 5a - 4) - (a^3 - 2a)]$
    $- [(a^3 + 2a^2 + 2) - (1 + a + a^2)]$
37. $2y - \{4x - 1 - [3y - x + (y - x + 4)]\}$
38. $9 + \{x - [y + 3(x - 3) - 2(y - 5)]\}$
39. $4a + \{b - 5 - [(a - b) + 2(a - 3)]\}$
40. $z - \{-4x - [3 + 3(z - x) - 4(z + 2)]\}$

In Exercises 41–45, fill in the parentheses.

41. $13x^2 - 5x + 2 = 13x^2 - ($ $)$
42. $-10a^2 + 4a - 7 = -($ $)$
43. $4 - [(7 + x) + 3 + x]$
    $= 4 - [(7 + x) - ($ $)]$
44. $-3x + 2y + x^2 - 7$
    $= -(3x$ $) - ($ $+ 7)$

45. $4x^2 - 6xy - 4x + 2$
$= 4x^2 - 2(\qquad)$

In Exercises 46–50, find the mistake(s), if any.
Find the correct answer.

46. $(y + 2) - (y - 3) = -1$

47. $-(x^2 + 3x - 7) - (2x^2 - 3x + 5)$
$= -3x^2 + 2$

48. $(xy + 3x^2 + y^2)$
$- (4xy + 4x^2 + 3y^2 + x - 1)$
$= -3xy - x^2 + 4y^2 - x - 1$

49. $-3(x^2y - 2x + 4) = -3x^2y - 6x - 12$

50. $(-y^2 + 2y - 3) - 2(5y^2 - y + 1)$
$= -6y^2 + 3y - 4$

In Exercises 51–60, what must be added to the first polynomial to produce the second polynomial?

51. $2; 2xy$

52. $-3x; -3xy$

53. $7a + 2b; 21a - 3b$

54. $4x - 3y; 10x - y$

55. $4x - 3y + 1; 5x + 2y - 3$

56. $4x - 2xy; x^2 + 2x + 4xy$

57. $3x^4 - 5x^2 + 7; 0$

58. $4xy + 7y^2 - 5x^2; 0$

59. $2x^2y^2 - xy + x + 2y + 3;$
$3x^2y^2 + 2xy - y$

60. $3x^2y - xy^2 + y + x - 2;$
$2xy^2 + xy + x - 1$

61. Subtract $12x^2 - 4x + 5$ from the sum of $10x^2 - 7x + 15$ and $5x^2 - 3x + 1$.

62. Subtract the sum of $x^2 - 4$ and $-3x^2 + 1$ from $5x^2 - 6x + 3$.

63. Find the perimeter of a geometric figure whose sides are $3a - 4$, $5a - 5$, and $8a + 3$.

64. Find the perimeter of a square with sides $x^2 + 3x - 5$.

*65. The monthly salary paid to a trucker is $4d + 7$ and a bus driver $3d - 1$, where d represents the number of days worked. Find the combined salary of the trucker and bus driver in terms of d.

*66. In order to make a picture frame from a square piece of plywood with sides equal to $3x$, Wayne cuts out a square hole with sides $2x$. Find a polynomial that gives the area of the hole and a polynomial that gives the area of the remaining plywood. See Figure 2.7.

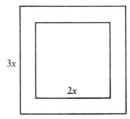

Figure 2.7

*67. Find polynomials for the perimeter and area of the polygon shown in Figure 2.8.

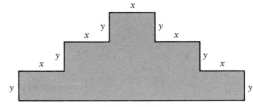

Figure 2.8

2.4 PROPERTIES OF EXPONENTS

Before we can discuss the multiplication of polynomials, we must examine some properties of exponents.

The Product of Powers

Recall that a^3, read "*a* to the third power," stands for $a \cdot a \cdot a$, and a^2, read "*a* to the second power," stands for $a \cdot a$. Thus,

$$5 \text{ factors}$$
$$a^3 \cdot a^2 = \overbrace{(a \cdot a \cdot a) \cdot (a \cdot a)} = a^5 = a^{3+2}$$
$$\underbrace{}_{3 \text{ factors}} \quad \underbrace{}_{2 \text{ factors}}$$

In multiplying these two powers of the **same base**, we could have found the new exponent by retaining the base and adding the exponents of the factors. In general, for positive integer exponents m and n,

$$m + n \text{ factors}$$
$$a^m \cdot a^n = \overbrace{(a \cdot a \cdots a) \cdot (a \cdot a \cdots a)} = a^{m+n}$$
$$\underbrace{}_{m \text{ factors}} \quad \underbrace{}_{n \text{ factors}}$$

This result suggests the following rule for multiplying powers of the same base:

Rule of Exponents for Multiplication

For all positive integers m and n, $a^m \cdot a^n = a^{m+n}$.

EXAMPLE 1 $7^4 \cdot 7^5 = 7^{4+5} = 7^9$

EXAMPLE 2 $(-5)^3 \cdot (-5)^{10} = (-5)^{3+10} = (-5)^{13}$

EXAMPLE 3   Recall that when no exponent is written, the exponent is understood to be 1. Thus, $6 = 6^1$.

$$6^2 \cdot 6 \cdot 6^5 = 6^{2+1+5} = 6^8$$

PROBLEM 1   Can we use the rule of exponents for multiplication on the product $5^3 \cdot (-5)^6$?

Answer   No! The rule of exponents for multiplication applies *only* when the bases of the powers are the *same*. In this problem, 5^3 has base 5 and $(-5)^6$ has base (-5). ■

The rule of exponents for multiplication, together with commutative and associative laws, may be used to multiply monomials (a polynomial with one term).

EXAMPLE 4   Multiply $(7xy^2) \cdot (4x^3y^4)$.

Solution $(7xy^2) \cdot (4x^3y^4) = (7 \cdot 4)(x \cdot x^3)(y^2 \cdot y^4)$

$$= 28x^{1+3}y^{2+4}$$
$$= 28x^4y^6 \quad \blacksquare$$

PROBLEM 2 Simplify the following: $(4x^2y)(2xy^2) + (6x^3y)y^2$.

Answer Recalling our order of operations, we perform multiplication before addition:

$$(4x^2y)(2xy^2) + (6x^3y)y^2 = 8x^3y^3 + 6x^3y^3$$
$$= 14x^3y^3 \quad \blacksquare$$

The Power of a Product

Consider the following example.

EXAMPLE 5 $(4a)^3 = (4a) \cdot (4a) \cdot (4a) = (4 \cdot 4 \cdot 4) \cdot (a \cdot a \cdot a) = 4^3a^3$

In general, for every positive integral exponent m, we have

$$(ab)^m = \overbrace{(ab)(ab) \cdots (ab)}^{m \text{ factors}}$$
$$= \underbrace{(a \cdot a \cdots \cdot a)}_{m \text{ factors}}\underbrace{(b \cdot b \cdots \cdot b)}_{m \text{ factors}}$$
$$= a^m b^m$$

These results may be summarized as follows:

Rule of Exponents for a Power of a Product
For every positive integer m, $(ab)^m = a^m b^m$.

EXAMPLE 6 $(-5y)^3 = (-5)^3(y)^3 = -125y^3$ and $(3xy)^2 = 3^2x^2y^2 = 9x^2y^2$

What happens if we raise a power to another power? Consider the following example.

EXAMPLE 7 $(a^3)^2 = a^3 \cdot a^3 = a^{3+3} = a^6 = a^{3 \cdot 2}$

In general,

$$\overbrace{(a^m)^n = \underbrace{(a^m)(a^m) \cdots (a^m)}_{n \text{ factors}} = a^{\overbrace{m+m+ \cdots +m}^{n \text{ terms}}} = a^{mn}}$$

We can summarize this result as follows:

Rule of Exponents for a Power of a Power
For all positive integers m and n, $(a^m)^n = a^{mn}$.

EXAMPLE 8 $(4^3)^5 = 4^{3 \cdot 5} = 4^{15}$

EXAMPLE 9 $(3x^2y^4)^3 = 3^3(x^2)^3(y^4)^3$ [by power of a product]

$\qquad\qquad\qquad = 27x^6y^{12}$ [by power of a product]

PROBLEM 3   Simplify the following expression: $(-x)^2x + (2x^2)(-2x^2)^3 + (2x^2)^4$.

Answer $(-x)^2x + (2x^2)(-2x^2)^3 + (2x^2)^4 = x^2 \cdot x + (2x^2)(-2)^3(x^2)^3 + 2^4(x^2)^4$

$\qquad\qquad\qquad\qquad\qquad\qquad\qquad = x^3 + (2x^2)(-8)x^6 + 16x^8$

$\qquad\qquad\qquad\qquad\qquad\qquad\qquad = x^3 + (2)(-8)x^2x^6 + 16x^8$

$\qquad\qquad\qquad\qquad\qquad\qquad\qquad = x^3 - 16x^8 + 16x^8$

$\qquad\qquad\qquad\qquad\qquad\qquad\qquad = x^3$ ∎

EXERCISES   2.4

In Exercises 1–40, perform the indicated operations and simplify. Assume that m and n are positive integers.

1. $3 \cdot 3^5$
2. $5^2 \cdot 5^4$
3. $(-7)^3(-7)^5$
4. $(-2) \cdot (-2)^6$
5. $x^2 \cdot x^7$
6. y^3y^4
7. $-a^4 \cdot a^4$
8. $(-a)^5(-a)^4$
9. $3^m \cdot 3^4$
10. $2^n \cdot 2$
11. $4^{2n} \cdot 4^3$
12. $a^{n+1} \cdot a$
13. $b^{n+4} \cdot b^n$
14. $x^{m-3} \cdot x^6$
15. $y^{n-2} \cdot y^4$
16. $5^n \cdot 5^{3n}$
17. $3^{2m} \cdot 3^{4m}$
18. $(-a)^n(-a)^{n+2}$
19. $b^{n+1} \cdot b^{n+3}$
20. $(x + y)(x + y)^5$
21. $(y - x)^4(y - x)^3$
22. $(3x - y)^3(3x - y)^5$

23. $-2(a + 4b)^2(a + 4b)^5$
24. $(xy)(y^3)$
25. $(x^2y)(y^4)$
26. $(3xy^2)(x^3)$
27. $(5a^3)(2a^2b^3)$
28. $(-a^3b^2)(a^2b)$
29. $b^2(-3a^2b)$
30. $(3xy)(-x^3y)$
31. $(5x^2y)(-3x^2z)$
32. $(-7x^2y^4)(-x^4y^2)$
33. $(-4xy)(-3xy)(-5x^2)$
34. $(-x^2y)(xy^3)(x^2y)$
35. $(2x^2y)(-3x^3y^2)(-y^3)$
36. $(2ab^3)(-a^2b^4)(-bc^2)$
37. $(3ab^2)(-3a^2b^4)(-3a^5b)$
38. $(x^4y)(-8x^3y)(-2xy^2z^3)$
39. $(6xy^5z)(-9yz^3)(-xy^4)$
40. $(-x^4y^2)(-3xy^3z)(4xz^5)$

In Exercises 41–50, simplify by finding each indicated product and then combining like terms.

41. $y(y^3) + (-2y^2)y^2$

42. $(-x^3)(x^4) + (4x^2)(-x)^5$

43. $3a^4(a^3b) - 4ab(-a)^6$

44. $-9a^2b(-b^4) + b^3(-2a^2b^2)$

45. $x^3 + (2x^2)(3x) - (5x)(4x^2)$

46. $(xy)(-y) + (3x)(y^2) - (5x)(-y)^2$

47. $(-2xyz)(3x^2yz) + (-5y^2)(2xz^2)(-x^2)$

48. $(8a^2bc^3)(-2ab^2c)$
 $+ (-3ac^2)(-5b^2a^2)(bc^2)$

*49. $-(3^2a^5b^3)(5ab^7)(-2c)$
 $+ (2^3b^6a^3c)(-b^2a^2)(ab^2)$

*50. $(5^2x^3y^2z)(-2x^4y^5z^4)$
 $- (-z^3x^2y^3)(4^2x^5z^2y^4)$

In Exercises 51–80, find the indicated products, given that m and n are positive integers.

51. $(3^2)^3$

52. $(4^3)^2$

53. $[(-5)^2]^4$

54. $[(-7)^4]^5$

55. $(-2^3)^2$

56. $(-6^4)^3$

57. $(-2x^2)^3$

58. $(-3x^2)^3$

59. $-2y(3yx)^2$

60. $-3a(5ab)^2$

61. $(-a)(-4a)^3$

62. $(-a)(-2a)^4$

63. $(8a)(2ab)^3$

64. $(3c)(4ac)^2$

65. $(2a)(-5a^2c)^2$

66. $(-3d)(5d^2e)^2$

67. $-(-x^2y^3z)^4$

68. $-(-x^3yz^2)^3$

69. $-x^5y^3(-xy^2)^3$

70. $5x^3(3x^3y)^3$

71. $(2xy)^2(x^2y)^3$

72. $(x^{3n}y^{n+1})(x^3y)^n$

73. $(y^n)^2(y^n)^3$

74. $(x^2y)^3(-2xy^3z)^2(-2yz^2)$

75. $(xyz^2)^2(3x^2y^3)^3(-x^2z)^4$

76. $[(x + 2y)^3]^4$

77. $[(x - y)^4]^5$

78. $[a^2b(a^2 + b^2)^3][ab^2(a^2 + b^2)^2]^3$

*79. $[-a^2b^3(a^2 - b^2)^2]^2[7b^2a(a^2 - b^2)^3]$

*80. $[3x^2y^2(2x - 3y)^4]^2[-xy^4(2x - 3y)^5]^3$

In Exercises 81–85, simplify by finding each indicated product and then combining like terms.

81. $(-x^3)^2(x^4y) + x^2y(-x^4)^2$

82. $3x^4(-2xy^2)^3 + 7x(-x^2y^2)^3$

83. $(-5^2x^3y)^2(-y^4) + (-3x^3y^3)^2$

*84. $4x^4y(-3xy^3)^3 - (-7x^2y^4)^2(-x^3y^2)$

*85. $(-a^2)^3(3ab^2c)^6 - (-a^2c)^2(a^2b^3c)^4$

2.5 MULTIPLICATION OF POLYNOMIALS

Multiplying a Polynomial by a Monomial

Now that we know the rules of exponents for multiplication and the distributive law of numbers, we can multiply any polynomial by a monomial.

EXAMPLE 1

$$3x(4x^2 + 5) = (3x)(4x^2) + (3x)(5) \qquad \text{[distributive law]}$$

$$= 12x^3 + 15x$$

Geometrically this result is illustrated in Figure 2.9, in which the largest rectangle is made up of the two smaller rectangles (the area of the largest rectangle is equal to the sum of the areas of the smaller two rectangles).

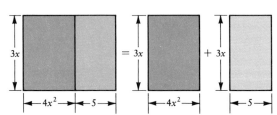

Figure 2.9

EXAMPLE 2

$$-2xy^2(2x^2 + 3xy - 5x^2y - xy^2 + 4y^2)$$

$$= (-2xy^2)(2x^2) + (-2xy^2)(3xy) + (-2xy^2)(-5x^2y)$$

$$+ (-2xy^2)(-xy^2) + (-2xy^2)(4y^2)$$

$$= -4x^3y^2 - 6x^2y^3 + 10x^3y^3 + 2x^2y^4 - 8xy^4$$

Rule for Multiplication of a Polynomial by a Monomial

In general, to multiply a polynomial by a monomial, use the distributive law: multiply each term of the polynomial by the monomial, and then add the products.

PROBLEM 1   A car travels for 5 hours at a rate of $(x + 25)$ miles per hour, then for 3 hours at a rate of $(x + 40)$ miles per hour. Find the total distance traveled by the car.

Answer   In order to find the total distance traveled by the car, we need to know the formula that relates distance, rate, and time:

$$\text{distance} = \text{rate} \cdot \text{time}$$

Thus, the total distance is given by the sum of the distance traveled at the rate of $(x + 25)$ miles per hour and the distance traveled at the rate of $(x + 40)$ miles per hour. See Figure 2.10. Using the formula $D = R \cdot T$, we find that the distance D_1 traveled at $(x + 25)$ miles per hour is

$$D_1 = (x + 25)\frac{\text{miles}}{\text{hour}} \cdot 5 \text{ hours}$$

$$= 5(x + 25) \text{ miles}$$

Similarly,

$$D_2 = (x + 40)\frac{\text{miles}}{\text{hour}} \cdot 3 \text{ hours}$$

$$= 3(x + 40) \text{ miles}$$

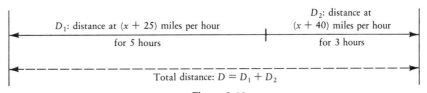

D_1: distance at $(x + 25)$ miles per hour

for 5 hours

D_2: distance at $(x + 40)$ miles per hour

for 3 hours

Total distance: $D = D_1 + D_2$

Figure 2.10

Hence, the total distance, $D = D_1 + D_2$, is

$$D = 5(x + 25) \text{ miles} + 3(x + 40) \text{ miles}$$
$$= (5x + 125) \text{ miles} + (3x + 120) \text{ miles}$$
$$= (8x + 245) \text{ miles} \ \blacksquare$$

Multiplying Two Polynomials

To find the product $A(x + 4)$, we used the distributive law of multiplication: $A(x + 4) = Ax + 4A$. We are now ready to find the product of two polynomials. For example, consider the product $(2x + 3)(x + 4)$. To find such a product, we treat $(2x + 3)$ as a number to be multiplied by the binomial $(x + 4)$ and apply the distributive law. We then apply the distributive law to each of the products and simplify:

$$(2x + 3)(x + 4) = (2x + 3) \cdot x + (2x + 3) \cdot 4 \qquad \text{[distributive law]}$$
$$= 2x \cdot x + 3 \cdot x + 2x \cdot 4 + 3 \cdot 4 \qquad \text{[distributive law]}$$
$$= 2x^2 + 3x + 8x + 12$$
$$= 2x^2 + 11x + 12$$

Figure 2.11 geometrically illustrates this product. Some people find it more convenient to set up the multiplication of polynomials in vertical form, and to multiply from left to right.

$$
\begin{array}{r}
2x + 3 \\
\underline{x + 4} \\
2x^2 + 3x \qquad \longleftarrow \text{this is } x(2x + 3) \\
\underline{ 8x + 12} \qquad \longleftarrow \text{this is } 4(2x + 3) \\
2x^2 + 11x + 12
\end{array}
$$

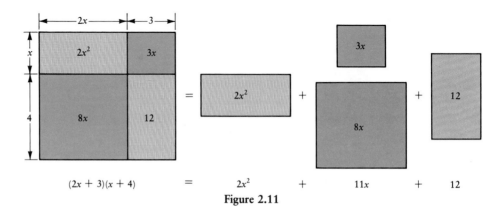

$(2x + 3)(x + 4) \qquad = \qquad 2x^2 \qquad + \qquad 11x \qquad + \qquad 12$

Figure 2.11

We now state a rule for multiplying two polynomials.

> ### Rule for Multiplying Two Polynomials
>
> To multiply two polynomials, use the distributive law: multiply each term of one polynomial by each term of the other, and add the products.

EXAMPLE 3 Multiply $x^2 + 3xy - 2y^2$ by $2xy - 2y^2$.

Solution

$$x^2 + 3xy - 2y^2$$
$$\underline{2xy - 2y^2}$$
$$2x^3y + 6x^2y^2 - 4xy^3 \qquad \longleftarrow \text{this is } 2xy \text{ times } (x^2 + 3xy - 2y^2)$$
$$\underline{\qquad\quad - 2x^2y^2 - 6xy^3 + 4y^4} \quad \longleftarrow \text{this is } -2y^2 \text{ times } (x^2 + 3xy - 2y^2)$$

Adding: $2x^3y + 4x^2y^2 - 10xy^3 + 4y^4$ ■

EXAMPLE 4 Multiply $(3x^2 + 2xy - 4y^2)$ by $(3x + 2y^2 - 5y)$.

Solution

$$3x^2 + 2xy - 4y^2$$
$$\underline{3x + 2y^2 - 5y}$$
$$9x^3 + 6x^2y - 12xy^2$$
$$+ 6x^2y^2 + 4xy^3 - 8y^4$$
$$\underline{- 15x^2y - 10xy^2 \qquad\qquad\qquad + 20y^3}$$
$$9x^3 - 9x^2y - 22xy^2 + 6x^2y^2 + 4xy^3 - 8y^4 + 20y^3 \quad ■$$

Note. Problems in multiplication can be tested by substituting convenient numerical values for the variables. It is best to use values larger than 1, since with 1 the exponents are not tested because any finite power of 1 is 1.

Test of Example 4, by letting $x = 3$ and $y = 2$:

$$\begin{aligned} 3x^2 + 2xy - 4y^2 &= 27 + 12 - 16 = 23 \\ 3x + 2y^2 - 5y &= 9 + 8 - 10 = 7 \end{aligned} \Big\} 23 \cdot 7 = 161$$

$$9x^3 - 9x^2y - 22xy^2 + 6x^2y^2 + 4xy^3 - 8y^4 + 20y^3$$
$$= 243 - 162 - 264 + 216 + 96 - 128 + 160 = 161$$

The work is *probably* correct since the product of the values of the two factors equals the value of the product, in this case, 161.

The product of two binomials occurs quite often, so we now present a very efficient way of multiplying two binomials called the FOIL technique.

FOIL Technique	Example
To multiply two binomials:	To multiply $(A + B)(C + D)$:
1. F—Multiply the **First** terms.	1. First terms: A and C Product: AC
2. O—Multiply the **Outside** terms.	2. Outside terms: A and D Product: AD
3. I—Multiply the **Inside** terms.	3. Inside terms: B and C Product: BC
4. L—Multiply the **Last** terms.	4. Last terms: B and D Product: BD
5. Add these products and simplify, if possible.	5. $(A + B)(C + D)$ $\quad = AC + AD + BC + BD$

EXAMPLE 5 Multiply $(3x + 7)(2x - 5)$.

Solution

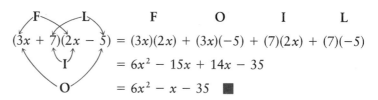

$$F \quad O \quad I \quad L$$
$$(3x + 7)(2x - 5) = (3x)(2x) + (3x)(-5) + (7)(2x) + (7)(-5)$$
$$= 6x^2 - 15x + 14x - 35$$
$$= 6x^2 - x - 35 \quad \blacksquare$$

Special Products

The products of certain binomials occur often enough to deserve special attention. Consider the following example.

EXAMPLE 6 Expand (multiply) $(x + 5)^2$.

Solution Recall that $(x + 5)^2 = (x + 5)(x + 5)$. Using the FOIL technique, we have $(x + 5)^2 = (x + 5)(x + 5) = x \cdot x + 5x + 5x + 5 \cdot 5 = x^2 + 10x + 25$. Note that the product, $x^2 + 10x + 25$, can be found using the terms x and 5. The product, $x^2 + 10x + 25$, is the square of the first term, x, plus twice their product, $5x$, plus the square of the second term. That is, $(x + 5)^2 = (x)^2 + 2 \cdot (5x) + (5)^2 = x^2 + 10x + 25$. $\blacksquare$

Special Product 1
$(A + B)^2 = A^2 + 2AB + B^2$

Proof $(A + B)^2 = (A + B)(A + B) = A^2 + AB + AB + B^2 = A^2 + 2AB + B^2$

EXAMPLE 7   Expand $(3x + 4)^2$.

Solution   This expression fits the form in Special product 1 with $A = 3x$ and $B = 4$. Thus,

$$(3x + 4)^2 = (3x)^2 + 2 \cdot 3x \cdot 4 + (4)^2 = 9x^2 + 24x + 16 \quad \blacksquare$$

EXAMPLE 8   Expand $[x + y + z]^2$.

Solution   In this problem we shall group two of the terms and treat them as a single term. Thus, $[x + y + z]^2 = [(x + y) + z]^2$. We now can use Special product 1 with $A = (x + y)$ and $B = z$, and

$$[x + y + z]^2 = [(x + y) + z]^2 = (x + y)^2 + 2 \cdot (x + y) \cdot z + z^2$$
$$= (x + y)^2 + 2z(x + y) + z^2$$

Using Special product 1 on $(x + y)^2$ and the distributive law on $2z(x + y)$, we obtain

$$= x^2 + 2xy + y^2 + 2xz + 2yz + z^2$$

Hence,

$$[x + y + z]^2 = x^2 + y^2 + z^2 + 2xy + 2xz + 2yz \quad \blacksquare$$

CAUTION

In general, $(A + B)^2 \neq A^2 + B^2$. For example, if we let $A = 1$ and $B = 1$, then

$$(1 + 1)^2 = (2)^2 = 4$$

where

$$1^2 + 1^2 = 1 + 1 = 2$$

Similar to the proof for Special product 1, we can show that

Special Product 2

$$(A - B)^2 = A^2 - 2AB + B^2$$

EXAMPLE 9 Expand $(4x - 3y)^2$.

Solution Using Special product 2 with $A = 4x$ and $B = 3y$, we have

$$(4x - 3y)^2 = (4x)^2 - 2(4x)(3y) + (3y)^2 = 16x^2 - 24xy + 9y^2$$

Be careful! $B = 3y$, *not* $-3y$. ■

We now state a formula that says that the product of the sum and the difference of two numbers is equal to the difference of their squares.

Special Product 3

$$(A + B)(A - B) = A^2 - B^2 \qquad \text{(Difference of two squares)}$$

EXAMPLE 10 Expand $(2x + 7y)(2x - 7y)$.

Solution Letting $A = 2x$ and $B = 7y$, we have

$$(2x + 7y)(2x - 7y) = (2x)^2 - (7y)^2 = 4x^2 - 49y^2 \quad ■$$

EXAMPLE 11 Multiply $(2x + 1)(2x - 1)(4x^2 + 1)$.

Solution Using Special product 3 on the first two factors, we obtain

$$(2x + 1)(2x - 1)(4x^2 + 1) = (4x^2 - 1)(4x^2 + 1)$$

Applying Special product 3 again,

$$= 16x^4 - 1 \quad ■$$

PROBLEM 2 A hollow square has dimensions, in inches, as shown in Figure 2.12. Find its area.

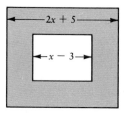

Figure 2.12

Answer Recall that the area of a square is given by the square of the length of a side. In this problem, the area of the hollow square is given by the area of the large square with side $(2x + 5)$ inches *minus* the area of the small square with side

$(x - 3)$ inches. Thus,

Area of hollow square = $[(2x + 5) \text{ in.}]^2 - [(x - 3) \text{ in.}]^2$

$$= (2x + 5)^2 \text{ sq in.} - (x - 3)^2 \text{ sq in.}$$

$$= (4x^2 + 20x + 25) \text{ sq in.} - (x^2 - 6x + 9) \text{ sq in.}$$

Removing parentheses yields

$$= [4x^2 + 20x + 25 - x^2 + 6x - 9] \text{ sq in.}$$

$$= (3x^2 + 26x + 16) \text{ sq in.} \quad \blacksquare$$

We now list five other special products that you should verify by doing the multiplication.

Other Special Products

4. $(A + B)(A^2 - AB + B^2) = A^3 + B^3$ (Sum of two cubes)
5. $(A - B)(A^2 + AB + B^2) = A^3 - B^3$ (Difference of two cubes)
6. $(A + B)^3 = A^3 + 3A^2B + 3AB^2 + B^3$
7. $(A - B)^3 = A^3 - 3A^2B + 3AB^2 - B^3$
8. $(Ax + By)(Cx + Dy) = ACx^2 + (AD + BC)xy + BDy^2$

EXAMPLE 12   Use Special products 4–7 to perform each multiplication:

(a) $(x + 3)(x^2 - 3x + 9)$ (b) $(2x - 5y)(4x^2 + 10xy + 25y^2)$

(c) $(4m + n)^3$ (d) $(5r - 2s)^3$

Solution   (a) Using Special product 4, with $A = x$ and $B = 3$, we have

$$(x + 3)(x^2 - 3x + 9) = x^3 + (3)^3 = x^3 + 27$$

(b) Using Special product 5, with $A = 2x$ and $B = 5y$, we obtain

$$(2x - 5y)(4x^2 + 10xy + 25y^2) = (2x^3) - (5y)^3 = 8x^3 - 125y^3$$

(c) Special product 6, with $A = 4m$ and $B = n$, yields

$$(4m + n)^3 = (4m)^3 + 3(4m)^2(n) + 3(4m)(n)^2 + (n)^3$$

$$= 64m^3 + 48m^2n + 12mn^2 + n^3$$

(d) Letting $A = 5r$ and $B = 2s$, Special product 7 gives

$$(5r - 2s)^3 = (5r)^3 - 3(5r)^2(2s) + 3(5r)(2s)^2 - (2s)^3$$

$$= 125r^3 - 150r^2s + 60rs^2 - 8s^3 \quad \blacksquare$$

EXERCISES 2.5

In Exercises 1–32, find each of the products.

1. $2(x - y)$
2. $-3(y - 4x)$
3. $x(x - 3y)$
4. $y(2x + 4y)$
5. $7x(x - 2)$
6. $-4x(3x - 2y)$
7. $\frac{1}{2} \cdot x(6x - 8)$
8. $\frac{2x}{3}(9x - 3y)$
9. $3x^2(2x - 1)$
10. $xy(x^2 - y^2)$
11. $5y^2(x - 2xy)$
12. $-7x^2(4 - 3x)$
13. $5(x^2 - x + 1)$
14. $-3(7 - y + y^2)$
15. $4(4x^2 + 3x - 7)$
16. $x^2(2x^2 - x - 1)$
17. $-2x^2(x^2 - 2xy - y^2)$
18. $x^3(2x^2 - 4x - 2)$
19. $2y^4(y^3 + 2y^2 - 3)$
20. $y^2(y^2 - 2xy - 3x^2)$
21. $(y^2 - 5y + 9)(-3y^2)$
22. $(1 - 2x - 3x^2)(-2x^4)$
23. $-3a^2b(-6a^3 + 5ab - 2b^2 + 4)$
24. $-5ab^2(-4a^2b^3 + 3a^3b^2 - 5ab + 1)$
25. $-4x^3y^2(x^4y - x^3y^2 + 2x^2y^4 - y^6)$
26. $-2x^2y^4(x^5y + 4x^4y^3 - x^2y^5 - y^6)$
27. $-12xyz(-12xyz + 4xyz^2 - 3x^2yz^3 - xy^2z)$
*28. $2xyz(7xy + 5xz - 3yz - 4x^2 - 2xy^2 + 5z^2)$
*29. $6x^2yz^2(-2x^2 + 3xy + xyz - 4xy^2z + 5xyz^2)$
*30. $-2x^2y^3z(4x^3 - 10xyz - x^2y^2z + 2x^2y^3z + z^2 - 5)$
*31. $xyz^2(x^ny^mz^p - 3xy^2z^n + 2x^my^nz)$
*32. $x^ay^bz^c(-xyz + 3x^2y^3z^3 - 2x^ny^nz^n)$

In Exercises 33–40, write a polynomial expression that represents the answer. Then simplify the expression.

33. A truck travels $(3x + 7)$ km per hr. Find the distance it travels in:
    (a) 2 hr   (b) 10 hr   (c) h hr
    (d) x hr

34. A 747 jet airplane flew at x mph for 3 hr, then at $(2x + 5)$ mph for another 5 hr. Find the polynomial that represents the total distance traveled by the plane.

35. A train traveled at 175 mph for h hr, then at 190 mph for $(h + 3)$ hr. Find the total distance traveled by the train.

36. Express the area of a rectangle whose length is $7x - 2y$ and whose width is $3x$.

37. How many square inches of cardboard are used to make a poster that measures $4x$ in. on each side?

38. A rectangular box is $7x$ cm long, $2x$ cm wide, and $(x + 3)$ cm deep. What is the volume of the box in terms of x?

*39. Find the surface area of the six faces of the rectangular box given in Exercise 38.

*40. A woman can row a boat $(x + 2)$ mph in still water. She rowed down a stream with a current of 3 mph for 3/4 hr. Then she rowed up a river against a current of 2 mph for 1/4 hr. Find a polynomial that represents the total distance she rowed.

In Exercises 41–121, multiply.

41. $(x + 2)(x - 3)$
42. $(x + 3)(x + 4)$
43. $(x - 1)(x - 2)$
44. $(y + 2)(y + 3)$
45. $(y - 3)(y - 1)$
46. $(x - 3)(x + 4)$
47. $(x - 3)(x - 4)$
48. $(x - 1)(x + 2)$
49. $(y + 3)(y - 2)$
50. $(y + 2)(y - 1)$
51. $(x - 5)(x - 2)$
52. $(x^2 - 2)(x^2 - 3)$
53. $(x^2 - 6)(x^2 - 5)$
54. $(y^2 + 6)(y^2 - 4)$
55. $(y^2 + 3)(y^2 + 6)$
56. $(x + 5z)(x - 2z)$
57. $(z^2 - 4)(z^2 - 6)$
58. $(x - y)(x + 5y)$
59. $(x + 2a)(x - a)$
60. $(x + 2y)(x + y)$
61. $(a - 3x)(a + 6x)$
62. $(x^2 - 7)(x^2 + 6)$
63. $(x - 3y)(x - 2y)$
64. $(x + y)(x + 3y)$
65. $(x^2 - 4)(x^2 - 5)$
66. $(2x - 3)(4x + 5)$
67. $(3x - 2)(4x + 1)$
68. $(5x - 3)(2x - 5)$
69. $(3y + 4)(3y + 2)$
70. $(x - 4)(4x + 3)$
71. $(3x + 5)(x - 5)$
72. $(2x - y)(x + 3y)$
73. $(4x^2 - 3)(2x^2 - 3)$
74. $(x - 6)(2x + 1)$
75. $(x - 2)(3x - 5)$
76. $(4a - x)(2a + 3x)$
77. $(3x + 2)(x + 1)$
78. $(2y - 3)(2y - 5)$
79. $(x - 2y)(3x - y)$

80. $(4y^2 - 3)(y^2 + 3)$

81. $(x + 3y)(3x - y)$

82. $(x - y)(4x - 5y)$

83. $(6x - 5)(x + 1)$

84. $(x - 3y)(x - 4y)$

85. $(2x - 3y)(2x - 3y)$

86. $(x - 2y)(x - 2y)$

87. $(3y^2 + 2)(3y^2 + 4)$

88. $(5x - 2y)(5x - 3y)$

89. $(7x - y)(x + y)$

90. $(a - x)(a^2 + ax + x^2)$

91. $(a - b)(a^2 - 2ab + b^2)$

92. $(4a - 3b)(16a^2 - 24ab + 9b^2)$

93. $(2x - 1)(1 + 4x^2 - 4x)$

94. $(x^2 + xy + y^2)(x - y)$

95. $(a^2 + x^2 + ax)(a^2 + x^2 - ax)$

96. $(5x^2 - 3x + 1)(2x - 3)$

97. $(2b^2 - 4ab + 3a^2)(4a - 3b)$

98. $(2x - 3xy + y)(y + 2x + 3xy)$

99. $(2a^2 - 3a + 1)(1 + 3a + 2a^2)$

100. $(3a - 5 + 4b)(4b + 3a + 5)$

101. $(8x^2y - 3xy^2 + y^3)(4x - 5y)$

102. $(3a^2 - 4ab + 2b^2)(5a - 2b)$

103. $(a^2 - 6a - 4)(3a - 2)$

104. $(4x^2 - x + 2)(x - 2)$

105. $(a - 2b + c)(a - 2b - c)$

106. $(a + b + c)(a + b - c)$

107. $(6x^2 + 3x - 4)(2x - 3)$

108. $(9x^2 - 6x + 4)(3x + 2)$

109. $(4a^2 + 6ab + 9b^2)(2a - 3b)$

110. $(y^2 - 3y + 4)(y^2 - 2y - 1)$

111. $(2a - 3 + c)(c + 2a + 3)$

112. $(x - 3y - 2z)(x + 2z - 3y)$

113. $(x^2 - xy + y^2 - 1)(x + y)$

114. $(x^2 + xy + y^2 - 1)(x - y)$

*115. $(2x^2 + y^2 - 2z^2)(2x^2 + y^2 + 2z^2)$

*116. $(81x^2 - 16y^2 - 25z^2)(81x^2 + 16y^2 + 25z^2)$

*117. $(x^2 + 3x + 2)(x^2 + 7x + 12)$

*118. $(x^3 - 3x^2)(x^3 - 6x^2 + 9x)$

*119. $(x^2 - y^2)(x^4 + x^2y^2 + y^4)$

*120. $(x^{2n} - 2x^n + 1)(x^n - 1)$

*121. $(x^{2n+2} + y^m)(x^{2n+2} - y^m)$

In Exercises 122–158, use special product formulas to find the product.

122. $(x + 5)^2$ 123. $(2x - 5)^2$

124. $(2x + 5y)^2$ 125. $(2x - 5y)^2$

126. $(x^2 - 2)^2$ 127. $(2x^2 + 1)^2$

128. $(1 - x^3)^2$ 129. $(4 - 5x^3)^2$

130. $(2x^2 - 3y)^2$ 131. $(2ax - 3by)^2$

132. $(2x^3 + 3xy^2)^2$

133. $(3^2 - 2^2)^2$

134. $(40 - 1)^2$ 135. $(20 + 1)^2$

136. $(50 - 1)^2$ 137. $(100 - 1)^2$

138. $(x + 2)(x - 2)$

139. $(3x + 4)(3x - 4)$

140. $(7x - 8y)(7x + 8y)$

141. $(11x^2 + 7y)(11x^2 - 7y)$

142. $(xy^3 + 2z^3)(xy^3 - 2z^3)$

143. $(4x + 5y^2)(4x - 5y^2)$

144. $(8x^4 - 3y^3)(8x^4 + 3y^3)$

145. $[(x + 2) + y][(x + 2) - y]$

*146. $[(x - y) + z][(x - y) - z]$

*147. $(2a + b + c)(2a + b - c)$

*148. $(a + b + 4)(a + b - 4)$

*149. $(a^2 - b^2 - ab)(a^2 + b^2 + ab)$
$= [a^2 - (b^2 + ab)][a^2 + (b^2 + ab)] = ?$

*150. $(10 + 2x + 3y)(10 - 2x - 3y)$

*151. $(3 - x + y)(3 + x + y)$

152. $98 \times 102 = (100 - 2)(100 + 2) = ?$

153. 97×103 154. 44×36

155. 702×698 156. 87×93

157. 1105×1095 158. 1001×999

159. Find the area of a rectangle that is c ft longer than it is wide. Hint: Let x be the width.

160. Find the volume of a rectangular box $(2x + 1)$ ft wide, $(3x - 1)$ ft long, and $(x + 2)$ ft deep.

161. Given that the volume of a cylinder is $\pi r^2 h$, where r is the radius and h is the altitude, find the volume of a right cylinder if the altitude is h ft and the radius $(2h - 3)$ ft.

162. Find a polynomial that expresses the product of three consecutive integers if the smallest is x.

*163. Recall that the area of a circle is $\frac{1}{4}\pi d^2$, where d is the diameter of the circle. Find the area of the ring that has dimensions shown in Figure 2.13.

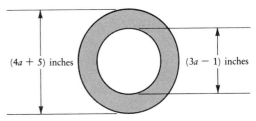

$(4a + 5)$ inches $(3a - 1)$ inches

Figure 2.13

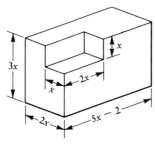

Figure 2.14

*164. Find the volume of the block in Figure 2.14.

165. A train can travel at a constant rate of $(x + 25)$ mph. Find the distance it can travel in $(3x - 1)$ hr.

In Exercises 166–171, use Special products 4–8 to perform each multiplication.

166. $(2 - x)(4 + 2x + x^2)$

167. $(x + 2y)(x^2 - 2xy + 4y^2)$

168. $(3x^2 + 2y)^3$ 169. $(5y - 4x^2)^3$

*170. $(x^m + 2y^n)^3$

*171. $[1 + x + y][(1 + x)^2 - (1 + x)y + y^2]$

*172. Obtain formulas for the expansion of
(a) $(A + B)^4$ *(b) $(A - B)^4$

2.6 DIVISION BY MONOMIALS

We begin this section by discussing another law of exponents that will be useful in the division of polynomials. Recall from arithmetic that if the numerator and denominator have common numerical factors, then such common factors could be cancelled. For example,

$$\frac{35}{40} = \frac{7 \cdot \cancel{5}}{8 \cdot \cancel{5}} = \frac{7}{8}$$

This principle also applies to variables that represent numbers. For example,

$$\frac{a^6}{a^2} = \frac{\cancel{aa}aaaa}{\cancel{aa}} = aaaa = a^4 \qquad (a \neq 0)$$

Notice that we could have found the exponent of the quotient by retaining the base and subtracting the exponent in the denominator from the exponent in the numerator. That is,

$$\frac{a^6}{a^2} = a^{6-2} = a^4$$

This example suggests that, in general,

Rule of Exponents for Division
For all positive integers m and n, $\dfrac{a^m}{a^n} = a^{m-n}$ $(a \neq 0)$.

EXAMPLE 1   (a) $\dfrac{x^{13}}{x^7} = x^{13-7} = x^6$ (c) $\dfrac{x^{2n+3}}{x^{n+1}} = x^{(2n+3)-(n+1)}$

             $= x^{2n+3-n-1}$

(b) $\dfrac{y^{10}}{y^7} = y^{10-7} = y^3$ $= x^{n+2}$

Note that if $m = n$ in the rule of exponents for division, we have

$$\frac{a^n}{a^n} = 1 \qquad (a \neq 0)$$

Note that we have not discussed the case where the exponent of the denominator is larger than the exponent of the numerator. This will be discussed in a later chapter.

Now consider the problem $\left(\dfrac{a}{b}\right)^2$. We know that $\left(\dfrac{a}{b}\right)^2 = \dfrac{a}{b} \cdot \dfrac{a}{b} = \dfrac{a^2}{b^2}$. Thus, $\left(\dfrac{a}{b}\right)^2 = \dfrac{a^2}{b^2}$. Similarly, $\left(\dfrac{a}{b}\right)^3 = \dfrac{a^3}{b^3}$. In general, if a and b are numbers $(b \neq 0)$ and n is a positive integer, then,

---

Power of a Quotient

If n is a positive integer, $\left(\dfrac{a}{b}\right)^n = \dfrac{a^n}{b^n}$ $(b \neq 0)$.

EXAMPLE 2   (a) $\left(\dfrac{x}{4}\right)^3 = \dfrac{x^3}{4^3} = \dfrac{x^3}{64}$

(b) $\left(\dfrac{2x}{y}\right)^4 = \dfrac{(2x)^4}{y^4}$ [by power of a quotient]

$= \dfrac{2^4 x^4}{y^4}$ [by power of product]

$= \dfrac{16x^4}{y^4}$

We are now ready to use the preceding rules to divide polynomials. We begin by showing how to divide a monomial by a monomial.

Dividing a Monomial by a Monomial

Procedure for Division of Monomials

1. Divide their numerical coefficients.
2. Divide variable factors that are powers having the same base.
3. Multiply the quotients obtained in steps 1 and 2.

EXAMPLE 3 If $x \neq 0$, $y \neq 0$, and $z \neq 0$,

$$\frac{42x^3y^6z^4}{7xy^3z^2} = \left(\frac{42}{7}\right)x^{3-1}y^{6-3}z^{4-2} = 6x^2y^3z^2$$

Note that $(7xy^3z^2)(6x^2y^3z^2) = 42x^3y^6z^4$, which verifies the solution.

EXAMPLE 4 If $a \neq 0$ and $b \neq 0$, and n is a positive integer,

$$\frac{a^{3n}b^{n+4}}{a^2b^n} = a^{3n-2}b^{n+4-n} = a^{3n-2}b^4$$

EXAMPLE 5
$$\frac{18(x+4)^3(y^2)^4}{-6(x+4)y^5} = \frac{18(x+4)^3y^8}{-6(x+4)y^5} \qquad \text{[by power of a power]}$$

$$= \left(\frac{18}{-6}\right)(x+4)^{3-1}y^{8-5} \qquad \text{[by rule of exponents for division]}$$

$$= -3(x+4)^2y^3$$

EXAMPLE 6 Find the quotient, given that none of the variables are zero:

$$\frac{(-3x^2y^6z^4)(-2xy^2z^3)^3}{(-4x^2y^3z^2)^2}$$

Solution
$$\frac{(-3x^2y^6z^4)(-2xy^2z^3)^3}{(-4x^2y^3z^2)^2} = \frac{(-3x^2y^5z^4)(-2)^3(x)^3(y^2)^3(z^3)^3}{(-4)^2(x^2)^2(y^3)^2(z^2)^2}$$

$$= \frac{(-3x^2y^6z^4)(-8)x^3y^6z^9}{16x^4y^6z^4}$$

$$= \frac{(-3)(-8)x^{2+3}y^{6+6}z^{4+9}}{16x^4y^6z^4}$$

$$= \frac{24x^5y^{12}z^{13}}{16x^4y^6z^4}$$

$$= \left(\frac{24}{16}\right)x^{5-4}y^{12-6}z^{13-4}$$

$$= \frac{3}{2} \cdot xy^6z^9 \quad \blacksquare$$

EXAMPLE 7 Find the quotient where $a \neq 0$, $b \neq 0$, and n is a positive integer:

$$\frac{a^n(a^{n+1}b^{2n-1})^3}{(a^{2n-1}b^{3n-2})^2}$$

Solution

$$\frac{a^n(a^{n+1}b^{2n-1})^3}{(a^{2n-1}b^{3n-2})^2} = \frac{a^n(a^{n+1})^3(b^{2n-1})^3}{(a^{2n-1})^2(b^{3n-2})^2}$$

$$= \frac{a^n \cdot a^{3n+3}b^{6n-3}}{a^{4n-2}b^{6n-4}}$$

$$= \frac{a^{4n+3}b^{6n-3}}{a^{4n-2}b^{6n-4}}$$

$$= a^{(4n+3)-(4n-2)}b^{(6n-3)-(6n-4)}$$

$$= a^{4n+3-4n+2}b^{6n-3-6n+4}$$

$$= a^5b \quad \blacksquare$$

Dividing a Polynomial by a Monomial

Recall from the rule for adding fractions that

$$\frac{4+5}{9} = \frac{4}{9} + \frac{5}{9}$$

In general, for all real numbers a, b_1, b_2, . . . , b_n, we have

$$\frac{b_1 + b_2 + \cdots + b_n}{a} = \frac{b_1}{a} + \frac{b_2}{a} + \cdots + \frac{b_n}{a}$$

This property of fraction extends to the division of a polynomial by a monomial.

Procedure for Dividing a Polynomial by a Monomial

To divide a polynomial by a monomial, divide each term of the polynomial by the monomial.

EXAMPLE 8 Divide $4x^2 - 6x$ by $2x$.

Solution
$$\frac{4x^2 - 6x}{2x} = \frac{4x^2}{2x} - \frac{6x}{2x} = 2x - 3 \quad \blacksquare$$

EXAMPLE 9 Divide and simplify $\dfrac{9x^2y - 18xyz + 36xy^2}{9xy}$.

Solution $\dfrac{9x^2y - 18xyz + 36xy^2}{9xy} = \dfrac{9x^2y}{9xy} - \dfrac{18xyz}{9xy} + \dfrac{36xy^2}{9xy} = x - 2z + 4y \quad \blacksquare$

EXAMPLE 10 Divide and simplify $\dfrac{14(x+y)^5 - 6(x+y)^3 + 4(x+y)^2}{-2(x+y)}$.

Solution $\dfrac{14(x+y)^5 - 6(x+y)^3 + 4(x+y)^2}{-2(x+y)}$

$$= \frac{14(x+y)^5}{-2(x+y)} - \frac{6(x+y)^3}{-2(x+y)} + \frac{4(x+y)^2}{-2(x+y)}$$

$$= -7(x+y)^4 + 3(x+y)^2 - 2(x+y) \quad \blacksquare$$

Just as in arithmetic, we can check our result by multiplying the result by the divisor $-2(x+y)$ and hope to obtain the dividend $14(x+y)^5 - 6(x+y)^3 + 4(x+y)^2$. Therefore, using the distributive law and exponent laws, we have

$$[-2(x+y)][-7(x+y)^4 + 3(x+y)^2 - 2(x+y)]$$

$$= (-2)(-7)(x+y)(x+y)^4 + (-2)(3)(x+y)(x+y)^2$$
$$+ (-2)(-2)(x+y)(x+y)$$

$$= 14(x+y)^5 - 6(x+y)^3 + 4(x+y)^2$$

EXAMPLE 11 Divide $x^{3n+1} - x^{3n} + x^{3n-1}$ by x^{n-2}, where n is a natural number.

Solution $\dfrac{x^{3n+1} - x^{3n} + x^{3n-1}}{x^{n-2}} = \dfrac{x^{3n+1}}{x^{n-2}} - \dfrac{x^{3n}}{x^{n-2}} + \dfrac{x^{3n-1}}{x^{n-2}}$

$$= x^{(3n+1)-(n-2)} - x^{3n-(n-2)} + x^{(3n-1)-(n-2)}$$

$$= x^{3n+1-n+2} - x^{3n-n+2} + x^{3n-1-n+2}$$

$$= x^{2n+3} - x^{2n+2} + x^{2n+1} \quad \blacksquare$$

PROBLEM 1 The area of a rectangle is $(30x^2 + 12x)$ square feet. Find the length of the rectangle if its width is $(6x)$ feet.

Answer The area of a rectangle is length times width. Thus, the

$$\text{width} = \frac{\text{area of rectangle}}{\text{length of rectangle}}$$

That is,

$$\text{width} = \frac{(30x^2 + 12x) \text{ square feet}}{(6x) \text{ feet}}$$

$$= \left[\frac{30x^2}{6x} + \frac{12x}{6x} \right] \text{ feet}$$

$$= (5x + 2) \text{ feet} \quad \blacksquare$$

EXERCISES 2.6

In Exercises 1–15, use the rule of exponents for division to simplify.

1. $\dfrac{7^3}{7^2}$

2. $\dfrac{8^{14}}{8^{10}}$

3. $\dfrac{5^{20}}{5^{11}}$

4. $\dfrac{10^{13}}{10^4}$

5. $\dfrac{14^{16}}{14^{16}}$

6. $\dfrac{x^9}{x^6}$

7. $\dfrac{y^7}{y^5}$

8. $\dfrac{(z-1)^{10}}{(z-1)^6}$

9. $\dfrac{x^{n+4}}{x^{n+1}}$

10. $\dfrac{x^{n2}}{x}$

11. $\dfrac{a^{3n}}{a^n}$

12. $\dfrac{y^{2n+4}}{y^{n-3}}$

13. $\dfrac{a^{n2+n-1}}{a^{n+1}}$

14. $\dfrac{b^{3n2+2n+1}}{b^{n2+n+1}}$

15. $\dfrac{(x+2)^{3n2+4n-2}}{(x+2)^{n2-2}}$

In Exercises 16–25, use the rule of power of a quotient to simplify.

16. $\left(\dfrac{x}{7}\right)^2$

17. $\left(\dfrac{3}{y}\right)^4$

18. $\left(\dfrac{x}{y}\right)^5$

19. $\left(\dfrac{2x}{3}\right)^3$

20. $\left(\dfrac{5}{7y}\right)^4$

21. $\left(\dfrac{4x}{9y}\right)^2$

22. $\left(\dfrac{x+2}{x-3}\right)^{10}$

23. $\left[\dfrac{x+21}{3(x-5)}\right]^3$

*24. $\left[\dfrac{4x}{5(x-1)}\right]^n$, n a positive integer

*25. $\left[\dfrac{7(x+3)}{(x-5)}\right]^n$, n a positive integer

*26. If we let $m = n$ in the rule of exponents for division, we get $a^n/a^n = a^{n-n} = a^0$. What meaning must we give to a^0 in order for this result to be consistent with the rules of arithmetic? Are there any conditions on a?

In Exercises 27–51, find the quotient. None of the variables are zero and m and n are positive integers.

27. $\dfrac{12x^3y}{6xy}$

28. $\dfrac{-18xy^3}{6xy^2}$

29. $\dfrac{15x^2y^2}{-5xy^2}$

30. $\dfrac{-30xyz^2}{-15yz}$

31. $\dfrac{-100x^5y^4}{-10xy^3}$

32. $\dfrac{-144x^4y^3z^2}{12xy^2z}$

33. $\dfrac{225x^3y^6}{25xy^4}$

34. $\dfrac{17a^3b^4c^5d^6}{-17abcd}$

35. $\dfrac{-51cx^9y}{17cx^7}$

36. $\dfrac{44x^5y^6}{-44x^5y^6}$

37. $\dfrac{55x^9y^8}{-5x^8y}$

38. $\dfrac{-55xy^3z}{10y^2z}$

39. $\dfrac{22x^5y^3z}{-xyz}$

40. $\dfrac{13x^4y^5z^3}{-y^4z^2}$

41. $\dfrac{65xy^5z^8}{13xz^6}$

42. $\dfrac{-21^4ab^4cd^4}{21^3abcd^3}$

43. $\dfrac{16x^my^4}{-4xy^m}$

44. $\dfrac{40x^{n+1}y^mz^2}{-10x^ny^{m-1}z}$

45. $\dfrac{-100x^{5n}y^{5n-1}z^6}{25xy^4z^5}$

46. $\dfrac{-4a^{m+n}b^{m+n}z^4}{-2a^nb^mz^2}$

47. $\dfrac{(2a^3b^4)(-5a^6b)}{5a^2b^3(a^3)^2}$

48. $\dfrac{2(3a^2b^4c^3)^3}{(3ab^5c^3)^2}$

49. $\dfrac{(2x^4y^3z^2)^4}{(-2xy^3z)(-x^2y^2z^2)^2}$

*50. $\dfrac{x(x^ny^{n+1})^2}{y(xy^2)^n}$

*51. $\dfrac{(x^{m+1})^m(y^{3m-1})^2}{(x^{m-1})^my^{2m+1}}$

In Exercises 52–71, divide and simplify.

52. $12y^4 - 16y^3z + 20y^5$ by $2y^2$; by $-4y^3$

53. $-3xy^4 + 6xy^5 - 9xy^6$ by xy^4; by $-3xy^2$

54. $7(32)^2 - 5(32) + 14(32)^3$ by $-(32)$

55. $4(-7)^3 - 10(-7)^2 + 13(-7)$ by (-7)

56. $125xyz^2 - 5x^2yz - 15xyz$ by $5xz$

57. $16x^3y^2z - 20x^2y + 24x^2y^2$ by $-4x^2y$

58. $12x^3 - 6x^2 + 2x$ by $2x$

59. $-16(xy)^2 + 8x^2y^3 - 24xy^2$ by $8xy$

60. $18a^8 - 4a^7 + 7a^5 - 5a^3$ by a^3

61. $x^4 - 1.5x^3 + 0.5x^2$ by $0.5x^2$

62. $3y^6 + 2y^4 - y^2$ by $0.5y$

63. $24x^5y^4 - 2x^4y^3 + 6x^3y^2$ by $2x^3y$

64. $8x - 16xy - 24x^2 + 12x^3 - 48x^4y$ by $4x$

*65. $x(1 + x)^4 - xy(1 + x)^3 + 7x(1 + x)^2$ by $x(1 + x)$; by x; by $(1 + x)$

*66. $24(x - y)^3 - 12(x - y)^2 - 6(x - y)$ by $-6(x - y)$

*67. $3(a - b)(x + y)^5 - 6(a - b)(x + y)^{10}$ by $-3(a - b)(x + y)^5$; by $(x + y)^2$

*68. $ay^n + by^{n-1} - cy^{n-2} + dy^{n-3} - ey^{n-4}$ by y^{n-4}

*69. $(1 - y)y^m - (1 - y)^2y^{m-1} + (1 - y)^3y^{m-2}$ by $(1 - y)y^m$

*70. $y^{2n} - y^{2n-2} + y^{2n-4} - y^{2n-6}$ by y^{n-6}

71. The area of a rectangle is $(15x^2 + 25x^3)$ square feet. Find the length of the rectangle if the width is $(5x^2)$ feet.

Since distance = rate · time, we can show that rate = distance/time and time = distance/rate.

72. If $30x^2 + 20x$ represents the distance in miles traveled by a woman, find the number of miles per hour (rate) she travels if she travels for:
    (a) 6 hr   (b) x hr   (c) $10x$ hr

73. If $(15x^3 - 45x^2)$ km represents the distance traveled by a train, find the number of hours the train travels if it goes:
    (a) 5 km/hr   (b) x km/hr
    (c) x^2 km/hr   (d) $15x^2$ km/hr

74. If $(18x + 81y)$ dollars represents the cost of manufacturing 9 widgets, find the cost of manufacturing 1 widget.

2.7 DIVISION OF A POLYNOMIAL BY A POLYNOMIAL

In Example 5 of Section 2.5, we saw that

$$(3x + 7)(2x - 5) = 6x^2 - x - 35$$

Using division, it follows that

$$\frac{6x^2 - x - 35}{3x + 7} = 2x - 5$$

However, we would like to obtain this result without using the first equation. There is a procedure for dividing polynomials that is similar to the long division process in arithmetic. When we divide a polynomial p by a polynomial d, we obtain a quotient q and a remainder r. That is,

$$\frac{p}{d} = q + \frac{r}{d} \quad \text{or} \quad p = qd + r$$

The expressions q and r are both polynomials, and the degree of the remainder r must be less than the degree of the divisor d. The procedure for dividing two polynomials is illustrated in the following example.

EXAMPLE 1   Divide $10 + 6x^2 - x$ by $x - 2$.

Solution

Procedure	Solution
Step 1. Arrange the terms of both the divisor and dividend in descending powers of one variable. If a power of that variable is missing, write the term with a zero coefficient.	$x - 2\overline{)6x^2 - x + 10}$
Step 2. Divide the first term of the dividend by the first term of the divisor to obtain the first term of the quotient.	$\dfrac{6x}{x - 2\overline{)6x^2 - x + 10}}$
Step 3. Multiply the whole divisor by the quotient obtained in step 2, and subtract this product from the dividend.	$\begin{array}{r} 6x \\ x - 2\overline{)6x^2 - x + 10} \\ \underline{6x^2 - 12x} \\ 11x + 10 \end{array}$

Procedure	Solution
Step 4. Repeat steps 2 and 3 until the remainder is 0 or until the degree of the remainder polynomial is less than the degree of the divisor polynomial in the chosen variable. *Step 5.* If the remainder is not 0, express the answer as follows: $$\frac{\text{dividend}}{\text{divisor}} = \text{quotient} + \frac{\text{remainder}}{\text{divisor}}$$	$$\begin{array}{r} 6x\ + 11 \\ x - 2\overline{)6x^2 -\quad x + 10} \\ \underline{6x^2 - 12x\quad\quad} \\ 11x + 10 \\ \underline{11x - 22} \\ 32 \end{array}$$ $$\frac{6x^2 - x + 10}{x - 2}$$ $$= 6x + 11 + \frac{32}{x - 2}$$

EXAMPLE 2   Divide $11x - 6x^2 + x^3 - 12$ by $x - 4$.

Solution   Rearranging the terms of the dividend in descending order of exponents, we proceed as follows:

$$\begin{array}{r} x^2 - 2x\ +\ 3 \\ x - 4\overline{)x^3 - 6x^2 + 11x - 12} \\ \text{Subtract} \longrightarrow \underline{x^3 - 4x^2\quad\quad\quad\quad} \\ - 2x^2 + 11x \\ \text{Subtract} \longrightarrow \underline{- 2x^2 +\ 8x\quad} \\ 3x - 12 \\ \text{Subtract} \longrightarrow \underline{3x - 12} \\ 0 \longleftarrow \text{Remainder} \end{array}$$

Thus

$$(11x - 6x^2 + x^3 - 12) \div (x - 4) = x^2 - 2x + 3$$

Note. If the remainder is 0, the divisor is called a factor of the dividend.

Hence, in Example 2, $x - 4$ is a factor of $x^3 - 6x^2 + 11x - 12$. ■

EXAMPLE 3 $(46xy + 16x^2 + 10y^2) \div (2x + 5y)$

Solution   Rearranging the dividend in descending order of exponents of x (we could have chosen y), we have

$$\begin{array}{r} 8x\ + 3y \\ 2x + 5y\overline{)16x^2 + 46xy + 10y^2} \\ \underline{16x^2 + 40xy\quad\quad\quad} \\ 6xy + 10y^2 \\ \underline{6xy + 15y^2} \\ -\ 5y^2 \longleftarrow \text{Remainder} \end{array}$$

Thus,

$$\frac{16x^2 + 46xy + 10y^2}{2x + 5y} = 8x + 3y + \frac{-5y^2}{2x + 5y}$$

Since the remainder is not 0, $(2x + 5y)$ is not a factor of $(16x^2 + 46xy + 10y^2)$. ■

EXAMPLE 4 $\dfrac{x^3 - 8}{x - 2}$

Solution Note that in the dividend $x^3 - 8$, there are no x^2 and x terms.

$$
\begin{array}{r}
x^2 + 2x + 4 \\
x - 2 \overline{) x^3 + 0x^2 + 0x - 8} \\
\underline{x^3 - 2x^2} \\
2x^2 + 0x \\
\underline{2x^2 - 4x} \\
4x - 8 \\
\underline{4x - 8} \\
0
\end{array}
$$

Thus, $x - 2$ is a factor of $x^3 - 8$ and

$$\frac{x^3 - 8}{x - 2} = x^2 + 2x + 4 \quad ■$$

PROBLEM 1 $[(x + 2y)^2 + 3(x + 2y) - 4] \div [(x + 2y) + 4]$

Answer We can simplify the division by making the substitution $A = x + 2y$. The dividend and divisor take the form $(A^2 + 3A - 4) \div (A + 4)$.

$$
\begin{array}{r}
A \; - 1 \\
A + 4 \overline{) A^2 + 3A - 4} \\
\underline{A^2 + 4A} \\
-\; A - 4 \\
\underline{-\; A - 4}
\end{array}
$$

Thus, the quotient is $A - 1$. Replacing A by the expression $x + 2y$, the quotient becomes $x + 2y - 1$. Hence,

$$\frac{[(x + 2y)^2 + 3(x + 2y) - 4]}{[(x + 2y) + 4]} = x + 2y - 1 \quad ■$$

EXERCISES 2.7

In Exercises 1–80, divide.

1. $(x^2 + 5x + 4) \div (x + 1)$

2. $(x^2 + 9x + 14) \div (x + 7)$

3. $(63 + x^2 + 16x) \div (x + 9)$

4. $(-63 + x^2 - 2x) \div (x + 7)$
5. $(y^3 - 3y^2 + 3y - 1) \div (y - 1)$
6. $(6y^2 + 11y + 3) \div (3y + 1)$
7. $(12x^2 + 7xy - 12y^2) \div (4x - 3y)$
8. $(8x^4 - 8x^2 - 6) \div (2x^2 - 3)$
9. $(5x^2 - 6x - 4) \div (x - 1)$
10. $(8x^2 + 14xy - 15y^2) \div (4x - 3y)$
11. $(6x^2 - 13xy + 5y^2) \div (3x - 5y)$
12. $(4x^3 - 5x^2 + 2x - 16) \div (x - 2)$
13. $(53x^2 + 15x^3 - 30x - 8) \div (3x - 2)$
14. $(8x^2 + 10xy - 3y^2) \div (4x - y)$
15. $(9x^2 - 6xy + y^2) \div (3x - y)$
16. $(4x^2 - 9y^2) \div (2x + 3y)$
17. $(y^2 - yx - 2x^2) \div (y - 2x)$
18. $(12x^4 + 15 - 29x^2) \div (4x^2 - 3)$
19. $(y^4 + y^2x^2 + x^4) \div (y^2 + x^2 - yx)$
20. $(6x^2 - 7y^2 + 11xy) \div (3x + 7y)$
21. $(16x^4 - 8x^2y^2 + y^4) \div (4x^2 - y^2)$
22. $(3x^4 - 13x^2y^2 + 12y^4) \div (3x^2 - 4y^2)$
23. $(2x^3 - 33xy^2 + 20y^3 + 5x^2y)$
    $\div (2x - 5y)$
24. $(4x^3 - 3x + 1) \div (2x - 1)$
25. $(14x^3 + 17x^2y + 9y^3) \div (2x + 3y)$
26. $(a^3 - a^2 + a - 1) \div (a + 1)$
27. $(a^4 - 7a^3 + 8a^2 + 28a - 48) \div (a - 3)$
28. $(x^4 - 6x^2y^2 + 8y^4 - x^3y + 4xy^3)$
    $\div (x - 2y)$
29. $(24x^4 - 1 + 4x^3 - 2x) \div (2x - 1)$
30. $(x^3 - 6x^2 + 7x + 4) \div (x^2 - 2x - 1)$
31. $(x^3 - 3x^2y + 3xy^2 - y^3)$
    $\div (x^2 - 2xy + y^2)$
32. $(4a^3 - 13a - 6) \div (2a^2 - 2 - 3a)$
33. $(a^4 + a^2b^2 + b^4) \div (a^2 + ab + b^2)$
34. $(6z^4 + 13z - 11z^3 - 10 - z^2)$
    $\div (3z^2 - 5 - z)$
35. $(a^3 + 27) \div (a + 3)$
36. $(8a^3 + 6a - 12a^2 - 1) \div (2a - 1)$
37. $(4x^3 - 9x^2 - 4 + 4x) \div (x - 2)$
38. $(12a^3 - 17a - 12a^2 + 16)$
    $\div (6a^2 + 3a - 4)$
39. $(a^2 - 6ab + 9b^2 - 4c^2) \div (a + 2c - 3b)$
40. $(3x^3 - 20x^2 + 8) \div (3x - 2)$
41. $(8x^3 + y^3) \div (4x^2 - 2xy + y^2)$
42. $(x^4 - 7x^2y^2 + 9y^4) \div (x^2 - xy - 3y^2)$
43. $(4x^4 - 5x^2 + 1) \div (2x^2 - 3x + 1)$
44. $(a^4 + 3ab^2 - 4b^4) \div (a^2 + 2b^2)$
45. $(27a^3 + 4ab^2 - 8b^3)$
    $\div (9a^2 + 6ab + 4b^2)$
46. $(4a^3 - 7ab^2 + 3b^3) \div (2a^2 - 3ab + b^2)$
47. $(10a^3 - 21a^2 + 11a - 3) \div (2a - 3)$

48. $(8x^3 + 20x^2y + 6xy^2 + 15y^3)$
    $\div (4x^2 + 3y^2)$
49. $(12x^3 - 25x^2y + 20xy^2 - 6y^3)$
    $\div (3x^2 - 4xy + 2y^2)$
50. $(8a^3x^3 - 1) \div (4a^2x^2 + 2ax + 1)$
51. $(4a^4 - 4a^2b^2 + 2a + b^4) \div (2a^2 - b^2)$
52. $(x^2 - 2xy + y^2 - z^2) \div (x - y + z)$
53. $(64x^6 - 8) \div (4x^2 - 2)$
54. $(15x^3 - 26x^2y + 18xy^2 - 4y^3)$
    $\div (3x^2 - 4xy + 2y^2)$
55. $(27a^3 - 54a^2 + 36a - 8)$
    $\div (9a^2 - 12a + 4)$
56. $(x^3 + 6x^2y + 12xy^2 + 8y^3) \div (x + 2y)$
57. $(125a^3 - 75a^2b + 15ab^2 - b^3)$
    $\div (5a - b)$
58. $(x^4 - y^4) \div (x^3 + x^2y + xy^2 + y^3)$
59. $(y^4 - 16) \div (y^2 + 4)$
60. $(9x^2 - 5x - 5x^3 + x^4 - 4)$
    $\div (x^2 - 3x + 4)$
61. $(32a^3b - 52a^2b^2 + 19ab^3 - 5b^4)$
    $\div (8a^2b - 3ab^2 + b^3)$
62. $(64x^3 - 144x^2y + 108xy^2 - 27y^3)$
    $\div (4x - 3y)$
63. $(x^2 - 6xy + 9y^2 - 4) \div (x + 2 - 3y)$
64. $(4x^2 + 4xy + y^2 - 1) \div (2x + y - 1)$
65. $(27a^6 + 8b^3) \div (9a^4 + 6a^2b + 4b^2)$
66. $(9x^2 - 12xy + 6xz - 4yz + 4y^2 + z^2)$
    $\div (3x - 2y + z)$
67. $(x^4 - 16) \div (x^3 - 2x^2 + 4x - 8)$
68. $(3a^3 - 22a^2 - 5a + 4) \div (3a - 1)$
69. $(16a^4 - 9a^2b^2 + b^4) \div (4a^2 - ab - b^2)$
70. $(x^4 - 2x^2y^2 + y^4) \div (x^2 + 2xy + y^2)$
71. $(2b^3 - 3b^2 + 7b + 24) \div (b^2 - 3b + 8)$
72. $(a^3 - ax^2 - a^2x + x^3) \div (a + x)$
73. $(3x^3 - xy^2 - 6x^2y + 2y^3) \div (3x^2 - y^2)$
74. $(2a^3 - 3a^3b + 15ab^2 - 7b^3) \div (2a - b)$
75. $(2x^3 - 7x^2y + 9y^3) \div (2x^2 - xy - 3y^2)$
76. $(15x^3 - 38x^2 + 21x + 4) \div (3x - 4)$
77. $(a^4 - 13a^2c^2 + 4c^4) \div (a^2 + 3ac - 2c^2)$
78. $(2x^3 - 3x^2y + 4xy^2 - 5y^3) \div (x - 2y)$
79. $(12a^4 - 10a^3b + 8a^2b^2 - 6ab^3 + 4b^4)$
    $\div (a^2 + b^2)$
*80. $(x^{5m} - y^{5m}) \div (x^m - y^m)$

In Exercises 81–85, use the technique of substitution to find the quotient.

81. $[(x + y)^2 - (a - b)^2]$
    $\div [(x + y) - (a - b)]$
82. $[(a - b)^2 - 4(a - b) + 3] \div [(a - b) - 3]$
83. $[a^2(x - 2y)^2 + 4a(x - 2y) + 4]$
    $\div [a(x - 2y) + 2]$

84. $[12(2a + b)^2 - 2(2a + b) - 30]$
$\div [3(2a + b) - 5]$

*85. $[(x + y)^3 - (m + n)^3]$
$\div [(x + y) - (m + n)]$

86. If the area of a rectangle is
$(8x^3 - 18x^2 + 13x - 6)$ sq ft and one side
is $(2x - 3)$ ft, find the other side.

87. One factor of $5x^2 - 22x + 21$ is $x - 3$.
Find the other factor.

88. One factor of $8y^3 - 64$ is $2y - 4$. Find the
other factor.

*89. Find the number k for which $x - 1$ is a factor of $3x^2 + 3x + k$. (*Hint:* Recall that a
polynomial is a factor of another polynomial
if and only if the remainder is zero.)

*90. Find the number k for which $3x - 5$ is a
factor of $3x^3 - 14x^2 + 18x + k$.

2.8 SYNTHETIC DIVISION

Let us consider a simplified method of dividing a polynomial $p(x)$ by a divisor
of the form $x - k$. This method determines the coefficients of the quotient and
the value of the remainder. The method is called **synthetic division** and consists
of working with only the coefficients of the variable.

EXAMPLE 1 Consider the division of $x^3 - 6x^2 + 20$ by $x - 3$. Using the ordinary process
of long division, we obtain:

$$
\begin{array}{r}
x^2 - 3x - 9 \quad \longleftarrow \text{Quotient} \\
x - 3{\overline{\smash{\big)}\,x^3 - 6x^2 + 0x + 20}} \quad \longleftarrow \text{Dividend} \\
\underline{x^3 - 3x^2} \\
-3x^2 + 0x \\
\underline{-3x^2 + 9x} \\
-9x + 20 \\
\underline{-9x + 27} \\
-7 \quad \longleftarrow \text{Remainder}
\end{array}
$$

Divisor $\longrightarrow$

where the term $0x$ has been inserted in the dividend so that *all* powers of x are
taken into account.

The first step in shortening the process is to write only the coefficients:

$$
\begin{array}{r}
1 - 3 - 9 \\
1 - 3{\overline{\smash{\big)}\,1 - 6 + 0 + 20}} \\
\underline{1 - 3} \\
-3 + 0 \\
\underline{-3 + 9} \\
-9 + 20 \\
\underline{-9 + 27} \\
-7
\end{array}
$$

Note that the numbers 1, -3, and -9, the coefficients of the quotient, have each been written three times. Thus, we can eliminate these repetitions. Also, since the divisor is a polynomial of the form $x - k$, the two coefficients in the far left position are always $1 - k$ and we can also eliminate the coefficient 1. Hence, the long division process can be written

$$
\begin{array}{r}
1 - 3 - 9 \\
-3\overline{)1 - 6 + 0 + 20} \\
\underline{- 3 + 9 + 27} \\
1 - 3 - 9 - 7
\end{array} \tag{1}
$$

Note that in this form the 1, -3, and -9 in the third line are the coefficients of the descending powers of the variable of the quotient, and -7 is the value of the remainder.

We now make one last modification in the ordinary long division process. In form (1), the numbers in the second row were subtracted from the corresponding numbers in the first row to obtain the third row. If the number $-k$ is replaced by k and the sign of every number of the second row is changed, we may *add* rather than subtract. Thus, the final form is

$$
\begin{array}{r|rrrr}
3 & 1 & -6 & 0 & 20 \\
& & 3 & -9 & -27 \\
\hline
& 1 & -3 & -9 & -7
\end{array}
$$

Synthetic Division

To divide a polynomial $p(x)$ by $x - k$:

1. Write the coefficients of the given polynomial $p(x)$, including negative signs, in order of descending powers of x. Include a zero coefficient for each missing power of x.

2. Place the number k in a box at the left end of the row of coefficients listed in step 1.

3. Bring down the first coefficient of row one, multiply it by k, add the product to the second coefficient in row one, and bring down their sum.

4. Repeat multiplication and addition until all coefficients of row one have been used.

5. The numbers in the third row, except for the last one, are the coefficients of the quotient. The last number in the third row is the remainder.

Synthetic division is shown diagrammatically in Figure 2.15.

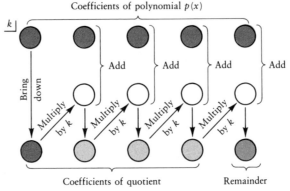

Coefficients of polynomial $p(x)$

Add   Add   Add   Add

Bring down   Multiply by k Multiply by k Multiply by k Multiply by k

Coefficients of quotient   Remainder

Figure 2.15

EXAMPLE 2   Divide $4x^3 - 13x^2 + 16x + 11$ by $x - 2$.

Solution   Here $x - k = x - 2$; hence, $k = 2$.

$$
\begin{array}{r|rrrr}
2 & 4 & -13 & 16 & 11 \\
  &   & 8 & -10 & 12 \\
\hline
  & 4 & -5 & 6 & 23
\end{array}
$$

Quotient $= 4x^2 - 5x + 6$; remainder $= 23$. Recall that

$$\frac{\text{dividend}}{\text{divisor}} = \text{quotient} + \frac{\text{remainder}}{\text{divisor}}$$

Thus, we have

$$\frac{4x^3 - 13x^2 + 16x + 11}{x - 2} = 4x^2 - 5x + 6 + \frac{23}{x - 2}$$

Note that the remainder equals the value of the given polynomial (the dividend) when k is substituted for x. That is, $4(2)^3 - 13(2)^2 + 16(2) + 11 = 23$. ∎

EXAMPLE 3   Divide $x^3 - 3x^2 + 4x - 5$ by $x + 3$.

Solution   Here $x - k = x + 3$; hence, $-k = 3$ and $k = -3$.

$$
\begin{array}{r|rrrr}
-3 & 1 & -3 & 4 & -5 \\
   &   & -3 & 18 & -66 \\
\hline
   & 1 & -6 & 22 & -71
\end{array}
$$

Quotient $= x^2 - 6x + 22$; remainder $= -71$. Thus,

$$\frac{x^3 - 3x^2 + 4x - 5}{x + 3} = x^2 - 6x + 22 + \frac{-71}{x + 3}$$

Again, if we substitute $x = -3$ into our given polynomial $x^3 - 3x^2 + 4x - 5$, we get the remainder:

$$(-3)^3 - 3(-3)^2 - 4(-3) - 5 = -27 - 27 - 12 - 5 = -71 \quad \blacksquare$$

EXAMPLE 4 Divide $x^6 - 4$ by $x + 2$.

Solution Here $x - k = x + 2$; hence, $-k = 2$ and $k = -2$. Since the x^5, x^4, x^3, x^2, and x terms are missing, we insert zero coefficients for each of these terms:

$$
\begin{array}{r|rrrrrrr}
-2 & 1 & 0 & 0 & 0 & 0 & 0 & -4 \\
   &   & -2 & 4 & -8 & 16 & -32 & 64 \\
\hline
   & 1 & -2 & 4 & -8 & 16 & -32 & 60
\end{array}
$$

Quotient $= x^5 - 2x^4 + 4x^3 - 8x^2 + 16x - 32$; remainder $= 60$. Thus,

$$\frac{x^6 - 4}{x + 2} = x^5 - 2x^4 + 4x^3 - 8x^2 + 16x - 32 + \frac{60}{x + 2}$$

Also, the value of $x^6 - 4$ when $x = -2$ equals the remainder:

$$(-2)^6 - 4 = 64 - 4 = 60 \quad \blacksquare$$

EXAMPLE 5 Divide $4x^5 - 3x^4 - 5x^3 + 2$ by $x + 1$.

Solution Here $x - k = x + 1$; hence, $-k = 1$ and $k = -1$. Since x^2 and x terms are missing, we insert zero coefficients:

$$
\begin{array}{r|rrrrrr}
-1 & 4 & -3 & -5 & 0 & 0 & 2 \\
   &   & -4 & 7 & -2 & 2 & -2 \\
\hline
   & 4 & -7 & 2 & -2 & 2 & 0
\end{array}
$$

Quotient $= 4x^4 - 7x^3 + 2x^2 - 2x + 2$; remainder $= 0$. Thus,

$$\frac{4x^5 - 3x^4 - 5x^3 + 2}{x + 1} = 4x^4 - 7x^3 + 2x^2 - 2x + 2$$

That is,

$$4x^5 - 3x^4 - 5x^3 + 2 = (x + 1)(4x^4 - 7x^3 + 2x^2 - 2x + 2)$$

and we see that $(x + 1)$ is a factor of the given polynomial $4x^5 - 3x^4 - 5x^3 + 2$. Also note that the remainder equals 0 and the value of $4x^5 - 3x^4 - 5x^3 + 2$ when $x = -1$ equals the remainder:

$$4(-1)^5 - 3(-1)^4 - 5(-1)^3 + 2 = -4 - 3 + 5 + 2 = 0 \quad \blacksquare$$

EXERCISES 2.8

In Exercises 1–16, find the quotient and remainder using synthetic division.

1. $(3x^3 - x^2 + 4x + 8) \div (x + 2)$
2. $(x^4 + 2x^3 + x^2 - 3x - 5) \div (x + 1)$
3. $(2x^3 - 5x^2 - 6) \div (x - 3)$
4. $(x^5 + 3x^2 - 4x + 10) \div (x - 2)$
5. $(x^3 + 8) \div (x + 2)$
6. $(x^5 - 1) \div (x + 1)$
7. $(x^5 - 2x^3 + 3x^2 - 2) \div (x - 1)$
8. $(x^3 - x^2 + x - 1) \div (x - 1)$
9. $(2x^3 - 6x^2 + x - 5) \div (x - 6)$
10. $(x^3 + 6x^2 + 6x + 2) \div (x + 2)$
11. $(2x^4 - x^2 + 3x - 5) \div (x + 3)$
12. $(x^5 - 2x^4 - 4x^2 + 2) \div (x - 1)$
13. $(x^5 + 1) \div (x + 1)$
14. $(4x^4 + 2x^2 - 1) \div (x - \frac{1}{2})$
15. $(3x^3 + ax^2 + a^2x + 2a^3) \div (x - a)$
16. $(2x^3 + ax^2 + a^2x - 2a^3) \div (x + a)$

2.9 THE REMAINDER THEOREM AND THE FACTOR THEOREM

In this section we discuss two very important and useful theorems in algebra.

> ### The Remainder Theorem
>
> If p is a polynomial defined by $p(x)$, and $p(x)$ is divided by $x - k$, then the remainder is given by $p(k)$. That is, the remainder is the value obtained by substituting the number k for x in the expression $p(x)$.

Proof   Recall that by the definition of division, we have

$$\frac{\text{dividend}}{\text{divisor}} = \text{quotient} + \frac{\text{remainder}}{\text{divisor}}$$

Then, dividing $p(x)$ by $x - k$ yields

$$\frac{p(x)}{x - k} = q(x) + \frac{R}{x - k} \tag{2}$$

where $q(x)$ is the quotient and R is the constant remainder. Multiplying both sides of equation (2) by $x - k$ gives

$$p(x) = (x - k)\, q(x) + R$$

Since this is an identity, it must be true for all values of x, including $x = k$. Substituting k for x yields

$$p(k) = (k - k)\, q(k) + R = 0\, q(k) + R = R$$

Hence,

$$p(k) = R \quad \blacksquare$$

EXAMPLE 1 Without dividing, find the remainder when $x^3 + 4x^2 - x + 5$ is divided by $x - 2$.

Solution Here $p(x) = x^3 + 4x^2 - x + 5$ and $x - k = x - 2$; hence, $k = 2$:

$$p(2) = (2)^3 + 4(2)^2 - 2 + 5$$
$$= 8 + 16 - 2 + 5$$
$$= 27$$

The remainder theorem states that remainder $= p(2)$. Thus, remainder $R = 27$. ■

EXAMPLE 2 Find the values of m so that $x^3 - 2x^2 + mx + 4$ will leave a remainder of -2 when divided by $x - 3$.

Solution Here $p(x) = x^3 - 2x^2 + mx + 4$; $x - k = x - 3$, or $k = 3$; and $R = -2$. Substituting $x = 3$ in $p(x)$, we get

$$p(3) = (3)^3 - 2(3)^2 + m(3) + 4 = 27 - 18 + 3m + 4 = 3m + 13$$

However, the remainder theorem states that $p(3) = R$, the remainder; thus,

$$3m + 13 = -2$$
$$3m = -15$$
$$m = -5$$

The remainder theorem can be used with the method of synthetic division to find the value of a polynomial at various real numbers. ■

EXAMPLE 3 Given $p(x) = x^4 - 3x^3 - 11x^2 + 3x + 10$, find $p(5)$.

Solution We use synthetic division to find the remainder R, which is the same as $p(5)$ when the polynomial is divided by $x - 5$:

$$
\begin{array}{r|rrrrr}
5 & 1 & -3 & -11 & 3 & 10 \\
  &   & 5 & 10 & -5 & -10 \\
\hline
  & 1 & 2 & -1 & -2 & 0 \\
\end{array}
$$

Thus, $p(5) = 0$, and we see that 5 is a root of the given polynomial. ■

For certain polynomials it is easier to substitute the value in the expression $p(x)$ than to use synthetic division. The choice usually depends on the polynomial—the powers that appear or the coefficients involved.

A direct consequence of the remainder theorem is the following theorem.

> ### The Factor Theorem
>
> Let p be a polynomial defined by $p(x)$. Then $x - k$ is a factor of $p(x)$ if and only if $p(k) = 0$.

Proof If $x - k$ is a factor of $p(x)$, then

$$p(x) = (x - k)q(x),$$

and

$$p(k) = (k - k)q(x) = 0 \cdot q(x) = 0.$$

Conversely, if $p(k) = 0$, then by the remainder theorem the remainder $R = 0$ and $p(x) = (x - k)\, q(x) + R$ can be written

$$p(x) = (x - k)\, q(x)$$

Thus, $x - k$ is a factor of $p(x)$. ∎

EXAMPLE 4 Using the factor theorem, determine whether $x + 1$ is a factor of $x^3 + 1$.

Solution We have $p(x) = x^3 + 1$ and $x - k = x + 1$; hence $k = -1$:

$$p(-1) = (-1)^3 + 1 = -1 + 1 = 0$$

Since $p(-1) = 0$, $x + 1$ is a factor of $x^3 + 1$. ∎

EXAMPLE 5 Use the factor theorem to prove that $x + a$ is a factor of $x^n - a^n$, if n is an even integer.

Solution We have $p(x) = x^n - a^n$, where n is an even integer. Also, $x - k = x + a$; hence, $k = -a$. Thus,

$$p(-a) = (-a)^n - a^n = a^n - a^n = 0$$

since n is even. So, $x + a$ is a factor of $x^n - a^n$, if n is an even integer. ∎

EXAMPLE 6 Determine m so that $x - 4$ is a factor of $x^3 - mx^2 + 8$.

Solution Here $p(x) = x^3 - mx^2 + 8$ and $x - k = x - 4$; hence, $k = 4$. Thus,

$$p(4) = (4)^3 - m(4)^2 + 8 = 64 - 16m + 8 = 72 - 16m$$

If $x - 4$ is a factor, then $p(4) = 0$. So, we have

$$72 - 16m = 0$$

$$m = \frac{9}{2} \quad \blacksquare$$

EXERCISES   2.9

In Exercises 1–8, use the remainder theorem to find the remainder.

1. $(x^3 + 2x^2 - 3x + 5) \div (x - 3)$
2. $(x^4 + x^3 + x^2 + 2) \div (x + 2)$
3. $(4x^3 + x^2 - 3x - 1) \div (x - 1)$
4. $(x^5 - x^2 + 8) \div (x + 1)$
5. $(x^3 + 27) \div (x - 3)$
6. $(x^3 - a^3) \div (x - a)$
7. $(x^4 - 6x^3 + x^2 - 9) \div (x + 1)$
8. $(4x^5 - 3x^4 - 5x^3 + 2) \div (x - 1)$
9. Find the value of m so that $x^3 + mx^2 + 2x - m$ will leave a remainder of 7 when divided by $x + 3$.
10. Find the value of m so that $3x^3 + 2x^2 - mx - 6$ will leave a remainder of -20 when divided by $x + 2$.
*11. Find a condition involving m so that $x^3 - 3x - 6$ will leave a remainder of 4 when divided by $x - m$.
*12. Find a condition involving m so that $2x^3 + x - 5$ will leave a remainder of -2 when divided by $x + m$.

In Exercises 13–24, use the factor theorem to determine whether the first polynomial is a factor of the second polynomial.

13. $x + 3; \quad x^4 + 3x^3 - x^2 - 3x$
14. $x - 2; \quad x^4 - 5x^2 + 5x - 6$
15. $x - 1; \quad x^9 + 2x^8 + 3x^7 - 6$
16. $x + 6; \quad x^4 + 7x^3 - 39x - 18$
17. $x + 3; \quad x^4 + 3x^3 - 2x^2 - x + 15$
18. $x + 1; \quad x^{99} + 1$
19. $x - 4; \quad x^5 + 1024$
20. $x - 2; \quad x^8 - 256$
21. $x - \dfrac{1}{3}; \quad 6x^4 - 5x^3 + 10x^2 - 1$
22. $x + \dfrac{1}{2}; \quad 4x^3 - 12x^2 + 7x + 1$
23. $x + m; \quad x^4 - m^4$
24. $x + m; \quad x^5 - m^5$
25. Determine m so that $x - 2$ will be a factor of $x^3 + 2x^2 + mx - 2$.
26. Determine m so that $x + 1$ will be a factor of $x^3 + 3x^2 - x - 6m$.
*27. If n is a positive integer, prove that $x - a$ is a factor of $x^n - a^n$.
*28. If n is a positive odd integer, prove that $x + a$ is a factor of $x^n + a^n$.
*29. If n is a positive even integer, prove that $x + a$ is not a factor of $x^n + a^n$.
*30. If n is a positive integer, prove that $x + a$ is a factor of $x^{2n} - a^{2n}$.

Chapter 2   SUMMARY

TERMS AND RULES

1. If a is any real number and n is a positive integer, then $a^n = \underbrace{a \cdot a \cdot a \cdot \ldots \cdot a}_{n \text{ factors}}$, and a is called the base and n is called the exponent.
2. An algebraic expression which involves *only* addition, multiplication, division, and in which all variables occur only as non-negative integral powers is called a rational expression.
3. Any rational expression in which *no* variable occurs in a denominator is called a polynomial. A polynomial in a single variable x:

$$a_n x^n + a_{n-1} x^{n-1} + \cdots + a_1 x + a_0$$

4. The **degree of a term** is the sum of the exponents of the variables in that term.
5. The **degree of a polynomial** is the degree of the term(s) with the highest degree.
6. *Removal of Parentheses.* When an expression within parentheses is preceded by a:

(a) positive sign, +, the parentheses may be removed without making any change in the expression.
(b) negative sign, −, the parentheses may be removed if the sign of *every* term within the parentheses is changed.

ADDITION, SUBTRACTION, MULTIPLICATION, AND DIVISION OF POLYNOMIALS

1. To **add** (**subtract**) two polynomials, add (subtract) like terms.
2. For all positive integers m and n, $a^m \cdot a^n = a^{m+n}$.
3. For every positive integer m, $(ab)^m = a^m b^m$.
4. For all positive integers m and n, $(a^m)^n = a^{mn}$.
5. To **multiply** two polynomials, multiply one polynomial by each term of the other polynomial and add the products.
6. FOIL: *First Outside Inside Last*

$$(a + b)(c + d) = ac + ad + bc + bd$$

7. *Special Products*
   (1) $(A + B)^2 = A^2 + 2AB + B^2$
   (2) $(A - B)^2 = A^2 - 2AB + B^2$
   (3) $(A + B)(A - B) = A^2 - B^2$
       [Difference of two squares]
   (4) $(A + B)(A^2 - AB + B^2)$
       $= A^3 + B^3$ [Sum of two cubes]
   (5) $(A - B)(A^2 + AB + B^2) = A^3 - B^3$
       [Difference of two cubes]

(6) $(A + B)^3 = A^3 + 3A^2B + 3AB^2 + B^3$
(7) $(A - B)^3 = A^3 - 3A^2B + 3AB^2 - B^3$
(8) $(Ax + By)(Cx + Dy)$
    $= ACx^2 + (AD + BC)xy + BDy^2$

8. For all positive integers m and n,

$$\frac{a^m}{a^n} = a^{m-n} \qquad (a = 0).$$

9. If n is a positive integer,

$$\left(\frac{a}{b}\right)^n = \frac{a^n}{b^n} \qquad (b = 0).$$

10. To divide one polynomial by another, follow the division procedure.
11. *The Remainder Theorem.* If p is a polynomial defined by $p(x)$, and $p(x)$ is divided by $x - k$, then the remainder is given by $p(k)$.
12. *The Factor Theorem.* Let p be a polynomial defined by $p(x)$. Then $x - k$ is a factor of $p(x)$ if and only if $p(k) = 0$.

Chapter 2 EXERCISES

In Exercises 1–12, determine which expressions are polynomials. If the expression is a polynomial, state its degree.

1. $3x$
2. $4x^2 + 1$
3. $\dfrac{1}{x}$
4. $\dfrac{x + z^3}{x + z}$
5. $x^2 + 2xy + y^2$
6. 10
7. $3x^2y^3$
8. $\dfrac{5x - y}{2}$
9. $\dfrac{x^2 - 4x + 4}{x + 3}$
10. $xyz + 3y^2$
11. $(x + y)(x - y)$
12. $x^2yz^3 + 2y^4 + z^8$

In Exercises 13–25, if $x = 3$, $y = 4$, and $z = 10$, find the value of each of the polynomials.

13. $2xyz^2$
14. $(2xyz)^2$
15. $3x^2 - y^2$
16. $3(x^2y^2 - z^2)$
17. $(3xy^2 - z)^2$
18. $(3x + y - z)^2$
19. $3x + (y + z)^2$
20. $(xy)^2(z - y)^2$
21. $[xy(z - y)^2]$
22. $(x + y)^2(z - y)^2$
23. $[z + (x + y)][z - (x + y)]$
24. $(x + y)^3$
25. $x^3 + 3x^2y + 3xy^2 + y^3$

In Exercises 26–30, if $x = 1$, $y = 3$, $z = 5$, and $w = 0$, find the value of each of the algebraic expressions.

26. $x^4 + 4x^3y + 6x^2y^2 - 4xy^3 + y^4$

27. $\dfrac{12x^3 - y^2}{3x^2} + \dfrac{2z^2}{x + y^2} - \dfrac{x + y^2 + z^3}{5y^3}$

28. $(z^2 - y^2 - x^2) \div (xyz - y^2)$

29. $(z - y - x)^2 \div (xyz - 3)^2$

30. $\dfrac{10xy + zw}{(w + 2xy - z)^2} + \dfrac{(2z)^2}{x^2 + y} - \dfrac{8x^2}{2y^2}$

In Exercises 31–35, find the perimeter and area of each form.

SAMPLE 1 Find the perimeter and area of the cross given in Figure 2.16. The unit of measure is inches.

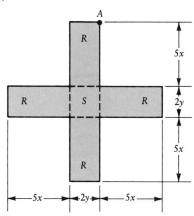

Figure 2.16

Solution (a) Starting at point A and moving in a clockwise direction, we have

perimeter $= (5x + 5x + 2y + 5x + 5x + 2y$

$+ 5x + 5x + 2y$

$+ 5x + 5x + 2y)$ in.

$= (40x + 8y)$ in.

(b) The total area given by adding the areas of the four rectangles, R, and the square, S, is

$$A = 4R + S$$

The area of each rectangle $= [(5x)\text{ in.}][(2y)\text{ in.}]$
$= (10xy)$ sq in.

The area of the square $= [(2y)\text{ in.}]^2$
$= 4y^2$ sq in.

Thus,

total area $= 4R + S = 4[(10xy)$ sq in.$]$
$+ 4y^2$ sq in.
$= 40xy$ sq in. $+ 4y^2$ sq in.
$= (40xy + 4y^2)$ sq in.

31.

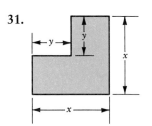

32.

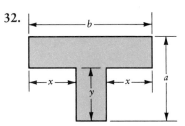

33.

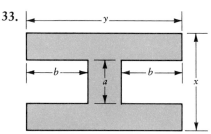

34.

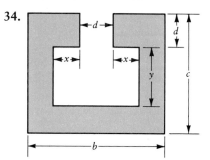

35.

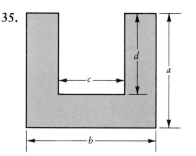

In Exercises 36–45, write a polynomial that represents the number characterized.

SAMPLE 2 Write a polynomial that represents the area of a triangle in which the length of the altitude is 4 ft greater than the length of the base (see Figure 2.17).

Solution: Let x = number of feet in the base. Then $x + 4$ = number of feet in the altitude.

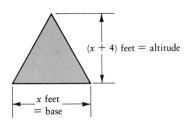

$(x + 4)$ feet = altitude

x feet = base

Figure 2.17

$$\text{Area} = \frac{1}{2}(\text{base})(\text{altitude}) = \frac{1}{2}[x \text{ ft}][(x + 4) \text{ ft}]$$

$$= \frac{1}{2}x(x + 4) \text{ sq ft}$$

$$= \frac{1}{2}(x^2 + 4x) \text{ sq ft}$$

36. Write a polynomial that represents the area of a rectangle whose length is 9 ft longer than its width.

37. Write a polynomial that represents the area of a triangle in which the length of the base is 7 in. shorter than the length of its altitude.

38. The area of a trapezoid equals one-half the sum of the two bases times the altitude. Write a polynomial that represents the area of a trapezoid whose bases are x ft and $(x + 5)$ ft, respectively, and whose altitude is $4x$ ft.

39. Write a polynomial that represents the total distance traveled by a car that goes for 4 hr at $(60 + x)$ mph and then for 5 hr at $(90 + x)$ mph.

*40. A plane traveled for 3 hr at an average airspeed of v mph with an average tail wind of 30 mph. Then the plane traveled against a wind averaging 10 mph for 2 hr. Write a polynomial that represents the total land distance traveled.

41. A girl is x years old now. Her father is four times as old as the girl will be 5 years from now. Write a polynomial that represents the father's present age.

42. A child has $4x - 3$ nickels and $5x + 1$ dimes. Write a polynomial that represents the total number of cents the child has.

43. A woman earns x dollars per day and her son y dollars per day. How many dollars did they both earn in a month if the woman worked 25 days and the son worked 19 days?

*44. If $13x$ books cost $156x^4$ dollars, how much does one book cost?

*45. If $3x^2 - 2x$ represents the cost of x television sets, what is the cost of one television set?

In Exercises 46–60, perform indicated operations.

46. $(8x - 3y) + (-5x + 2y)$

47. $(21x + 13y) + (9y + 11z)$

48. $(2xy - 8xyz - 15yz)$
$+ (-12xy + 15xyz - 20yz)$

49. $(26a + 10x + 14y + z)$
$+ (-12a + 15x - 20z)$

50. $(3x^2 + xy + z) + (2z - 4x^2 - xy)$

51. $(-2ab + b^3 + a^2) - (2a^2 - b^3 + 4ab)$

52. $(15x^2y + 5xy^2 - 3xy + 2)$
$+ (5xy + x^2 - y^2)$

53. $(7x^3 - 5y^3 + 2xy) - (6x^3 + 4y + 3xy)$

54. $(2x^2 - 3y^2 + x) + (5a^2 + 2b^2 + y)$

55. $(3x^2y - 5xy^2 + 7xy)$
$+ (4x^3 + 8x^2y + 3xy^2)$
$+ (-2xy^2 + 5xy + 3)$

56. $(3x^2 - 17y^2) + (4x^2 + 9y^2)$
$+ (-3a^2 + 5b^2)$

57. $(7x^4 - 3a^2 + 5xy) - (2x^4 + 6a^2 - 2xy)$
$+ (3x^4 + 2a^2 + 4xy)$

58. $(4xy + x^2 - 3y^4) + (x^2 + y^2 - 2xy)$
$- (x^3 - y^2 + 7xy)$

59. $(2x^2 - 3y^2) - (4xy + y^2)$
$- (5x^2 + 7xy + 11y^2)$

60. $(x^3 - 2y^3 + 3x^2y) + (4xy^2 - 5x^3 + 2x^2y)$
$- (3x^2y + 3xy^2 + y^3)$

61. A woman saved $(6x - 7y)$ dollars the first year, $(4x + 10y)$ dollars the second year, $(3y - 5x)$ dollars the third year, and $(x + y)$ dollars the fourth year. What was her total amount saved during the four years?

62. Find the perimeter of a hexagon having sides of lengths $4x + y$, $2x - y$, $5x + 6y$, $x + y$, $x - y$, and $8x + 9y$.

In Exercises 63–72, remove the parentheses and simplify.

63. $3x + 4y + (2x - y)$

64. $7x - [2x - (6 - x)]$

65. $4x - 7y - (3x - 4y) + x + 3y$

66. $3x - [x - (2x + 4)]$

67. $8x - [4y - (-3x + 3y) + x]$

68. $2y + 3[y - 2(y + 1) + 1]$

69. $-7y + \{6z - [2y - z + (8z - 4y) + y]\}$

70. $4(3y - 2x) - [y - x - (y - x)]$

71. $11xy - 3x^2y$
$- [5xy^2 - (6xy + 2x^2y) + xy] + 3xy^2$

72. $2[6x - (2x - y)]$
$+ 3[-(x + 4y) + 2(x - y)]$

In Exercises 73–90, find the product.

73. $2x^3z^3(18z - 19x + 16a^2b^2)$

74. $5axz^2(16xy - 7axy + 9x^2z^3 - 8az)$

75. $-7abx(12x^3 - 5x^2y + 6x^2 - 2bx)$

76. $5a^3cz^5(19b^2x^5 - 13ac^3d^5 + 18b)$

77. $8x^3y^2z(10x^3y^2z - 9xyz^2 + 3x^2y^3z^2$
$+ 2x^2yz^3)$

78. $-2abx(5ab^2x - 7bx + 6a^3b^4x - 2ax)$

79. $6a^2uv(3az^2w^2 - 4a^2w^2u + 5a^3u^2v^2)$

80. $(-3a^3x^2)$
$\times (-12bcx^3 - 5ac^2x^2 + 3a^3bc - 2abx)$

81. $4abx(-5ab^2x + 2a^2bx^3 - 5b^2cx - x^3)$

82. $6a^9b^{30}m^2(-2b^{19}x^{11} + 3m^{25}x^{23})$

83. $-8aby(5a^5b - 6by^5 + 7ay^5)$

84. $(12x^3p^6q)$
$\times (x^4pq - 7pqx^2 + 5pq^2x - x^9p^9q^9)$

85. $4abx^{19}y^{21}z(14e^3y^2 - 8a^5d^{19}e^{21}x^9y^2z^4)$

86. $-axyz(12 - 5x^4z^4 + 2ayz)$

87. $-x^2yz^2(xyz - xyz^2 + x^2yz - x^2yz^2)$

88. $6ab^5x(5bx - 4ax - 3 - 2b^2)$

89. $7ax^4y^2(12a^2xz - 5a^2xyz^3 + 4a^5b^2y^3z^3)$

90. $-10b^6x^6y^6(5x^4y^4 - 7b^4y^4 - 6b^4x^4)$

In Exercises 91–125, multiply.

91. $(x + y)(x + y)$ 92. $(b + x)(b + x)$

93. $(m + n)(p + q)$ 94. $(h + k)(x - y)$

95. $(c - d)(c + d)$ 96. $(r - s)(r + s)$

97. $(x^2 + y^2)(x^2 + y^2)$

98. $(a^2 + ax + x^2)(a - x)$

99. $(c^2 + 2cd + d^2)(c + d)$

100. $(x^2 - x + 1)(x + 1)$

101. $(m^3 + m^2n + mn^2 + n^3)(m - n)$

102. $(8x^4 + 3y^2)(8x^4 - 3y^2)$

103. $(a^6 + a^4 + a^2 + 1)(a^2 - 1)$

104. $(5a^3xy^2 - 3a^4x)(5a^3xy^2 + 3a^4x)$

105. $(p^3 + q^3)(p + q)$

106. $(3x^2 + 2y^2)(2x^2 + 3y^2)$

107. $(3a^4 - 12a^3x + 2ax^3 - 5x^4)(7a - 2x)$

108. $(7a^5x - 2b^3y + zx)(5a^2 + 3b^4 - y^2)$

109. $(11b^2c - 5a^3y^2 + 2z)(14a^3m + yz)$

110. $(15c^2d^3 - 5ef + 2cd)(16ef + 4c^2y - z)$

111. $(25a^3 + 6a^4b + 7a^5b^2)(2a - 3b)$

112. $(7a - 3b)(-5a + 2b)$

113. $(3a^2 - 4ax + 5x^2)(7a^2 - 2ax - 3x^2)$

114. $(m^4 - n^4)(m^4 + n^4)$

115. $(3x^5 - 2x + 5)(4x^2 - x + 6)$

116. $(81c^4 + 27c^3y + 9c^2y^2 + 3cy^3 + y^4)$
$\times (3c - y)$

117. $(1024b^5 - 256b^4c + 64b^3c^2 - 16b^2c^3$
$+ 4bc^4 - c^5)(4b + c)$

118. $(r^5s^5 + r^4s^4 + rs + 1)$
$\times (-r^3s^3 + r^2s^2 - rs + 1)$

119. $(a^7 - 2a^6y + 3a^5y^2 - 4a^4y^3 + 5a^3y^4$
$- 6a^2y^5 + 7ay^6 - 8y^7) \cdot (a^2 + 2ay + y^2)$

120. $(x + 1)(x - 2)(x + 3)(x - 4)$

121. $(x + a)(x + b)(x + c)$

122. $(a^5 - x^5)(a + x)(a^4 - a^3x + a^2x^2 - ax^3$
$+ x^4)$

123. $(a^2 - x^2)(a^2 + ax + x^2)(a^2 - ax + x^2)$

124. $(a + b)(a - b)(a^2 + b^2)$

125. $(x + y)(y + z)(z + w)$

In Exercises 126–168, use the special product formulas to find the product.

126. $(x + 1)^2$ 127. $(x - 5)^2$

128. $(x + 2)(x + 2)$ 129. $(9 - y)^2$

130. $(10 + a)^2$

131. $(xy - 4)(xy - 4)$

132. $(ab + 2)(ab + 2)$

133. $(x^2 - 5)^2$ 134. $(c^2 + 3)^2$

135. $(2a + 1)^2$ 136. $(2x + 1)^2$

137. $(5 + 3z)^2$ 138. $(1 - 2p)^2$

139. $(5m + 2n)^2$

140. $(x + 7y)^2$ 141. $(x^2 - 5y)^2$

142. $(x^3 + 6y^2)^2$ 143. $\left(\dfrac{1}{2}xy + 6z\right)^2$

144. $\left(\dfrac{1}{3}a^2 - 9b\right)^2$ 145. $(20 - 1)^2$

146. $(50 - 1)^2$ 147. $(x + 3)(x - 3)$

148. $(y - 4)(y + 4)$ 149. $(x + y)(x - y)$

150. $(x + a)(x - a)$ 151. $(xy + 5)(xy - 5)$

152. $(bc + 3)(bc - 3)$

153. $(7 - 3m)(7 + 3m)$

154. $(4 - 5y)(4 - 5y)$

155. $\left(\dfrac{a}{x} - \dfrac{c}{y}\right)\left(\dfrac{a}{x} + \dfrac{c}{y}\right)$

156. $(8m^4 - an^3)(8m^4 + an^3)$

157. $[(a + 2) - c][(a + 2) + c]$

158. $[x - y + z][x - y - z]$

159. $[x^2 - y^2 - xy][x^2 + y^2 + xy]$

160. $[8 + 2x + 3y][8 - 2x - 3y]$
161. $(x^2 - 3y)^3$ 162. $(2x + 5y)^3$
163. $(3x^2 + 4y)^3$ 164. $(y^2 - x^2)^3$
165. $(x + 4y)(x^2 - 4xy + 16y^2)$
166. $(3x + 5y)(9x^2 - 15xy + 25y^2)$
167. $(x - 4y)(x^2 + 4xy + 16y^2)$
168. $(3x - 2y)(9x^2 + 6xy + 4y^2)$

In Exercises 169–182, find the quotient.

169. $(ax + bx + cx) \div x$
170. $(20a^2c^2 - 45a^2cx + 10a^3xy) \div (5a^2)$
171. $(15a^3by - 9a^2cy^3 + 12a^2x^3y^5) \div (3a^2y)$
172. $(36a^2b^2x^2y - 16x^5y^2 - 20x^3y^4)$
     $\div (-4x^2y)$
173. $(4a^2x^3 + 7a^3x^4 - 9a^4x^5) \div (-ax^2)$
174. $(6a^4x^3m^2 - 8a^3bx^3m^2 + 10a^3cx^3m^2$
     $- 4a^3dx^3m^2) \div (2a^3x^3m^2)$
175. $(-15a^4x^3y^4 - 9a^4x^6y^2 + 6a^8x^3y^2)$
     $\div (-3a^4x^3y^2)$
176. $(75a^2c^3x^2 + 25a^2c^3x - 15ac^2x^3yz)$
     $\div (5c^2ax)$
177. $(6a^4cm - 9a^4cm^2 + 12a^4cm^3) \div (-3a^3c)$
178. $(7ac^5x + 7ac^5y - 7ac^5z - 7ac^5d)$
     $\div (7ac^5)$
179. $(4am^4n^3 - 27am^3n^4 + 63m^3n^5y)$
     $\div (9m^3n^3)$
180. $(36p^2q^2rs^3 - 72p^2q^2rt^3 + 84pq^2rx^2)$
     $\div (-12pq^2)$
181. $(42am^3r^3st^2 - 63bc^2r^3st^2 + 105r^3s^3t^2x)$

182. $(126a^4x^2y - 30a^3b^2x^2y + 18ax^3y^2)$
     $\div (6x^2y)$

In Exercises 183–202, find the quotient and remainder.

183. $(a^2 + 2ax + x^2) \div (a + x)$
184. $(a^4 + 2a^2x^2 + x^4) \div (a^2 + x^2)$
185. $(m^4 + m^3n + mn^3 + n^4) \div (m + n)$
186. $(16x^4 - 81y^4) \div (2x - 3y)$
187. $(x^3 + y^3) \div (x + y)$
188. $(96x^9y^2 + 12x^6y^4 - 24x^5y^6$
     $- 48x^5y^3 - 6x^2y^5 + 12xy^7)$
     $\div (6x^5y - 3xy^2)$
189. $(312x^4y^2 - 32x^3y^3 + 20x^3y^2$
     $- 160x^2y^4 + 88x^2y^3 - 12x^2y^2)$
     $\div (12x^2y + 8xy^2 - 2xy)$
190. $(a^3 - b^3) \div (a + b)$
191. $(1 + 4x) \div (1 - 6x)$
192. $(56y^2 - 4ay - a^2) \div (7y - a)$
193. $(20a^2b - 25a^3 - 18b^3 + 27ab^2)$
     $\div (-5a + 6b)$
194. $(-3cd^3 + 4c^3d - 9c^2d^2 + 6c^4 + 2d^4)$
     $\div (-cd + 3c^2 - 2d^2)$
195. $(ay + by + ax + bx) \div (a + b)$
196. $(3x^4 - 8x^2y^2 + 3x^2z^2 + 5y^4 - 3y^2z^2)$
     $\div (x^2 - y^2)$
197. $(6x^2y - 12xyz + 6yz^2) \div (2xy - 2yz)$
*198. $(ax + cx + ay + bx + bz + by + cz$
     $+ cy + az) \div (x + y + z)$
*199. $(a^3b^2 + a^2b^3 + a^2b + ab^2)$
     $\div (a^2b^2 + ab)$
*200. $(x^5 + 32a^5) \div (x + 2a)$
*201. $(n^2 + x^2) \div (n^2 - x^2)$
*202. $(a^{40} - b^4) \div (a^{10} - b)$

3 Linear Equations and Inequalities in One Variable

One of the major concerns of algebra is solving equations. In this chapter, we shall use the techniques discussed in the preceding chapters to solve linear (first-degree) equations and linear inequalities in one variable. All variables will represent real numbers.

3.1 EQUATIONS: CONDITIONAL AND IDENTITIES

If x is a variable, then $3x + 1$, $4(x - 3)$, $10 - \dfrac{x}{2}$, and 0 represent what we have defined in Chapter 2 as algebraic expressions. If we substitute specific numerical values for the variable x, the algebraic expression yields a particular number. In this chapter, we are interested in "equating" two algebraic expressions.

DEFINITION

An algebraic equation, or simply an equation, in the variable x is a statement that expresses the equality between two algebraic expressions.

EXAMPLE 1   The following are equations:

(a) $4(x - 3) = 4x - 12$

(b) $x + 1 = x + 9$

(c) $3x + 1 = 0$

(d) $10 - \dfrac{x}{2} = 6$

The expressions on either side of the equals sign are referred to as the left-hand side of the equation and the right-hand side of the equation.

$$\underbrace{5x + 4}\ = \underbrace{2x + 1}$$

Left-hand side   Right-hand side

If certain numerical values are substituted for the variable x in equations, true statements result. However, other numerical values may result in false statements. Any value of the variable that makes the equation a true statement is called a solution or root of the equation.

EXAMPLE 2   The equation $5x + 4 = 2x + 1$ is a true statement only when we substitute -1 for x; that is, let $x = -1$. Thus, -1 is the solution of the equation.

There are two important types of equations: identities and conditional equations.

DEFINITION

An identity is an equation that is true for all possible values of the variable.

EXAMPLE 3   (a) $x + 5 = 5 + x$

(b) $4(x - 3) = 4x - 12$

(c) $\dfrac{x}{x} = 1$ (provided $x \neq 0$)

are all examples of identities.

Other examples of identities were discussed in Chapter 1. For instance, the properties $ab = ba$, $a(bc) = (ab)c$, and $a(b + c) = ab + ac$ are also identities, since they are true for all possible values of the variables.

DEFINITION

A conditional equation is an equation that is true only for certain values of the variable.

EXAMPLE 4   (a) $3x + 1 = 0$ is a conditional equation because it is true when x is replaced by $-\frac{1}{3}$, and it is false for any other value of x.

(b) $10 - \dfrac{x}{2} = 6$ is a conditional equation because it is true only for the value $x = 8$.

PROBLEM 1   Which of the following are identities and which are conditional equations?

(a) $5 + x = 9$ (b) $5x + 1 - x = 4x + 1$ (c) $3(2 + x) = 9$

Answer   (a) $5 + x = 9$ is a conditional equation because $x = 4$ is the only value that makes the equation true.

(b) $5x + 1 - x = 4x + 1$
$\qquad 4x + 1 = 4x + 1$ is an identity.

(c) $3(2 + x) = 9$
$\qquad 6 + 3x = 9$ is a conditional equation because $x = 1$ is the only value that makes the equation true.   ■

PROBLEM 2   Determine which of the given numbers is a solution of the given equation.

(a) $2x + 3 = x + 4$, for 0, 1

(b) $3(x - 2) = 5 - (x - 1)$, for 3, -1

(c) $x + 2 = x - 1$, for -2, 0

Answer   (a) For $x = 0$, the left-hand side of the equation becomes 3, while the right-hand side becomes 4. Since $3 \neq 4$, 0 is not a solution. For $x = 1$, both sides of the equation become 5; thus, $x = 1$ is a solution.

(b) For $x = 3$, the equation becomes $3 = 3$, and $x = 3$ is a solution. For $x = -1$ we have $-9 \neq 7$, so $x = -1$ is not a solution.

(c) For $x = -2$, we have $0 \neq -3$, and for $x = 0$, we have $2 \neq -1$. Thus, neither $x = -2$ nor $x = 0$ are solutions. Note that the equation $x + 2 = x - 1$ has no solutions since no value of the variable x makes the equation a true statement.   ■

EXERCISES   3.1

Which of the equations in Exercises 1–28 are identities and which are conditional equations?

1. $3x + 4 = 4 + 3x$
2. $2x = x + x$
3. $x + 7 = 3$
4. $x - 4 = 1$
5. $4x - x = 3x$
6. $3x + 1 = 6$
7. $x - 1 = 9$
8. $2 - x = x$
9. $4x + 3 = 15$
10. $9x + 2 = 7x + 2x + 1 + 3x$
11. $3x + 5x - 2x = 7x$
12. $3x + 5x - 2x = 7$
13. $9x + x - 4 = -4 + 10x$
14. $6x - 2x + 4x + 1 = 1 + 8x$
15. $5(2 + x) = 10 + 5x$
16. $3x - 6 = 3(x - 2)$

17. $2(x - 1) = 2x - 1 + x$

18. $3(2x + 4) = 12 + 5x$

19. $\dfrac{7x - 2x + 4x}{6} = \dfrac{9x}{6}$

20. $\dfrac{5x + 3x - 8x}{2} = 0$

21. $3(x + 2) - (x - 4) = x$

22. $x - 2(x - 2) = 2 - (x - 2)$

23. $x + 3(x - 4) = 2 + 2(x + 4)$

24. $2(3x + 1) - 3(2x - 1) = 5$

25. $(3x + 1)^2 - 1 = 3x(3x + 2)$

26. $(2x + 3)^2 - 9 = 4x(x + 3)$

27. $(2x - 3)(x + 2) = x(x + 1) + x(x - 6)$

28. $x(x - 1) + 2(x - 1) - x^2 = 0$

In Exercises 29–50, determine whether the given number is a solution of the equation.

29. $x + 4 = 6$, for $x = 2$

30. $y + 9 = 12$, for $y = 3$

31. $y - 5 = 7$, for $y = 2$

32. $x - 5 = 12$, for $x = 7$

33. $x + 7 = 8$, for $x = 1$

34. $3x + 2 = -4$, for $x = -2$

35. $2y - 3 = 5$, for $y = 1$

36. $4x - 10 = 2x + 2$, for $x = 6$

37. $9y - 9 + 3y = 15$, for $y = 2$

38. $300x - 250 = 50x + 750$, for $x = 4$

39. $17x - 7x = x + 18$, for $x = -2$

40. $2.5x + 0.5 = 1.5x + 1.5$, for $x = 3$

41. $9x - 19 + x = 11$, for $x = 3$

42. $x + 2x + 3 - 4x = 5x - 9$, for $x = 4$

43. $2x + 3x - 4 = 5x + 6x - 16$, for $x = 2$

44. $75y - 150 = 80y - 300$, for $y = 30$

45. $3.3y + 2.7y - 4.6 = 7.4$, for $y = 2$

46. $2x - 3x + 4x - 5 = 6x - 7x + 15$, for $x = -5$

47. $(4x + 6) - 2x = (x - 6) + 24$, for $x = 10$

48. $2(3 + 4x) - 5 = 5 - 12x$, for $x = 1$

49. $x - (4x - 8) + 9 + (6x - 8)$
    $= 9 - x + 24$, for $x = 6$

50. $[2x - (3x - 4) + 5x - 6] + 10x$
    $= (12x - 12) + 36$, for $x = 13$

3.2 EQUIVALENT LINEAR EQUATIONS

Solving an equation involves finding all values of the variable that make the equation true. Such values are called solutions or roots of the equation. Certain equations can be solved easily by inspection. For example, the equation $7 - x = 2$ has $x = 5$ as its solution. However, consider the equation $2(3x - 5) = 5(x - 2)$; the solution of this equation is not as easy to find by inspection.

In order to solve such equations, we introduce the idea of equivalent equations.

DEFINITION

Two equations are said to be equivalent if and only if they have exactly the same solutions.

EXAMPLE 1   The equations $x + 3 = 7$ and $10 - x = 6$ are equivalent equations since each has 4 as its solution.

EXAMPLE 2   The equations $4x + 1 = 5$ and $2x + 3 = 1$ are *not* equivalent since 1 is the only solution to the first equation and -1 is the only solution to the second equation.

EXAMPLE 3   The equations $3x + 1 = 10$, $3x = 9$, and $x = 3$ are equivalent equations because each has 3 as its solution.

We can solve a more complicated linear equation by finding a simpler equivalent equation. The following properties of equality can be used to reduce to a simple equivalent equation.

Rules That Yield Equivalent Equations

Given an equation, we may
1. Add the same quantity to each side
2. Subtract the same quantity from each side
3. Multiply each side by the same *nonzero* quantity
4. Divide each side by the same *nonzero* quantity

In each of these cases, the resulting equation will be equivalent to the original equation.

EXAMPLE 4   Solve the equation $x - 5 = 4$ and check the result.

Solution

$$x - 5 = 4 \qquad \text{[given equation]}$$
$$x - 5 + 5 = 4 + 5 \qquad \text{[adding 5 to both sides]}$$
$$x = 9 \qquad \text{[simplifying]} \quad \blacksquare$$

Check

$$(9) - 4 \overset{?}{=} 5 \qquad \text{[substituting 9 for } x \text{ in given equation]}$$
$$5 \overset{\checkmark}{=} 5 \qquad \text{[simplifying]}$$

EXAMPLE 5   Solve the equation $x + 3 = 7$ and check the result.

Solution

$$x + 3 = 7 \qquad \text{[given equation]}$$
$$x + 3 - 3 = 7 - 3 \qquad \text{[subtracting 3 from both sides]}$$
$$x = 4 \qquad \text{[simplifying]} \quad \blacksquare$$

Check

$$(4) + 3 \overset{?}{=} 7 \qquad \text{[substituting 4 for } x \text{ in given equation]}$$
$$7 \overset{\checkmark}{=} 7 \qquad \text{[simplifying]}$$

EXAMPLE 6   Solve $\dfrac{x}{2} = 5$ and check the result.

Solution

$$\frac{x}{2} = 5 \qquad \text{[given equation]}$$

$$2 \cdot \frac{x}{2} = (2) \cdot 5 \qquad \text{[multiplying both sides by 2]}$$

$$x = 10 \qquad \text{[simplifying]} \quad \blacksquare$$

Check $\qquad \frac{(10)}{2} \overset{?}{=} 5 \qquad \text{[substituting 10 for } x \text{ in given equation]}$

$$5 \overset{\checkmark}{=} 5 \qquad \text{[simplifying]}$$

EXAMPLE 7 Solve $3x = 9$ and check the result.

Solution $\qquad 3x = 9 \qquad \text{[given equation]}$

$$\frac{3x}{3} = \frac{9}{3} \qquad \text{[dividing both sides by 3]}$$

$$x = 3 \qquad \text{[simplifying]} \quad \blacksquare$$

Check $\qquad 3 \cdot (3) \overset{?}{=} 9 \qquad \text{[substituting 3 for } x \text{ in given equation]}$

$$9 \overset{\checkmark}{=} 9 \qquad \text{[simplifying]}$$

EXAMPLE 8 Solve $\dfrac{-3}{2}x = 6$ and check the result.

Solution $\qquad \left(-\frac{3}{2}\right)x = 6 \qquad \text{[given equation]}$

$$\left(-\frac{2}{3}\right)\left(-\frac{3}{2}\right)x = \left(-\frac{2}{3}\right) \cdot 6 \qquad \left[\text{multiplying both sides by } \left(-\frac{2}{3}\right)\right]$$

$$x = -4 \qquad \text{[simplifying]} \quad \blacksquare$$

Check $\qquad \left(-\frac{3}{2}\right)(-4) \overset{?}{=} 6 \qquad \text{[substituting 4 for } x \text{ in given equation]}$

$$6 \overset{\checkmark}{=} 6 \qquad \text{[simplifying]}$$

EXAMPLE 9 Solve $-x = 4$.

Solution $\qquad -x = 4 \qquad \text{[given equation]}$

$$\frac{-x}{-1} = \frac{4}{-1} \qquad \text{[dividing both sides by } -1]$$

$$x = -4 \qquad \text{[simplifying]} \quad \blacksquare$$

EXERCISES 3.2

In Exercises 1–27, solve and check the result.

1. $x - 5 = 13$
2. $x - 3 = 7$
3. $x + 8 = 12$
4. $x + 4 = 6$
5. $17 = x - 9$
6. $x + 5 = 9$
7. $x + 20 = 36$
8. $5 = x - 2$
9. $4 = x + 6$
10. $10 = 7 + x$
11. $x - 5 = -9$
12. $1 = 3 + x$
13. $4 = x - 3$
14. $x + 7 = 4$
15. $3 = x + 12$
16. $2 = x + 1$
17. $5 + x = -9$
18. $-5 = -7 + x$
19. $x - 8 = -4$
20. $8 = x + 12$
21. $x + .7 = .3$
22. $x - 1.4 = .6$
23. $-.4 = x + .6$
24. $x + 2\frac{1}{2} = 3$
25. $x - 3\frac{1}{4} = -4$
26. $-\frac{1}{3} = x - 1\frac{5}{6}$
27. $4\frac{1}{5} = x + 5\frac{7}{10}$
28. If $x + 7 = 10$, find the value of $5x$.
29. If $x - 2 = -3$, find the value of $7x$.
30. If $.3 = x - 1.7$, find the value of $x + 10$.
31. If $17 = x + 12$, find the value of $x + 5$.
32. If $41 = x + 41$, find the value of $2x - 1$.
33. If $x - 1\frac{1}{4} = 2\frac{3}{4}$, find the value of $\frac{1}{16}x$.
34. If $x - 8\frac{3}{4} = 6\frac{1}{2}$, find the value of $x - \frac{1}{4}$.
35. If $2\frac{1}{3} + x = \frac{1}{2}$, find the value of $1 - x$.

In Exercises 36–71, solve and check the result.

36. $2x = 12$
37. $5x = 15$
38. $9x = 36$
39. $3x = -9$
40. $4x = -16$
41. $15x = -60$
42. $-88 = 11x$
43. $42 = 7x$
44. $-2x = 8$
45. $-4x = 20$
46. $4x = 9$
47. $2x = -5$
48. $-6 = 9x$
49. $-17x = 51$
50. $-7x = 7.7$
51. $3x = -2.1$
52. $-7x = -35$
53. $-4x = -24$
54. $-x = 27$
55. $-x = 41$
56. $-x = -5$
57. $-x = -13$
58. $\frac{1}{3}x = 4$
59. $\frac{x}{25} = -4$
60. $\frac{x}{13} = -2$
61. $\frac{x}{2} = -.5$
62. $\frac{x}{9} = 0.15$
63. $\frac{2}{5}x = -10$
64. $\frac{5}{6}x = 5$
65. $24 = \frac{3}{4}x$
66. $\frac{2}{3}x = -3$
67. $\frac{3}{5}x = \frac{6}{8}$
68. $\frac{-3}{4} = \frac{12}{9}x$
69. $-\frac{4}{3} = \frac{4}{3}x$
70. $\frac{x}{-8} = 3\frac{1}{4}$
71. $\frac{x}{7} = 1\frac{1}{14}$
72. If $4x = 28$, find the value of $3x$.
73. If $-3x = 36$, find the value of $x + 12$.
74. If $\frac{x}{3} = 5$, find the value of $\frac{x}{15}$.
75. If $\frac{5}{7}x = 20$, find the value of $2x - 48$.

3.3 GENERAL SOLUTION OF LINEAR EQUATIONS

When the addition or subtraction rule is used to solve an equation, the same quantity is added to or subtracted from both sides of the equation. Thus, subtracting 7 from both sides of the equation $x + 7 = 10$ eliminates the 7 on the left side of the equation and gives $x = 10 - 7$. The result would be the same as if we transposed the 7 from the left side of the original equation to the right side and changed its sign. In general, we can state the following rule.

Transposition Rule

Any term may be transposed from one side of an equation to the other side, provided that its algebraic sign is changed.

EXAMPLE 1   Using the transposition rule, the equation

$$3x - 7 = 9x + 5$$

becomes

$$3x - 9x = 5 + 7$$

The process of transposing terms from one side of the equation to the other can simplify the solving of many equations. By the process of transposition, we can move all the terms containing the variable (unknown) to one side of the equation and all other terms to the other side of the equation. This results in like terms on each side of the equation, and such terms can be combined. Thus, we can reduce the original equation to an equivalent equation of the form

$$ax = b$$

where a and b represent the numbers obtained in combining terms. The solution of this equation is easily obtained by dividing both sides by the number a, provided $a \neq 0$.

DEFINITION

Any equation that can be reduced to the form $ax = b$, $a \neq 0$, is called a first-degree equation in one variable or more simply, a linear equation in one variable.

We can now state a general method for solving linear equations.

Method for Finding the Solution of Linear Equations

1. Clear fractions by multiplying each term by the least common denominator, if necessary.
2. Use the distributive law to remove parentheses, if they exist.
3. Transpose all terms containing the unknown to one side of the equation and all other terms to the other side of the equation.
4. Combine like terms on each side of the equation.
5. Divide both sides of the equation by the coefficient of the unknown.

EXAMPLE 2   Solve the equation $3x - 7 = 9x + 5$ and check the result.

Solution
$$3x - 7 = 9x + 5 \qquad \text{[given equation]}$$
$$3x - 9x = 5 + 7 \qquad \text{[transposing terms]}$$

$$-6x = 12 \qquad \text{[combining like terms]}$$

$$\frac{-6x}{-6} = \frac{12}{-6} \qquad \text{[dividing by coefficient of unknown]}$$

$$x = -2 \quad \blacksquare$$

Check
$$3 \cdot (-2) - 7 \stackrel{?}{=} 9(-2) + 5 \qquad \text{[substituting } -2 \text{ for } x]$$

$$-6 - 7 \stackrel{?}{=} -18 + 5 \qquad \text{[simplifying]}$$

$$-13 \stackrel{\checkmark}{=} -13$$

EXAMPLE 3 Solve the equations $3x - 4(3x - 5) = 12 - x$ and check the result.

Solution
$$3x - 4(3x - 5) = 12 - x \qquad \text{[given equation]}$$

$$3x - 12x + 20 = 12 - x \qquad \text{[removing parentheses]}$$

$$3x - 12x + x = 12 - 20 \qquad \text{[transposing terms]}$$

$$-8x = -8 \qquad \text{[combining like terms]}$$

$$\frac{-8x}{-8} = \frac{-8}{-8} \qquad \text{[dividing by coefficient of unknown]}$$

$$x = 1 \quad \blacksquare$$

Check
$$3(1) - 4[3(1) - 5] \stackrel{?}{=} 12 - (1) \qquad \text{[substituting 1 for } x]$$

$$3 - 4[3 - 5] \stackrel{?}{=} 12 - 1 \qquad \text{[simplifying]}$$

$$3 + 8 \stackrel{?}{=} 11 \qquad \text{[simplifying]}$$

$$11 \stackrel{\checkmark}{=} 11$$

Some equations that initially involve higher powers of the unknown reduce to linear equations.

EXAMPLE 4 Solve $(x - 5)^2 = 4 + (x + 3)(x - 2)$.

Solution
$$(x - 5)^2 = 4 + (x + 3)(x - 2) \qquad \text{[given equation]}$$

$$x^2 - 10x + 25 = 4 + x^2 + x - 6 \qquad \text{[removing parentheses]}$$

$$x^2 - 10x - x^2 - x = 4 - 6 - 25 \qquad \text{[transposing terms]}$$

$$-11x = -27 \qquad \text{[combining like terms]}$$

$$\frac{-11x}{-11} = \frac{-27}{-11} \qquad \text{[dividing by coefficient of unknown]}$$

$$x = \frac{27}{11} \quad \blacksquare$$

The following example further illustrates the point that some equations that do not look like linear equations initially can be transformed to linear equations. One such type of equation involves denominators. To solve such equations, we multiply each side of the equation by the least common denominator (LCD) and solve the resulting linear equation as usual.

EXAMPLE 5 Solve $\dfrac{x}{3} - \dfrac{4x + 1}{2} = x - \dfrac{5}{6}$ and check the result.

Solution

$$\dfrac{x}{3} - \dfrac{4x + 1}{2} = x - \dfrac{5}{6} \qquad \text{[given equation]}$$

$$6\left(\dfrac{x}{3} - \dfrac{4x + 1}{2}\right) = 6\left(x - \dfrac{5}{6}\right) \qquad \text{[multiplying both sides by LCD]}$$

$$6\left(\dfrac{x}{3}\right) - \dfrac{6(4x + 1)}{2} = 6x - 6\left(\dfrac{5}{6}\right) \qquad \text{[distributive law]}$$

$$2x - 3(4x + 1) = 6x - 5 \qquad \text{[simplifying]}$$

$$2x - 12x - 3 = 6x - 5 \qquad \text{[removing parentheses]}$$

$$2x - 12x - 6x = -5 + 3 \qquad \text{[transposing terms]}$$

$$-16x = -2 \qquad \text{[combining like terms]}$$

$$\dfrac{-16x}{-16} = \dfrac{-2}{-16} \qquad \text{[dividing by coefficient of unknown]}$$

$$x = \dfrac{1}{8} \qquad \text{[simplifying]} \quad \blacksquare$$

Check

$$\dfrac{\left(\dfrac{1}{8}\right)}{3} - \dfrac{4\left(\dfrac{1}{8}\right) + 1}{2} \stackrel{?}{=} \dfrac{1}{8} - \dfrac{5}{6} \qquad \left[\text{substituting } \dfrac{1}{8} \text{ for } x\right]$$

$$\dfrac{1}{24} - \dfrac{\dfrac{1}{2} + 1}{2} \stackrel{?}{=} \dfrac{1}{8} - \dfrac{5}{6}$$

$$\dfrac{1}{24} - \dfrac{\dfrac{3}{2}}{2} \stackrel{?}{=} \dfrac{1}{8} - \dfrac{5}{6}$$

$$\dfrac{1}{24} - \dfrac{3}{4} \stackrel{?}{=} \dfrac{1}{8} - \dfrac{5}{6}$$

$$\dfrac{1}{24} - \dfrac{18}{24} \stackrel{?}{=} \dfrac{3}{24} - \dfrac{20}{24}$$

$$-\dfrac{17}{24} \stackrel{\checkmark}{=} -\dfrac{17}{24}$$

EXERCISES   3.3

Solve and check the result.

1. $7x + 1 = 6x + 5$
2. $3 - 3x = 3 - 2x$
3. $4x - 2 = -8x + 7$
4. $10x + 3 = 6x + 5$
5. $-10x - 5 = 5x - 14$
6. $4x + 6 = 3x + 4$
7. $x - 11 = 7 + 2x$
8. $4x - 2 = -8x + 7$
9. $10x + 3 = 6x + 5$
10. $9x - 4 = 3x$
11. $3x - 4 = 2x - 1$
12. $7x + 10 = 3x + 50$
13. $4x + 20 = 5x + 9$
14. $8x + 56 = 14x + 26$
15. $x + 4 = 9x + 4$
16. $9x - 3 = 2x + 46$
17. $7x - 4 = 5x - x + 35$
18. $5x + 9 - 4x = 51 - 5x$
19. $9x - 2x + 8 = 4x + 38$
20. $12x - 5 = 8x - x + 50$
21. $6x - 12 - x = 9x + 53$
22. $3x - 5x - 12 = 7x - 88 - 5$
23. $5 - 3x - 18 = x - 1 + 8x$
24. $5 - 6x - 11x = 13 - 11x - 4x$
25. $2x + 3x - 4 = 5x + 6x - 16$
26. $4x + 6 - 2x = x - 6 + 24$
27. $3x + 4 - 6x = 6 + 5x$
28. $8x - x - 8 = 5 - 6x$
29. $4x + 8 - 12x = 7 - 5x$
30. $10 + 13x - 4x = 7 - 3x$
31. $4x - 3 = 3(6 - x)$
32. $7x = -2(x + 9)$
33. $(2x - 9) - (x - 3) = 0$
34. $4 - (2 - x) = 2 + 2x$
35. $(1 - x) - 6 = 3(7 - 2x)$
36. $6(x - 5) = 15 + 5(7 - 2x)$
37. $(x - 1) - (x + 4) = 2x - 5$
38. $5(x + 2) - 4(x + 1) - 3 = 0$
39. $3(8x - 2) - 3(1 - x) + 8 = 8$
40. $3(4x - 3) + 2(3x + 4) = 23$
41. $6(x - 4) - 7(x - 2) = 0$
42. $7(2x - 1) - 3(4x - 5) = 0$
43. $-6(2x + 3) - 2x = 38$
44. $3x + (2x - 5) = 13 - 2(x + 2)$
45. $2(x - 3) - 13 = 17 - 3(x + 2)$
46. $2(x + 1) - 3x = 3(2x + 3)$

47. $2 - 7(x - 1) = 3(x - 2) - 5(x + 3)$
48. $5(x + 1) + 6 = 3(4 + x) + 2x - x$
49. $4(3x + 4) - 4(x - 3)$
    $= 6(x + 1) - 2(8 + 6x)$
50. $-(4x + 3) + 6(2x + 1) - 3$
    $= -3(4x - 3) - (6x + 10)$
51. $15x - [3 - (4x + 4) - 57] = (2 - x)$
52. $9[x - (3x - 1) + 5] = x + 3$
53. $x - (4x - 8) + 9 + (6x - 8)$
    $= 9 - x + 24$
54. $2x + 2 = -2[x - (2x + 4) + 3] + x$
55. $[2x - (3x - 4) + 5x - 6] + 10x$
    $= 12x + 24$
56. $-(7x + 2) = 4[2x - (3 - x) + 5]$
57. $4x - (12x - 24) + 38x - 38 = 0$
58. $-2[x + 3(5 - x)] - 5(x - 7) = 0$
59. $6[x - 2(2x + 3) + 1]$
    $= -3(5 + 6x) - x$
60. $-3[2(x - 5) - x] = 2[3x - 5(-3 + x)]$
61. $x^2 + 51 = (x + 3)(x + 2)$
62. $(x + 1)^2 = x^2 + 7$
63. $x(x + 1) = (x - 5)(x + 2)$
64. $(x + 2)(x + 3) = x(x + 2)$
65. $(x - 3)(x + 2) = (x - 2)(x + 4)$
66. $(2x + 1)(3x - 2) = (6x - 1)(x + 4)$
67. $(3x + 2)(x - 3) - (3x^2 + 2x - 1) = 4$
68. $(2x + 1)(x - 7) = 2(x^2 - 3) + 25$
69. $(x + 2)^2 - (x - 3)^2 = 7$
70. $(x + 3)^2 - (x + 2)^2 = 1$
71. $(2x - 1)^2 - 4(x - 2)^2 = 0$
72. $(x + 1)^2 = (x + 5)(x - 2)$
73. $(x - 1)^2 + (x + 2)^2 - 2x^2 = 1$
74. $(x - 1)(x - 5) - x^2 = -11$
75. $x(x + 3) - 7 = 2x^2 - x(x + 2)$
76. $(x - 2)^2 - (x - 4)^2 = 4x - x$
77. $-2(x^2 + 2x - 7)$
    $= 5 - (2x - 3)(x + 5)$
*78. $3(3x + 2)(2x + 7)$
    $= 6(17x + 7) + 2(9x^2 - 21x + 10)$
*79. $(x + 3)(x + 1) - (x + 2)(2x + 2)$
    $+ (x + 2)(x + 3) = 0$
*80. $6x(x - 3) + (3 + x)(x - 3)(3 + x)$
    $= 6(3 + x)(x - 3) + x^2(3 + x)$

81. $\dfrac{x}{2} + 4 = \dfrac{1}{3}$ 82. $\dfrac{x}{2} - x = 12$

83. $5 - \dfrac{x}{4} = x - 1$

84. $\dfrac{x - 8}{6} = 3 - 2x$

85. $\dfrac{2x}{3} - 1 = \dfrac{x - 2}{2}$

86. $6x + 1 = \dfrac{3}{4} - \dfrac{x}{3}$

87. $\dfrac{x}{2} + \dfrac{x}{3} + \dfrac{x}{4} = 78$

88. $\dfrac{x}{4} + \dfrac{x}{6} + \dfrac{x}{8} = 26$

89. $\dfrac{x}{3} - \dfrac{7}{6} + \dfrac{2x}{5} = \dfrac{3x}{4}$

90. $\dfrac{5x}{7} - x - \dfrac{5}{3} = \dfrac{x}{7} + \dfrac{1}{21}$

91. $\dfrac{x + 3}{5} = \dfrac{3x + 2}{4} - 1$

92. $\dfrac{x + 1}{4} - \dfrac{2x - 9}{10} = \dfrac{3}{2}$

93. $\dfrac{x}{3} - \dfrac{4x + 1}{2} + x = \dfrac{5}{6}$

94. $\dfrac{3x - 6}{4} - \dfrac{x + 6}{6} = 5 - \dfrac{2x}{3}$

95. $\dfrac{x + 1}{3} - \dfrac{x + 3}{5} = x - \dfrac{3x - 2}{3}$

96. $\dfrac{x}{2} + \dfrac{3x + 1}{5} = \dfrac{x + 3}{10}$

97. $\dfrac{2x + 3}{4} = 1 + \dfrac{3x + 2}{12}$

98. $\dfrac{x + 1}{15} - \dfrac{6x + 2}{5} = \dfrac{-(3x + 5)}{3}$

*99. $\dfrac{x + 9}{4} = 2 + \dfrac{3}{14}(2x - 3)$

*100. $x - \dfrac{1}{3}(x + 1) + \dfrac{1}{5}(x + 3) = \dfrac{3x - 2}{2}$

3.4 LITERAL EQUATIONS

An equation in which at least one of the coefficients or terms is represented by a letter other than the unknown is called a **literal equation**. The procedures used to solve literal equations are the same as those for linear equations with numerical coefficients. That is,

1. Clear fractions by multiplying each term by the least common denominator.
2. Remove parentheses, if any.
3. Transpose all terms containing the unknown to one side and all other terms to the opposite side.
4. Combine like terms.
5. Divide both sides of the equation by the coefficient of the unknown.

EXAMPLE 1   Solve the equation $5x - 2a = 2x + 4a$ for x and check the result.

Solution

$$5x - 2a = 2x + 4a \qquad \text{[given equation]}$$

$$5x - 2x = 4a + 2a \qquad \text{[transposing terms]}$$

$$3x = 6a \qquad \text{[combining like terms]}$$

$$\frac{3x}{3} = \frac{6a}{3} \qquad \text{[dividing by coefficient of } x\,]$$

$$x = 2a \qquad \text{[simplifying]} \quad \blacksquare$$

Check

$$5(2a) - 2a \overset{?}{=} 2(2a) + 4a \qquad \text{[substituting } 2a \text{ for } x\text{]}$$

$$10a - 2a \overset{?}{=} 4a + 4a \qquad \text{[simplifying]}$$

$$8a \overset{\checkmark}{=} 8a$$

EXAMPLE 2   Solve for x: $9b + a + 3x = 4a$.

Solution

$$9b + a + 3x = 4a \qquad \text{[given equation]}$$

$$3x = 4a - a - 9b \qquad \text{[transposing terms]}$$

$$3x = 3a - 9b \qquad \text{[combining like terms]}$$

$$\frac{3x}{3} = \frac{3a}{3} - \frac{9b}{3} \qquad \text{[dividing by coefficient of } x\text{]}$$

$$x = a - 3b \qquad \blacksquare$$

Check

$$9b + a + 3(a - 3b) \overset{?}{=} 4a \qquad \text{[substituting } (a - 3b) \text{ for } x\text{]}$$

$$9b + a + 3a - 9b \overset{?}{=} 4a \qquad \text{[simplifying]}$$

$$4a \overset{\checkmark}{=} 4a$$

EXAMPLE 3   Solve for x: $\dfrac{2(mx + 2m^2)}{3} = \dfrac{mx - 2m}{2}$.

Solution

$$\frac{2(mx + 2m^2)}{3} = \frac{mx - 2m}{2} \qquad \text{[given equation]}$$

$$6\left[\frac{2(mx + 2m^2)}{3}\right] = 6\left[\frac{mx - 2m}{2}\right] \qquad \begin{array}{l}\text{[multiplying both sides by}\\ \text{least common denominator]}\end{array}$$

$$2[2(mx + 2m^2)] = 3[mx - 2m]$$

$$4mx + 8m^2 = 3mx - 6m \qquad \text{[removing parentheses]}$$

$$4mx - 3mx = -8m^2 - 6m \qquad \text{[transposing terms]}$$

$$mx = -2m(4m + 3) \qquad \text{[combining like terms]}$$

$$\frac{mx}{m} = \frac{-2m(4m + 3)}{m} \qquad \text{[dividing by coefficient of } x\text{]}$$

$$x = -2(4m + 3) \qquad \blacksquare$$

EXAMPLE 4   Solve for x: $x(x - a) + a = x(x + a) + b$

Solution

$$x(x - a) + a = x(x + a) + b \qquad \text{[given equation]}$$

$$x^2 - ax + a = x^2 + ax + b \qquad \text{[removing parentheses]}$$

$$x^2 - ax - x^2 - ax = b - a \qquad \text{[transposing terms]}$$

$$-2ax = b - a \qquad \text{[combining like terms]}$$

$$\frac{-2ax}{-2a} = \frac{(b - a)}{-2a} \qquad \text{[dividing by coefficient of } x\text{]}$$

$$x = \frac{(b - a)}{-2a} \quad \blacksquare$$

Other examples of literal equations are the many formulas that arise in mathematics, science, engineering, business, and other disciplines. Examples of such formulas are:

Examples of Literal Equations	
Perimeter of a rectangle = 2 × (length + width)	$[P = 2(\ell + w)]$
Distance = rate × time	$D = rt$
Circumference of a circle = 2π × radius	$C = 2\pi r$
Simple interest = principal × rate × time	$I = prt$
Area of a rectangle = length × width	$A = \ell w$
Area of a triangle = $\frac{1}{2}$ × base × height	$A = \frac{1}{2}bh$
Area of a circle = π × radius squared	$A = \pi r^2$

Sometimes it is not as important to have definite values for a particular variable in a formula as it is to have a definite expression of that variable in terms of all the other variables. For example, the formula $A = p + prt$ (another formula for simple interest) states A (amount due) in terms of p (principal), r (rate), and t (time in years). However, suppose we wish to find t in terms of A, p, and r:

EXAMPLE 5 Solve $A = p + prt$, for t.

Solution

$$p + prt = A \qquad \text{[given equation]}$$

$$prt = A - p \qquad \text{[transposing terms]}$$

$$t = \frac{A - p}{pr} \qquad \text{[dividing by coefficient of } t\text{]} \quad \blacksquare$$

EXAMPLE 6 The area of a trapezoid is given by the formula

$$A = \frac{1}{2}h(\ell + c)$$

where h is the height, and ℓ and c are the bases of the trapezoid. Solve the given formula for ℓ and check the result.

Solution

$$A = \frac{1}{2}h(\ell + c)$$ [given equation]

$$2A = h(\ell + c)$$ [clearing fraction, multiplying each side by 2]

$$2A = h\ell + hc$$ [removing parentheses]

$$2A - hc = h\ell$$ [transposing terms]

$$\frac{2A - hc}{h} = \ell$$ [dividing by h, the coefficient of ℓ] ∎

Check

$$A \stackrel{?}{=} \frac{1}{2}h\left[\left(\frac{2A - hc}{h}\right) + c\right]$$ [substituting result for ℓ]

$$A \stackrel{?}{=} \frac{1}{2}h\left(\frac{2A - hc}{h}\right) + \frac{1}{2}hc$$ [removing brackets]

$$A \stackrel{?}{=} \frac{1}{2}(2A - hc) + \frac{1}{2}hc$$ [cancelling h]

$$A \stackrel{?}{=} \frac{1}{2} \cdot 2A - \frac{1}{2}hc + \frac{1}{2}hc$$ [removing parentheses]

$$A \stackrel{\checkmark}{=} A$$ [simplifying]

EXAMPLE 7 $F = \frac{9}{5}C + 32$ is the formula for converting Celsius temperature to degrees Fahrenheit. For instance, 20° Celsius is equal to

$$F = \frac{9}{5}(20) + 32 = 36 + 32 = 68$$

68° Fahrenheit. Find a formula for converting Fahrenheit temperature to Celsius temperature.

Solution Such a formula can be found by solving $F = \frac{9}{5}C + 32$ for C.

$$F = \frac{9}{5}C + 32$$ [given equation]

$$5F = 9C + 160$$ [multiplying each term by 5]

$$5F - 160 = 9C$$ [transposing terms]

$$\frac{5F - 160}{9} = C$$ [dividing by 9, coefficient of C]

$$\frac{5}{9}(F - 32) = C$$ [simplifying] ∎

EXERCISES 3.4

In Exercises 1–50, solve the equations for x, and check.

1. $mx = 5$

2. $4x = d$

3. $x - 5 = n$

4. $x + c = s$

5. $\dfrac{x}{a} = b$

6. $\dfrac{x}{5} = a$

7. $2x - a = 5a$

8. $3x - 5 = 4m$

9. $ax + 7 = b$

10. $ax + b = r$

11. $2(a + x) = b$

12. $x - 5a = 15$

13. $x + 4m = p$

14. $ax - n = m$

15. $7x + mn = p$

16. $ax + b = c - b$

17. $a + 3b - 2x = b$

18. $5x + 2mp = 6np + x$

19. $7x + 2n = 5m - n$

20. $ax + 3b = c + 3b$

21. $mx - 6np = 3mx - 2np$

22. $3(x + m) = x - 5m$

23. $a(x + b) = 2ax - 7ab$

24. $mx - 6n = 3(m - n)$

25. $nx - mp = p(m + r)$

26. $3x - 8n = 5x + 4n$

27. $2x + 4m = 3x + m$

28. $a + 2x = 4a - 3x$

29. $6x - 7m = 3x + 5m$

30. $4x - n = 7x + 8n$

31. $mx - 3p + 2n = p + 5n$

32. $a - 3b + 6x = 3x - 6a$

*33. $7mp + 4n + 4px = 6px - 2r + 4n$

*34. $7(x - m) + 4(2x + m) - 2m = 0$

*35. $3(x - 2m) + 4(m + 2x) = 9m$

*36. $3 + 4(3x - a) + 5(a - 3x) = 10 - 6x$

*37. $3(2a - x) = 6(3a - x) + 2(x - 4)$

*38. $2(px - pm) - p(2m - x) = 0$

*39. $7b(ax - 5ac) = 5a(bx - 3bc)$

*40. $2x^2 + 2m + x$
$= 16m^2 + x^2 + (x - 4m)(x + 4m)$

41. $\dfrac{2x + 3a}{7} = \dfrac{x + a}{3}$

42. $\dfrac{1}{2}(x - 6m) = \dfrac{1}{3}(6m - x)$

43. $\dfrac{3x + 2a}{3} = \dfrac{x + 4a}{6}$

44. $\dfrac{6x - 5m}{2} = \dfrac{5x + 2m}{8} + x$

45. $\dfrac{6x - 7m}{7} = \dfrac{5x + 7m}{14} + x$

46. $\dfrac{1}{2}x - \dfrac{1}{3}p = \dfrac{1}{4}x - \dfrac{1}{5}p$

47. $\dfrac{m - x - 2}{4} = \dfrac{2 - x}{3}$

48. $\dfrac{2x + 3m}{9} = \dfrac{x - m}{2}$

49. $\dfrac{2x - a}{c} = a$

50. $\dfrac{m + 4x}{n} = p$

51. $p = a + b + c$ (perimeter of a triangle with sides a, b, and c). Solve for b.

52. $A = \dfrac{bh}{2}$ (area of triangle with base b and height h). Solve for h.

53. $A + B + C = 180°$ (sum of angles of a triangle where A, B, and C are the three angles). Solve for A.

54. $P = 2(\ell + w)$ (perimeter of a rectangle with length ℓ and width w). Solve for w.

55. $A = \ell w$ (area of a rectangle with length ℓ and width w). Solve for ℓ.

56. $A = \dfrac{1}{2}h(\ell + c)$ (area of a trapezoid with height h and bases ℓ and c). Solve for c.

57. $V = \ell wh$ (volume of a rectangular parallelepiped with length ℓ, width w, and height h). Solve for w.

58. $p = nS$ (perimeter of a regular polygon with n sides, each of length S). Solve for n.

59. $S = 180(n - 2)$ (sum of angles of a regular polygon with n sides). Solve for n.

60. $p = 2a + b$ (perimeter of an isosceles triangle). Solve for a.

61. $C = 2\pi r$ (circumference of a circle of radius r). Solve for r.

62. $V = \dfrac{1}{3}abh$ (volume of a pyramid with rectangular base). Solve for h.

63. $V = \dfrac{1}{3}\pi r^2 h$ (volume of a cone with circular base and height h). Solve for h.

64. $V = \pi r^2 h$ (volume of a circular cylinder with height h). Solve for h.

65. $S = 2\pi rh$ (lateral surface area of a right circular cylinder). Solve for r.

66. $D = RT$ (distance equals rate times time). Solve for R; T.

67. $y = mx + b$ (slope-intercept equation of a line). Solve for x.
68. $W = Fd$ (work equals force F times distance d). Solve for d.
69. $A = p(1 + rt)$ (simple interest when p is principal invested, r is rate of interest, and t is the time). Solve for t.
70. $I = \dfrac{E}{R}$ (electrical current in amperes with voltage E and resistance R). Solve for E.
71. $W = I^2 R$ (power in watts when I is current and R is resistance in ohms). Solve for R.
72. $S = \dfrac{1}{2}gt^2$ (distance traveled by a freely falling object with g the pull of gravity and t the time in seconds). Solve for g.
73. $v = v_0 + at$ (velocity where v_0 is the initial velocity, a is acceleration due to gravity, and t is time). Solve for t.
74. $y = P - At$ (book value where A is annual depreciation, P is purchase price, and t is time in years). Solve for A.
75. $V = c - \left(\dfrac{c}{t}\right)n$ (undepreciated value of an object at the end of n years where c is original cost and t is the number of years of linear depreciation). Solve for n.

3.5 INEQUALITIES AND THEIR PROPERTIES

So far we have considered only the solution of equations or equalities. We now turn our attention to the solution of linear equalities. First, recall the two symbols of inequalities:

$$< \text{ means } \textit{is less than} \qquad \text{(for example, } 2 < 5\text{)}$$

$$> \text{ means } \textit{is greater than} \quad \text{(for example, } 7 > 4\text{)}$$

Note that the inequality symbol *points* to the smaller of the two numbers, and *opens wide* to the larger.

Before learning how to solve inequalities, we first discuss some new properties of real numbers. Given two real numbers a and b and their point representation on the real number line, then one and only one of the following situations happens:

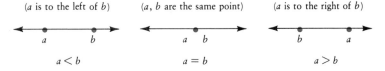

This ordering property of real numbers is called the trichotomy law.

Trichotomy Law

If a and b are real numbers, then exactly one of the following is true: $a < b$, $a = b$, or $a > b$.

	Algebraic Meaning	Geometric Meaning
1. $a < b$	a is less than b	a is to the left of b
2. $a = b$	a is equal to b	a is the same point as b
3. $a > b$	a is greater than b	a is to the right of b

EXAMPLE 1   Of the following statements, only the third is true:

$$-5 > -3 \qquad -5 = -3 \qquad -5 < -3$$

If we consider the three different points a, b, and c, we see that if a lies to the left of b and b lies to the left of c, then a lies to the left of c. Using our inequality notation, if $a < b$ and $b < c$, then $a < c$. Similarly, if c lies to the right of b and b lies to the right of a, then c lies to the right of a. That is, if $c > b$ and $b > a$, then $c > a$. This figure illustrates the transitive law of inequalities.

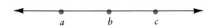

Transitive Law of Inequalities

If a, b, and c are real numbers, then

1. If $a < b$ and $b < c$, then $a < c$.
2. If $c > b$ and $b > a$, then $c > a$.

EXAMPLE 2   Since $4 < 8$ and $8 < 12$, then $4 < 12$.

Geometrically:

Since 4 is to the left of 8 and 8 is to the left of 12, then 4 is to the left of 12.

EXAMPLE 3   Since $2 > -4$ and $-4 > -7$, then $2 > -7$.

Geometrically:

Since 2 is to the right of -4 and -4 is to the right of -7, then 2 is to the right of -7.

The following properties of inequalities are fundamental in solving inequalities.

Addition and Subtraction Properties of Inequalities

When the same number is added to or subtracted from both sides of an inequality, the order of the inequality remains unchanged. That is, for a, b, and c real numbers:

If $a < b$, then $a + c < b + c$ and $a - c < b - c$.
If $a > b$, then $a + c > b + c$ and $a - c > b - c$.

EXAMPLE 4   Since $3 < 6$, then $3 + 2 < 6 + 2$, or $5 < 8$.

Geometrically: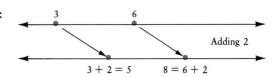

Adding 2

EXAMPLE 5 Since $-1 < 4$, then $-1 - 2 < 4 - 2$, or $-3 < 2$.

Geometrically: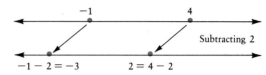

Subtracting 2

EXAMPLE 6 Since $2 > 0$, then $2 + 1 > 0 + 1$, or $3 > 1$.

Geometrically: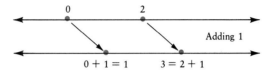

Adding 1

EXAMPLE 7 Since $-3 > -5$, then $-3 - 1 > -5 - 1$, or $-4 > -6$.

Geometrically: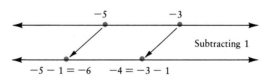

Subtracting 1

Multiplication and Division Properties of Inequalities

When both sides of an inequality are multiplied or divided by a positive number, the order of the inequality remains unchanged. That is,

1. If $c > 0$ and $a < b$, then $ac < bc$ and $\dfrac{a}{c} < \dfrac{b}{c}$.

2. If $c > 0$ and $a > b$, then $ac > bc$ and $\dfrac{a}{c} > \dfrac{b}{c}$.

When both sides of an inequality are multiplied or divided by a negative number, the order of the inequality is reversed. That is,

3. If $c < 0$ and $a < b$, then $ac > bc$ and $\dfrac{a}{c} > \dfrac{b}{c}$.

4. If $c < 0$ and $a > b$, then $ac < bc$ and $\dfrac{a}{c} < \dfrac{b}{c}$.

EXAMPLE 8   Since $-3 < -1$, then $(-3)(4) < (-1)(4)$, or $-12 < -4$.

Geometrically:

EXAMPLE 9   Since $4 > 2$, then $\dfrac{4}{2} > \dfrac{2}{2}$, or $2 > 1$.

Geometrically:

EXAMPLE 10   Since $\dfrac{1}{3} < \dfrac{5}{6}$, then $(-6)\left(\dfrac{1}{3}\right) > (-6)\left(\dfrac{5}{6}\right)$, or $-2 > -5$.

Geometrically:

EXAMPLE 11   Since $6 > 3$, then $\dfrac{6}{-3} < \dfrac{3}{-3}$, or $-2 < -1$.

Geometrically:

EXERCISES   3.5

In Exercises 1–25, insert the symbol $<$ or $>$ in the box so that the resulting statement will be true.

1. 5 ☐ 9   **2.** -4 ☐ -5

3. 0 ☐ 6   **4.** $\dfrac{1}{3}$ ☐ $\dfrac{5}{8}$

5. 3 ☐ -2 **6.** -1 ☐ 0
7. -2.3 ☐ -4.5 **8.** 3 ☐ -3
9. 0.1 ☐ 0.0986
10. Since $3 > 1$, then $3 + 5$ ☐ $1 + 5$.
11. Since $-5 < 4$, then $-5 - 2$ ☐ $4 - 2$.
12. Since $10 > 7$,   then $10 - (-2)$ ☐ $7 - (-2)$.
13. Since $-3 < -1$, then $(-3)(5)$ ☐ $(-1)(5)$.
14. Since $0 > -2$, then $(0)(-4)$ ☐ $(-2)(-4)$.
15. Since $14 > 7$, then $(14) \div (7)$ ☐ $(7) \div (7)$.

16. Since $10 > 9$, then $(10) \div (-6)$ ☐ $(9) \div (-6)$.
17. If $x + 3 < 5$, then $x + 3 - (3)$ ☐ $5 - (3)$.
18. If $x > 7$, then $9x$ ☐ $9(7)$.
19. If $4x < 16$, then $(4x) \div (4)$ ☐ $(16) \div 4$.
20. If $-5x > 25$, then $(-5x) \div (-5)$ ☐ $(25) \div (-5)$.
21. If $-\dfrac{1}{3}x < 4$, then $(-3)\left(-\dfrac{1}{3}x\right)$ ☐ $(-3)(4)$.
22. If $x > -8$ and $-8 > y$, then x ☐ y.
23. If $x < 7$ and $7 < y$, then x ☐ y.
24. If $x < 13$ and $y > 13$, then x ☐ y.
25. If $x > y$ and $z < y$, then x ☐ z.
In Exercises 26–30, given that $a > b$, determine if

the following statements are true. If the statement is false, give a counterexample.

26. $-a < -b$ **27.** $a^2 < b^2$

28. $a + 2 < b + 2$ **29.** $6a > 6b$

30. $a - b < 0$

31. If $m - n = 2$, then what inequality symbol can be placed in the box to make $m \;\Box\; n$ a true statement?

32. Do the same as in Exercise 31: If $x - y = -3$, then $x \;\Box\; y$.

*33. If $x > 0$, then what can be said about x^n, where n is a positive integer?

*34. If $x < 0$, then what can be said about x^n
(a) if n is a positive even integer;
(b) if n is a positive odd integer.

In Exercises 35–40, what can be said about the numbers a and b if it is known that

35. $ab > 0$ **36.** $ab < 0$

37. $\dfrac{a}{b} > 0$ **38.** $\dfrac{a}{b} < 0$

39. $a^2 b > 0$ **40.** $\dfrac{a}{b^2} < 0$

*41. Prove that if $0 < x < 1$, then $x^2 < x$.

*42. Prove that if $x > 0$, then $\dfrac{1}{x} > 0$.

*43. Prove that if $x > 1$, then $0 < \dfrac{1}{x} < 1$.

*44. Suppose $x > 0$, $y > 0$, and $x < y$. Show that $\dfrac{-1}{x} < \dfrac{-1}{y}$.

3.6 SOLVING FIRST-DEGREE INEQUALITIES IN ONE VARIABLE

Before discussing the solution of inequalities, we introduce two other symbols of inequality that occur in problems:

$$\leq \text{ means } \textit{is less than or equal to}$$

$$\geq \text{ means } \textit{is greater than or equal to}$$

For example,

$$5 \geq 4 \text{ and } 5 \geq 5 \text{ are both true statements}$$

and

$$6 \leq 9 \text{ and } 6 \leq 6 \text{ are both true statements}$$

The addition, subtraction, multiplication, and division properties of inequalities are still true when $<$ is replaced by $\leq$ and when $>$ is replaced by $\geq$. Now that we have the basic properties of inequalities, we can solve inequalities in one variable. The **solution set** of a first-degree inequality is the set of all numbers that make the inequality a true statement. **Equivalent inequalities** are those inequalities that have the same solution set.

Inequalities that are true for all permissible values of the variable are called **absolute inequalities**. For example, if x is any real number,

$$x^2 \geq 0$$

is an absolute inequality. Inequalities that are true only for certain values of the variable are called **conditional inequalities**. For example,

$$x - 1 > 0$$

is true only for $x > 1$.

As with first-degree equations, to find the solution set of an inequality, we will use the properties of inequalities to transform the given inequality into a simpler equivalent inequality.

EXAMPLE 1 Solve the inequality $2x + 7 < 15$ and graph the solution.

Solution

$$2x + 7 < 15 \qquad \text{[given inequality]}$$

$$2x + 7 - 7 < 15 - 7 \qquad \text{[subtracting 7 from both sides]}$$

$$2x < 8 \qquad \text{[combining like terms]}$$

$$\frac{2x}{2} < \frac{8}{2} \qquad \text{[dividing both sides by 2. Note that division by a positive number does not change the sense of the inequality.]}$$

$$x < 4 \ \blacksquare$$

We can use the real number line to graph the solution to inequalities in one variable. For Example 1, the graph of $x < 4$ consists of all points to the left of 4 on the real number line. The open circle on the graph indicates that all points up to, but *not* including, 4 are in the solution set.
Geometrically:

Using set-builder notation, this solution can be written $\{x \mid x < 4\}$.

EXAMPLE 2 Find all values of x that satisfy $2x - 2 < 4x + 6$.

Solution

$$2x - 2 < 4x + 6 \qquad \text{[given inequality]}$$

$$2x - 2 - 4x < 4x + 6 - 4x \qquad \text{[subtracting } 4x \text{ from both sides]}$$

$$-2 - 2x < 6 \qquad \text{[combining like terms]}$$

$$2 - 2 - 2x < 6 + 2 \qquad \text{[adding 2 to both sides]}$$

$$-2x < 8 \qquad \text{[combining like terms]}$$

$$\frac{-2x}{-2} > \frac{8}{-2} \qquad \text{[dividing both sides by } -2 \text{ and } \textit{changing} \text{ the order of the inequality]}$$

$$x > -4 \ \blacksquare$$

Geometrically:

Using set-builder notation, this solution can be written $\{x \mid x > -4\}$.

EXAMPLE 3 Find all values of x that satisfy $2(x + 1) \geq 3(x - 2)$.

Solution

$$2(x + 1) \geq 3(x - 2) \qquad \text{[given inequality]}$$

$$2x + 2 \geq 3x - 6 \qquad \text{[removing parentheses]}$$

$$2x + 2 - 2 \geq 3x - 6 - 2 \qquad \text{[subtracting 2 from both sides]}$$

$$2x \geq 3x - 8 \qquad \text{[combining like terms]}$$

$$2x - 3x \geq 3x - 8 - 3x \qquad \text{[subtracting } 3x \text{ from both sides]}$$

$$-x \geq -8 \qquad \text{[combining like terms]}$$

$$\frac{-x}{-1} \leq \frac{-8}{-1} \qquad \text{[dividing by } -1 \text{ and changing the order of the inequality]}$$

$$x \leq 8 \quad \blacksquare$$

Geometrically:

Solid circle

Note that the solid circle at the point represents 8, since 8 is a solution to the problem. Using set-builder notation, the solution set is $\{x \,|\, x \leq 8\}$.

EXAMPLE 4 Solve the inequality $-\dfrac{x}{2} < 3x - \dfrac{(2x + 3)}{4}$.

Solution

$$-\frac{x}{2} < 3x - \frac{(2x + 3)}{4} \qquad \text{[given inequality]}$$

$$4\left(-\frac{x}{2}\right) < 4\left[3x - \frac{(2x + 3)}{4}\right] \qquad \text{[multiplying both sides by 4. The sense of the inequality remains the same.]}$$

$$-2x < 12x - (2x + 3) \qquad \text{[simplifying]}$$

$$-2x < 12x - 2x - 3 \qquad \text{[removing parentheses]}$$

$$-2x < 10x - 3 \qquad \text{[combining like terms]}$$

$$-10x - 2x < -10x + 10x - 3 \qquad \text{[subtracting } 10x \text{ from both sides]}$$

$$-12x < -3 \qquad \text{[combining like terms]}$$

$$\frac{-12x}{-12} > \frac{-3}{-12} \qquad \text{[dividing by } -12 \text{ and changing the order of the inequality]}$$

$$x > \frac{1}{4} \quad \blacksquare$$

Thus, the solution set is $\left\{x \,\middle|\, x > \dfrac{1}{4}\right\}$.

Geometrically:

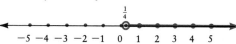

EXAMPLE 5   Solve the inequality $-5 < 3x + 1 \le 10$.

Solution

$$-5 < 3x + 1 \le 10 \qquad \text{[given inequality]}$$

$$-5 - 1 < 3x + 1 - 1 \le 10 - 1 \qquad \text{[subtracting 1 from all quantities]}$$

$$-6 < 3x \le 9 \qquad \text{[combining like terms]}$$

$$\frac{-6}{3} < \frac{3x}{3} \le \frac{9}{3} \qquad \text{[dividing each quantity by 3]}$$

$$-2 < x \le 3 \quad \blacksquare$$

Thus, the solution set is $\{x \mid -2 < x \le 3\}$.
Geometrically:

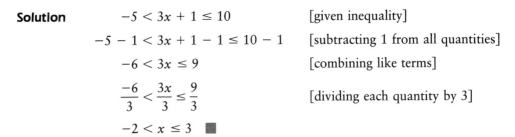

Open circle

Solid circle

$-5 \quad -4 \quad -3 \quad -2 \quad -1 \quad 0 \quad 1 \quad 2 \quad 3 \quad 4 \quad 5$

EXAMPLE 6   The formula that converts Celsius degrees to Fahrenheit degrees is

$$F = \frac{9}{5}C + 32$$

Find the range of values of C if F is between 10 and 60.

Solution   We are given the inequality $10 < F < 60$. Substituting the formula for F in this inequality, we obtain

$$10 < \frac{9}{5}C + 32 < 60$$

$$10 - 32 < \frac{9}{5}C + 32 - 32 < 60 - 32 \qquad \text{[subtracting 32 from all quantities]}$$

$$-22 < \frac{9}{5}C < 28 \qquad \text{[simplifying]}$$

$$\frac{5}{9}(-22) < \frac{5}{9}\left(\frac{9}{5}C\right) < \frac{5}{9}(28) \qquad \left[\text{multiplying all quantities by } \frac{5}{9}\right]$$

$$-\frac{110}{9} < C < \frac{140}{9} \quad \blacksquare$$

Thus, the values of C should be between $-\dfrac{110}{9}$ and $\dfrac{140}{9}$.

EXERCISES   3.6

In Exercises 1–10, determine if the given inequality is an absolute inequality or a conditional inequality. If it is a conditional inequality, give an

example showing that the inequality can be false.

1. $x^2 \ge 0$ **2.** $x + 3 \ge 5$

3. $3x - 1 \le 2x + 4$ **4.** $-x^2 \le 0$

5. $x^2 + 1 \geq 0$

6. $x \geq 0$

7. $\sqrt{x} > 0$

8. $(x - 3)^2 \geq 0$

9. $-(x + 4)^2 \leq 0$

10. $-3 \leq 2x - 3 < 5$

In Exercises 11–80, find the solution set of the given inequality and graph each solution.

11. $x - 7 > 0$ **12.** $x + 4 < 0$

13. $3x \geq 9$ **14.** $-14x \leq 56$

15. $12x < 11x - 1$ **16.** $9x < 10x - 2$

17. $3x + 7 < 0$ **18.** $5 - 2x \geq 0$

19. $x + 3 \leq 5x + 7$

20. $-3x - 4 \geq 5x + 4$

21. $3 - 5x \leq 6 - 8x$

22. $12 - x > 6 - 2x$

23. $x + 3 - 4x > 2x + 23$

24. $-14 + 2x < x + 7 - 5x$

25. $2x > x - 3$ **26.** $3x < x + 2$

27. $(7x + 1) - x < 5x + 3$

28. $(7x + 1) \geq 6x + 3$

29. $7x \leq 5x$ **30.** $-10x \geq 6x$

31. $3(x + 5) < x + 7$

32. $2(x + 1) \geq 3(x - 4)$

33. $6 + x \leq 2(x + 2)$

34. $6 - 8x < 3(7 - x)$

35. $5(x + 3) + 4 < x - 1$

36. $6 + 3(x + 2) \geq x + 6$

37. $-6(x + 2) < 2(3x - 5)$

38. $-5(12 - 3x) < 16(x + 3)$

39. $3 - (x + 4) > 5 + x$

40. $9 - x \geq 7 + (x - 2)$

41. $3[1 - (x - 2)] \geq 2 - 5(1 - x)$

42. $-4[1 - (1 + x)] \leq 24 + 6(2x - 3)$

43. $x(2 - x) - 8 \geq -(x^2 + x + 4)$

44. $x(3 - 2x) - 4 < 2[x(5 - x) - 1]$

*45. $x(3x - 2) > x[3(x + 1) - 2]$

*46. $-(5 + x + x^2) \leq 8 + x(4 - x)$

*47. $(x - 3)^2 \geq (x + 1)^2$

*48. $-(x + 5)^2 + 1 < -(x^2 - 6x + 9)$

*49. $(x + 2)^2 - (x - 5)^2 < 0$

*50. $10 \leq (x - 1)^2 - (x + 1)^2$

51. $\dfrac{x}{2} \leq 13$ **52.** $-\dfrac{x}{4} > 3$

53. $-\dfrac{2x}{3} > 9$ **54.** $\dfrac{3x}{4} \leq 5$

55. $\dfrac{x}{2} - 5 > \dfrac{x}{4} + 3$

56. $\dfrac{3}{4}x + 1 \leq \dfrac{5}{8}x - 2$

57. $-6 - \dfrac{3}{5}x > -\dfrac{2}{5}x + 1$

58. $\dfrac{7}{8}x - 6 \geq \dfrac{1}{8}x + 4$

59. $\dfrac{5x}{4} - \dfrac{x}{3} \leq 2$ **60.** $\dfrac{2x}{7} + \dfrac{3x}{5} > 1$

61. $\dfrac{x - 2}{4} + \dfrac{x}{6} \geq -1$

62. $\dfrac{x - 1}{10} - \dfrac{x}{25} - 7 \leq 0$

63. $\dfrac{5 + x}{12} - \dfrac{x + 3}{4} \leq 1$

64. $\dfrac{2 + x}{6} - \dfrac{x - 3}{9} \geq -7$

65. $\dfrac{2x + 1}{5} - \dfrac{2 - x}{3} > 2$

66. $\dfrac{1}{2}(3x - 1) + \dfrac{x}{5} < 7x + 10$

67. $4 - \dfrac{3}{2}x > \dfrac{13}{8} - \dfrac{1}{6}(4x - 3)$

68. $\dfrac{x - 3}{2} - \dfrac{3(3 - x)}{10} + \dfrac{7x - 6}{4}$
$< \dfrac{x + 10}{3} - \dfrac{3 - 16x}{20}$

69. $\dfrac{2x^2 - 5x + 3}{6} - \dfrac{4 - x}{12} + \dfrac{5 - x^2}{3}$
$\geq \dfrac{2x - 1}{9}$

70. $\dfrac{1}{2} + \dfrac{1}{4}[(x^2 - 1)x + 2] > \dfrac{3}{2}x + \dfrac{x^3 - 1}{4}$

71. $-2 < x + 1 < 5$

72. $-5 < x + 1 < 7$

73. $-4 < x - 4 < 4$

74. $0 < x + 3 \leq 8$

75. $2 \leq 4 - x \leq 9$

76. $-2 \leq 8 - x < 3$

77. $-12 < 2 - 7x \leq 2$

78. $-13 \leq 5 - 6x \leq 5$

*79. $3x - 2 \leq 4x + 1 \leq 3x + 1$

*80. $-2x + 1 \leq -x - 7 \leq -2x + 3$

In Exercises 81–90, use inequalities to describe the collection of numbers graphed.

81.

3   10

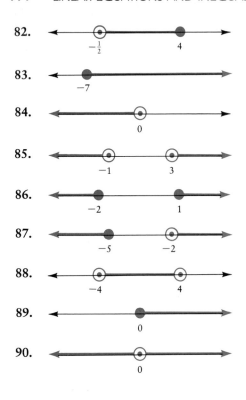

82.

83.

84.

85.

86.

87.

88.

89.

90.

*91. Show that if $a \neq b$, then $a^2 + b^2 > 2ab$. (*Hint:* Use the absolute identity $(a - b)^2 \geq 0$.)

*92. Prove that if $a \neq b$ are two positive real numbers, then $\dfrac{a}{b} + \dfrac{b}{a} > 2$. (*Hint:* Use the results of Exercise 91.)

*93. If $a^2 < b^2$, does it follow that $a < b$? Use numerical examples to illustrate the answer.

*94. Prove that for all real numbers a where $a \neq 0$, $a^2 > 0$. (*Hint:* Consider cases (**a**) $a > 0$ and (**b**) $a < 0$.)

*95. Suppose $a < b < 0$ and $0 < c < d$. Is $ac < bd$? Is $bd < ac$?

*96. Prove that if $a > 0$, then $a + \dfrac{1}{a} \geq 2$. (*Hint:* Use the absolute inequality $(a - 1)^2 \geq 0$.)

*97. Show by example that even though $a > b$ and $c > d$, it need not be true that $a - c > b - d$.

*98. Under what conditions does the inequality $\dfrac{1}{a} \leq \dfrac{1}{b}$ imply that $b \leq a$?

3.7 LINEAR EQUATIONS AND LINEAR INEQUALITIES INVOLVING ABSOLUTE VALUES

We begin this section by recalling the definition of absolute value given in Chapter 1:

$$|x| = x \qquad \text{if } x \text{ is positive or if } x \text{ is zero}$$
$$|x| = -x \qquad \text{if } x \text{ is negative}$$

This definition can be written using inequalities:

DEFINITION

$$|x| = \begin{cases} x \text{ if } x \geq 0 \\ -x \text{ if } x < 0 \end{cases}$$

For example,

$$|3| = 3 \qquad |-3| = -(-3) = 3 \qquad |0| = 0$$

The definition of absolute value can be applied to solving equations. Consider the following examples.

EXAMPLE 1 Solve $|x - 2| = 4$.

Solution Since the values of x are unknown at this time, we don't know whether $x - 2 \geq 0$ or $x - 2 < 0$. Thus, we must consider both possibilities.

Case 1. Suppose $x - 2 \geq 0$. Then, $|x - 2| = x - 2$ and our equation becomes

$$|x - 2| = 4$$
$$x - 2 = 4$$
$$x = 6$$

Case 2. Suppose $x - 2 < 0$. Then, $|x - 2| = -(x - 2)$ and we have

$$|x - 2| = 4$$
$$-(x - 2) = 4$$
$$-x + 2 = 4$$
$$-x = 2$$
$$x = -2$$

Thus, the solutions to the equation $|x - 2| = 4$ are $x = 6$ and $x = -2$. Checking our answer, we have:

$$|6 - 2| \overset{?}{=} 4 \qquad |-2 - 2| \overset{?}{=} 4$$
$$|4| \overset{?}{=} 4 \qquad |-4| \overset{?}{=} 4$$
$$4 \overset{\checkmark}{=} 4 \qquad 4 \overset{\checkmark}{=} 4 \quad \blacksquare$$

EXAMPLE 2 Solve the equation $|3x + 4| = 6$.

Solution *Case 1.* If $(3x + 4) \geq 0$, then $|3x + 4| = 3x + 4$, and

$$|3x + 4| = 6$$
$$3x + 4 = 6$$
$$3x = 2$$
$$x = \frac{2}{3}$$

Case 2. If $(3x + 4) < 0$, then $|3x + 4| = -(3x + 4)$, and

$$|3x + 4| = 6$$
$$-(3x + 4) = 6$$
$$-3x - 4 = 6$$

$$-3x = 10$$

$$x = -\frac{10}{3}$$

Therefore, the solutions are $x = \frac{2}{3}$ and $x = -\frac{10}{3}$. ■

In order to solve inequalities involving absolute values, we recall that $|x|$ represents the distance between the origin and the point representing x on the real number line. For example, the inequality $|x| < 3$ represents all points *at most three units from the origin* on the real number line.

Therefore, $|x| < 3$ is equivalent to $-3 < x < 3$.

The inequality $|x| > 3$ consists of all points *more than three points from the origin* on the real number line.

Thus, $|x| > 3$ is equivalent to the inequalities $x > 3$ or $x < -3$.

In general, we have

Inequalities Involving Absolute Values

Given a positive number a,

$$|x| < a \text{ is equivalent to } -a < x < a$$

$$|x| > a \text{ is equivalent to } x > a \text{ or } x < -a$$

EXAMPLE 3   Solve $|x - 8| < 5$ and graph the solution set.

Solution $|x - 8| < 5$ is equivalent to

$$-5 < x - 8 < 5$$

$$3 < x < 13 \qquad \qquad \text{[adding 8 to each expression]} \quad ■$$

EXAMPLE 4   Solve $|2x - 1| \leq 7$ and graph the solution set.

Solution $|2x - 1| \leq 7$ is equivalent to

$$-7 \leq 2x - 1 \leq 7$$

$$-6 \leq 2x \leq 8 \qquad \text{[adding 1 to each expression]}$$

$$-3 \leq x \leq 4 \qquad \text{[dividing each expression by 2]} \quad \blacksquare$$

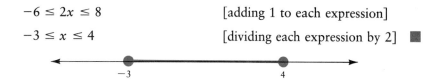

EXAMPLE 5 Solve $|3x - 5| > 2$ and graph the solution set.

Solution $|3x - 5| > 2$ is equivalent to the inequalities

$$3x - 5 > 2 \quad \text{or} \quad 3x - 5 < -2$$

$$3x > 7 \qquad\qquad 3x < 3$$

$$x > \frac{7}{3} \qquad\qquad x < 1 \quad \blacksquare$$

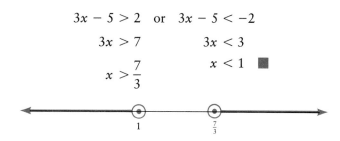

EXAMPLE 6 Solve $\left|\dfrac{2x + 1}{3}\right| \geq 4$ and graph the solution set.

Solution $\left|\dfrac{2x + 1}{3}\right| \geq 4$ is equivalent to the inequalities

$$\frac{2x + 1}{3} \geq 4 \quad \text{or} \quad \frac{2x + 1}{3} \leq -4$$

$$2x + 1 \geq 12 \qquad 2x + 1 \leq -12$$

$$2x \geq 11 \qquad\qquad 2x \leq -13$$

$$x \geq \frac{11}{2} \qquad\qquad x \leq -\frac{13}{2} \quad \blacksquare$$

CAUTION
Note that the graphs in Exercises 5 and 6 are disjoint segments and they are described separately. Inequalities describing disjoint segments must always be written separately. *Do not try* to describe disjoint segments using a single inequality. For example, do *not* try to write the answer to Exercise 6 as $-\frac{13}{2} > x > \frac{11}{2}$, because there are no values of x that satisfy this inequality.

EXAMPLE 7 Solve $|x + 1| = 4x - 1$.

Solution *Case 1.* If $(x + 1) \geq 0$, then $|x + 1| = x + 1$ and we have

$$|x + 1| = 4x - 1$$
$$x + 1 = 4x - 1$$
$$-3x = -2$$
$$x = \frac{2}{3}$$

Case 2. If $(x + 1) < 0$, then $|x + 1| = -(x + 1)$ and we get

$$|x + 1| = 4x - 1$$
$$-(x + 1) = 4x - 1$$
$$-x - 1 = 4x - 1$$
$$-5x = 0$$
$$x = 0 \quad \blacksquare$$

Because the expression without absolute value signs, $4x - 1$, has the unknown x in it, we must always check the results:

Check

$$x = \frac{2}{3}$$
$$\left| \frac{2}{3} + 1 \right| \stackrel{?}{=} 4\left(\frac{2}{3}\right) - 1$$
$$\left| \frac{5}{3} \right| \stackrel{?}{=} \frac{8}{3} - 1$$
$$\frac{5}{3} \stackrel{\checkmark}{=} \frac{5}{3}$$

$$x = 0$$
$$|0 + 1| \stackrel{?}{=} 4(0) - 1$$
$$|1| \stackrel{?}{=} -1$$
$$1 \neq -1$$

Thus, the solution to $|x + 1| = 4x - 1$ is only $x = \frac{2}{3}$.

EXERCISES 3.7

In Exercises 1–20, solve each equation and check the results.

1. $|x| = 4$
2. $|x| = 10$
3. $|x| = \frac{1}{4}$
4. $|x| = \frac{2}{5}$
5. $|x| = 0$
6. $|x + 3| = 2$
7. $|x - 5| = 3$
8. $|5 - x| = \frac{1}{2}$
9. $|7 - x| = \frac{3}{4}$
10. $|2x - 4| = 5$
11. $|3x + 1| = 8$
12. $|4x + 2| = 8$
13. $|-4x - 1| = 3$
14. $|5x - 6| = 0$
15. $|9 - 2x| = 0$
16. $|x - 2| = x - 3$

17. $|x + 1| = 2x + 3$

18. $|x + 1| = \frac{1}{2}x + 3$

19. $|x - 2| = \frac{1}{2}x + 1$

20. $|3x + 2| = 2x + 1$

In Exercises 21–55, solve and graph the solution set.

21. $|x| < 2$ **22.** $|x| > 1$

23. $|x| \leq 6$ **24.** $|x| \geq 8$

25. $|x| > 0$ **26.** $|x| \geq 0$

27. $|3x| < 9$ **28.** $|4x| \leq 17$

29. $\left|\dfrac{x}{2}\right| > 6$ **30.** $\left|\dfrac{x}{3}\right| \geq 7$

31. $|x + 3| < 5$ **32.** $|x - 2| \geq 4$

33. $|3 - x| \leq 10$ **34.** $|2 + x| > 1$

35. $|x - 7| \leq 0$

36. $|8 - x| \geq 0$ **37.** $|-x| \geq 2$

38. $|-x| \leq 4$ **39.** $|-x + 5| \leq |-2|$

40. $|-x + 3| \geq |-6|$ **41.** $|x + 1| - 2 \leq 5$

42. $|x - 3| + 7 \geq 10$

43. $|2x - 5| < 1$ **44.** $|4x + 3| < 6$

45. $|5x + 1| \geq 3$ **46.** $|3x + 2| \geq 1$

47. $|2x + 7| \leq 3$ **48.** $|4x - 1| < 2$

49. $|4 - 3x| > 1$ **50.** $|7x + 2| \geq 4$

51. $|5x + 4| \geq 1$ **52.** $\left|\dfrac{x}{2} + 5\right| \geq 3$

53. $\left|4 - \dfrac{x}{3}\right| \geq 1$ **54.** $|3x - 1| + 2 \geq 7$

55. $|2x + 3| + 5 \leq 8$

*56. Solve $|x| < \delta$, where δ (delta) is a positive real number.

*57. Solve $|y| < \varepsilon$, where ε (epsilon) is a positive real number.

*58. Solve $|y - 4| < \varepsilon$.

*59. Solve $|x - 3| < \delta$.

In Exercises 60–73, write the given inequality as an absolute-value inequality.

60. $-2 < x < 2$ **61.** $-4 \leq x \leq 4$

62. $-7 < x + 1 < 7$ **63.** $-8 \leq x - 4 \leq 8$

64. $-3 < x < 1$

65. $0 \leq x \leq 5$

66. $1.1 \leq x \leq 1.2$

67. $-0.6 < x < -0.5$

68. $x > 5$ or $x < -5$

69. $x > 7$ or $x < -7$

70. $x - 1 \geq 3$ or $x - 1 \leq -3$

71. $4 - x \geq 6$ or $4 - x \leq -6$

*72. $3 - \delta < x < 3 + \delta$

*73. $5 - \varepsilon \leq x \leq 5 + \varepsilon$

*74. Is $|ab| = |a||b|$?

*75. Give a numerical example showing that $|a + b| \neq |a| + |b|$, where a and b are any real numbers.

*76. For what numbers a and b, if any, is it true that $|a| + |b| = 0$?

*77. Prove that $a^2 = |a|^2$ for every real number a. (*Hint:* Consider cases $a \geq 0$ and $a < 0$, and the definition of absolute value.)

Chapter 3 SUMMARY

EQUATIONS

1. An **equation** is a mathematical statement that expresses the equality between two algebraic expressions.
2. An **identity** is an equation that is true for all possible values of the variables involved.
3. A **conditional equation** is an equation that is true only for certain values of the variables.

SOLVING LINEAR EQUATIONS

1. *Equivalence.* Two equations are said to be **equivalent** if and only if they have exactly the same solutions.

2. *Rules that Yield Equivalent Equations.* Given an equation, we may add, subtract, and multiply by a nonzero quantity and divide by a nonzero quantity. In each of these cases, the resulting equation will be equivalent to the original equation.

3. *Method for Finding the Solution of Linear Equations.*
   (1) Clear fractions by multiplying each term by the least common denominator.
   (2) Use the distributive law to remove parentheses.

(3) Transpose all terms containing the unknown to one side of the equation and all other terms to the other side of the equation.

(4) Combine like terms on each side of the equation.

(5) Divide both sides of the equation by the coefficient of the unknown.

INEQUALITIES

$<$ means *is less than*

$>$ means *is greater than*

$\leq$ means *is less than or equal to*

$\geq$ means *is greater than or equal to*

1. *Trichotomy Law.* If a and b are real numbers, then exactly one of the following is true:

$$a < b \qquad a = b \qquad a > b$$

2. *Transitive Law of Inequalities.* If a, b, and c are real numbers, then:

If $a < b$ and $b < c$, then $a < c$.

If $a > b$ and $b > c$, then $a > c$.

3. *Addition and Subtraction Properties of Inequalities.* For a, b, and c real numbers, then:

If $a < b$,

then $a + c < b + c$ and $a - c < b - c$.

If $a > b$,

then $a + c > b + c$ and $a - c > b - c$.

4. *Multiplication and Division Properties of Inequalities.* For a, b, and c real numbers, then:

(1) If $c > 0$ and $a < b$,

then $ac < bc$ and $\dfrac{a}{c} < \dfrac{b}{c}$.

(2) If $c > 0$ and $a > b$,

then $ac > bc$ and $\dfrac{a}{c} > \dfrac{b}{c}$.

(3) If $c < 0$ and $a < b$,

then $ac > bc$ and $\dfrac{a}{c} > \dfrac{b}{c}$.

(4) If $c < 0$ and $a > b$,

then $ac < bc$ and $\dfrac{a}{c} < \dfrac{b}{c}$.

LINEAR EQUATIONS AND LINEAR INEQUALITIES INVOLVING ABSOLUTE VALUES

1. $|x| = \begin{cases} x, & \text{if } x \geq 0 \\ -x, & \text{if } x < 0 \end{cases}$

2. If a represents a real number, $a \geq 0$, then: $|x| = a$ is equivalent to $x = -a$ and $x = a$.

$|x| < a$ is equivalent to $-a < x < a$.

$|x| > a$ is equivalent to $x < -a$ or $x > a$.

Chapter 3 EXERCISES

In Exercises 1–10, determine whether each equation is an identity or a conditional equation.

1. $4x + 3 = 7$ 2. $2x + 5 = 5 + 2x$
3. $-7(x + 2) = -7x - 14$
4. $3(2x - 1) = 3(1 - 2x)$
5. $-x + 3 = 2x + 4$
6. $(x + 2)^2 - 4x = x^2 + 4$
7. $5x + 2 = 3x + 5$
8. $x(x - 3) - 3x = (x - 3)^2 - 9$
9. $(x + 7)(x + 2) - x^2 - 9x = 14$

10. $-3(5 + 3x) + 8(3 + x)$
    $= 5(5 - x) - 4(4 - x)$

In Exercises 11–125, solve each equation.

11. $3x - 5 = 7$ 12. $4x - 9 = 27$
13. $2x + 13 = 29$
14. $7x + 11 = 39$
15. $5x = 14 - 2x$
16. $7x = 40 - 3x$
17. $5x = 9x - 32$
18. $9x + 40 = 17$

19. $3x + 2 = -5 + x$

20. $11x - 9 = 26 + 6x$

21. $7x + 3 = 3x - 5$

22. $3x - 2 = 10x - 16$

23. $26 - x = 6x + 5$

24. $9x + 8 = 63 - 2x$

25. $3x - 17 = -7x + 43$

26. $5x - 16 = 29 - 4x$

27. $3 + 7x = 19 - x$

28. $8 + 3x = 4 - 5x$

29. $22 - x = 7 - 6x$

30. $9 - 4x = 16 + 3x$

31. $10 - 3x = 7 + 3x$

32. $25 + 3x = 3 - 8x$

33. $17 - 19x = 5 + 17x$

34. $23 - 13x = 16 + x$

35. $3x - 5 + 2x = 11 + x$

36. $5x + 8 - 3x = 6x - 2$

37. $4x - 11 + 3x - 7 - x = 0$

38. $7x + 8 - 3x + 7 = 3 - 2x$

39. $9 - 3x - 2 + 8x = 11 - 2x + 17$

40. $5x - 7 + 2x - 8 = 4x - 3 + 2x - 9$

41. $\dfrac{x}{2} = \dfrac{9}{4}$

42. $\dfrac{x}{3} = \dfrac{7}{6}$

43. $\dfrac{x}{7} = \dfrac{-4}{3}$

44. $\dfrac{2x}{3} = \dfrac{9}{2}$

45. $\dfrac{3x}{4} = \dfrac{-18}{5}$

46. $\dfrac{7x}{15} = \dfrac{14}{5}$

47. $-\dfrac{5x}{6} = \dfrac{8}{9}$

48. $\dfrac{2x}{11} = \dfrac{-9}{4}$

49. $\dfrac{4}{7} = \dfrac{2x}{3}$

50. $\dfrac{5}{9} = \dfrac{7x}{6}$

51. $\dfrac{x}{2} + \dfrac{x}{3} = 10$

52. $\dfrac{x}{2} + \dfrac{2x}{3} = \dfrac{14}{3}$

53. $\dfrac{x}{4} - \dfrac{x}{5} = \dfrac{3}{4}$

54. $\dfrac{x}{7} - \dfrac{x}{5} = \dfrac{11}{3}$

55. $x + 5 = \dfrac{x}{2}$

56. $x - 7 = \dfrac{x}{3}$

57. $2x = 21 - \dfrac{x}{3}$

58. $3x + \dfrac{x}{3} = 14 - \dfrac{2}{3}$

59. $2x - \dfrac{x}{6} + \dfrac{x}{3} = 15 - \dfrac{x}{4} - \dfrac{1}{2}$

60. $3x - 25 = \dfrac{x}{2} + x - 12 + \dfrac{x}{5}$

61. $2x + 20 - \dfrac{x}{6} = \dfrac{x}{4} + 1$

62. $\dfrac{x}{3} - 2 + x + \dfrac{x}{4} = 19 - \dfrac{3x}{4}$

63. $3(x + 1) = 5(x - 1)$

64. $13(2 + x) = 9x - 10$

65. $18 - 3(x + 2) = 5(x + 4)$

66. $1 - 4(x + 1) = x - 2(x + 3)$

67. $6 - 5(x - 5) = 9 + 2(x - 3)$

68. $3(x - 3) + 8 = 9 - 4(x - 1)$

69. $3(x - 5) - 15 = 5(2 - x)$

70. $4(x - 2) + 5(x - 4) = 6 + 7(x - 4)$

71. $6(x - 5) - 1 = 13(x - 4)$

72. $5(x + 2) - 3(x - 1) = 7 - 10(x - 3)$

73. $6(2x - 1) - (6x + 14) + 13 = 0$

74. $2(3 - 2x) - 1 = 16(2x + 1) + 3(2x - 6)$

75. $3(2x - 7) + 5(3x - 11)$
$= 5x - 19 - 7(2x - 9)$

76. $5(2x + 3) + 17 - 9(7 - 2x)$
$= 4(2x - 2) - 3(3 - 2x)$

77. $4(2x - 5) + 7(3x - 8) = 5(4x - 10) + 1$

78. $8 - [4x - (3x - 1) + 1] = 2$

79. $4x - 25 - 5[10 - (2x - 5) + 6x] = 40$

80. $3x - 2[2x + 5 - 3(x + 4) - 3]$
$= 4(x + 3) + 4$

81. $6x - 7[4x - 5(2x - 3) - 3] + 8(3x - 1)$
$- 10x - 1 = 0$

82. $2x - 3\{5x - 4[3x - 7(x - 2) - 5]$
$- 6\} - 4 = 0$

83. $2\{8x - 6[7 + 4(x - 3) - 5(x - 2) - 2]$
$-11\} = 9x - 1$

84. $5\{3x - 2[1 - 3(2x - 5) + 4(x - 2)$
$- 2] - 2\} = -x$

85. $\dfrac{4x + 6}{9} = \dfrac{3x + 5}{7}$

86. $\dfrac{1}{3}(5x + 2) = \dfrac{1}{5}(7x - 1)$

87. $\dfrac{5 - 4x}{29} + \dfrac{x - 1}{7} = 0$

88. $\dfrac{x + 1}{2} - \dfrac{4x - 5}{3} = \dfrac{5x + 9}{3}$

89. $\dfrac{6x + 2}{3} - \dfrac{2x + 1}{4} = \dfrac{4x + 3}{5} - \dfrac{5}{12}$

90. $\dfrac{3x + 2}{5} - \dfrac{x}{3} = \dfrac{x + 8}{10} + \dfrac{7x + 3}{15}$

91. $\dfrac{4x - 3}{3} - \dfrac{10x - 9}{6} = \dfrac{6x - 5}{4} - \dfrac{14x - 1}{12}$

92. $\dfrac{2x + 1}{3} - \dfrac{3x - 1}{11} = \dfrac{x + 1}{5} + \dfrac{4x - 7}{9}$

93. $\dfrac{1}{3}(3x + 1) - \dfrac{1}{7}(4x + 1) + \dfrac{1}{6}(2x + 10) = 1$

94. $\frac{1}{12}(5x + 3) + \frac{1}{3}(2x - 3) - \frac{1}{11}(7x - 1)$
$+ \frac{1}{7}(5 - 3x) = 0$

95. $\frac{1}{5}(4x + 5) - \frac{1}{6}(5x - 3) = \frac{1}{8}(7x - 4)$
$+ \frac{1}{12}(24 - 11x)$

96. $\frac{1}{4}(3x + \frac{1}{2}) - \frac{1}{3}(5x - \frac{1}{2}) + \frac{1}{6}(7x + \frac{1}{2})$
$- \frac{1}{11}(9x - \frac{1}{2}) + 1 = 0$

97. $\frac{1}{4}(5x - \frac{1}{4}) - \frac{1}{5}(3x + \frac{1}{4}) - \frac{1}{2}(7x - \frac{3}{4})$
$+ \frac{3}{2}(2x - \frac{1}{2}) = 0$

98. $\frac{1}{4}\{3 + \frac{1}{4}[\frac{1}{4}(x + 3) + x - 3]\} = x - 4$

99. $\frac{1}{4}\{x - \frac{1}{2}[\frac{1}{3}(2x - 3) + 2] + \frac{7}{2}\} - 2(x - 1)$
$= 0$

100. $\frac{1}{3}\{3x - \frac{1}{2}[x - \frac{1}{4}(2x - 3) + \frac{3}{2}] - 1\}$
$+ 3 = 0$

101. $x(x + 3) = (x + 2)(x + 5) - 18$

102. $(x + 2)^2 = (x + 1)(x + 3) + x - 2$

103. $2(x + 4)(x + 7) = (x + 5)(2x + 9) + 8$

104. $3(x - 1)(3x - 8) = (3x + 2)^2 - 10$

105. $(x + 3)^2 + (x - 5)^2 = 2(x - 4)^2 - 22$

106. $(2x + 3)(3x + 2) = 6(x + 1)^2 - 4x + 5$

107. $(4x - 5)^2 + (3x + 2)^2 = (5x + 3)^2 + 20$

108. $(2x - 5)(x - 2) + (7x - 6)(x - 4)$
$= (3x - 8)^2 - 15$

109. $3(2x + 7)(x - 1) - 2(3x - 1)(2x - 1)$
$= 6 - (6x - 1)(x - 5)$

110. $2(4 - 3x)(5 + 2x) - 3(x + 3)(2 - x)$
$= (3x - 10)(14 - 3x) - 4$

111. $3(7 + 2x)(5 - x) + (2x - 9)(3 - 5x)$
$= 27 - (13 - 4x)^2$

112. $4(x + 8)(3 - 2x) - (5 - x)^2$
$= 27 - (10 + 3x)^2$

113. $2(5 - 3x)(6 - x) + 3(x + 2)^2$
$= (3x - 4)^2 - 4$

114. $\frac{1}{2}(8 - 3x)(5 - 2x)$
$= \frac{1}{10}(4 - 5x)(7 - 6x) - 2$

115. $\frac{1}{3}(7 - x)(6x - 5)$
$= 10x + \frac{1}{4}(25 - 8x)(x + 8)$

116. $\frac{1}{5}(x - 3)(5x + 4)$
$= \frac{1}{4}(2x - 3)(2x - 5) - \frac{3}{4}$

117. $\frac{1}{7}(2x + 10)(9 - 7x)$
$= \frac{1}{3}(8 + x)(1 - 6x) - 31$

118. $\frac{1}{4}(11 - 2x)(3x - 9)$
$= 15 - \frac{3}{2}(5 - x)(8 - x)$

119. $\frac{1}{3}(3x + 2)(x - 4) + \frac{1}{2}(x - 2)^2$
$= \frac{1}{6}(3x - 10)^2 + \frac{4}{3}$

120. $\frac{1}{4}(5 - 2x)(3 - 2x) + \frac{1}{3}(x - 7)^2$
$= \frac{2}{3}(x + 1)(2x - 9) - \frac{23}{12}$

121. $(\frac{1}{2}x + 3)^2 = (\frac{1}{2}x + 2)(\frac{1}{2}x + 4) + \frac{1}{2}x - 1$

122. $(\frac{1}{2}x + 1)(\frac{1}{2}x + 3) = (x - 2)(\frac{1}{4}x + 1) + 8$

123. $(\frac{1}{2}x + \frac{1}{4})(x + \frac{1}{3}) = \frac{1}{2}(x + 1)^2 - \frac{11}{4}$

124. $(\frac{1}{2}x - 2)(\frac{1}{2}x - 3)$
$= (x + 2)(5 - \frac{1}{4}x) + (x + 2)(\frac{1}{2}x - 2)$

125. $(\frac{2}{3}x + 9)(\frac{1}{3}x + \frac{15}{2})$
$= (\frac{2}{3}x + 11)(\frac{1}{3}x + 5) + 8$

In Exercises 126–140, solve for x.

126. $x - a = 3a$

127. $x - 5c = 20$

128. $x + 10a = a + 8$

129. $cx = -7c$

130. $dx = d^3$

131. $2ax = -14ab$

132. $a + b = \frac{x}{2}$

133. $3x - a = -13a$

134. $\frac{x}{c} + c = -c$

135. $4(x + a) = 7(x - 2a)$

136. $\frac{x}{3} + a = c - 4a$

137. $\frac{x}{18} = \frac{a}{6} + b$

138. $\frac{x}{5} - \frac{c}{3} = \frac{d}{30}$

139. $3(5 - x) = 8(x - a)$

140. $\frac{x}{4} + \frac{x}{3} + a = 0$

In Exercises 141–170, solve each inequality for x.

141. $5x - 8 < 0$

142. $4x + 5 \geq 0$

143. $7 - 3x > -2$

144. $3 - 2x > -1$

145. $-5x + 3 \leq 18$

146. $-3x - 1 \leq -7$

147. $3(5 - x) \geq 2(x + 1)$

148. $3(1 - 2x) \geq 7x$

149. $3(7 + x) > 2(15 + x)$

150. $5 - (4x - 6) > (2x - 3) - (8x + 2)$

151. $(3x + 1) - (4x + 2) \leq (4x - 2)$
$+ (6 - 3x)$

152. $(4 + x)^2 > (x - 3)(x - 2)$

153. $(x + 1)^2 \geq (x + 2)^2$

154. $(x + 1)(x - 2) \geq (x - 1)(x + 3)$

155. $(x + 2)(x - 2) \geq (x + 2)^2$

156. $\frac{x + 6}{2} \leq \frac{7}{4}$

157. $\frac{x}{5} + 2 < 4$

158. $x + 1 < \dfrac{4(2x + 1)}{3}$

159. $\dfrac{3(2x + 1)}{-2} + 2 < 2$

160. $\dfrac{x}{2} + \dfrac{x - 2}{3} < 2x - \dfrac{1}{12}$

161. $\dfrac{x - 1}{3} + \dfrac{1}{2} \le \dfrac{x}{2} - 1$

162. $-3 < 4x + 1 < 9$

163. $4 < 2x - 5 < 13$

164. $-2 \le x - 3 \le 4$

165. $-3 \le 5 - 2x \le 7$

166. $-5 < 3x + 1 < 5$

167. $-7 \le 1 + 2x \le 7$

168. $-1 < \dfrac{x + 4}{6} \le \dfrac{1}{3}$

169. $3x - 1 \le 6x - 2 \le 3x + 8$

170. $9x + 2 > 7x > 9x - 4$

In Exercises 171–180, solve for x.

171. $|x + 3| = 7$

172. $|x - 3| = 5$

173. $|-x + 4| = 8$

174. $|5 - x| = 10$

175. $|2x + 1| = 11$

176. $|5x + 6| = 13$

177. $|7 - 6x| = 3$

178. $|3 - 4x| = 9$

179. $\left| \dfrac{x}{3} - 2 \right| = 5$

180. $\left| \dfrac{2x - 5}{-3} \right| = 2$

In Exercises 181–190, solve each inequality for x. Graph the solution.

181. $|x - 1| < 2$

182. $|x + 4| < 1$

183. $\left| x - \dfrac{1}{2} \right| \le \dfrac{1}{5}$

184. $|2x - 5| \le 0$

185. $|3x - 2| < 4$

186. $|5x + 3| \le 0$

187. $\left| \dfrac{-x}{3} - \dfrac{5}{2} \right| < \dfrac{1}{3}$

188. $\left| \dfrac{2x + 3}{8} \right| \le 1$

189. $\left| 5 - \dfrac{x}{3} \right| \le 2$

190. $\left| \dfrac{-5 + x}{-3} \right| < 6$

In Exercises 191–200, solve for x. Graph the solution.

191. $|x - 4| > 3$

192. $|2 + x| > 7$

193. $\left| x - \dfrac{1}{4} \right| \ge \dfrac{1}{2}$

194. $|2x + 7| \ge 3$

195. $|3x - 9| \ge 2$

196. $|4 - 5x| > 14$

197. $\left| \dfrac{x}{3} + 4 \right| > 6$

198. $\left| 5 - \dfrac{x}{2} \right| \ge 7$

199. $|-4x + 7| \ge 0$

200. $|3 - 5x| > 0$

4 Problem Solving Involving Linear Equations and Inequalities

Linear equations and inequalities can be used to solve a variety of real-life problems. Problems related to concepts such as length, area, volume, uniform motion, mixture, investments, and other areas can often be reduced to linear equations and linear inequalities. However, very few of the problems that arise in mathematics-related courses (physics, chemistry, and so on) or in a chosen occupation will be given in equation form. Most of these problems will either be stated in words or they will arise as questions in actual situations. In such cases, it will be necessary to define the problem, as well as solve it. Thus, the ability to analyze and solve such problems is one of the most important skills we can learn in algebra. Although no general procedure can be given for obtaining the required equations, methods used are illustrated by a number of examples. Before discussing particular types of word problems, we shall look at some problem-solving techniques.

4.1 PROBLEM-SOLVING TECHNIQUES

In this section, we discuss three useful ideas in solving word problems: dimensional analysis, translating key words and statements into algebraic form, and an outline for dealing with word problems.

A. Dimensional Analysis and Conversion of Units

The application of mathematics to a physical problem requires more than operating with numbers and equations. Choosing the proper dimensions to represent the various physical quantities is the first important step. Any algebraic or numerical procedure will be useless if a wrong combination of dimensions is used in an equation relating physical quantities.

What do we mean by **dimensions**? Suppose you were asked "Do you have 7?" Your response would probably be "Seven what?" Most physical problems involve quantities expressed in terms of numbers and dimensions (sometimes called **units**) such as 7 feet, 7 hours, 7 miles per hour, or 7 pounds per square inch. While the most common dimensions are measures of length, mass, and time, there are several different units that may be used to represent each of these dimensions. For example, time can be measured in hours, minutes, or seconds.

As we learned in Chapter 2, we can add or subtract only like algebraic terms. The same will hold for units, such as

$$3 \text{ minutes} + 6 \text{ minutes} = 9 \text{ minutes}$$

However, we cannot perform the addition

$$3 \text{ minutes} + 6 \text{ hours} = 9 \text{ ?}$$

even though both represent units of time. Thus, *it is very important that both sides of an equation have an equality of units*. This balance of units often requires conversion of units. For example, 6 hours = 360 minutes; thus, 3 minutes + 6 hours = 3 minutes + 360 minutes = 363 minutes.

As the preceding example illustrated, we shall often be required to convert a quantity of a given unit to an equivalent value in terms of another related unit. While there are many tables of conversion factors, we shall concentrate on the most commonly used conversions. If the conversion factors are known, quantities expressed in terms of one unit can be converted to an equivalent value in terms of another unit.

EXAMPLE 1   Convert a length of 5 yards to its equivalent in feet.

Solution   Each yard is equivalent to 3 feet, or "3 feet per yard." The word *per* means divided by. Thus, 3 feet per yard can be represented by the fraction 3 feet/1 yard. However, the number 1 is usually not written.

$$? \text{ feet} = 5 \text{ yards}$$
$$? \text{ feet} = (5 \text{ yards})\left(3 \frac{\text{feet}}{\text{yard}}\right) = 15 \text{ feet}$$

Note that we have treated the units as algebraic quantities and the yard units have cancelled each other. ■

Conversion of Units
To convert units: **1.** Find a conversion factor relating the units to be converted with the units you want.

> 2. Express the conversion factor as a fraction, with the units to be converted in the denominator.
> 3. Multiply the original number by the conversion factor.

EXAMPLE 2 Convert 10.16 centimeters (cm) to inches (in.).

Solution We want the final units to be inches.

$$? \text{ in.} = 10.16 \text{ cm}$$

The conversion factor 2.54 cm = 1 in. can be written with the unit to be converted (cm) in the denominator:

$$\frac{1 \text{ in.}}{2.54 \text{ cm}} = 1$$

Since the value of this fraction is 1, we may multiply the number whose units we want to convert by this fraction without changing the value of that number:

$$10.16 \text{ cm} \cdot \frac{1 \text{ in.}}{2.54 \text{ cm}} = \frac{10.16}{2.54} \text{ in.} = 4 \text{ in.} \quad \blacksquare$$

Some conversions require several conversion factors before the desired units can be obtained. However, be sure to select conversion factors so that at least one preceding unit is cancelled.

EXAMPLE 3 An engine part moves at the rate of 255 ft/min. Find the rate in inches per second.

Solution Since

$$\left(\frac{\text{ft}}{\text{min}}\right)\left(\frac{\text{in.}}{\text{ft}}\right)\left(\frac{\text{min}}{\text{sec}}\right) = \frac{\text{in.}}{\text{sec}}$$

we have

$$\left(255\frac{\text{ft}}{\text{min}}\right)\left(12\frac{\text{in.}}{\text{ft}}\right)\left(\frac{1 \text{ min}}{60 \text{ sec}}\right) = 51\frac{\text{in.}}{\text{sec}}$$

Note that we used the reciprocal of a familiar conversion factor. If there are 60 sec/min, then there is $\frac{1}{60}$ min/sec or 1 min/60 sec. $\quad \blacksquare$

EXAMPLE 4 Determine the conversion factor necessary to convert centimeters per second to miles per hour.

Solution

$$\left(\frac{\text{cm}}{\text{sec}}\right)\left(\frac{\text{in.}}{\text{cm}}\right)\left(\frac{\text{ft}}{\text{in.}}\right)\left(\frac{\text{mi}}{\text{ft}}\right)\left(\frac{\text{sec}}{\text{min}}\right)\left(\frac{\text{min}}{\text{hr}}\right) = \frac{\text{mi}}{\text{hr}} \quad \blacksquare$$

Often quantities are expressed in terms of more than one unit, and some of the units may be raised to powers. For example, the quantity "7 pounds per square inch" can be written 7 lb/in.^{2}. When a ratio of units is raised to some power, the units are also raised to the power:

$$\left(3\,\frac{\text{ft}}{\text{yd}}\right)^2 = (3)^2\left(\frac{\text{ft}}{\text{yd}}\right)^2 = 9\,\frac{\text{ft}^2}{\text{yd}^2}$$

EXAMPLE 5 Given a cube (Figure 4.1) with sides of 5 in., find the volume of the cube.

Solution

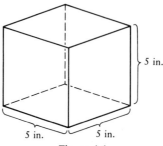

5 in.

5 in.   5 in.

Figure 4.1

The volume of a cube is given by cubing the length of one of its sides. Thus,

$$V = (5 \text{ in.})^3 = 5^3 \text{ in.}^3 = 125 \text{ in.}^3 \quad \blacksquare$$

EXAMPLE 6 The acceleration of a free-falling body due to gravity is approximately 32.2 ft/sec^2. Express acceleration in cm/sec^2.

Solution Performing the unit conversion first, we get

$$\left(\frac{\text{ft}}{\text{sec}^2}\right)\left(\frac{\text{in.}}{\text{ft}}\right)\left(\frac{\text{cm}}{\text{in.}}\right) = \frac{\text{cm}}{\text{sec}^2}$$

Thus,

$$\left(32.2\,\frac{\text{ft}}{\text{sec}^2}\right)\left(12\,\frac{\text{in.}}{\text{ft}}\right)\left(2.54\,\frac{\text{cm}}{\text{in.}}\right) = 981.456\,\frac{\text{cm}}{\text{sec}^2} \quad \blacksquare$$

The following are some useful tables of equivalent units.

Table 4.1. English–English Equivalents

1 mile (mi) = 5280 feet (ft)

1 yard (yd) = 3 feet

1 foot = 12 inches (in.)

1 pound (lb) = 16 ounces (oz)

1 ton = 2000 pounds

1 gallon (gal) = 4 quarts (qt)

1 quart = 2 pints (pt)

Table 4.2. Metric–Metric Equivalents

1 kilometer (km) = 1000 meters (m)

1 meter = 100 centimeters (cm)

1 centimeter = 10 millimeters (mm)

1 liter (ℓ) = 1000 cm^3 (cc)

1 kilogram (kg) = 1000 grams (g)

Table 4.3. English–Metric Conversions

1 inch = 2.54 centimeters

1 mile = 1609.3 meters

1 quart = 0.946 liters

1 mile per hour (mph) = 0.447 meter/sec

Table 4.4. Metric–English Conversions

1 centimeter = 0.3937 inches

1 meter = 1.0936 yards

1 kilometer = 0.6214 miles

1 liter = 0.2642 gallons

EXERCISES 4.1(A)

In Exercises 1–18, perform the indicated operations and simplify if possible. Do not change any units.

1. $(\frac{1}{3}$ hr$)(54$ mi/hr$)$
2. $(18$ ft$)(\frac{1}{3}$ yd/ft$)$
3. $(4$ cm$)(5$ g/2 cm$)$
4. $(12$ sec$)(3$ ft/4 sec$^2)$
5. $(1000$ atoms/in.$^3)(2$ in./cm$)^3$
6. $(17$ ft$^2)(12$ in./ft$)^2$
7. 8 m + 13 m
8. 22 ft^2 + 14 ft^2
9. 17 cm^3 + 13 cm^3
10. 2 tons + 1000 lb
11. 3 min + 45 sec
12. 7 yd^2 + 10 yd^3
13. $(8$ mi/hr$)(24$ hr/day$)$
14. $(36$ g/cm$)(2.54$ cm/in$)$
15. $(38$ ft/min$) \div (12$ ft$)$
16. $(60$ ℓ/sec$) \div (12$ ℓ$)$
17. $(32$ ft/sec$) \div (4$ sec$)$
18. $(7$ lb/in.$^3) \div (14$ in.^{3}/lb$)$

In Exercises 19–36, perform the indicated conversions.

19. 42 cm, change to m
20. 6 hr, change to sec
21. 8 lb, change to oz
22. 720 sec, change to hr
23. 2 mi, change to in.
24. 100 ft, change to yd
25. 9 ft 5 in., change to in.
26. 90 oz, change to lb
27. 200 m, change to ft
28. 7000 g, change to kg
29. 50 m^2, change to cm^2
30. 45 ft^3, change to in.^{3}
31. 44 ft/sec, change to mi/hr
32. 12 dollars/day, change to cents/hr
33. 60 mi/hr, change to ft/sec
34. 10 kg/m, change to g/cm
35. 186,000 mi/sec (speed of light), change to mi/yr (let 1 yr = 365 days)
36. 1100 ft/sec (speed of sound), change to mi/yr
37. A student is building a bookcase that has four shelves, each 2 ft 7 in. long. What is the total length of the material needed to build the shelves?
38. Butter costs $3.25 per pound. What is the cost of $2\frac{1}{2}$ pounds of butter?
39. Jogging 1 mi requires 150 calories. How many miles must you jog to burn off 750 calories?
40. The distance from Los Angeles to San Francisco is 405 mi. What is the cost of a trip from Los Angeles to San Francisco if gasoline costs $1.20/gal and the car gets 15 mpg?

B. Translation from English to Algebra

Algebra may be viewed as a language, different from English. Therefore, before solving algebraic problems that are stated in words, it is helpful to learn how to translate English statements into equations and inequalities. A statement given in English may seem very confusing, whereas a similar algebraic statement may be quite clear. For example, the answer to the statement "By how much does x exceed a?" may not be so familiar as the numerical statement "By how much does 19 exceed 13?" The answer to the numerical statement is 6, which is obtained by subtracting 13 from 19. Similarly, we can express the statement "By how much does x exceed a" by the algebraic statement $x - a$.

If the meaning of any statement involving words and letters is not clear, try substituting numbers for the letters and then reexamine the statement.

EXAMPLE 7   If x is an even integer, what is the next larger even integer?

Solution   Numerical statement: "If 6 is an even integer, what is the next larger even integer?" The answer is 8, which is obtained by adding 2 to 6. Thus, the next larger even integer following x is $x + 2$. ■

While it is not possible to give any easy rules for translating English statements to algebraic statements, certain key words and phrases occur quite often. The following statements point out several such key words and phrases.

1. *Statement*: A number is five more than another.
   Translation: Let the letter n represent the smaller number. Then $n + 5$ is the larger number.
   Note: The phrase *more than* translates to $+$.
2. *Statement*: A number is six less than another.
   Translation: If n is the larger number, then the smaller number is $n - 6$.
   Note: The phrase *less than* translates to $-$.
3. *Statement*: A number is four times another.
   Translation: If we let n represent one number, then the other is $4n$.
   Note: The word *times* indicates multiplication.
4. *Statement*: A number is one-third of another.
   Translation: Let n be one number, then $\frac{1}{3}n$ is the other.
5. *Statement*: A number is seven more than twice another.
   Translation: Let n represent the smaller number, then the larger is $2n + 7$.
6. *Statement*: The sum of a number and six is fourteen.
   Translation: Let x represent the number. Then $x + 6 = 14$.
   Note: The word *sum* indicates addition and the word *is* translates to $=$.
7. *Statement*: The difference of a number and five is three.
   Translation: Let x represent the number. Then $x - 5 = 3$.
   Note: The word *difference* indicates subtraction. Also, the phrase *difference of a number and five* translates to $x - 5$, not to $5 - x$.
8. *Statement*: The cube of a number is negative.
   Translation: Let y stand for the number. Then $y^3 < 0$.
   Note: The phrase *is negative* translates to *is less than 0*, and is written < 0.
9. *Statement*: Six more than a number squared is greater than ten.
   Translation: Let x represent the number. Then *the number squared* is written x^2; and *six more than the number squared* translates to $x^2 + 6$. Thus, the statement translates to $x^2 + 6 > 10$.
   Note: *is greater than* translates to $>$.
10. *Statement*: The sum of a number and that number squared is less than or equal to twelve.
    Translation: Let x represent the number. Then $x + x^2 \leq 12$.
    Note: *is less than or equal to* translates to $\leq$.
11. *Statement*: The product of two consecutive natural numbers is 30.
    Translation: Let x represent the first natural number. Then $x + 1$ represents the next natural number and we have $x(x + 1) = 30$.
    Note: The word *consecutive* translates to *follow each other in order*.
12. *Statement*: The sum of the cubes of two numbers is 29.

Translation: Let x represent one number and y the other. Then $x^3 + y^3 = 29$.

13. *Statement*: The cube of the sum of two numbers is 29.

   Translation: Let the numbers be represented by x and y, respectively. Then their sum is $x + y$, and $x + y^{\frac{3}{}}$ is their sum cubed. Thus, $x + y^{\frac{3}{}} = 29$. *Note*: The *sum of cubes* is not the same as the *cube of the sum*; that is, $x^3 + y^3 \neq x + y^{\frac{3}{}}$.

While it is not possible to give examples of all translations, the following table contains a list of words and phrases that occur quite often.

Table 4.5

English Word or Phrase	Algebraic Translation
Sum, plus, more, increased by, . . .	+
Difference, minus, less, decreased by, . . .	−
Equals, is, results, gives, . . .	=
Times, product, multiplied by, . . .	·
Divide by, quotient, . . .	÷

EXERCISES 4.1(B)

In Exercises 1–30, write an algebraic phrase for the given English phrase.

1. Four times x
2. Six times y
3. One-fifth times a
4. Two-thirds times b
5. The product of 8 and x
6. The product of 7 and y
7. The product of x and y
8. The product of a and b
9. The sum of 4 and x
10. The sum of 6 and y
11. x increased by 27
12. y increased by 14
13. Five decreased by x
14. Ten decreased by y
15. 9 less x
16. 13 less y
17. x less 9
18. y less 13
19. The quotient of x and y
20. The quotient of a and b
21. The sum of x and $x + 4$
22. By how much does $x + 7$ exceed 4?
23. Three less than twice x
24. What number is 10% larger than x?
25. The difference between x and y divided by six
26. Four times the sum of a and b

27. The quotient of x and two increased by the product of x and y.
28. The average of x, y, and z.
29. The square of the sum of a and b.
30. The sum of the squares of a and b, respectively.

In Exercises 31–50, express in algebraic symbols.

31. If Skip is 17 years old, how old was he x years ago?
32. Mark is 21 years old; how old will he be in y years?
33. What was Lucy's age five years ago, if she will be x years old in three years?
34. Represent two consecutive natural numbers starting with:
    (a) 3   (b) x (c) $x + 2$ (d) $4x$
35. Represent two consecutive odd integers starting with:
    (a) -15 (b) x (c) $x - 3$ (d) $2x + 5$
36. The number of days in x weeks and y days.
37. The number of weeks in 3 years.
38. The number of minutes in x hours.
39. If a board, L feet long, is broken into two parts, one of which is three times as long as the

other, how long is the larger piece in terms of L and the length of the shorter piece?

40. The larger of two numbers is four times the smaller. What is their sum?
41. How many cents in x dollars and y dimes?
42. A number decreased by 14 is 32.
43. Five less than twice a number is the same as the number increased by 16.
44. Seven times a number is five greater than the number.

45. The sum of three consecutive whole numbers is 15.
46. The sum of two squared numbers is 13.
47. The square of the sum of two numbers is 13.
48. A number divided by five is greater than or equal to four.
49. One-third of an unknown decreased by seven is negative.
50. Three times the first of four consecutive odd integers is three less than the sum of the third and the fourth.

C. Outline for Solving Word Problems

While there is no foolproof method for solving word problems, the following outline should be helpful.

Outline for Solving Word Problems

1. *Read the problem very carefully*, at least twice. Try and determine the type of problem: uniform motion, age problem, geometric, investment, coin, mixture, ... Even sparsely stated problems usually contain a great deal of information.
2. Examine the problem statement to determine exactly *what is to be found*. Look for phrases such as:

    What is . . . ?   Find the . . .
    How much . . . ?   How long . . . ?
    At what rate . . . ?

3. *Let a variable represent the unknown.* Make sure to specify *units* of measure.
4. *Draw a picture* if appropriate. Label any dimensions such as lengths, widths, distances, angles, and so on, that are given in the problem or that can be expressed in terms of the assigned variable.
5. Express any other unknown quantities in terms of the assigned variable.
6. List any formulas that might be useful.
7. Set up an equation or inequality relating the unknown quantities with the known quantities.
8. Solve the equation.
9. Check that the solution to the equation satisfies the conditions of the stated problem. The equation formed in step 8 may not correctly represent the stated problem. Thus, while the answer may satisfy the equation, it may not satisfy the stated problem.

EXAMPLE 8   A problem says: "Find the rate of the first car." Label the unknown.

Solution   If we choose miles per hour (mph) as our unit, we write:

Let $x =$ the rate in mph of the first car

The following statements are not sufficient for labeling the unknown.

(a) Let $x =$ first car   This does not refer to the *rate* of the car.

(b) Let $x =$ rate of car   Which car?

(c) Let $x =$ rate of first car   Units are missing. ■

EXAMPLE 9   A problem contains the question "How many pounds of coffee A and coffee B must be blended to produce 40 lb of coffee C?"

Solution   Despite the fact that there are two unknown quantities, the number of pounds of coffee A and coffee B, we need only one variable to determine the unknown quantities. If we let x equal the number of pounds of coffee A, then $40 - x$ equals the number of pounds of coffee B. ■

Sometimes the equation can be obtained quite easily from the problem statement.

EXAMPLE 10   Seven times a number, increased by ten, is twenty-four.

Solution   Let $x =$ the number. Then the equation given in the problem statement is:

$$\underbrace{7x}_{\text{seven times a number}} + \underbrace{10}_{\text{increased by ten}} = \underbrace{24}_{\text{is twenty-four}}$$

Solving for x,

$$7x + 10 = 24$$
$$7x = 14$$
$$x = 2$$

The number is 2. ■

However, deriving the equation from the problem statement may require a little more skill.

EXAMPLE 11   A problem contains the following sentence: "A rocket leaves Earth and travels to a distant planet at a rate of 25,000 mph and returns to Earth at a rate of 22,000 mph." The implied equation is:

distance going = distance returning

Frequently, the relationships between the quantities in a problem may not be given explicitly. The reader is expected to know or to be able to find the necessary relationships.

EXAMPLE 12 One of the acute angles in a right triangle is twice the other acute angle. Find the acute angles.

Solution To solve this problem, we need to know that the nonacute angle is 90 degrees and that the sum of the three angles in any triangle (in Euclidean geometry) is 180 degrees.

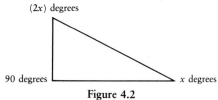

Figure 4.2

If we let x = number of degrees in the smaller acute angle in Figure 4.2,

then $2x$ = number of degrees in the larger acute angle

Thus, we have

90 degrees + x degrees + $(2x)$ degrees = 180 degrees

Solving for x,

$$90 + x + 2x = 180$$
$$3x = 90$$
$$x = 30$$

Hence, the smaller acute angle equals 30 degrees and the larger acute angle equals 60 degrees. ■

As mentioned previously, checking an answer by substituting it into the derived equation is not sufficient. The derived equation may be wrong. Thus, the answer should satisfy the problem statement.

EXAMPLE 13 In solving for the length of a rectangle, represented by x feet, suppose the derived equation is

$$4x + 9 = 7$$

Solving for x, we obtain $x = -\frac{1}{2}$. While $x = -\frac{1}{2}$ satisfies the equation, the answer does not satisfy the problem statement; the length cannot be a negative number.

When checking your answer, be sure to:

1. Check the units. For example, if you are solving for a rate and the answer is given in terms of miles, then you know something is wrong. Units associated with rates are ratios such as mi/hr, ft/sec, gal/min, and so on.
2. In geometric problems, check your answer with the figure. For example, the sum of the angles of a triangle in Euclidean geometry cannot be greater than 180 degrees, measures of distances such as length, width, and height are not negative, and right triangles must satisfy the Pythagorean theorem.
3. Use common sense and your own experience. For example, a man running 700 mph, mixtures containing 150 percent copper, and time being negative are not appropriate answers.

4.2 NUMBER PROBLEMS

EXAMPLE 1 The sum of two numbers is 28, and their difference is 4. Find the numbers.

Solution Let
$$x = \text{the smaller number}$$
Then
$$x + 4 = \text{the larger number}$$

Since the sum of the two numbers is 28, we have

$$x + x + 4 = 28$$

Solving,
$$2x + 4 = 28$$
$$2x = 24$$
$$x = 12$$

Thus, the smaller number is 12, and the larger is $x + 4 = 12 + 4 = 16$. ∎

Check
$$16 + 12 = 28 \quad \text{and} \quad 16 - 12 = 4$$

EXAMPLE 2 The sum of three numbers is 40. If the second number is twice as large as the first and the third number is 5 greater than the second number, find the numbers.

Solution   Let x = the first number

Then $2x$ = the second number

and $2x + 5$ = the third number

The sum being 40 gives the equation

$$x + 2x + (2x + 5) = 40$$

that is, $$5x + 5 = 40$$

$$5x = 35$$

$$x = 7$$

Thus, the first number is 7, the second number is $2x = 2 \cdot 7 = 14$, and the third number is $2x + 5 = 14 + 5 = 19$. ■

Check $$7 + 14 + 19 = 40$$

EXAMPLE 3   Divide 70 into two parts such that one-fourth the larger part exceeds one-half the smaller part by 1.

Solution   Although this problem deals with two parts of 70, we need only one variable. From the problem statement, we know

$$\frac{1}{4} \text{ (larger part)} = \frac{1}{2} \text{ (smaller part)} + 1$$

Let x = the smaller part

Then $70 - x$ = the larger part

and we have $$\frac{1}{4}(70 - x) = \frac{1}{2}(x) + 1$$

Multiplying both sides by 4 gives

$$70 - x = 2x + 4$$

$$66 = 3x$$

$$22 = x$$

Thus, the smaller part is 22 and the larger part is $70 - x = 70 - 22 = 48$. ■

Check $$\frac{1}{4}(48) = 12$$

$$\frac{1}{2}(22) = 11 \quad \text{and} \quad 12 = 11 + 1$$

EXAMPLE 4   Find three consecutive odd integers whose sum is 5 more than twice the largest.

Solution   Let $\qquad\qquad\qquad\qquad\qquad\qquad x =$ the smallest odd integer

Then $\qquad\qquad\qquad\qquad\qquad x + 2 =$ the middle odd integer

and $\qquad\qquad\quad (x + 2) + 2 = x + 4 =$ the largest odd integer

The problem states that

$$(\text{smallest}) + (\text{middle}) + (\text{largest}) = 2(\text{largest}) + 5$$

Substituting the symbols gives the equation:

$$x + (x + 2) + (x + 4) = 2(x + 4) + 5$$
$$3x + 6 = 2x + 8 + 5$$
$$3x + 6 = 2x + 13$$
$$x = 7$$

If $x = 7$, then $x + 2 = 9$ and $x + 4 = 11$. Thus, the three consecutive odd integers are 7, 9, and 11. ■

Check   Their sum is $7 + 9 + 11 = 27$, which is 5 more than twice the largest, $2 \cdot 11 = 22$.

EXAMPLE 5   A number consists of three digits. The middle digit is twice the hundreds digit, and exceeds the units digit by 2. If the number is 28 times the sum of its digits, find the number.

Solution   First we consider a numerical example. The value of the three-digit number 123 can be written $123 = (100) \cdot 1 + (10) \cdot 2 + (1) \cdot 3$, since 1 is in the hundreds position, 2 in the tens position, and 3 in the units position. So, the three-digit number whose digits from left to right are x, y, z can be written $xyz = 100x + 10y + z$.

Since the digits in our problem are related, we can use one variable.

Let $\qquad\qquad\qquad\qquad x =$ the hundreds digit

Then $\qquad\qquad\qquad\quad 2x =$ the tens digit

and $\qquad\qquad\qquad 2x - 2 =$ the units digit

The value of the number consisting of these digits is

$$100 \cdot (x) + 10 \cdot (2x) + (2x - 2)$$

Given that this number is 28 times the sum of the digits, we have the equation

$$100 \cdot (x) + 10 \cdot (2x) + (2x - 2) = 28\{x + 2x + (2x - 2)\}$$

$$100x + 20x + 2x - 2 = 28\{5x - 2\}$$

$$122x - 2 = 140x - 56$$

$$54 = 18x$$

$$3 = x$$

Thus, the hundreds digit is 3, the tens digit is $2x = 2(3) = 6$, and the units digit is $2x - 2 = 2(3) - 2 = 4$. Hence the three-digit number is 364. ■

Check   The sum of the digits is $3 + 6 + 4 = 13$ and $28 \cdot (13) = 364$.

EXAMPLE 6   The sum of the digits of a certain two-digit number is 12. If we reverse the digits and multiply by $\frac{4}{7}$, we get the original number. What is the original number?

Solution   There are four unknown quantities: the units digit, the tens digit, the value of the number as it stands, and the value when the digits are reversed. However, all four can be expressed in terms of the units digit or the tens digit.

Let $x =$ the units digit

Then $12 - x =$ the tens digit

The value of the required number is $10(12 - x) + x$, and the value with digits reversed is $10x + (12 - x)$. The problem statement gives the equation

$$10(12 - x) + x = \frac{4}{7}[10x + (12 - x)]$$

Solving, we obtain

$$10(12 - x) + x = \frac{4}{7}[10x + (12 - x)]$$

$$120 - 9x = \frac{4}{7}[9x + 12]$$

multiplying both sides by 7 gives

$$840 - 63x = 4[9x + 12]$$

$$840 - 63x = 36x + 48$$

$$-99x = -792$$

$$x = 8$$

If $x = 8$, then $12 - x = 4$. Thus, the units digit is 8 and the tens digit is 4. Hence, the required number is 48. ■

EXERCISES 4.2

In Exercises 1–20, find the number(s) satisfying the given condition.

1. If twice a certain number is decreased by 5, the result is 25.
2. If five times a certain number is increased by 4, the result is 29.
3. One-fourth a number decreased by 16 equals 8.
4. Two-thirds of a number increased by 11 is equal to 17.
5. If three times a certain number is increased by 9, the result is 30.
6. Seven times a certain number decreased by 10 is equal to 53.
7. One number exceeds another by 5 and their sum is 33.
8. The difference between two numbers is 8. If 2 is added to the larger, the result will be three times the smaller.
9. Two numbers whose sum is 58 and difference is 28.
10. If 288 is added to a certain number, the result will be equal to three times the amount by which the number exceeds 12.
11. One number is four times another. Three times the larger number decreased by five times the smaller is 21.
12. Find two numbers differing by 10, whose sum is equal to twice their difference.
13. Find three consecutive numbers such that the first increased by twice the second and by three times the third is 98.
14. Find three consecutive numbers whose sum is 84.
15. Find three consecutive even numbers such that five times the middle one shall exceed the sum of the other two by 24.
16. Find a number such that if you multiply it by 3 and take 2 away, the result is five times as great as if you divide the number by 3 and add 2.

17. Find three consecutive numbers such that if three times the largest is added to twice the smallest and the sum decreased by the middle number, the result is 33.
18. Find four consecutive odd integers such that twice the second number added to the third number is 9 less than twice the sum of the other two numbers.
19. The difference between the squares of two consecutive numbers is 27.
20. Find four consecutive numbers such that the product of the first two is 50 less than the product of the other two.
21. The sum of the digits of a two-digit number is 14. If the order of the digits is reversed, then the number is increased by 18. What is the number?
22. The sum of the digits of a two-digit number is 12. When the digits are interchanged, the number increases by 75%. Find the number.
23. In a three-digit number, the tens digit is twice the hundreds digit and the units digit is 8. Find the number if it is given that the number is 4 less than twelve times the sum of its digits.
*24. In a three-digit number, the tens digit is one and one-half times the units digit, and the difference between the tens digit and the units digit is equal to three times the hundreds digit. Find the number if the number is equal to 4 more than twelve times the sum of its digits.
*25. A two-digit number has the units digit 6 more than the tens digit. If the number is added to four times the product of its digits, the result is 3 more than the square of the sum of the digits. Find the number.
*26. The sum of the digits of a certain number of two digits is a. If the order of the digits is reversed, the number is increased by b. What is the number?

4.3 MOTION PROBLEMS (UNIFORM MOTION)

If an automobile travels for 5 hr at a uniform rate of 30 mph, the distance traveled is found by multiplying the rate by the time; that is,

$$30\frac{mi}{hr} \cdot 5 \; hr = 150 \; mi$$

If we travel 300 mi in 6 hr, our average rate of speed is given by dividing the distance by the time; that is,

$$\frac{300 \; mi}{6 \; hr} = 50 \frac{mi}{hr}$$

If we travel 400 mi at 40 mph, the time for the trip is determined by dividing the distance by the rate; that is,

$$400 \; mi \div 40\frac{mi}{hr} = 400 \; mi \cdot \frac{1 \; hr}{40 \; mi} = 10 \; hr$$

These relationships between distance, rate, and time can be represented by the triangle in Figure 4.3.

Figure 4.3

$$distance = rate \cdot time$$

$$rate = \frac{distance}{time} \qquad time = \frac{distance}{rate}$$

When setting up motion problems, there are several questions that should be considered.

1. Are there two distances that are equal?
2. Have the two objects traveled the same distance?
3. Is the distance going equal to the distance returning?
4. Is the sum (difference) of two distances equal to a constant?

Note. The same questions can be asked about time.

EXAMPLE 1 A truck leaves San Francisco for Los Angeles at 1:00 PM. At 2:00 PM a second truck leaves San Francisco heading for Los Angeles and traveling 10 mph faster than the first truck. If the second truck overtakes the first at 5:00 PM, what is the average speed of each truck?

Solution We begin by drawing the diagram in Figure 4.4.

Let d = distance traveled, r = rate (speed) traveled, and t = time traveled. As shown in the diagram, both trucks have traveled the *same distance* by the time the second truck overtakes the first truck. Then

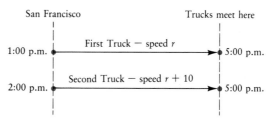

Figure 4.4

time traveled by first truck = 4 hours (5:00 − 1:00)

rate traveled by first truck = r mph

distance traveled by first truck: $d = $ rate $\cdot$ time $= r\dfrac{\text{mi}}{\cancel{\text{hr}}} \cdot 4 \, \cancel{\text{hr}}$

$$= 4r \text{ mi}$$

time traveled by second truck = 3 hr (5:00 − 2:00)

rate traveled by second truck = $(r + 10)$ mph

distance traveled by second truck: $d = (r + 10) \dfrac{\text{mi}}{\cancel{\text{hr}}} \cdot 3 \, \cancel{\text{hr}}$

$$= 3(r + 10) \text{ mi}$$

We can summarize our information in a table:

	r	t	$d = r \cdot t$
1st truck	r mph	4 hr	$4r$ mi
2nd truck	$(r + 10)$ mph	3 hr	$3(r + 10)$ mi

Since the two distances are equal, we have

$$4r \text{ mi} = 3(r + 10) \text{ mi}$$

Solving, we obtain

$$4r = 3(r + 10)$$
$$4r = 3r + 30$$
$$r = 30$$

Then, $r + 10 = 40$. Thus, the first truck travels at 30 mph and the second truck travels at 40 mph. ∎

EXAMPLE 2 A freight train and a passenger train travel toward each other from towns 630 miles apart. The freight train travels at a rate of 60 mph and the passenger train at a rate of 90 mph. If the freight train starts half an hour sooner than the passenger train, how many hours will the passenger train travel before it meets the freight train?

Solution Again, we start with a diagram (see Figure 4.5).

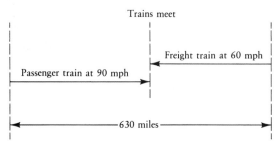

Figure 4.5

Let $\quad t =$ hours the passenger train travels to meet freight train

$$t + \frac{1}{2} = \text{hours the freight train travels to meet passenger train}$$

Then $\quad$ distance passenger train travels $= 90\dfrac{\text{mi}}{\text{hr}} \cdot t \text{ hr} = 90t \text{ mi}$

$$\text{distance freight train travels } = 60\frac{\text{mi}}{\text{hr}} \cdot \left(t + \frac{1}{2}\right)\text{hr}$$

$$= 60\left(t + \frac{1}{2}\right)\text{mi}$$

Summarizing our information in a table, we get

	r	t	$d = r \cdot t$
Passenger train	90 mph	t hr	$90t$ mi
Freight train	60 mph	$\left(t + \frac{1}{2}\right)$ hr	$60\left(t + \frac{1}{2}\right)$ mi

Together the trains travel the entire 630 mi. Thus, the sum of their distances must equal 630 mi. That is,

$$(90t) \text{ mi} + \left\{60\left(t + \frac{1}{2}\right)\right\} \text{ mi} = 630 \text{ mi}$$

Solving, we obtain

$$90t + 60\left(t + \frac{1}{2}\right) = 630$$

$$90t + 60t + 30 = 630$$
$$150t = 600$$
$$t = 4$$

The passenger train travels for 4 hr and the freight train travels for $4\frac{1}{2}$ hr before they meet. ■

EXAMPLE 3   A boat travels 60 km upstream at 15 km/hr and returns to its dock at 20 km/hr. Find the average speed for the round trip.

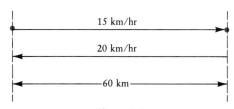

Figure 4.6

Solution   We first draw the diagram in Figure 4.6.

Let r = average speed in km/hr

Then   time upstream = $\dfrac{\text{distance}}{\text{rate}} = \dfrac{60 \text{ km}}{15 \text{ km/hr}} = 4$ hr

time downstream = $\dfrac{60 \text{ km}}{20 \text{ km/hr}} = 3$ hr

Then,

total time = time upstream + time downstream
$$= 4 \text{ hr} + 3 \text{ hr}$$
$$= 7 \text{ hr}$$

and

total distance = distance upstream + distance downstream
$$= 60 \text{ km} + 60 \text{ km}$$
$$= 120 \text{ km}$$

Thus,

average speed = $\dfrac{\text{total distance}}{\text{total time}} = \dfrac{120 \text{ km}}{7 \text{ hr}} = 17\dfrac{1}{7} \dfrac{\text{km}}{\text{hr}}$ ■

Note. The average speed is *not* the average of the speeds,

$$17\frac{1}{7}\frac{\text{km}}{\text{hr}} \neq \frac{15\frac{\text{km}}{\text{hr}} + 20\frac{\text{km}}{\text{hr}}}{2} = 17\frac{1}{2}\frac{\text{km}}{\text{hr}}$$

EXAMPLE 4 A bus traveling at 50 mph leaves Oakland at 8:00 AM heading for Seattle. At the same time, another bus leaves Seattle heading for Oakland at a rate of 40 mph. How far from Oakland do the two buses pass each other, going in opposite directions, if the distance between Oakland and Seattle is 900 miles?

Solution We first draw the diagram in Figure 4.7. Since both buses leave at the same time, the amount of time traveled is the same for both buses.

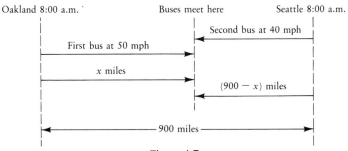

Figure 4.7

Let $\qquad x$ = number of miles traveled by the first bus

Then $\qquad (900 - x)$ = number of miles traveled by the second bus

time for the first bus is t = $\dfrac{x \text{ mi}}{50 \text{ mi/hr}} = \dfrac{x}{50}\text{hr}$

time for the second bus is $t = \dfrac{(900 - x) \text{ mi}}{40 \text{ mi/hr}} = \dfrac{(900 - x)}{40}\text{ hr}$

	d	r	$t = \dfrac{d}{r}$
First bus	x mi	50 mph	$\dfrac{x}{50}$ hr
Second bus	$(90 - x)$ mi	40 mph	$\dfrac{(900 - x)}{40}$ hr

Equating times gives

$$\frac{x}{50}\text{ hr} = \frac{900 - x}{40}\text{ hr}$$

Solving, we obtain

$$\frac{x}{50} = \frac{900}{40} - \frac{x}{40}$$

$$\frac{x}{50} + \frac{x}{40} = \frac{900}{40}$$

$$90x = 45,000$$

$$x = 500$$

Thus, the buses meet 500 miles from Oakland. ∎

EXERCISES 4.3

1. A plane flies 980 km in 3.5 hr against a head-wind blowing at 20 km/hr. How fast would the plane be flying in still air?

2. Two women travel in opposite directions, starting from the same place, and at the same time. The first travels twice as many miles an hour as the second. If at the end of 5 hr they are 45 mi apart, how many miles per hour does each travel?

3. Two hikers, Gilbert and Linda, start from the same place and travel in the same direction. Gilbert has a 6 hr headstart. If Gilbert travels 2 mph and Linda travels 4 mph, how long will it take Linda to overtake Gilbert?

4. Two rowers are a certain distance apart. They travel toward each other, one starting 3 hr before the other and rowing at a rate of 4 mph. The second rower travels at a rate of 6 mph. When the rowers meet, it is found that each has traveled the same distance. How long did each travel?

5. Pete and Lynn travel on bicycles from the same place, in opposite directions, Pete traveling 4 mph faster than Lynn. After 5 hr, they are 120 mi apart. Find the rate of each biker.

6. A postal truck leaves its station and heads for Chicago, averaging 40 mph. An error in the mailing schedule is spotted, and 24 min after the truck leaves, a car is sent to overtake the truck. If the car averages 50 mph, how long will it take to catch the postal truck?

7. A helicopter leaves the deck of a destroyer and travels east at a rate of 128 mph. The carrier travels east at a rate of 28 mph. When will the helicopter pass out of range of communication with the destroyer if the helicopter's radio has a range of 500 mi?

8. A speed boat can cover a distance in 4 hr with the wind, but can travel only two-thirds of the way back in the same time. If the speed boat travels 110 mph in still air, what is the wind velocity?

9. A sailing vessel making 4 mph has been at sea 12 hr when she is pursued by a motor-boat traveling at 20 mph. How far behind is the motorboat at the end of $2\frac{1}{2}$ hr?

10. Mark Willard drove 3 hr and 20 min at a uniform speed. Engine trouble forced him to drive at two-fifths of his original speed for an hour. At the end of this time, he had driven a distance of 168 mi. What was Willard's original speed?

11. A pilot flies from New York to Boston against a 20 mph headwind, and returns to New York with the aid of a 20 mph tailwind. Her plane can go 100 mph in still air, and the total trip took 5 hr flying time. Find the distance between New York and Boston.

*12. Wayne ran the 440-yd race in 59.2 sec, and Bruce finished second with a time of 60.4 sec. Assuming that both ran at a uniform speed throughout the race, how far back was Bruce when Wayne crossed the finish line?

4.4 AGE PROBLEMS

EXAMPLE 1   A father is six times as old as his son, and in four years he will be four times as old. What are their present ages?

Solution   Let $\qquad\qquad\qquad x =$ the son's present age in years

Then $\qquad\qquad\qquad 6x =$ the father's present age in years

In four years, their respective ages will be:

$$x + 4 = \text{son's age in years, after four years}$$

and $\qquad\qquad 6x + 4 =$ father's age in years, after four years

From the problem statement, we know that in four years the father will be four times the son's age. That is,

$$6x + 4 = 4(x + 4)$$
$$6x + 4 = 4x + 16$$
$$2x = 12$$
$$x = 6$$

Thus, the son's present age is 6 years and the father's age is $(6x)$ years = 36 years.   ■

Check   If the father is presently 36 years old and the son is 6 years old, in four years they will be 40 years old and 10 years old, respectively. At that time, the father's age will be four times the age of the son; that is, 40 years = $4 \cdot (10 \text{ years})$.

EXAMPLE 2   Kim is now one-half the age of Brad. Seven years ago the sum of their ages was 28. What are their present ages?

Solution   The following table may be useful in solving this example.

	Kim	Brad	Equation
If Kim is now x years,	x years	$(2x)$ years	
then 7 years ago	$(x - 7)$ years	$(2x - 7)$ years	$(x - 7)$ years $+ (2x - 7)$ years $= 28$ years

Thus, we solve the equation:

$$(x - 7) + (2x - 7) = 28$$
$$3x - 14 = 28$$
$$3x = 42$$
$$x = 14$$

Hence, Kim is 14 years old and Brad is 28 years old. ■

Check   If Kim is 14 years old now, then he was 7 years old seven years ago, and if Brad is presently 28 years old, then he was 21 years old seven years ago. At that time, the sum of their ages was 7 years + 21 years = 28 years.

EXERCISES 4.4

1. The sum of Bob's age and Ed's age is 49 years. Four years from now Bob will be twice as old as Ed. Find their present ages.
2. Mary is twice as old as Jim, and seven years ago the sum of their ages was equal to Jim's present age. Find their present ages.
3. A father is four times as old as his son. In 24 years he will only be twice as old. Find their ages.
4. David is 25 years older than Skip, and David's age is as much above 20 as Skip's age is below 85. Find their ages.
5. A man has four children, each child being three years older than the next one. If the sum of their ages is 38, how old is the eldest?
6. A boy is half as old as his sister. Three years ago the girl was three and one-half times as old as her brother. What are their present ages?
7. Kathy's age is one-fourth that of Sherry. In six years Sherry will be twice as old as Kathy. How old is Kathy?
8. Tad is 11 years younger than Ric, and in two years Tad will be half as old as Ric. How old is Tad?
9. Ron is two years older than Pam and five years older than Adrienne. In five years the sum of the ages of Pam and Adrienne will be one and one-half times that of Ron. Find their present ages.
10. In seven years, a mother will be three times as old as her daughter. If she is now five and one-third times as old as her daughter, what is the mother's present age?

11. A man is 30 years older than his son, and six years ago he was six times as old as his son. How long is it since he was seven times as old as his son?
12. A mother is presently four times as old as her son. If both she and her son live 20 years longer, she will then be twice as old as her son. What are their present ages, and in how many years will the mother be three times as old as the son?
13. A man's age is twice the sum of the ages of his two sons, one of whom is three years older than the other. In seven years the sum of the son's ages will be five-sevenths of their father's age. Find their present ages.
14. A mother's present age is twice the sum of the ages of her son and daughter. The son is three years older than the daughter. Six years ago the mother was six times as old as her son. Find their present ages.
*15. At present a father's age is five years more than three times the age of his son. Four years ago the father was four times as old as his son. Find their present ages; find also how long it will be until the son is one-half as old as his father.
*16. Two persons M and N are x and y years old, respectively. Is there a time when M was or when M will be s times as old as N, and if so, when? Discuss the result for various values of x, y, and s.

4.5 COIN, STAMP, AND CURRENCY PROBLEMS

In dealing with coin, stamp, and currency problems, we must distinguish between the **number** of coins, stamps, and currencies, and their **value**.

EXAMPLE 1

(a) 6 nickels have a value of $6 \cdot (5 \text{ cents}) = 30$ cents

(b) 4 dimes have a value of $4 \cdot (10 \text{ cents}) = 40$ cents

(c) 3 quarters have a value of $3 \cdot (25 \text{ cents}) = 75$ cents

(d) 2 dollars have a value of $2 \cdot (100 \text{ cents}) = 200$ cents

(e) 5 (3-cent stamps) have a value of 15 cents

In general, for coins and currency, we have

x nickels have a value of $5x$ cents
x dimes have a value of $10x$ cents
x quarters have a value of $25x$ cents
x dollars have a value of $100x$ cents
$\vdots$

When working with coins, stamps, or currency, there are two important ideas to remember.

1. $\left(\begin{array}{c} \text{number of} \\ \text{coins} \end{array} \right) \cdot \left(\begin{array}{c} \text{number of} \\ \text{cents in} \\ \text{each coin} \end{array} \right) = \left(\begin{array}{c} \text{value of} \\ \text{coins in cents} \end{array} \right)$

2. Express all denominations in the same units.

EXAMPLE 2   A toy savings bank holds nickels, dimes, and quarters. If it contains four times as many dimes as nickels, and three more quarters than dimes, and if the total value of its contents is $6.55, how many of each coin are there in the bank?

Solution   Let

x = the number of nickels

Then

$4x$ = the number of dimes

and

$4x + 3$ = the number of quarters

Putting our information in the form of a table, we get:

	Nickels	Dimes	Quarters	Collection
Number of coins	x	$4x$	$4x + 3$	
Value of each coin	5 cents	10 cents	25 cents	
Total value	$(5x)$ cents	$(40x)$ cents	$(4x + 3)25$ cents	655 cents

Equating values, we have

$$\left(\begin{matrix}\text{value of}\\\text{nickels}\end{matrix}\right) + \left(\begin{matrix}\text{value of}\\\text{dimes}\end{matrix}\right) + \left(\begin{matrix}\text{value of}\\\text{quarters}\end{matrix}\right) = \left(\begin{matrix}\text{total}\\\text{value}\end{matrix}\right)$$

That is,

$$(5x) \text{ cents} + [(4x)(10)] \text{ cents} + [(4x + 3)(25)] \text{ cents} = 655 \text{ cents}$$
$$(5x) \text{ cents} + \quad (40x) \text{ cents} + \quad (100x + 75) \text{ cents} = 655 \text{ cents}$$

Solving for x, we obtain

$$5x + 40x + 100x + 75 = 655$$
$$145x + 75 = 655$$
$$145x = 580$$
$$x = 4$$

If $x = 4$, then $4x = 16$ and $4x + 3 = 19$; hence, we have 4 nickels, 16 dimes, and 19 quarters. ■

Check

$$\begin{aligned}4 \text{ nickels} &= \quad 20 \text{ cents} = \$0.20\\16 \text{ dimes} &= 160 \text{ cents} = \quad 1.60\\19 \text{ quarters} &= 475 \text{ cents} = \quad \underline{4.75}\\&\qquad\qquad\qquad\qquad \$6.55\end{aligned}$$

EXAMPLE 3 A special delivery letter requires \$2.20 postage. If we have only 20¢ stamps and 15¢ stamps, and if we use just 12 stamps, then how many of each kind of stamp is required?

Solution Let

$$x = \text{number of 20¢ stamps required}$$

Then

$$12 - x = \text{number of 15¢ stamps required}$$

In table form, we have:

	20¢ Stamps	15¢ Stamps	Collection
Number of stamps	x	$12 - x$	
Value of each stamp	20 cents	15 cents	
Total value	$(20x)$ cents	$(12 - x)15$ cents	220 cents

Then

$$(\text{value of 20¢ stamps}) + (\text{value of 15¢ stamps}) = \$2.20$$

That is,

$$(x)(20 \text{ cents}) + (12 - x)(15 \text{ cents}) = (2.20)(100 \text{ cents})$$

$$(20x) \text{ cents} + [(12 - x)(15)] \text{ cents} = 220 \text{ cents}$$

Solving for x, we obtain

$$20x + (12 - x)15 = 220$$

$$20x + 180 - 15x = 220$$

$$5x = 40$$

$$x = 8$$

If $x = 8$, then $12 - x = 4$; hence, we have

$$8 \text{ required 20¢ stamps}$$

$$4 \text{ required 15¢ stamps} \quad \blacksquare$$

Check $(8)(20 \text{ cents}) = 160 \text{ cents} = \$1.60$

$(4)(15 \text{ cents}) = 60 \text{ cents} = \underline{0.60}$

$\phantom{(4)(15 \text{ cents}) = 60 \text{ cents} = }\$2.20$

EXAMPLE 4   The admission price at a rock concert is \$6.00 for the balcony and \$8.50 for loge. For one evening performance, the cashier's receipts were \$3409. If the cashier sold 479 tickets, how many of each kind of ticket did he sell?

Solution   Let $ x = $ the number of \$6 tickets sold

Then $ (479 - x) = $ the number of \$8.50 tickets sold

	\$6 tickets	\$8.50 tickets	all tickets
Number of tickets	x	$479 - x$	
Value of each ticket	6 dollars	8.5 dollars	
Total value	$(6x)$ dollars	$(479 - x)8.5$ dollars	3409 dollars

Then

(receipts from $6 tickets) + (receipts from $8.50 tickets) = $3409

That is,

$(6x)$ dollars + $[(479 - x)8.5]$ dollars = 3409 dollars

Solving for x, we obtain

$$6x + (479 - x)8.5 = 3409$$
$$6x + 4071.5 - 8.5x = 3409$$
$$-2.5x = -662.5$$
$$x = 265$$

If $x = 265$, then $479 - x = 214$; hence, the cashier sold

265 tickets at $6

214 tickets at $8.50.   ■

Check

$$(265)(\$6) = \$1590$$
$$(214)(\$8.50) = \underline{\$1819}$$
$$= \$3409$$

EXERCISES 4.5

1. A jar contains $4.45 in dimes and quarters. If there are 8 less dimes than quarters, then how many coins of each type does the jar contain?
2. Craig has $2.25 in nickels and dimes. If there are 6 more dimes than nickels, then how many coins of each type does Craig have?
3. A savings bank contains nickels, dimes, and quarters. If the number of dimes exceeds twice the number of nickels by 3 and the number of quarters is 4 less than five times the number of nickels, and if the total value of the coins is $17.30, then how many of each coin are in the bank?
4. Scott pays a $2.30 bill with nickels, dimes, and quarters. He uses three times as many quarters as nickels, and five more dimes than nickels. How many coins of each kind does he use to pay the bill?
5. The local movie theater charges an admission fee of $5 for adults and $2.50 for children. One day's gross is $265. If the number of adult tickets sold was 100 less than children

tickets, find the number of adult tickets and children tickets sold.
6. Jack bought 80 stamps for $12.80. Some were 4¢ stamps and some were 20¢ stamps. How many of each kind did he buy?
7. Mrs. Launer asked Daniel to deposit $990 in the bank for her. There were exactly 70 bills, consisting of $10 bills and $20 bills. How many of each kind of bill did he deposit?
8. A football game played for charity has tickets selling for $5, $10, and $25. The hope is to raise $2415 from the sale of tickets. If there are twice as many $5 tickets as $25 tickets and five more $10 tickets than $5 tickets, how many tickets of each type are there?
9. Is it possible to have $4.50 in dimes and quarters and have three times as many quarters as dimes?
10. Is it possible to spend exactly $4.52 for 50 stamps consisting of 4¢ stamps and 10¢ stamps?

4.6 INVESTMENT PROBLEMS

When we borrow money, we pay a fee called interest for the use of the money. When we lend money, we receive interest. Many of the transactions that occur in business deal with borrowing and lending money, each with a stipulated rate of interest. The rate of interest can be viewed as a measure of the cost of borrowing money or a measure of the return of lending money. Consider the following examples.

EXAMPLE 1 If $12,000 is invested at an annual interest rate of 6%, then the simple interest at year's end will be

$$I = (\$12,000) \cdot (.06) = \$720$$

In general,

$$\text{simple interest} = (\text{principal})(\text{rate})(\text{time})$$

$$\text{or}$$

$$I = P \cdot r \cdot t$$

where I is the simple interest, P is the principal (the amount invested), r is the rate of interest per period, and t is the number of periods. If the period $t = 1$ year, then $I = P \cdot r$, where r is the annual interest rate (yearly rate).

EXAMPLE 2 A teacher invests part of $15,000 at 8% annual interest rate and the remainder at 6% annual interest rate. If she expects a total return of $1105 a year on her investments, how much has she invested at each rate?

Solution Let $\qquad x$ = the amount in dollars invested at 8%

Then $\qquad 15,000 - x$ = the amount in dollars invested at 6%

The relationship used to set up an equation is:

$$(8\% \text{ investment return}) + (6\% \text{ investment return}) = \text{total return}$$

We now calculate the investment returns:

	Principal	• Rate	• Time	= Interest
8% portion	x dollars	.08/year	1 year	$0.08x$ dollars
6% portion	$(15,000 - x)$ dollars	.06/year	1 year	$0.06(15,000-x)$ dollars

Since the total interest is the sum of the interest from the two parts, we have

$$0.08x \text{ dollars} + 0.06(15{,}000 - x) \text{ dollars} = 1105 \text{ dollars}$$

Solving for x, we obtain

$$0.08x + 0.06(15{,}000 - x) = 1105$$
$$0.08x + 900 - 0.06x = 1105$$
$$0.02x = 205$$
$$2x = 20{,}500$$
$$x = 10{,}250$$

If $x = 10{,}250$, then $15{,}000 - x = 4750$. Therefore, $10{,}250$ was invested at 8%, and $4750 was invested at 6%. ■

Check   The simple interest on $10,250 for one year at 8% is $(0.08)(\$10{,}250) = \$820$. The simple interest on $4750 for one year at 6% is $(0.06)(\$4750) = \$285$. Total interest $= \$820 + \$285 = \$1105$.

EXAMPLE 3   C. T. Draper has $20,000 to invest. He invests $8000 at 6% and $7000 at 4%. At what rate should he invest the remainder in order to have a total yearly income of $1000?

Solution   Let $x =$ the rate of interest to be earned on the remaining investment. Consider the following table.

	Principal	•	Rate	•	Time	=	Simple Interest
Portion at 6%	$8000		.06/year		1 year		$480.00
Portion at 4%	$7000		.04/year		1 year		$280.00
Portion at x%	$5000		x/year		1 year		$(5000x)

Since the total interest is to be $1000, we have

$$\$480 + \$280 + \$5000x = \$1000$$

Solving, we obtain

$$480 + 280 + 5000x = 1000$$
$$760 + 5000x = 1000$$
$$5000x = 240$$
$$x = 0.048$$

Thus, the remainder ($5000) should be invested at a rate of 4.8% in order to have a total yearly income of $1000. ■

Check 4.8% of $5000 = $240 and $480 + $280 + $240 = $1000.

EXAMPLE 4 A digital clock cost a dealer $204. At what price should he mark the watch so that he can give a discount of 15% from the marked price and still make a profit of 20% on the selling price?

Solution Let x = the marked price in dollars

Then $x - .15x = .85x$ = the selling price in dollars

For this problem we use the relationship

$$\text{Selling price} = \text{cost} + \text{profit}$$
$$.85x \text{ dollars} = 204 \text{ dollars} + [(.20)(.85x)] \text{ dollars}$$

Solving, we obtain

$$.85x = 204 + [(.20)(.85x)]$$
$$.85x = 204 + .17x$$
$$.68x = 204$$
$$x = 300$$

Thus, the marked price should be $300. ■

Check The marked price = $300 and the selling price is 15% less:

$$\text{selling price} = \$300 - [(15\%)(\$300)] = \$300 - \$45 = \$255$$
Since $\quad \text{profit} = \text{selling price} - \text{cost}$
$$= \$255 - \$204$$
$$= \$51$$

And the profit of $51 is 20% of the selling price of $255.

EXERCISES 4.6

1. A woman invested $8500, part at 4% and part at 5%. Her yearly income on the investments was $400. How much did she invest at each rate?
2. A man had $12,000 invested, part at 5% and part at 6%. If his yearly income from these

investments was $680, how much had he invested at each rate?
3. Mr. Carlson has $3000 invested at 8% and $4000 at 7%. How much must he invest at 5% to make his total annual income be 6% of his total investment?

4. From three investments, Mrs. Beach receives $224 interest. The amount invested at 6% is $1000 more than the amount invested at 7%, and the amount invested at 5% is $400 more than the amount invested at 7%. How much did she invest at each rate?

5. George inherits $5200 and invests part of it at 6% and the remainder at 8%. If the total yearly income from both investments is $362, what is the amount of each investment?

6. Dr. Bell's $4000 invested in stocks yields her 2% more than her $3200 invested in bonds. If her total income is $440, what is the rate of interest on each investment?

7. Mr. Hill has invested $6000 at one rate of interest, another $4000 at half that rate. If his yearly income is $360, what are the two rates?

8. Shirley divides $6800 between two investments. On the first she gains 10% and on the second she loses 5%. If her net gain is $305, how much does she invest in each investment?

9. Mrs. Smith had invested part of her $6000 at 3% and the remainder at 5%. A reversal of these investments would have resulted in a return of $48 less. How much did she invest at 3%?

10. Ken invests $550 at 6% simple interest, while Ron invests $600 at $4\frac{2}{3}$% simple interest. How many years will it take for the principal plus the accrued interest to be the same in each case?

11. A chair cost a furniture dealer $26. What must her selling price be to give her 20% of the selling price as overhead and 15% profit?

12. At a discount store, the merchant wishes to give a discount of 10% of his marked price. If a belt cost him $3.25, what must his marked price be in order that he receive a margin of 25% of his marked price?

13. A dealer paid $1500 for a diamond. How much must he receive for it in order to cover its cost, $50 for registration fee, 10% of cost as profit, and 15% of the selling price as overhead?

14. Carol paid $11,500 for a lot 3 years ago. Now she wishes to sell it at a price that will return her the cost, simple interest at 5% for three years on the cost, a profit of 5% of the selling price, and enough additional to pay 10% of the selling price to the agent who makes the sale. What must the selling price be?

15. Adrienne pays $684 yearly on borrowed money for which she used her house as collateral. She borrowed part through a first mortgage at 13% and the rest, which is $3000 less than the first part, by means of a second mortgage at 11%. How much did she borrow on the first mortgage?

*16. A man invested p dollars by paying a certain amount for a piece of land, fives times as much for an apartment, and investing m dollars less in stocks than he paid for the land and the apartment. How much did he invest in each?

*17. If P dollars are to be invested, part at 8% and part at 5%, how much must be invested at each rate so that the total income will be 6% of the P dollars?

*18. A woman has P dollars invested at r%, and Q dollars invested at s%. How many dollars must she invest at t% to make her income n% of her investment?

4.7 MIXTURE PROBLEMS

Many applications from chemistry, medicine, nursing, pharmacy, and business involve what are called mixture problems. These applications involve the blending of ingredients of different cost to obtain a specified mixture, or the mixing of solutions containing different concentrations of ingredients to obtain a solution containing a specified concentration of ingredients.

Mixture Problems Involving Cost

Many problems deal with the mixing of ingredients that have different cost. For instance, we might mix an inexpensive tea with an expensive tea to obtain a blend. In order to sell the new blend, we must be able to find the cost of the blend before we can set a selling price.

These types of mixture problems can be solved using a few simple principles:

1. The sum of the weights of each ingredient must equal the weight of the obtained mixture.
2. The cost of each ingredient in the mixture is its price per unit weight multiplied by its weight in the mixture.
3. The value of the mixture is its total cost divided by its total weight.

EXAMPLE 1 An herbalist has some tea worth $1.40 a pound and some worth $1.60 a pound. How much of each must he mix together to produce a blended mixture of 40 pounds worth $1.52 a pound?

Solution Let

$$x = \text{the number of pounds of \$1.40 tea}$$

Then

$$40 - x = \text{the number of pounds of \$1.60 tea}$$

A diagram such as in Figure 4.8 is helpful:

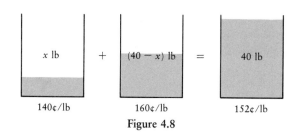

Figure 4.8

In solving these problems, we express the value of each ingredient in the same unit of money, such as cents.

$$\text{Value of \$1.40 tea} = 140\,\frac{\text{cents}}{\text{lb}} \cdot x\,\text{lb} = 140x \text{ cents}$$

$$\text{Value of \$1.60 tea} = 160\,\frac{\text{cents}}{\text{lb}} \cdot (40 - x)\,\text{lb} = 160(40 - x) \text{ cents}$$

$$\text{Value of blended mixture} = 152\,\frac{\text{cents}}{\text{lb}} \cdot 40\,\text{lb} = 6080 \text{ cents}$$

Using the principle

$$\begin{bmatrix} \text{value of \$1.40 tea} \\ \text{in the mixture} \end{bmatrix} + \begin{bmatrix} \text{value of \$1.60 tea} \\ \text{in the mixture} \end{bmatrix} = \begin{bmatrix} \text{value of} \\ \text{total mixture} \end{bmatrix}$$

we obtain

$$140x \text{ cents} + [160(40 - x)] \text{ cents} = 6080 \text{ cents}$$

Solving for x,

$$140x + 160(40 - x) = 6080$$
$$140x + 6400 - 160x = 6080$$
$$-20x = -320$$
$$x = 16$$

If $x = 16$, then $40 - x = 24$. Thus, the mixture should be 16 lb of $1.40 tea and 24 lb of $1.60 tea. ■

Check   (a) 16 lb + 24 lb = 40 lb

(b) 16 ɭb $\cdot$ 1.40 $\dfrac{\text{dollars}}{\text{ɭb}}$ = 22.40 dollars

24 ɭb $\cdot$ 1.60 $\dfrac{\text{dollars}}{\text{ɭb}}$ = $\dfrac{38.40 \text{ dollars}}{60.80 \text{ dollars}}$ = total cost

(c) $\dfrac{60.80 \text{ dollars}}{40 \text{ lb}}$ = 1.52 $\dfrac{\text{dollars}}{\text{lb}}$

Mixture Problems Involving Percentages

Some mixture problems involve solutions containing different concentrations of ingredients. The principles used in mixture problems involving percentages are:

1. Number of units of solution times the percentage of an ingredient A equals the number of units of ingredient A.
2. The number of units in the mixture equals the sum of the number of units in each solution.
3. The number of units of ingredient A in the mixture equals the sum of the number of units of ingredient A in each solution.

EXAMPLE 2   A biochemist has 30 oz of 4% acid solution. How much 20% acid solution must she add to bring the total solution up to 10% strength?

Solution   Let x = the number of ounces of 20% solution to be added

Then $(30 + x)$ = the number of ounces in the resulting solution

Before solving, draw a diagram such as in Figure 4.9.

4% of 30 oz = the number of ounces of pure acid in the original
4% solution

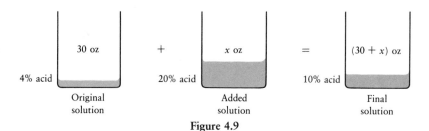

Figure 4.9

20% of x oz = the number of ounces of pure acid that will be added

10% of (30 + x) oz = the number of ounces of pure acid in the final
10% solution

Using the relationship

$$\begin{pmatrix} \text{pure acid in} \\ \text{original solution} \end{pmatrix} + \begin{pmatrix} \text{pure acid in} \\ \text{added solution} \end{pmatrix} = \begin{pmatrix} \text{pure acid in} \\ \text{resulting solution} \end{pmatrix}$$

we obtain

$$(.04)(30) \text{ oz} + (.20)(x) \text{ oz} = (.10)(30 + x) \text{ oz}$$

Solving for x,

$$(.04)(30) + (.20)(x) = (.10)(30 + x)$$
$$1.2 + .20x = 3 + .10x$$
$$.10x = 1.80$$
$$x = 18$$

Thus, 18 oz of the 20% acid solution must be added. ■

Check   4% of 30 oz = 1.2 oz pure acid in original solution

20% of 18 oz = 1.6 oz pure acid in added solution

Adding these two, we get 1.2 oz + 3.6 oz = 4.8 oz in the final solution; and
(4.8 oz) ÷ (48 oz) = .10 = 10%, the strength of the final solution.

EXAMPLE 3   Chemically pure hydrochloric acid contains about 5% water. How much water
must be added to 1 gal of chemically pure hydrochloric acid to make a solution
that contains 50% acid?

Solution   Let x = the number of gallons of water to be added

Then   (x + 1) = the number of gallons in the resulting solution

A diagram such as in Figure 4.10 is helpful.

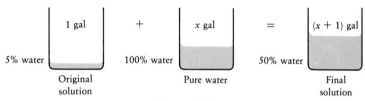

Figure 4.10

5% of 1 gal = the number of gallons of water in original solution

100% of x gal = the number of gallons of water in added solution

50% of $(x + 1)$ gal = the number of gallons of water in final solution

Using the relationship

$$\left(\begin{array}{c} \text{water in} \\ \text{original solution} \end{array}\right) + \left(\begin{array}{c} \text{water in} \\ \text{added solution} \end{array}\right) = \left(\begin{array}{c} \text{water in} \\ \text{final solution} \end{array}\right)$$

we get

$$(.05)(1) \text{ gal} + (1.00)(x) \text{ gal} = (.50)(x + 1) \text{ gal}$$

Solving for x,

$$(.05)(1) + (1.00)x = (.50)(x + 1)$$
$$.05 + x = .50x + .50$$
$$.50x = .45$$
$$x = .9$$

Thus, .9 gal of water must be added. ■

Check

$$5\% \text{ of } 1 \text{ gal} = .05 \text{ gal of water in original solution}$$
$$100\% \text{ of } .9 \text{ gal} = .9 \text{ gal of water in added solution}$$

Adding, we get .05 gal + .9 gal = .95 gal of water in final solution; and
(.95 gal) ÷ (1.9 gal) = .50 = 50%, the strength of water in final solution.
Thus, 100% − 50% = 50% is the strength of the hydrochloric acid in the final
solution.

EXAMPLE 4 A doctor has 40 cc of a 2% tincture of iodine. If the iodine is boiled, alcohol
is evaporated away and the strength of the tincture is raised. How much alcohol
must be boiled away in order to raise the strength of the tincture to 8%?

Solution Since we have evaporation occurring in this problem, we shall be using subtraction.

Let $\quad\quad\quad\quad\quad x =$ the number of cc of alcohol evaporated

Then $\quad\quad (40 - x) =$ the number of cc in the resulting solution

A diagram such as in Figure 4.11 is helpful.

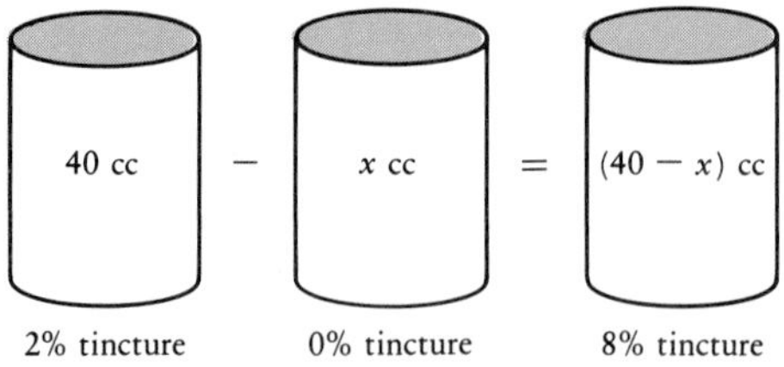

$$\underset{\text{2\% tincture}}{\boxed{40 \text{ cc}}} \quad - \quad \underset{\text{0\% tincture}}{\boxed{x \text{ cc}}} \quad = \quad \underset{\text{8\% tincture}}{\boxed{(40 - x) \text{ cc}}}$$

Figure 4.11

2% of 40 cc = the number of cc of iodine in original solution

0% of x cc = the number of cc of iodine in the alcohol
being boiled away

8% of $(40 - x)$ cc = the number of cc of iodine in the resulting solution

Using the relationship

$$\left(\begin{array}{c}\text{iodine in}\\\text{original solution}\end{array}\right) - \left(\begin{array}{c}\text{iodine in}\\\text{evaporating solution}\end{array}\right) = \left(\begin{array}{c}\text{iodine in}\\\text{resulting solution}\end{array}\right)$$

we get

$$(.02)(40) \text{ cc} - (.00)(x) \text{ cc} = (.08)(40 - x) \text{ cc}$$

Solving for x,

$$(.02)(40) - (.00)(x) = (.08)(40 - x)$$
$$.80 = 3.2 - .08x$$
$$.08x = 2.4$$
$$x = 30$$

Thus, 30 cc of alcohol must be evaporated in order to raise the strength of the tincture to 8%. ■

Check In the original tincture we had

$$(.02)(40) \text{ cc} = .80 \text{ cc of iodine}$$

After evaporating 30 cc of alcohol, we have 10 cc of solution left.

$$\frac{.80 \text{ cc}}{10 \text{ cc}} = .080 = 8\% \text{ concentration}$$

EXERCISES 4.7

1. How many pounds each of 35¢ candy and 50¢ candy must a grocer mix to make 100 pounds to sell at 40¢ per pound?
2. How much alcohol must be added to 4 gal of a 30% mixture of alcohol and water to make a 50% mixture?
3. How many pounds of copper must be added to 24 lb of an 80% copper alloy to produce an alloy that is 90% copper?
4. A storekeeper has two kinds of cookies, one worth 45¢ a pound and the other worth 65¢ a pound. How many pounds of each should she use to make a mixture of 60 lb selling for 57¢ a pound?
5. A radiator contains $7\frac{1}{2}$ gal of a 25% mixture of alcohol and water. How much of the mixture must be drained off and replaced by alcohol so that the resulting mixture will be a 50% mixture?
6. An alloy of copper and zinc weighing 40 lb contains 4.5 lb of zinc. How much zinc must be added to the alloy in order that one-sixth of the mass will be zinc?
7. A solution contains 30% acid. How much water must be added to 30 gal of it so that it will contain 20% acid?
8. How much distilled water must be added to 1 pint of a medical solution that contains 20% alcohol so that the resulting mixture will contain 15% of alcohol?
9. How much cream containing 15% butterfat may be removed from 100 lb of raw milk containing 4% butterfat so that the remaining mixture contains 3% butterfat?
10. If 4 gal of hot water and 3 gal of 10°C water are mixed together, the temperature of the mixture turns out to be 40°C. Find the temperature of the hot water used for the mixture.
*11. In manufacturing a certain kind of wine, three kinds of grapes are used. The first kind contains 35% more sugar than the third kind, and the second kind contains 20% more sugar than the third. How much of the first kind must be added to 100 lb of the third kind and 75 lb of the second kind in order to obtain a mixture containing 25% more sugar than the third kind?
*12. If a certain quantity of water is added to a quart of a solution, it contains 30% alcohol; if twice this quantity of water is added, it contains 20% alcohol. How much water is added the first time, the second time, and what percentage of alcohol was contained in the original solution?

4.8 GEOMETRIC PROBLEMS

The following table contains some useful facts from plane geometry.

Table 4.6. Facts From Plane Geometry

1. Complementary angles A and B
$$m\angle A + m\angle B = 90°$$

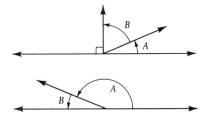

Supplementary angles A and B
$$m\angle A + m\angle B = 180°$$

2. Square with side of length a
Perimeter: $P = 4a$
Area: $A = a^2$

---

3. Rectangle with length a and width b
Perimeter: $P = 2(a + b)$
Area: $A = ab$

---

4. Triangle ABC with sides of length a, b,
and c, with base b and altitude h.
Perimeter: $P = a + b + c$
Area: $A = \dfrac{1}{2}bh$

Sum of angles: $m \angle A + m \angle B + m \angle C = 180°$

(a) *Isosceles Triangle*   **(b)** *Equilateral Triangle*
two congruent sides   three congruent sides
two congruent (base) angles three congruent angles (60°)
(c) Right triangle with hypotenuse c:
Pythagorean theorem: $a^2 + b^2 = c^2$

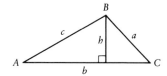

---

5. Circle of radius r or diameter $d = 2r$
Circumference (perimeter): $C = 2\pi r = \pi d$
Area: $A = \pi r^2$

---

6. Trapezoid with legs of length a and c,
bases of length b_1 and b_2, and
altitude h.
Perimeter: $P = a + b_2 + c + b_1$
Area: $A = \dfrac{1}{2}h(b_1 + b_2)$

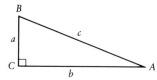

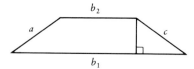

In solving geometric problems, drawings are very useful. Drawing to scale is
not necessary.

EXAMPLE 1 A 57-cm long board is to be cut into three pieces so that the second piece is three
times the length of the first piece, and the third piece is five times the length of
the second piece. Find the lengths of the three pieces of board.

Solution Let $x =$ the length, in cm, of the first piece

Then $3x$ = the length, in cm, of the second piece

and $5(3x)$ = the length, in cm, of the third piece

The diagram in Figure 4.12 helps illustrate.

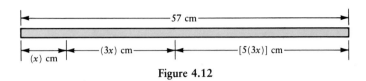

Figure 4.12

Using the relationship

(length of first piece) + (length of second piece) + (length of third piece)

= (total length)

we get

$$(x) \text{ cm} + (3x) \text{ cm} + [5(3x)] \text{ cm} = 57 \text{ cm}$$

Solving for x,

$$x + 3x + 5(3x) = 57$$
$$x + 3x + 15x = 57$$
$$19x = 57$$
$$x = 3$$

If $x = 3$, then $3x = 9$ and $5(3x) = 45$. Thus, the first piece is 3 cm long, the second is 9 cm long, and the third is 45 cm long. ■

Check   3 cm + 9 cm + 45 cm = 57 cm

EXAMPLE 2   The perimeter of a rectangle, whose length is $3\frac{1}{2}$ times its width, is 36 ft. Find the dimensions of the rectangle.

Solution   Let x = the width of the rectangle, measured in feet

Then $3\frac{1}{2}(x)$ = the length of the rectangle, measured in feet

See the diagram in Figure 4.13.

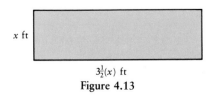

$3\frac{1}{2}(x)$ ft

Figure 4.13

Using the relationship

$$\text{perimeter of a rectangle} = 2 \cdot (\text{width} + \text{length})$$

we get

$$36 \text{ ft} = 2[x \text{ ft} + 3\frac{1}{2}(x) \text{ ft}]$$

$$36 \text{ ft} = (2x) \text{ ft} + (7x) \text{ ft}$$

Solving for x,

$$36 = 2x + 7x$$

$$36 = 9x$$

$$4 = x$$

If $x = 4$, then $3\frac{1}{2}(4) = 14$. Thus, the width of the rectangle is 4 ft and its length is 14 ft. ■

Check $2(4 \text{ ft} + 14 \text{ ft}) = 2(18 \text{ ft}) = 36 \text{ ft}$

EXAMPLE 3   Two rectangles of equal area are such that the longer side of the first rectangle is 2 ft shorter than the longer side of the second rectangle. If the widths of the rectangles are 6 ft and 5 ft, respectively, find the dimensions of the rectangles.

Solution   Let $x =$ the length of the longer side of the first rectangle, measured in ft

Then $x + 2 =$ the length of the longer side of the second rectangle, measured in ft

See the diagrams in Figure 4.14.

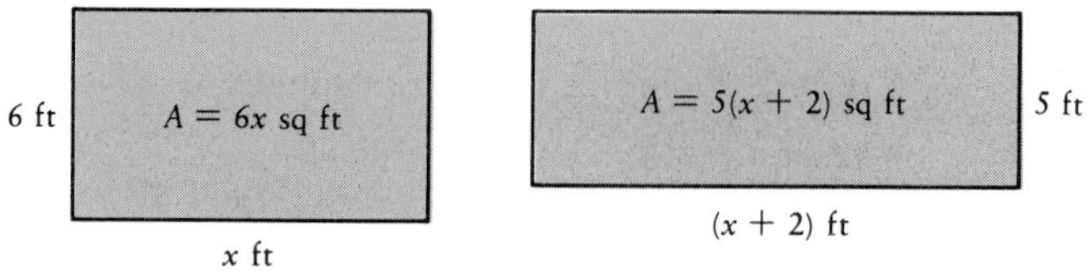

6 ft $A = 6x$ sq ft

$A = 5(x + 2)$ sq ft   5 ft

x ft

$(x + 2)$ ft

Figure 4.14

$$\text{area of first rectangle} = 6x \text{ sq ft}$$

$$\text{area of second rectangle} = 5(x + 2) \text{ sq ft}$$

Using the condition that the areas are equal, we get

$$6x \text{ sq ft} = 5(x + 2) \text{ sq ft}$$

Solving for x,

$$6x = 5(x + 2)$$

$$6x = 5x + 10$$

$$x = 10$$

If $x = 10$, then $x + 2 = 12$. Thus, the dimensions of the first rectangle are 10 ft by 6 ft. The dimensions of the second rectangle are 5 ft by 12 ft. ∎

Check

$$\text{area of first rectangle} = (10 \text{ ft})(6 \text{ ft}) = 60 \text{ sq ft}$$

$$\text{area of second rectangle} = (12 \text{ ft})(5 \text{ ft}) = 60 \text{ sq ft}$$

EXAMPLE 4   Find the radius of a circle if an increase of 3 in. in the radius increases the area by 33π in.²

Solution   Let $r =$ the length of the radius, measured in inches

Then $(r + 3) =$ the length of the radius of the larger circle, measured in inches.

See the diagrams in Figure 4.15.

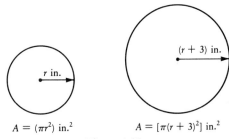

$A = (\pi r^2)$ in.² $A = [\pi(r + 3)^2]$ in.²

Figure 4.15

$$\text{the area of smaller circle} = \pi r^2 \text{ in.}^2$$

$$\text{the area of larger circle} = \pi(r + 3)^2 \text{ in.}^2$$

Using the fact that

$$(\text{area of smaller circle}) + (33\pi) \text{ in.}^2 = (\text{area of larger circle})$$

we get

$$(\pi r^2) \text{ in.}^2 + (33\pi) \text{ in.}^2 = [\pi(r + 3)^2] \text{ in.}^2$$

Solving for x,

$$\pi r^2 + 33\pi = \pi(r + 3)^2$$

dividing both sides by π,

$$r^2 + 33 = (r + 3)^2$$
$$r^2 + 33 = r^2 + 6r + 9$$
$$24 = 6r$$
$$4 = r$$

If $r = 4$, then $r + 3 = 7$. Thus, the radius of the smaller circle equals 4 in. and the radius of the larger circle equals 7 in. ■

Check
$$\text{area of smaller circle} = \pi(4 \text{ in.})^2 = 16\pi \text{ in.}^2$$
$$\text{area of larger circle} = \pi(7 \text{ in.})^2 = 49\pi \text{ in.}^2$$

and

$$49\pi \text{ in.}^2 - 16\pi \text{ in.}^2 = 33\pi \text{ in.}^2$$

EXAMPLE 5 Find the angle whose supplement and complement total 150°.

Solution Let $x =$ number of degrees in the angle

Then $(180 - x) =$ number of degrees in the supplementary angle

and $(90 - x) =$ number of degrees in the complementary angle

See Figure 4.16.

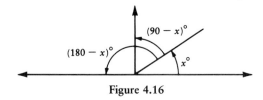

Figure 4.16

Since the sum of the supplementary and complementary angles is 150°, we have

$$(180 - x) \text{ degrees} + (90 - x) \text{ degrees} = 150 \text{ degrees}$$

Solving for x,

$$(180 - x) + (90 - x) = 150$$
$$270 - 2x = 150$$
$$-2x = -120$$
$$x = 60$$

Thus, the angle is 60°. ■

Check   supplementary angle is $(180 - x)° = 120°$

complementary angle is $(90 - x)° = 30°$

Hence, $120° + 30° = 150°$.

EXERCISES   4.8

1. A board 48 ft long is cut so that one piece is 14 ft longer than the other. Find the dimensions of the two pieces.

2. A piece of wire 6 ft long is cut into three pieces. The middle-sized piece is 6 in. shorter than one piece and 6 in. longer than the other. What is the length of the longest piece?

3. The length of a rectangle is 7 less than four times its width. If the perimeter is 116 in., find the dimensions of the rectangle.

4. A rectangle is twice as long as it is wide, and has a perimeter of 54 cm. What are its dimensions?

5. A pentagon has a perimeter of 51 in. Three sides are the same length and the two remaining sides are each 7 in. shorter. How long is one of the three equal sides?

6. The perimeter of an isosceles triangle is 45 cm. One side is 5 cm shorter than the sum of the other two sides. What is the length of the longest side?

7. A square has the same area as a rectangle whose length is 4 yd less and whose width is 12 yd greater than the side of the square. Find the area of each.

8. One side of a rectangle is 7 in. more than the other side. Find the side of a square whose area equals that of the rectangle, if the side of the square is 3 in. more than the shorter side of the rectangle.

9. When the sides of a square are increased by 1 m, the area of the new square is 41 sq yd more than that of the original square. Find the side of the first square.

10. A rectangle is 2 ft less in width and 1 ft more length than the side of a square. The area of the rectangle is 9 sq ft less than the area of the square. Find the length of the side of the square.

11. By removing a partition at one end, a square room is made into a rectangular one 6 ft longer, and the area is increased by 72 sq ft. Find the side of the square room.

12. There are two rectangles; the length of the first is 2 ft more than its width. The second rectangle is 3 ft longer and 2 ft narrower than the first. If the area of the first is 3 sq ft more than the second, find the width of the second rectangle.

13. The sum of four angles about a point is 360°. The second is twice the first; the third is three times the second; and the fourth is 10° greater than the first. Find the angles.

14. One of the two supplementary angles is three times the other. How many degrees are there in each angle?

15. One of two complementary angles added to 20° equals the other. Find the angles.

16. The number of degrees in the four angles about a point in a plane are $3x - 10$, $2x + 10$, $4x + 40$, and $6x + 20$, respectively. Find the four angles.

*17. The three angles of a triangle are in the ratio 1:2:3. Find the number of degrees in each angle.

*18. If two angles of a triangle are complements of each other and in the ratio of 2:3, find the three angles.

19. The radius of one circle is 2 cm more than that of a smaller one and its area is 113.4 sq cm more. Find the radius of the smaller circle.

20. One base of a trapezoid is 3 in. longer than the other base. If the height of the trapezoid is 14 in. and the area is 189 in.^{2}, what are the lengths of the bases?

4.9 MISCELLANEOUS WORD PROBLEMS

In this section, we give several types of examples of word problems not previously given in this chapter.

Work Problems

If a man can do a job in 7 days, he completes $\frac{1}{7}$ of the job in one day, $\frac{2}{7}$ of the job in 2 days, and $\frac{x}{7}$ of the job in x days.

> In general, if it takes n days to complete a job, then the part of the job completed in x days is represented by the fraction $\frac{x}{n}$.

EXAMPLE 1 If A can do a job in 9 days and B can do the same job in 14 days, how long would it take the two working together?

Solution Let x = the number of days it would take the two working together.

Then $\frac{x}{9}$ = the part of the job done by A

and $\frac{x}{14}$ = the part of the job done by B

Using the relationship

(part of job done by A) + (part of job done by B) = 1 job

we get

$$\frac{x}{9} + \frac{x}{14} = 1$$

$$14x + 9x = 126$$

$$23x = 126$$

$$x = 5\frac{11}{23}$$

Thus, it takes $5\frac{11}{23}$ days to complete the job if A and B are working together. ■

Check $$\frac{5\frac{11}{23}}{9} + \frac{5\frac{11}{23}}{14} = \frac{\frac{126}{23}}{9} + \frac{\frac{126}{23}}{14} = \frac{14}{23} + \frac{9}{23} = 1$$

EXAMPLE 2 A tank can be filled in 6 hr by one pipe, by a second pipe in 9 hr, and can be drained when full, by a third pipe, in 8 hr. How long would it take to fill the tank if it is empty, and if all pipes are in operation?

Solution Let $x =$ the number of hours the pipes are in operation.

Then $\dfrac{x}{6} =$ part of tank filled by first pipe

and $\dfrac{x}{9} =$ part of tank filled by second pipe

and $\dfrac{x}{8} =$ part of tank emptied by third pipe

Using the relationship

$$\begin{pmatrix} \text{part of tank} \\ \text{filled by first pipe} \end{pmatrix} + \begin{pmatrix} \text{part of tank} \\ \text{filled by second pipe} \end{pmatrix}$$

$$- \begin{pmatrix} \text{part of tank} \\ \text{emptied by third pipe} \end{pmatrix} = \begin{pmatrix} \text{full} \\ \text{tank} \end{pmatrix}$$

we get

$$\frac{x}{6} + \frac{x}{9} - \frac{x}{8} = 1$$

$$12x + 8x - 9x = 72$$

$$11x = 72$$

$$x = 6\frac{6}{11}$$

Thus, it takes $6\frac{6}{11}$ hours to fill the tank. ■

Check $\dfrac{6\frac{6}{11}}{6} + \dfrac{6\frac{6}{11}}{9} - \dfrac{6\frac{6}{11}}{8} = \dfrac{\frac{72}{11}}{6} + \dfrac{\frac{72}{11}}{9} - \dfrac{\frac{72}{11}}{8} = \dfrac{12}{11} + \dfrac{8}{11} - \dfrac{9}{11} = 1$

EXAMPLE 3 A woman can paint her garage in 9 hr and her daughter can do the same job in 16 hr. They start working together. After 4 hr the daughter leaves and the woman finishes the job alone. How many hours did the woman take to finish the job alone?

Solution Let $x =$ the number of hours it takes the woman to finish the job alone

Then $(x + 4) =$ the number of hours the woman works

and $4 =$ the number of hours the daughter works

Thus,

$$\frac{(x + 4) \text{ hr}}{9 \text{ hr}} = \text{part of job done by the woman}$$

and

$$\frac{4 \text{ hr}}{16 \text{ hr}} = \text{part of job done by the daughter}$$

Using the relationship

$$\left(\begin{array}{c} \text{part of job} \\ \text{done by woman} \end{array} \right) + \left(\begin{array}{c} \text{part of job} \\ \text{done by daughter} \end{array} \right) = 1 \text{ job}$$

we get

$$\frac{x + 4}{9} + \frac{4}{16} = 1$$

$$16(x + 4) + (4)(9) = (9)(16)$$

$$16x + 64 + 36 = 144$$

$$x = 2\frac{3}{4}$$

Thus, it takes the woman $2\frac{3}{4}$ hours to finish the job alone. ■

Check
$$\frac{2\frac{3}{4} + 4}{9} + \frac{4}{16} = \frac{\frac{11}{4} + 4}{9} + \frac{1}{4} = \frac{\frac{27}{4}}{9} + \frac{1}{4} = \frac{3}{4} + \frac{1}{4} = 1$$

Clock Problems

EXAMPLE 4 At what time between 4 o'clock and 5 o'clock is the minute hand of a watch 13 minutes in advance of the hour hand?

Solution Let x = the number of minutes after 4 o'clock when the minute hand is 13 minutes in advance of the hour hand (see Figure 4.17). That is, the minute hand will move over x minute spaces of the watch face in x minutes.

Since the minute hand moves 12 times as fast as the hour hand, the hour hand will move over $\frac{x}{12}$ minute spaces in x minutes.

At 4 o'clock the minute hand is 20 minute spaces behind the hour hand, and finally the minute hand is 13 minute spaces in advance. Thus, in x minutes, the minute hand has moved $20 + 13 = 33$ minute spaces more than the hour hand. Hence,

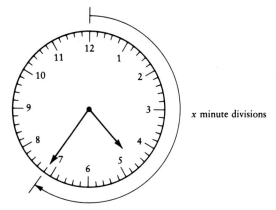

x minute divisions

Figure 4.17

$$x = \frac{x}{12} + 33$$

$$12x = x + 396$$

$$11x = 396$$

$$x = 36$$

Therefore, at 36 minutes past 4, the minute hand will be 13 minutes in advance of the hour hand. ■

Note. If the question had asked: "At what *times* between 4 and 5 o'clock will there be 13 minutes between the two hands?" we must also take into account the case where the minute hand is 13 minute spaces *behind* the hour hand. In this case, the minute hand gains $20 - 13 = 7$ minute spaces. Thus,

$$x = \frac{x}{12} + 7$$

$$12x = x + 84$$

$$x = 7\frac{7}{11}$$

Therefore, the times are $7\frac{7}{11}$ minutes past 4, and 36 minutes past 4.

Lever Problems

A lever is a rigid bar having one point of support called the fulcrum. If a weight or other force, w, is applied to the lever at a distance ℓ from the fulcrum, ℓ is called the lever arm of w, and the product $w\ell$ is called the moment of w with respect to the fulcrum (see Figure 4.18):

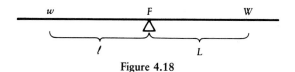

Figure 4.18

$$m = w\ell$$

Similarly, the moment of W with respect to the fulcrum is given by the product

$$m = WL$$

It can be shown that if two or more forces are applied to a lever in such a way that it is in balance or equilibrium, the *sum of the moments of the forces tending to turn the lever in one direction is equal to the sum of the moments of the forces tending to turn the lever in the opposite direction.*

For example, if w and W represent weights, and ℓ and L their respective distances from the fulcrum (see Figure 4.18), then the above principle is stated by the formula

$$\boxed{w\ell = WL}$$

Unless stated otherwise, we shall neglect the weight of the lever itself.

EXAMPLE 5 If an 8-lb weight is placed on a lever 10 ft away from the fulcrum, how much weight must be placed 4 ft from the fulcrum to keep the lever in balance (see Figure 4.19)?

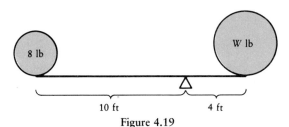

8 lb

W lb

10 ft   4 ft

Figure 4.19

Solution Using the formula $w\ell = WL$, where $w = 8$ lb, $\ell = 10$ ft, and $L = 4$ ft, we have

$$(8 \text{ lb})(10 \text{ ft}) = (W \text{ lb})(4 \text{ ft})$$
$$80 \text{ ft} \cdot \text{lb} = 4W \text{ ft} \cdot \text{lb}$$
$$20 = W$$

Thus, $W = 20$ lb. ■

Check $(8 \text{ lb})(10 \text{ ft}) = (20 \text{ lb})(4 \text{ ft}) = 80$ ft-lb

EXAMPLE 6 A girl weighing 45 lb is on one end of a seesaw, 6 ft from the fulcrum. On the other end of the seesaw, 5 ft from the fulcrum, is a boy weighing 75 lb. Where must a third child, weighing 60 lb, be placed in order to balance the seesaw?

Solution   In Figure 4.20, let points G and B indicate the positions of the first two children.

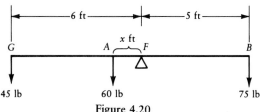

Figure 4.20

Let $x =$ the distance (in feet) of the third child, A, from the fulcrum

Then $m_G = (45 \cdot 6)$ ft-lb $=$ the moment of G with respect to the fulcrum

$\qquad m_A = (60 \cdot x)$ ft-lb $=$ the moment of A with respect to the fulcrum

$\qquad m_B = (75 \cdot 5)$ ft-lb $=$ the moment of B with respect to the fulcrum

Since the sum of the moments on one side is equal to the sum of the moments on the other side,

$$m_G \quad + \quad m_A \quad = \quad m_B$$

or $\qquad\qquad$ 270 ft-lb $+$ (60x) ft-lb $=$ 375 ft-lb

Solving for x,

$$270 + 60x = 375$$
$$60x = 105$$
$$x = 1\frac{3}{4}$$

Thus, the third child must be placed on the same side as G and $1\frac{3}{4}$ ft from the fulcrum. ∎

Check $\qquad m_G = 270$ ft-lb, $m_A = 60 \cdot 1\frac{3}{4} = 105$ ft-lb, and $m_B = 375$ ft-lb

$$270 \text{ ft-lb} + 105 \text{ ft-lb} = 375 \text{ ft-lb}$$

Note. How did we know that the third child should be placed on the side with G? The moment on the left, $m_G = 270$ ft-lb, is less than that on the right, $m_B = 375$ ft-lb, so the 60-lb weight must be placed at an unknown distance, x, to the left of the fulcrum.

EXAMPLE 7   A plank 18 ft long is to be used as a seesaw for Tracy and Kate. Tracy weighs 125 lb and Kate weighs 100 lb. If the girls are to balance one another, how far from the fulcrum must each sit (see Figure 4.21)?

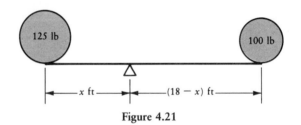

Figure 4.21

Solution   Let x = the number of feet Tracy is away from the fulcrum

Then $18 - x$ = the number of feet Kate is away from the fulcrum

$(125x)$ ft-lb = Tracy's moment with respect to the fulcrum

$[100(18 - x)]$ ft-lb = Kate's moment with respect to the fulcrum

Equating the moments on both sides of the fulcrum, we get

$$(125x) \text{ ft-lb} = [100(18 - x)] \text{ ft-lb}$$

Solving for x,

$$125x = 100(18 - x)$$
$$125x = 1800 - 100x$$
$$225x = 1800$$
$$x = 8$$

If $x = 8$, then $18 - x = 10$. Thus, Tracy sits 8 ft from the fulcrum and Kate sits 10 ft from the fulcrum.   ■

Check   Tracy's moment = $(125 \cdot 8)$ ft-lb = 1000 ft-lb
Kate's moment = $(100 \cdot 10)$ ft-lb = 1000 ft-lb

EXERCISES 4.9

1. A tank is filled by two pipes; the first can fill it in 4 hr and the second in 3 hr. How long will it take both pipes working together to fill the tank?

2. A battery is charged by three chargers; the first can charge it in 1 hr and 20 min, the second in 3 hr and 20 min, and the third in 5 hr. If all three chargers are working together, how long will it take to charge the battery?

3. A job can be done by Frank in 18 days and Bob in 21 days. In how many days can the job be done by both working together?

4. Skip can do a piece of work in 15 hr and Greg can do the same work in 25 hr. After Skip has worked a certain time, Greg completes the job, working 9 hr longer than Skip. How many hours did Skip work?

5. A water storage tank can be filled in 15 hr by two pipes, A and B, running together. After A has been running by itself for 5 hr, B is also turned on, and the tank is filled in 13 hr more. In how much time would it take each pipe working separately to fill the storage tank?

*6. At what time between 3 and 4 o'clock is the

minute hand one minute ahead of the hour hand?

*7. At what time between 8 and 9 o'clock do the hands of a watch point in the same direction? In the opposite direction?

*8. How soon after 4 o'clock are the hands of a clock at right angles?

*9. The hands of a clock are at right angles to each other at 3 o'clock. When are the hands next at right angles?

*10. The hands of a clock are together at noon. When will they be together again for the first time?

11. A child weighing 75 lb is seated on one end of a seesaw at a distance of 5 ft from the fulcrum. When a second child is seated $7\frac{1}{2}$ ft from the fulcrum, the seesaw balances. Find the weight of the second child.

12. One girl weighs 110 lb and another 120 lb. The first is seated 6 ft from the fulcrum. At what distance from the fulcrum must the other sit in order that the seesaw balance?

13. Carol and Bill together weigh 299 lb. They balance when seated 5 ft and 8 ft, respectively, from the fulcrum on opposite sides. What is the weight of each?

14. Two weights of 250 lb and 150 lb are placed 8 ft and 5 ft, respectively, from the fulcrum of a lever and on the same side. A third weight of 200 lb is placed 4 ft at the other side of the fulcrum, and a fourth weight is placed 6 ft from the fulcrum on the same side as the third weight. What is the amount of the fourth weight if the lever is in equilibrium?

*15. At what distance from the end of a 6-m crowbar must the fulcrum be placed so that a 600-lb weight may be lifted by a woman exerting a force of 120 lb on the other end of the crowbar?

4.10 WORD PROBLEMS INVOLVING LINEAR INEQUALITIES

Just as some word problems can be solved using linear equations, other word problems can be solved using linear inequalities.

EXAMPLE 1 If the temperature for a one-week period in Alaska ranged between $-58°F$ and $5°F$, what was the range in Centigrade degrees?

Solution The expression "ranged between $-58°F$ and $5°F$" can be expressed using the inequality $-58° \leq F \leq 5°$.

Recall that the formula relating Fahrenheit and Celsius is $F = \frac{9}{5}C + 32°$. Replacing F by this formula in the inequality yields

$$-58° \leq F \leq 5°$$

$$-58° \leq \frac{9}{5}C + 32° \leq 5°$$

solving for C,

$$-58° - 32° \leq \frac{9}{5}C + 32° - 32° \leq 5° - 32°$$

$$-90° \leq \frac{9}{5}C \leq -27°$$

$$\left(\frac{5}{9}\right)(-90°) \le \left(\frac{5}{9}\right)\left(\frac{9}{5}\right) C \le \left(\frac{5}{9}\right)(-27°)$$

$$-50° \le C \le -15° \quad \blacksquare$$

EXAMPLE 2 Scott earns $6 per hour and June earns $175 a week. How many hours per week must Scott work so that their combined weekly income is at least $370?

Solution Let x = the number of hours per week worked by Scott. Then (5 dollars/h̸r̸) × x h̸r̸ = $(5x)$ dollars is Scott's weekly income, and $(5x + 175)$ dollars is their combined weekly income. Since the combined weekly income must be "at least" $370, we solve the inequality:

$$(5x + 175) \text{ dollars} \ge 370 \text{ dollars}$$

Solving, we obtain

$$5x + 175 \ge 370$$

$$5x \ge 195$$

$$x \ge 39$$

Thus, Scott must work at least 39 hr per week. ■

Check If Scott must work at least 39 hr, then his weekly income is at least $5 \dfrac{\text{dollars}}{\text{h̸r̸}} \cdot 39$ h̸r̸ = $195, and $195 + $175 is at least $370.

EXAMPLE 3 Dave Bradford received grades of 88, 87, 86, and 92 on four algebra exams. Find the least grade that he must receive on a fifth exam in order to have an average of at least 90 on the five exams.

Solution Let x be Dave's score on the fifth exam. Then the average for these five exams is

$$\frac{88 + 87 + 86 + 92 + x}{5}$$

The problem requires that this average be at least 90. That is,

$$\frac{88 + 87 + 86 + 92 + x}{5} \ge 90$$

$$\frac{353 + x}{5} \ge 90$$

$$353 + x \ge 450$$

$$x \ge 97$$

Thus, Dave must receive at least 97 on the fifth exam in order to have a 90 average.   ■

EXAMPLE 4   If the perimeter of a rectangle with a width of 7 ft must be smaller than 36 ft, how large may the length be?

Solution   Let x = the number of feet in the length of the rectangle. See Figure 4.22.

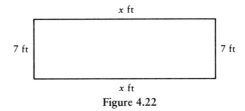

Figure 4.22

The perimeter of the rectangle is given by the expression

$$(2x) \text{ ft } + 14 \text{ ft } = (2x + 14) \text{ ft}$$

Since the perimeter must be smaller than 36 ft, we have

$$(2x + 14) \text{ ft} < 36 \text{ ft}$$

Solving for x,

$$2x + 14 < 36$$
$$2x < 22$$
$$x < 11$$

Thus, the length must be less than 11 ft.   ■

EXERCISES   4.10

1. The Fahrenheit (F) and Celsius (C) temperature scales are related by the formula $C = \frac{5}{9}(F - 32)$. If the temperature in a city ranges between $-10°C$ and $35°C$, find the range in Fahrenheit degrees.
2. Jerry's budget calls for him to spend no more than $120 for a pair of pants and a dress shirt. If the price of the pants was $20 more than three times the price of the shirt, find the maximum possible price of the shirt.
3. A student must have an average of 72 in order to pass a course. Her grades on the first three exams were 68, 67, and 73. What is the min-

imum grade she must receive on the fourth exam to pass the course?
4. The degree measure of angle M is twice the degree measure of angle N and one less than the degree measure of angle P. If the sum of the degree measures of the three angles is at least 61, which is the least possible degree measure of angle M?
5. A can of soda has 10 fewer calories than twice the number of calories in a can of grapefruit juice. Together they contain at least 185 calories. What is the smallest possible number of calories in the can of grapefruit juice?

6. A certain stock sells for $60 and yields an 11% annual dividend. How many years will it take for the stock to yield at least $58 in dividends?

7. A collection of 30 coins is made up of dimes and quarters. The total value of these coins is no more than $6. What is the maximum number of quarters that can be in the collection?

8. The two equal sides of an isosceles triangle are each 5 in. less than twice the base. If the perimeter of the triangle is at most 60 in., what is the maximum length of the base?

9. A broker invests $25,000, part at 10% and part at 15%. What is the least amount she can invest at 15% if she wishes an annual return of at least $2800?

10. A carpenter has 200 in. of material to make a record cabinet that has three compartments as shown in Figure 4.23. If the total length of the cabinet is to be 52 in., at most, how high can the cabinet be?

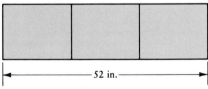

─ 52 in. ─

Figure 4.23

5 Factoring Polynomials

In Chapter 2, we discussed the operation of multiplying polynomials. For example, we multiplied the polynomials $(x + 3)$ and $(x - 5)$ and obtained the product $x^2 - 2x - 15$:

$$(x + 3)(x - 5) = x^2 - 2x - 15$$

In this chapter, we study the reverse process. That is, we would like to write the polynomial $x^2 - 2x - 15$ as the product of the polynomials $(x + 3)$ and $(x - 5)$. This process of finding two or more polynomials, each of lower degree than the given polynomial, whose product yields the given polynomial, is called factoring. The polynomials that form the products are called factors. Thus,

$$\underbrace{x^2 - 2x - 15}_{\text{product}} = \underbrace{(x + 3)(x - 5)}_{\text{factors}}$$

In this book, we shall restrict factorization to polynomials with integral coefficients. The factors must also have integral coefficients. If a polynomial can be factored, we say the polynomial is factorable. If a polynomial has no factors other than itself and 1, or its negative and -1, the polynomial is called prime. For example, the polynomial

$$x + y$$

is prime because the only factors it has are $\pm (x + y)$ and ± 1. Other examples of prime polynomials are 3, $4a + 5b$, and $y + 7$.

A polynomial is said to be factored completely if it is expressed as a product of prime polynomial factors.

5.1 FACTORING POLYNOMIALS BY COMMON FACTORS

The distributive property

$$a(b + c) = ab + ac$$

provides the primary bridge between multiplying polynomials and factoring polynomials. When trying to factor any algebraic expression, we shall first look for a common quantity occurring in *each term* of the given expression. This common factor may be a simple monomial quantity or a polynomial quantity. If such a quantity is present in each term, we remove that factor first, and divide this factor into each term of the expression to determine the other factor.

Common Monomial Factor

We have learned that to multiply a polynomial by a monomial we apply the distributive property for multiplication. For example,

$$7x(y + 2x - 3) = 7xy + 14x^2 - 21x$$

Conversely, given an expression such as $7xy + 14x^2 - 21x$, it can be written in factored form as $7x(y + 2x - 3)$.

EXAMPLE 1 Factor the expression $5x - 15$.

Solution Each of the two terms contains the common factor 5. Thus,

$$5x - 15 = 5(x - 3) \quad \blacksquare$$

EXAMPLE 2 Factor $2ax + 6ay - 10a$.

Solution Each term contains $2a$. Hence,

$$2ax + 6ay - 10a = 2a(x + 3y - 5) \quad \blacksquare$$

EXAMPLE 3 Factor $x^5 - x^4 + x^3$.

Solution Each term contains x^3. Thus,

$$x^5 - x^4 + x^3 = x^3(x^2 - x + 1) \quad \blacksquare$$

Note. If only x or x^2 had been factored out, the factorization would not have been complete. That is, the common factor should be the *greatest* factor that is common to all terms.

EXAMPLE 4   Factor $6a^3b^2 - 15a^2b^4 + 9a^4b$.

Solution   The highest number that is a factor of all the coefficients 6, 15, and 9 is 3. The highest power of a that is a factor of each term is a^2. The highest power of b that is a factor of each term is b. Thus, $3a^2b$ is the highest common factor and

$$6a^3b^2 - 15a^2b^4 + 9a^4b = 3a^2b(2ab - 5b^3 + 3a^2) \quad ■$$

EXAMPLE 5   Factor $a(x + y) + b(x + y)$.

Solution   If we think of $(x + y)$ as a single quantity, we may factor the common factor. Thus,

$$a(x + y) + b(x + y) = (x + y)(a + b) \quad ■$$

EXAMPLE 6   Factor $6(a + 2b)^2 + 8a(a + 2b) - 6b(a + 2b)$.

Solution   Each term has the number 2 and the quantity $(a + 2b)$. Thus, the common factor is $2(a + 2b)$:

$$6(a + 2b)^2 + 8a(a + 2b) - 6b(a + 2b)$$

$$= 2(a + 2b)[3(a + 2b) + 4a - 3b] \quad ■$$

Group Factoring

While all the terms of an algebraic expression may not contain a common factor, it is sometimes possible to group the terms in such a way that each grouping will have a common factor.

EXAMPLE 7   Factor the expression $ax + bx + ay + by$.

Solution   We see that there is no factor common to all terms. However, notice that the first pair of terms has x as a common factor, and the second pair of terms has y as a common factor. Thus, we may write

$$ax + bx + ay + by = x(a + b) + y(a + b)$$

The expression has been reduced to two terms having the common factor $(a + b)$. Hence,

$$ax + bx + ay + by = x(a + b) + y(a + b) = (a + b)(x + y)$$

These three expressions can be illustrated geometrically by Figures 5.1(a), (b), and (c), respectively, where ax represents the area of a rectangle of length a and width x, bx represents the area of a rectangle of length b and width x, and so on.

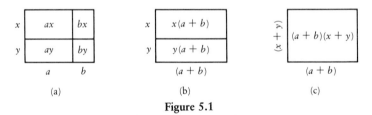

(a)   (b)   (c)

Figure 5.1

The same factorization of $ax + bx + ay + by$ may be obtained by an alternate grouping of the terms. That is, grouping the first and third terms, and grouping the second and fourth terms, we obtain

$$ax + bx + ay + by = ax + ay + bx + by$$
$$= a(x + y) + b(x + y)$$
$$= (x + y)(a + b)$$

These factors are the same as those obtained in the first grouping method, but the factors are in reverse order. The steps of this second method of grouping are illustrated geometrically in Figures 5.2(a), (b), and (c).

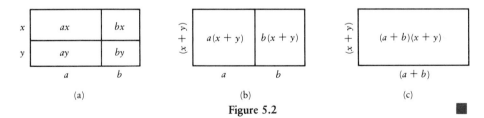

(a)   (b)   (c)

Figure 5.2

One property of real numbers that occurs frequently in factorization is

$$\boxed{A - B = -(B - A)}$$

EXAMPLE 8   Factor $2ax - 8a + 12 - 3x$.

Solution   Grouping the first two terms and the last two terms, we get

$$2ax - 8a + 12 - 3x = 2a(x - 4) + 3(4 - x)$$

The two terms do not appear to have any common factors. But

$4 - x = -(x - 4)$ and we have

$$2ax - 8a + 12 - 3x = 2a(x - 4) + 3(4 - x)$$
$$= 2a(x - 4) - 3(x - 4)$$
$$= (x - 4)(2a - 3)$$ ■

EXAMPLE 9   Factor $ab(x + 2) + (b - a)(x + 2) - x - 2$

Solution   Factoring (-1) out of the last two terms, we get

$$ab(x + 2) + (b - a)(x + 2) - x - 2$$
$$= ab(x + 2) + (b - a)(x + 2) - (x + 2)$$
$$= (x + 2)[ab + b - a - 1]$$
$$= (x + 2)[b(a + 1) - (a + 1)]$$

[factoring the second bracket]

$$= (x + 2)(a + 1)(b - 1)$$ ■

Some algebraic expressions contain more than four terms. In such expressions we must determine the grouping that will yield the factors.

EXAMPLE 10   Factor $8ay - 3x(3c - b) + 12cy - 4by - 6ax$.

Solution   Removing the parentheses, we obtain

$$8ay - 3x(3c - b) + 12cy - 4by - 6ax$$
$$= 8ay - 9cx + 3bx + 12cy - 4by - 6ax$$

Grouping the terms containing y and the terms containing x yields

$$8ay - 3x(3c - b) + 12cy - 4by - 6ax$$
$$= 8ay - 9cx + 3bx + 12cy - 4by - 6ax$$
$$= 8ay + 12cy - 4by - 9cx + 3bx - 6ax$$
$$= 4y(2a + 3c - b) - 3x(3c - b + 2a)$$
$$= 4y(2a + 3c - b) - 3x(2a + 3c - b)$$
$$= (2a + 3c - b)(4y - 3x)$$ ■

Note. Any algebraic expression should always be examined for monomial common factors before using other factoring methods.

EXAMPLE 11   Factor $4a^2b^2 - ab^3 + 4a^2 - ab$.

Solution   Each term has the common factor a:

$$4a^2b^2 - ab^3 + 4a^2 - ab = a[4ab^2 - b^3 + 4a - b]$$
$$= a[b^2(4a - b) + (4a - b)]$$
$$= a(4a - b)(b^2 + 1) \quad \blacksquare$$

EXERCISES 5.1

In Exercises 1–36, factor.

1. $x^2 + bx$
2. $ax + xb$
3. $ax + x$
4. $ab - b^2$
5. $a^2b - a^3$
6. $2ax - 2xb$
7. $10d - 20cd$
8. $bcx + bdx$
9. $a^3x - 2ax$
10. $14ax + 21bx$
11. $5ax^2 - 10x^3$
12. $ax^2c - ax^2d$
13. $a^2b + ab$
14. $2ab^3 - a^2b^2$
15. $5acx + 15cx$
16. $18ab^2c + 12a^2dc$
17. $22a^3bc - 33a^2c$
18. $x^3y^2 - xy^3$
19. $3x^4y^4 - 7x^5y^2$
20. $2a^2bd^2 + 3ab^3$
21. $17p^3q^3 - 51p^2q^2$
22. $39x^2 - 52$
23. $a^3 - a^2b + ab^2$
24. $-3ab - 2b^3c + ab^2$
25. $3a + 3b + 3c$
26. $-9x^3y^2 + 15x^2y^3 - 21xy^3$
27. $57x^6 - 19m^6x^2 + 38xy^6$
28. $6x^3 + 18x^2 + 30x$
29. $6ax^2 + 14ax - 2a$
30. $12a^7b^6 - 36a^6b^6 - 48a^3b^5c$
31. $35x^3y - 77x^2y - 63xy$
32. $a^2x^5 + a^2x^3y^2 + a^2xy^3$
33. $m + 23mx - 7x^2m$
34. $7x^3y^3z - 3xy^3z^3 - x^3yz^3$
35. $xyz - xzw - xyw$
36. $-a^5 - a^4 + a^3$

In Exercises 37–44, use factoring to find the value of:

37. $18 \times 61 + 18 \times 39$
38. $14 \times 72 - 14 \times 62$
39. $26 \times 42 - 37 \times 26$
40. $33 \times 26 - 33 \times 7$
41. $\dfrac{3}{8}$ of 142 + $\dfrac{3}{8}$ of 18
42. $\dfrac{5}{11}$ of 152 $-$ $\dfrac{5}{11}$ of 119
43. 9% of 35 + 9% of 65
44. 25% of 195 $-$ 25% of 31

In Exercises 45–95, factor.

45. $a(b - 2c) - 2(b - 2c)$
46. $3(x + 3y) + y(x + 3y)$
47. $2x^2(x - 1) - 3(x - 1)$
48. $y(y^2 + z^2) - z(y^2 + z^2)$
49. $2a(a^2 - 1) - 3(1 - a^2)$
50. $x^3(2x - 3) - (2x - 3)$
51. $m^2(m + 2h) - 4(m + 2h)$
52. $8z^3(2z + 1) + (2z + 1)$
53. $4x(x - 2y) + 3y(2y - x)$
54. $5a^2(a^2 + 4) - 3(a^2 + 4)$
55. $c^2(c - d) + d^2(d - c)$
56. $6x(x - 2) + 5(x - 2)$
57. $3m(m - 2n) - 2n(m - 2n)$
58. $2a^2(3a - 2b) - (2b - 3a)$
59. $x^4(x - 1) + 8(1 - x)$
60. $2x^2 - 3x - 2xy + 3y$
61. $a^3 - ab^2 + a^2b - b^3$
62. $3a - 12 + ac - 4c$
63. $3n - 3x + 4x^2 - 4nx$
64. $2a^2 - 5ac - 5bc + 2ab$
65. $5ax - 10x - a + 2$
66. $2y^2 + 6y + yz + 3z$
67. $20x + 4xy - y - 5$
68. $12x^2 + 2y - 8x - 3xy$
69. $m^3 - mn^2 - m^2n + n^3$
70. $xy - 2z^2 + 2xz - yz$
71. $3x^3 + 3x^2 - x - 1$
72. $a^4 + a^3b + ab^3 + b^4$
73. $x^2 - 2 - 3x^3 + 6x$
74. $8m^2 - 6m - 3n + 4mn$
75. $x^3 - x^2y - xy^2 + y^3$
76. $2a^3 + 2b^3 - 4a^2b - ab^2$
77. $3c^2 - 4cd - 3c + 4d$
78. $5y^3 + 10y^2z + 2z^2 + yz$
79. $6x^2 + 9x - 8xy - 12y$
80. $12x^4 - 6x^2 - 4x^3 + 2x$
81. $18c^3 - 6c^2d + 6c^2 - 2cd$
82. $2x^3y - 5x^2y - 6xy^2 + 15y^2$
83. $m^4 + m^3 + m + 1$

84. $6z^2 - 6z + 5xz - 5x$
85. $12a^3 - 3a^2c - 4ac^2 + c^3$
86. $4ax^2 + 12x^3 - 5ax - 15x^2$
87. $2ab^2 + ab - 2b^3 - b^2$
88. $12x - 16y + 6xy - 8y^2$
89. $5m^2 + 10mn - 3m - 6n$

90. $4m^2n^2 - mn^3 - 4m^2 + mn$
91. $a^2 + ab + ac - ad - bd - cd$
92. $ac + bd - (bc + ad)$
93. $a^2c - abd - abc + a^2d$
94. $ad + ce + bd + ae + cd + be$
95. $a^2 + cd - ab - bd + ac + ad$

In the next two sections we discuss the factorization of trinomials. We begin by factoring trinomials of the form $x^2 + bx + c$, and then proceed to the factorization of trinomials of the form $ax^2 + bx + c$, where a, b, and c are integers and $a \neq 0$. A trinomial of this form is called a second-degree trinomial or a quadratic trinomial. Finally, we cover the factorization of other types of trinomials.

5.2 FACTORING TRINOMIALS OF THE FORM $x^2 + bx + c$

Before considering the factoring of $x^2 + bx + c$, let us recall how the product of two binomials can result in a trinomial.

$$(x + 7)(x + 2) = x^2 + 9x + 14$$

$$(x - 7)(x - 2) = x^2 - 9x + 14$$

$$(x + 7)(x - 2) = x^2 + 5x - 14$$

$$(x - 7)(x + 2) = x^2 - 5x - 14$$

Each term of the above products was obtained as follows:

1. The first term of the trinomial is the product of the first terms of the binomials.
2. The second term of the trinomial is the algebraic sum of the product of the two outer terms and the product of the two inner terms.
3. The constant term of the trinomial is the product of the second terms of the binomials.

We now reverse the process and factor trinomials of the form $x^2 + bc + c$ into a product of binomials.

EXAMPLE 1   Factor $x^2 + 7x + 12$.

Solution   Since the first term in the trinomial is the product of the first terms of the binomials, we let x be the first term of each binomial:

$$(x \qquad)(x \qquad)$$

Since the constant term $(+12)$ is positive, both of its factors will have the same sign as the second term $(+7x)$ of the trinomial. Thus, both binomials have $(+)$ signs:

$$(x + \quad)(x + \quad)$$

We now look for factors of 12 whose sum is 7. Thus, the factors are 3 and 4, and we have

$$(x + 3)(x + 4)$$

Hence,

$$x^2 + 7x + 12 = (x + 3)(x + 4) \quad \blacksquare$$

EXAMPLE 2   Factor $x^2 - 5x + 6$.

Solution   Again, we let the first term of each binomial be x:

$$(x \quad)(x \quad)$$

Since the constant term $(+6)$ is positive, both of its factors will have the same sign as the second term $(-5x)$ of the trinomial. Thus, both binomials have $(-)$ signs:

$$(x - \quad)(x - \quad)$$

We now look for factors of 6 whose sum is 5. Such factors are 3 and 2, and we have

$$(x - 3)(x - 2)$$

Hence,

$$x^2 - 5x + 6 = (x - 3)(x - 2) \quad \blacksquare$$

The following rule can be used when the constant term is positive.

Rule for Factoring $x^2 + bx + c$ if the Constant Term Is Positive

If the constant term of the trinomial $x^2 + bx + c$ is positive, then look for two numbers whose product is c and whose sum is b, the coefficient of the middle term. Also, both of the numbers should be given the same signs as b.

EXAMPLE 3   Factor $x^2 - 6x + 8$.

Solution   The constant term $(+8)$ is positive. Thus, we are looking for two numbers whose product is 8 and whose sum is 6. These are 4 and 2. We assign each the same sign as (-6); that is, both are negative. Hence,

$$x^2 - 6x + 8 = (x - 4)(x - 2) \quad \blacksquare$$

EXAMPLE 4   Factor $15x + x^2 + 36$.

Solution   Rearranging the terms in descending powers of x, we obtain $x^2 + 15x + 36$. Since the constant term $(+36)$ is positive, we are looking for two numbers whose product is 36 and whose sum is 15. These are 12 and 3. Each number is positive because $(+15)$ is positive. Thus,

$$x^2 + 15x + 36 = (x + 12)(x + 3) \quad \blacksquare$$

We now consider factoring trinomials of the form $x^2 + bx + c$, where the constant term is negative.

EXAMPLE 5   Factor $x^2 + 4x - 12$.

Solution   The first term of each binomial factor is x:

$$(x \qquad)(x \qquad)$$

The constant term (-12) is negative, so its factors must have opposite signs, one positive and one negative:

$$(x + \qquad)(x - \qquad)$$

We are looking for factors such that their product is (-12) and their algebraic sum, which is numerically equal to their difference, is $(+4)$. Thus, the greater factor must be positive in order to give its sign to their sum. The required factors are $(+6)$ and (-2). Hence,

$$x^2 + 4x - 12 = (x + 6)(x - 2) \quad \blacksquare$$

The following rule can be used when the constant term is negative.

Rule for Factoring $x^2 + bx + c$ If the Constant Term Is Negative

If the constant term of the trinomial $x^2 + bx + c$ is negative, then look for two numbers whose product is c and whose difference is b, the coefficient of the middle term. Also, the larger number is given the same sign as b.

EXAMPLE 6   Factor $x^2 - 5x - 24$.

Solution The constant term (-24) is negative. Thus, we are looking for two numbers whose product is 24 and whose difference is 5. These numbers are 8 and 3, and the larger number 8 is given a negative sign. Thus,

$$x^2 - 5x - 24 = (x - 8)(x + 3) \quad \blacksquare$$

EXAMPLE 7 Factor $y^2 + 7y - 30$.

Solution $$y^2 + 7y - 30 = (y + 10)(y - 3) \quad \blacksquare$$

EXAMPLE 8 Factor $-x^2 + x + 2$

Solution When factoring an expression that contains $-x^2$, it is convenient to factor out the minus sign. Thus,

$$-x^2 + x + 2 = -(x^2 - x - 2)$$

and factoring $x^2 - x - 2$ we obtain

$$-x^2 + x + 2 = -(x^2 - x - 2) = -(x - 2)(x + 1) = (2 - x)(x + 1) \quad \blacksquare$$

EXERCISES 5.2

Factor.

1. $x^2 - 3x + 2$
2. $x^2 + 2x - 15$
3. $a^2 - 7a + 12$
4. $r^2 - r - 6$
5. $t^2 - 9t + 20$
6. $21 - 4h - h^2$
7. $p^2 - 10p + 21$
8. $a^2 - 8a + 15$
9. $b^2 + 2b - 15$
10. $18 + 7y - y^2$
11. $x^2 + 10x + 24$
12. $y^2 + 11y + 28$
13. $z^2 - 11z + 30$
14. $w^2 - 9w + 18$
15. $r^2 + 4r - 21$
16. $s^2 - 2s - 15$
17. $-m^2 - m + 12$
18. $-x^2 + 4x + 45$
19. $-y^2 + 15y - 56$
20. $-3x - x^2 + 40$
21. $11x - x^2 - 30$
22. $-2x - x^2 + 24$
23. $-a^2 + 5a + 36$
24. $-b^2 - 2b + 48$

5.3 FACTORING TRINOMIALS OF THE FORM $ax^2 + bx + c$

When the coefficient of x^2 is not 1, we can use the approach discussed in Section 5.2, but the number of possible combinations of factors is much greater. Some of the information we learned can still be useful for trinomials of the form $ax^2 + bx + c$, where $a > 0$. The following observations will be helpful.

Rules for Factoring Trinomials of the Form $ax^2 + bx + c$, $a > 0$

1. If the third term of the trinomial is positive, the second terms of its binomial factors both have the same sign, and this sign is the same as that of the middle term of the trinomial.
2. If the third term of the trinomial is negative, the second terms of its binomial factor have opposite signs.

As before, we begin by considering the product of two binomials:

$$2x + 5$$
$$x - 3$$
$$2x^2 - 6x$$
$$+ 5x - 15$$
$$2x^2 - x - 15$$

Note the following:

1. The first term in the product $(2x^2)$ is obtained by multiplying the first terms of the binomials. That is, $2x^2 = (2x)(x)$.
2. The last term in the product (-15) is obtained by multiplying the last terms of the binomials. That is, $-15 = (+5)(-3)$.
3. The middle term of the product $(-x)$ is the algebraic sum of the cross-products. That is, $-x = (2x)(-3) + (x)(5)$.

If the binomials are written horizontally, the middle term may be found by finding the algebraic sum of the outer products and the inner products.

$$(2x + 5)(x - 3)$$
$$(5x)$$
$$(-6x)$$
$$-x$$

We now use this example and the observations given above to help us factor trinomials.

EXAMPLE 1 Factor $3x^2 - 5x - 2$.

Solution The first term, $3x^2$, has $(3x)$ and x as factors. Thus, we write

$$3x^2 - 5x - 2 = (3x \qquad)(x \qquad)$$

Since the constant (-2) is negative, the factors must have opposite signs and the *difference* between the inner and outer products must be $(-5x)$. Thus, the greater cross-product must be negative. The possible factors of (-2) are $(-1)(2)$, $(1)(-2)$, $(-2)(1)$, and $(2)(-1)$. Using the factors $(-1)(2)$, we have

$$(3x - 1)(x + 2)$$

However, since $6x - x = 5x$, this combination fails to give the correct coefficient of the middle term. If we try the combination $(1)(-2)$, we obtain

$$(3x + 1)(x - 2) = 3x^2 - 5x - 2$$

Thus,

$$3x^2 - 5x - 2 = (3x + 1)(x - 2) \quad \blacksquare$$

EXAMPLE 2   Factor $2x^2 - 12x + 18$.

Solution   Since the constant term $(+18)$ is positive, both factors must have the same sign and they have the same sign as the middle term $(-12x)$. Thus, we have

$$(2x - \quad)(x - \quad)$$

We now find factors of 18 so that the *sum* of the inner and outer products give 12. Since $(2)(3) + (6)(1) = 12$, we have

$$2x^2 - 12x + 18 = (2x - 6)(x - 3) \quad \blacksquare$$

EXAMPLE 3 $10x^2 - x - 21$.

Solution   Factors of 10: 10 and 1   or   5 and 2
Factors of 21: 21 and 1   or   7 and 3

$$(5x + 7)(2x - 3)$$
$$(+14x)$$
$$(-15x)$$
$$-x$$

Thus, $10x^2 - x - 21 = (5x + 7)(2x - 3)$. $\blacksquare$

EXAMPLE 4   Factor $14 - 6x^2 - 17x$.

Solution   Rearranging the terms into descending order, we obtain $-6x^2 - 17x + 14$. Next, we factor out (-1) so that the resulting factor has a positive leading coefficient. Thus,

$$14 - 6x^2 - 17x = -6x^2 - 17x + 14 = -(6x^2 + 17x - 14)$$

Factors of 6: 6 and 1   or   3 and 2
Factors of 14: 14 and 1   or   7 and 2

$$(3x - 2)(2x + 7)$$
$$(-4x)$$
$$(+21x)$$
$$+17x$$

Thus, $14 - 6x^2 - 17x = -(6x^2 + 17x - 14) = -(3x - 2)(2x + 7) = (2 - 3x)(2x + 7)$. $\blacksquare$

Factoring Other Trinomials

EXAMPLE 5   Factor $28x^4y + 64x^3y - 60x^2y$.

Solution   First we look for any common factors. Each term has $4x^2y$ in common, and we have

$$28x^4y + 64x^3y - 60x^2y = 4x^2y(7x^2 + 16x - 15)$$
$$= 4x^2y(7x - 5)(x + 3) \blacksquare$$

EXAMPLE 6   Factor $56a^2 - 5ab - 25b^2$.

Solution $56a^2 - 5ab - 25b^2 = (7a - 5b)(8a + 5b)$ $\blacksquare$

EXAMPLE 7   Factor $6x^4 - 23x^2 + 20$.

Solution $6x^4 - 23x^2 + 20 = (3x^2 - 4)(2x^2 - 5)$ $\blacksquare$

EXAMPLE 8   Factor $24x^2 + 22xy - 21y^2$.

Solution $24x^2 + 22xy - 21y^2 = (2x + 3y)(12x - 7y)$ $\blacksquare$

EXAMPLE 9   Factor $(x + y)^2 + 5a(x + y) + 6a^2$.

Solution   Let $z = (x + y)$. Then the trinomial can be written

$$z^2 + 5az + 6a^2 = (z + 3a)(z + 2a)$$

Substituting $(x + y)$ for z, we get

$$(x + y)^2 + 5a(x + y) + 6a^2 = [(x + y) + 3a][(x + y) + 2a]$$
$$= (x + y + 3a)(x + y + 2a) \blacksquare$$

EXERCISES 5.3

Factor.

1. $3x^2 + 5x + 2$
2. $2x^2 + 5x + 2$
3. $2x^2 + 7x + 6$
4. $3x^2 + 8x + 4$
5. $2x^2 + 11x + 5$
6. $3x^2 + 10x + 3$
7. $3x^2 + 11x + 6$
8. $4x^2 + 11x - 3$
9. $3x^2 + x - 2$
10. $2x^2 + 3x - 2$
11. $2x^2 + 15x - 8$
12. $3x^2 - 19x - 14$
13. $6x^2 - 31x + 35$
14. $3x^2 + 19x - 14$
15. $4x^2 + x - 14$
16. $4x^2 + 8x - 5$
17. $21x^2 + 10x - 24$
18. $-a^2 + 19a - 88$
19. $-x^2 + 9x + 70$
20. $-y^2 + 19y + 42$
21. $-h^2 - h + 20$
22. $-x^2 + 18x + 144$
23. $-y^2 + 24y - 63$
24. $-b^2 - 20b - 51$

25. $2a^3 - 8a^2b + 8ab^2$

26. $3x^4 + 30x^3 + 72x^2$

27. $16a^2 - 24ax + 9x^2$

28. $3a^3 + 15a^2 - 18a$

29. $20x^2 + 11xy - 3y^2$

30. $16b^2 - 8bc + c^2$

31. $9a^2 - 6ax - 8x^2$

32. $6x^2 - 47xy - 8y^2$

33. $4x^4 + 15x^3 - 4x^2$

34. $16c^3 - 8c^2d - 3cd^2$

35. $16p^2q^2 + 14pq - 147$

36. $-68x^2y^2 + 48xy + 81$

37. $28ab - 44 + 57a^2b^2$

38. $-11mn - 28 + 24m^2n^2$

39. $-5y^2 - 21y^4 + 4$

40. $35c^2 - 12 - 8c^4$

41. $15a^4 + 29a^2 + 8$

42. $19x^2 + 8x^4 - 15$

43. $41a^2 + 14 - 28a^4$

44. $14(xy)^4 - 65(xy)^2 - 25$

45. $(x - y)^2 + 6(x - y) + 9$

46. $(x + y)^2 - 8(x + y) + 16$

47. $(x + y)^2 - 2(x + y) - 15$

48. $(a + b)^2 - 4(a + b) - 12$

49. $2(x + y)^2 + 5(x + y) + 2$

50. $4(x - y)^2 - 3a(x - y) - a^2$

51. $(x - y)^3 - (x - y)^2 - 2(x - y)$

5.4 SPECIAL FACTORING

There are several second-degree polynomials that occur quite often and factor easily. These polynomials are based on the special products presented in Chapter 2. We begin by listing those identities:

1. $\qquad (A + B)^2 = A^2 + 2AB + B^2$

2. $\qquad (A - B)^2 = A^2 - 2AB + B^2$

3. $\qquad (A + B)(A - B) = A^2 - B^2$

4. $(A + B)(A^2 - AB + B^2) = A^3 + B^3$

5. $(A - B)(A^2 + AB + B^2) = A^3 - B^3$

These products may be used as formulas for factoring.

Examples Using Identity 1: $(A + B)^2 = A^2 + 2AB + B^2$

EXAMPLE 1   Factor $x^2 + 6x + 9$.

Solution   The expression may be written in the form of Identity 1; that is,

$$(x)^2 + 2(3)(x) + (3)^2$$

If we let $A = x$ and $B = 3$, then by Identity 1 we get

$$x^2 + 6x + 9 = (x + 3)^2 \quad \blacksquare$$

EXAMPLE 2   Factor $36x^2 + 84xy + 49y^2$.

Solution   The expression may be written $(6x)^2 + 2(6x)(7y) + (7y)^2$. If we let $A = 6x$ and $B = 7y$ and use Identity 1, we obtain

$$36x^2 + 84xy + 49y^2 = (6x + 7y)^2 \quad \blacksquare$$

EXAMPLE 3 Factor $a^2 + 2ab + b^2 + 10(a + b) + 25$.

Solution Note that the first three terms can be factored using Identity 1. Thus, $a^2 + 2ab + b^2 = (a + b)^2$. Hence, the entire expression may be written

$$(a + b)^2 + 2(5)(a + b) + (5)^2$$

If we let $A = (a + b)$ and $B = 5$, then using Identity 1 again we get

$$a^2 + 2ab + b^2 + 10(a + b) + 25 = [(a + b) + 5]^2 = (a + b + 5)^2 \quad \blacksquare$$

Examples Using Identity 2: $(A - B)^2 = A^2 - 2AB + B^2$

EXAMPLE 4 Factor $y^2 - 14y + 49$.

Solution This expression may be written $(y)^2 - 2(7)y + (7)^2$. If we let $A = y$ and $B = 7$ and use Identity 2, we get

$$y^2 - 14y + 49 = (y - 7)^2 \quad \blacksquare$$

EXAMPLE 5 Factor $100a^2 - 160ab + 64b^2$.

Solution This expression may be written $(10a)^2 - 2(10a)(8b) + (8b)^2$. If we let $A = 10a$ and $B = 8b$, then the expression becomes

$$100a^2 - 160ab + 64b^2 = (10a - 8b)^2 \quad \blacksquare$$

EXAMPLE 6 Factor $81x^6 - 72x^3y^2 + 16y^4$.

Solution This expression may be written $(9x^3)^2 - 2(9x^3)(4y^2) + (4y^2)^2$. Letting $A = 9x^3$ and $B = 4y^2$ in Identity 2, we get

$$81x^6 - 72x^3y^2 + 16y^4 = (9x^3 - 4y^2)^2 \quad \blacksquare$$

EXAMPLE 7 Factor $9a^2 + 24ab + 16b^2 - 6a - 8b + 1$.

Solution The first three terms may be written

$$(3a)^2 + 2(3a)(4b) + (4b)^2 = (3a + 4b)^2 \qquad \text{[from Identity 1]}$$

The fourth and fifth terms may be written

$$-6a - 8b = -2(3a + 4b)$$

Thus, the entire expression may be written

$$\underbrace{9a^2 + 24ab + 16b^2}\; \underbrace{-\;6a - 8b} + 1 = (3a + 4b)^2 - 2(3a + 4b) + (1)^2$$

Letting $A = (3a + 4b)$ and $B = 1$ and using Identity 2, we obtain

$$9a^2 + 24ab + 16b^2 - 6a - 8b + 1 = [(3a + 4b) - 1]^2$$
$$= (3a + 4b - 1)^2 \quad \blacksquare$$

Examples Using Identity 3: $(A + B)(A - B) = A^2 - B^2$

The expression $A^2 - B^2$ is called the difference of two squares.

EXAMPLE 8   Factor $36x^2 - 81y^2$.

Solution   First we look for the greatest common factor:

$$36x^2 - 81y^2 = 9(4x^2 - 9y^2)$$

Using Identity 3, we obtain

$$36x^2 - 81y^2 = 9(4x^2 - 9y^2) = 9(2x + 3y)(2x - 3y) \quad \blacksquare$$

EXAMPLE 9   Factor $50x^6 - 32y^2$.

Solution   This expression is not at present the difference of two squares. However, if we factor the common factor 2 first, we get

$$50x^6 - 32y^2 = 2(25x^6 - 16y^2)$$
$$= 2[(5x^3)^2 - (4y)^2]$$
$$= 2(5x^3 + 4y)(5x^3 - 4y) \quad \blacksquare$$

EXAMPLE 10   Factor $(x^2 + y^2)^2 - (2xy)^2$.

Solution   If we let $A = (x^2 + y^2)$ and $B = 2xy$ in Identity 3, we get

$$(x^2 + y^2)^2 - (2xy)^2 = [(x^2 + y^2) + 2xy][(x^2 + y^2) - (2xy)]$$
$$= (x^2 + 2xy + y^2)(x^2 - 2xy + y^2)$$

But the first factor is an expression of the form given in Identity 1, and the second factor is of the form given in Identity 2. Thus,

$$(x^2 + y^2)^2 - (2xy)^2 = (x + y)^2(x - y)^2 \quad \blacksquare$$

It is sometimes necessary to group the terms in an expression before an identity can be used.

EXAMPLE 11 Factor $a^2 + b^2 - c^2 + 4c - 4 + 2ab$.

Solution Grouping the first, second, and last terms, and grouping the third, fourth, and fifth terms, we get

$$a^2 + b^2 - c^2 + 4c - 4 + 2ab = (a^2 + 2ab + b^2) - (c^2 - 4c + 4)$$
$$= (a + b)^2 - (c - 2)^2$$
$$= [(a + b) + (c - 2)][(a + b) - (c - 2)]$$
$$= (a + b + c - 2)(a + b - c + 2) \blacksquare$$

Examples Using Identity 4:
$$A^3 + B^3 = (A + B)(A^2 - AB + B^2)$$

EXAMPLE 12 Factor $27x^3 + 64$.

Solution The expression may be written as $(3x)^3 + (4)^3$. Letting $A = 3x$ and $B = 4$ and using Identity 4, we get

$$27x^3 + 64 = (3x + 4)[(3x)^2 - (3x)(4) + (4)^2]$$
$$= (3x + 4)(9x^2 - 12x + 16) \blacksquare$$

EXAMPLE 13 Factor $x^6 + y^6$.

Solution Writing x^6 as $(x^2)^3$ and y^6 as $(y^2)^3$, we see that

$$x^6 + y^6 = (x^2)^3 + (y^2)^3$$
$$= (x^2 + y^2)[(x^2)^2 - (x^2)(y^2) + (y^2)^2]$$
$$= (x^2 + y^2)(x^4 - x^2y^2 + y^4) \blacksquare$$

EXAMPLE 14 Factor $(a - b)^3 + c^3$.

Solution Let $A = (a - b)$ and $B = c$, and using Identity 4, we obtain

$$(a - b)^3 + c^3 = [(a - b) + c][(a - b)^2 - (a - b)(c) + c^2]$$
$$= (a - b + c)(a^2 - 2ab + b^2 - ac + bc + c^2) \blacksquare$$

Examples Using Identity 5:
$$A^3 - B^3 = (A - B)(A^2 + AB + B^2)$$

EXAMPLE 15 Factor $125x^3 - 8$.

Solution Writing the expression as $(5x)^3 - (2)^3$ and using Identity 5, we get

$$125x^3 - 8 = (5x - 2)[(5x)^2 + (5x)(2) + (2)^2]$$
$$= (5x - 2)(25x^2 + 10x + 4) \quad \blacksquare$$

EXAMPLE 16 Factor $343a^3 - 1$.

Solution
$$343a^3 - 1 = (7a)^3 - (1)^3$$
$$= (7a - 1)[(7a)^2 + (7a)(1) + (1)^2]$$
$$= (7a - 1)(49a^2 + 7a + 1) \quad \blacksquare$$

EXAMPLE 17 Factor $(x + 1)^3 - 27(x - 1)^3$.

Solution Let $A = (x + 1)$ and $B = 3(x - 1)$. Then

$$(x + 1)^3 - 27(x - 1)^3 = (x + 1)^3 - [3(x - 1)]^3$$
$$= [(x + 1) - 3(x - 1)][(x + 1)^2 + 3(x + 1)$$
$$\times (x - 1) + 9(x - 1)^2]$$
$$= (4 - 2x)(13x^2 - 16x + 7)$$
$$= 2(2 - x)(13x^2 - 16x + 7) \quad \blacksquare$$

EXAMPLE 18 Factor $x^3 + 2^3$ and expand $(x + 2)^3$.

Solution (a) $x^3 + 2^3 = (x + 2)(x^2 - 2x + 4)$

(b) $(x + 2)^3 = x^3 + 3 \cdot 2 \cdot x^2 + 3 \cdot 2^2 \cdot x + 2^3$
$$= x^3 + 6x^2 + 12x + 8 \quad \blacksquare$$

As Example 18 illustrates,

$$A^3 + B^3 \neq (A + B)^3$$

EXERCISES 5.4

In Exercises 1–24, factor using Identities 1 and 2.

1. $x^2 + 2ax + a^2$
2. $x^2 + 10x + 25$
3. $25x^2 + 10x + 1$
4. $a^2 - 2ay + y^2$
5. $y^2 - 20y + 100$
6. $81y^2 - 18y + 1$
7. $p^2q^2 - 2pqr + r^2$
8. $1 - 10c + 25c^2$
9. $m^2 + 14am + 49a^2$
10. $a^2x^2 + 2ax + 1$
11. $ab^2 - 2abc + ac^2$
12. $p^4 + 2p^2q^2 + q^4$

13. $4x^4 - 12x^2y^2 + 9y^4$
14. $a^4x^2 + 26a^2x + 169$
15. $x^4y^4 - 2cx^2y^2 + c^2$
16. $49b^4 - 14b^2y + y^2$
17. $x^4 + 4x^2y^2 + 4y^4$
18. $(a + b)^2 + 2(a + b) + 1$
19. $(a + 1)^2 - 2(a + 1)c + c^2$
20. $(m + n)^2x^2 + 2(m + n)x + 1$
21. $x^2 + 2xy + y^2 - 2(x + y) + 1$

22. $p^2 + 2pq + q^2 + 6(p + q) + 9$

23. $9(a + b)^2 x^2 + 12(a + b)x + 4$

24. $25 + 30k + 9k^2 + 12(3k + 5)m + 36m^2$

In Exercises 25–70, factor using Identity 3.

25. $x^2 - y^2$ 26. $a^2 - 1$

27. $m^2 - 4$ 28. $49 - y^2$

29. $(ab)^2 - 36$ 30. $25x^2 - y^2$

31. $1 - a^2 b^2$ 32. $16m^2 - 81n^2$

33. $(a^2)^2 - y^2$ 34. $169x^2 y^2 - 4$

35. $50y^2 - 2$ 36. $b^3 - b$

37. $9a^2 b^2 - c^2 d^2$ 38. $12 - 3m^2$

39. $m^4 - 4$ 40. $p^2 q^4 - t^4$

41. $121a^4 - 1$ 42. $x^4 y^6 - 25z^2$

43. $c^6 - 16d^2$ 44. $100 - k^8$

45. $9a^8 - 49b^4$ 46. $5a^2 - 20b^2$

47. $ab^2 - ad^2$ 48. $256x^2 y^4 - 81z^2$

49. $2 - 72m^4 n^6$ 50. $x^3 - xy^2$

51. $a^4 - a^2$ 52. $(a + b)^2 - 1$

53. $4(x - y)^2 - 9(xy)^2$ 54. $(a^2 + b)^2 - 25c^2$

55. $(x + y)^2 - (a - b)^2$

56. $3c^2 - 432(l + m)^2$

57. $ax^5 - axb^2$ 58. $x^4 - y^4$

59. $a^4 - 1$ 60. $a^3 b - ab^3$

61. $(a^2 + 2ab)^2 - b^4$

62. $(4p^2 - 4pq)^2 - q^4$

63. $a^2 + b^2 - c^2 + 4c - 4 + 2ab$

64. $9m^2 - 6mn + n^2 - 1$

65. $4x^2 y^2 - 9 + 4xyz + z^2$

66. $(c^2 + 4d^2)^2 - (4cd)^2$

67. $(p^2 + 5pq + q^2)^2 - 9p^2 q^2$

68. $16(x - y)^3 - (x - y)$

69. $(a - x)y^2 - 25(a - x)$

70. $(x^4 + 2x^2 y^2 + y^4) - x^2 y^2$

In Exercises 71–75, use factors to find the values.

71. $(134)^2 - (133)^2$ 72. $(372)^2 - (367)^2$

73. $(421)^2 - (179)^2$ 74. $(239)^2 - 121$

75. $(57 \cdot 6)^2 - (57 \cdot 2)^2$

In Exercises 76–100, factor using Identities 4 and 5.

76. $x^3 + y^3$ 77. $a^3 + 1$

78. $m^3 + (2n)^3$ 79. $(ab)^3 + y^3$

80. $64 + 27x^3$ 81. $m^3 - n^3$

82. $p^3 - 1$ 83. $(2a)^3 - 1$

84. $2x^3 - 2$ 85. $a^3 - 125$

86. $8x^3 - 27$ 87. $250 - 16m^3$

88. $3m^3 + 24$ 89. $xy^3 - x$

90. $a^4 + a$ 91. $(x^2)^3 - (y^2)^3$

92. $(a - 1)^3 + 8$

93. $(a + b)^3 - (a - b)^3$

94. $(x + y)^3 + (x - y)^3$

95. $c^6 - d^6$ 96. $x^2 a^6 - 729x^8$

97. $64 + a^6$ 98. $(8c^3)^2 - 1$

99. $128a^6 + 2$

100. $(2a - b)^3 - (2a)^3$

5.5 FACTORING OTHER POLYNOMIALS

EXAMPLE 1   Factor $a^3 + 3a^2 b + 3ab^2 + b^3$.

Solution   Grouping the first and last terms, we obtain $a^3 + b^3$, which may be factored using Identity 4 of Section 5.4. Also, the middle terms, $3a^2 b + 3ab^2$, have the common factor $3ab$. Thus,

$$a^3 + 3a^2 b + 3ab^2 + b^3 = a^3 + b^3 + 3a^2 b + 3ab^2$$

$$= (a + b)(a^2 - ab + b^2) + 3ab(a + b)$$

$$= (a + b)[(a^2 - ab + b^2) + 3ab]$$

$$= (a + b)(a^2 + 2ab + b^2)$$

$$= (a + b)(a + b)^2$$

$$= (a + b)^3 \quad \blacksquare$$

We have established a very important identity that occurs often in mathematics:

$$(a + b)^3 = a^3 + 3a^2b + 3ab^2 + b^3$$

EXAMPLE 2 Factor $a^3 - 3a^2b + 3ab^2 - b^3$.

Solution Using the procedure given in Example 1, we get

$$
\begin{aligned}
a^3 - 3a^2b + 3ab^2 - b^3 &= a^3 - b^3 - 3a^2b + 3ab^2 \\
&= (a - b)(a^2 + ab + b^2) - 3ab(a - b) \\
&= (a - b)[(a^2 + ab + b^2) - 3ab] \\
&= (a - b)(a^2 - 2ab + b^2) \\
&= (a - b)(a - b)^2 \\
&= (a - b)^3 \quad \blacksquare
\end{aligned}
$$

Thus,

$$(a - b)^3 = a^3 - 3a^2b + 3ab^2 - b^3$$

By adding or subtracting a perfect square, certain expressions take the form of the difference of two squares.

EXAMPLE 3 Factor $a^4 + a^2b^2 + b^4$.

Solution If the middle term had a coefficient of 2, the expression would be a perfect square. To this end, we add and subtract a^2b^2 to obtain

$$
\begin{aligned}
a^4 + a^2b^2 + b^4 &= a^4 + a^2b^2 + b^4 + a^2b^2 - a^2b^2 \\
&= a^4 + 2a^2b^2 + b^4 - a^2b^2 \\
&= (a^2 + b^2)^2 - a^2b^2 \\
&= [(a^2 + b^2) + ab][(a^2 + b^2) - ab] \\
&= (a^2 + b^2 + ab)(a^2 + b^2 - ab) \quad \blacksquare
\end{aligned}
$$

EXAMPLE 4 Factor $x^4 - 23x^2y^2 + y^4$.

Solution Adding and subtracting $25x^2y^2$, we get

$$
\begin{aligned}
x^4 - 23x^2y^2 + y^4 &= x^4 - 23x^2y^2 + y^4 + 25x^2y^2 - 25x^2y^2 \\
&= x^4 + 2x^2y^2 + y^4 - 25x^2y^2
\end{aligned}
$$

$$= (x^2 + y^2)^2 - 25x^2y^2$$
$$= [(x^2 + y^2) + 5xy][(x^2 + y^2) - 5xy]$$
$$= (x^2 + y^2 + 5xy)(x^2 + y^2 - 5xy) \quad \blacksquare$$

EXAMPLE 5   Factor $a^2x^2 - a^2y^2 - b^2x^2 + b^2y^2$.

Solution
$$a^2x^2 - a^2y^2 - b^2x^2 + b^2y^2 = a^2(x^2 - y^2) - b^2(x^2 - y^2)$$
$$= (x^2 - y^2)(a^2 - b^2)$$
$$= (x + y)(x - y)(a + b)(a - b) \quad \blacksquare$$

EXAMPLE 6 $ax^2 - ay^2 + ax + ay$.

Solution   First we remove the common factor a, and then use the grouping method to factor the resulting factor:

$$ax^2 - ay^2 + ax + ay = a[x^2 - y^2 + x + y]$$
$$= a[(x^2 - y^2) + (x + y)]$$
$$= a[(x + y)(x - y) + (x + y)]$$
$$= a(x + y)[(x - y) + 1]$$
$$= a(x + y)(x - y + 1) \quad \blacksquare$$

EXERCISES   5.5

Factor:

1. $a^3 - 6a^2b + 12ab^2 - 8b^3$
2. $a^3x^3 + 9a^2x^2b + 27axb^2 + 27b^3$
3. $1 - 3b + 3b^2 - b^3$
4. $x^3y^3 - 3x^2y^2 + 3xy - 1$
5. $y^4 + b^2y^2 + b^4$

6. $x^4 - 14x^2y^2 + y^4$
7. $x^3 - 2x^2 - x + 2$
8. $(x + y)^2 + x + y$
9. $3a^4 - 2a^3 + 3a^2 - 2a$
10. $9a^2 + 6a + 1 - 4b^2 - 2b - 1$

Chapter 5 SUMMARY

FACTORING POLYNOMIALS

1. Remove all common factors.
2. Depending on the number of terms in the remaining polynomial, factor as follows:
   (a) If the polynomial has two terms, try to factor as:
   > difference of two squares
   > difference of two cubes
   > sum of two cubes

(b) If the polynomial has three terms, try to factor as:
   > trinomial

(c) If the polynomial has four or more terms, try to factor by:
   > grouping

IDENTITIES

1. $(A + B)^2 = A^2 + 2AB + B^2$
2. $(A - B)^2 = A^2 - 2AB + B^2$
3. $A^2 - B^2 = (A + B)(A - B)$
4. $A^3 + B^3 = (A + B)(A^2 - AB + B^2)$

5. $A^3 - B^3 = (A - B)(A^2 + AB + B^2)$
6. $(A + B)^3 = (A^3 + 3A^2B + 3AB^2 + B^3)$
7. $(A - B)^3 = (A^3 - 3A^2B + 3AB^2 - B^3)$
8. $A - B = -(B - A)$

Chapter 5 EXERCISES

Factor:

1. $x^2 + xy$
2. $x^3 - x^2y$
3. $10x^3 - 25x^4y$
4. $x^3 - x^2y + xy^2$
5. $3a^4 - 3a^3b + 6a^2b^2$
6. $38a^3x^5 + 57a^4x^2$
7. $ax - bx - az + bz$
8. $2ax + ay + 2bx + by$
9. $6x^2 + 3xy - 2ax - ay$
10. $2x^4 - x^3 + 4x - 2$
11. $3x^3 + 5x^2 + 3x + 5$
12. $x^4 + x^3 + 2x + 2$
13. $y^3 - y^2 + y - 1$
14. $2ax^2 + 3axy - 2bxy - 3by^2$
15. $x^2 - 19x + 84$
16. $x^2 - 19x + 78$
17. $a^2 - 14ab + 49b^2$
18. $a^2 + 5ab + 6b^2$
19. $m^2 - 13mn + 40n^2$
20. $m^2 - 22mn + 105n^2$
21. $x^2 - 23xy + 132y^2$
22. $130 + 31xy + x^2y^2$
23. $132 - 23x + x^2$
24. $88 + 19x + x^2$
25. $65 + 8xy - x^2y^2$
26. $x^2 + 16x - 260$
27. $x^2 - 11x - 26$
28. $a^2b^2 - 3abc - 10c^2$
29. $x^4 - a^2x^2 - 132a^4$
30. $4x^2 + 23x + 15$
31. $12x^2 - 23xy + 10y^2$
32. $8x^2 - 38x + 35$
33. $12x^2 - 31x - 15$
34. $3 + 11x - 4x^2$
35. $6 + 5x - 6x^2$
36. $4 - 5x - 6x^2$
37. $5 + 32x - 21x^2$
38. $20 - 9x - 20x^2$
39. $25 - 64x^2$
40. $81p^4z^6 - 25b^2$
41. $(1811)^2 - (689)^2$
42. $(8133)^2 - (8131)^2$
43. $(24x + y)^2 - (23x - y)^2$
44. $(5x + 2y)^2 - (3x - y)^2$
45. $9x^2 - (3x - 5y)^2$
46. $16x^2 - (3x + 1)^2$
47. $a^6 + 729b^3$
48. $x^3y^3 - 512$
49. $500x^2y - 20y^3$
50. $(a + b)^4 - 1$
51. $(c + d)^3 + (c - d)^3$
52. $x^4y - x^2y^3 - x^3y^2 + xy^4$
53. $x^4 - 6x^2y^2 + y^4$
54. $ab(x^2 + 1) + x(a^2 + b^2)$
55. $a^3 + (a + b)x + bx^2$
56. $m^{12} + n^{12}$
57. $m^6 - n^3$
58. $x^4 - 7x^2y^2 + y^4$
59. $x^4 - 15x^2y^2 + 9y^4$
60. $a^9 - 64a^3 - a^6 + 64$

6 Fractional Expressions and Equations

Throughout this book we have assumed a knowledge of the rules for the operations with rational numbers. In this chapter we shall extend these ideas to rational expressions, which are quotients of polynomials.

6.1 RATIONAL EXPRESSIONS

In Chapter 1 we saw that a rational number can be represented in the form a/b, where a and b are integers and $b \neq 0$. The numeral a/b is also called a fraction whose numerator is a and whose denominator is b. For example, the fraction $\frac{3}{4}$ has 3 as its numerator and 4 as its denominator. However, a fraction has many equivalent names (numerals). The rational number $\frac{3}{4}$ may be named equivalently by the fractions

$$\frac{6}{8}, \quad \frac{15}{20}, \quad \frac{75}{100}, \quad \text{and} \quad \frac{9}{12}$$

One way to test whether two fractions are equivalent is to use the following fact: two fractions, a/b and c/d, are equivalent if and only if $ad = bc$. That is, we say

Rule for Equivalence of Fractions
$\dfrac{a}{b} = \dfrac{c}{d}$ if and only if $ad = bc$

EXAMPLE 1 The fraction $\frac{2}{3}$ is equivalent to $\frac{12}{18}$ because $2 \cdot 18 = 3 \cdot 12$. However, the fraction $\frac{3}{5}$ is *not* equivalent to $\frac{7}{10}$ because $7 \cdot 5 \neq 10 \cdot 3$.

In algebra, we can extend the idea of a fraction to include polynomials as possible numerators and denominators. Such fractions are called rational expressions.

DEFINITION

An expression that can be expressed in the form $\dfrac{P}{Q}$, where P and Q are polynomials and $Q \neq 0$, is called a rational expression.

Examples of rational expressions are:

$$\frac{2}{x}, \quad \frac{x-1}{2x+3}, \quad \frac{x^2-4}{x+2}, \quad \frac{x^2+x-2}{x-1}, \quad \frac{6x^3-12x^2}{1}$$

The property of equivalence can be extended to rational expressions. That is,

---

Rule for Equivalence of Rational Expressions

If $\dfrac{P}{Q}$ and $\dfrac{R}{S}$ are rational expressions, with $Q \neq 0$ and $S \neq 0$, then

$$\frac{P}{Q} = \frac{R}{S} \quad \text{if and only if} \quad P \cdot S = Q \cdot R$$

EXAMPLE 2 Determine which pairs of rational expressions are equivalent:

(a) $\dfrac{4x}{18x^2}$ and $\dfrac{2}{9x}$ (b) $\dfrac{1}{x-3}$ and $\dfrac{x+2}{x^2-x-6}$

(c) $\dfrac{x+2}{x+1}$ and $\dfrac{x}{x-1}$ (d) $\dfrac{a^2-9}{a+3}$ and $a-3$

Solution (a) Since $(4x)(9x) = (18x^2)(2)$, then $\dfrac{4x}{18x^2} = \dfrac{2}{9x}$.

(b) Since $(x-3)(x+2) = x^2 - x - 6$, then $\dfrac{1}{x-3} = \dfrac{x+2}{x^2-x-6}$.

(c) Since $(x+2)(x-1) \neq x(x+1)$, then $\dfrac{x+2}{x+1} \neq \dfrac{x}{x-1}$.

(d) First, we note that $a - 3$ can be written $\dfrac{a-3}{1}$. Since $(a^2 - 9)(1) = (a+3)(a-3)$, then $\dfrac{a^2-9}{a+3} = a - 3$. ∎

We now discuss ways to change a fraction into an equivalent fraction. One of the fundamental principles of fractions is that the value of a fraction is not changed if the numerator and denominator are both multiplied or divided by the same nonzero quantity. That is, if $\frac{a}{b}$ is a fraction and k is a real number, where $b \neq 0$ and $k \neq 0$, then

$$\frac{a}{b} = \frac{ak}{bk}$$

EXAMPLE 3

$$\frac{4}{7} = \frac{4 \cdot 2}{7 \cdot 2} = \frac{8}{14}$$

EXAMPLE 4

$$\frac{5}{6} = \frac{?}{42}$$

Solution

$$\frac{5}{6} = \frac{5 \cdot 7}{6 \cdot 7} = \frac{35}{42} \quad \blacksquare$$

As with equivalence, this fundamental principle extends to rational expressions.

Fundamental Principle of Fractions

For all polynomials P, Q, and K, where $Q \neq 0$ and $K \neq 0$, $\dfrac{P}{Q} = \dfrac{PK}{QK}$.

EXAMPLE 5

$$\frac{4x}{y} = \frac{?}{yx}$$

Solution

$$\frac{4x}{y} = \frac{4x \cdot x}{y \cdot x} = \frac{4x^2}{yx} \quad \blacksquare$$

EXAMPLE 6

$$\frac{x + 2}{x - 5} = \frac{?}{x^2 - 25}$$

Solution

$$\frac{x + 2}{x - 5} = \frac{(x + 2)(x + 5)}{(x - 5)(x + 5)} = \frac{x^2 + 7x + 10}{x^2 - 25} \quad \blacksquare$$

EXAMPLE 7

$$\frac{3a + b}{a + b} = \frac{?}{2a^2 + 5ab + 3b^2}$$

Solution   We begin by factoring the polynomial $2a^2 + 5ab + 3b^2$, obtaining $2a^2 + 5ab + 3b^2 = (2a + 3b)(a + b)$. Thus, the problem can be stated as

$$\frac{3a + b}{a + b} = \frac{?}{(a + b)(2a + 3b)}$$

Hence,

$$\frac{3a + b}{a + b} = \frac{(3a + b)(2a + 3b)}{(a + b)(2a + 3b)} = \frac{6a^2 + 11ab + 3b^2}{2a^2 + 5ab + 3b^2} \quad \blacksquare$$

In Examples 5–7, we multiplied the numerator and denominator by quantities to obtain equivalent rational expressions. However, we can also divide the numerator and denominator by nonzero quantities to obtain an equivalent, but reduced, rational expression. For example, dividing the numerator and denominator by 2, we see that $\frac{12}{16} = \frac{6}{8}$.

A fraction is said to be *reduced to lowest terms* when its numerator and denominator have no common factor other than 1. For example, the fractions $\frac{3}{5}, \frac{2}{7}$, and $\frac{-3}{14}$ are in lowest terms, while $\frac{6}{8}$ and $\frac{7}{21}$ are not in lowest form since each fraction has a common factor in its numerator and denominator; reduced to their lowest terms, we have $\frac{6}{8} = \frac{3}{4}$ (dividing numerator and denominator by 2) and $\frac{7}{21} = \frac{1}{3}$ (dividing numerator and denominator by 7).

This idea of reduction to lowest terms can be extended to rational expressions. The rational expressions $\dfrac{x}{x - 1}, \dfrac{x + 5}{x - 7}$, and $\dfrac{x^2 + 3}{x^3 - 5x + 9}$ are in lowest terms, since the numerator and denominator of each have no common factors. However, the rational expression $\dfrac{4x^2}{7xy}$ is *not* in lowest terms since x is a common factor in both the numerator and the denominator.

EXAMPLE 8 Reduce $\dfrac{4x^2}{7xy}$ to its lowest terms.

Solution Dividing numerator and denominator by x, we have $\dfrac{4x^2}{7xy} = \dfrac{4x}{7y}$. $\blacksquare$

When reducing a fraction, the division of the numerator and the denominator by a common factor can be denoted by a *cancellation*.

EXAMPLE 9

$$\frac{7(x - 6)}{8(x - 6)} = \frac{7(\overset{1}{\cancel{x - 6}})}{8(\underset{1}{\cancel{x - 6}})} = \frac{7}{8}$$

How to Reduce Fractions to Lowest Terms

1. Completely factor the numerator and the denominator.
2. Cancel all common factors that exist in both the numerator and the denominator.

EXAMPLE 10 Reduce $\dfrac{10x^3y^2}{12x^2y^3}$ to its lowest terms.

Solution Factoring the common factor $2x^2y^2$ from the numerator and denominator, we get

$$\frac{10x^3y^2}{12x^2y^3} = \frac{5x(2x^2y^2)}{6y(2x^2y^2)} = \frac{5x}{6y} \quad\blacksquare$$

EXAMPLE 11 Reduce $\dfrac{3x - 9y}{5x - 15y}$ to its lowest terms.

Solution Factoring the numerator and the denominator, we get

$$\frac{3x - 9y}{5x - 15y} = \frac{3\overset{1}{\cancel{(x - 3y)}}}{5\underset{1}{\cancel{(x - 3y)}}} = \frac{3}{5} \quad\blacksquare$$

Note. To economize time in writing, the 1's obtained as a result of cancellation may be omitted. *It should be remembered, however, that if all the factors cancel, the answer is 1, not 0.*

EXAMPLE 12 Reduce $\dfrac{x^3 - 4x}{x^3 + 5x^2 + 6x}$ to its lowest terms.

Solution Factoring the numerator and denominator, we have

$$\frac{x^3 - 4x}{x^3 + 5x^2 + 6x} = \frac{x(x^2 - 4)}{x(x^2 + 5x + 6)} = \frac{\cancel{x}\cancel{(x + 2)}(x - 2)}{\cancel{x}\cancel{(x + 2)}(x + 3)} = \frac{x - 2}{x + 3} \quad\blacksquare$$

EXAMPLE 13 Reduce $\dfrac{4a^3 + 8a^2 + 16a}{2a^3 - 16}$ to its lowest terms.

Solution
$$\frac{4a^3 + 8a^2 + 16a}{2a^3 - 16} = \frac{4a(a^2 + 2a + 4)}{2(a^3 - 8)} \qquad \text{[elementary factoring]}$$

$$= \frac{\overset{2}{\cancel{4}}\,a\cancel{(a^2 + 2a + 4)}}{\cancel{2}\,(a - 2)\cancel{(a^2 + 2a + 4)}} \qquad \text{[difference of cubes]}$$

$$= \frac{2a}{a - 2} \quad\blacksquare$$

Recall that $(a - b) = (-1)(b - a)$; thus, $\dfrac{a - b}{b - a} = -1$.

EXAMPLE 14 Reduce $\dfrac{x^2 - 5x + 6}{9 - x^2}$ to its lowest terms.

Solution

$$\frac{x^2 - 5x + 6}{9 - x^2} = \frac{(x - 2)(\overset{-1}{\cancel{x - 3}})}{(3 + x)(\cancel{3 - x})} = \frac{-(x - 2)}{(3 + x)} \quad \blacksquare$$

EXAMPLE 15 Reduce $\dfrac{xy - 2x + 3y - 6}{xy - 2x - 3y + 6}$ to its lowest terms.

Solution

$$\frac{xy - 2x + 3y - 6}{xy - 2x - 3y + 6} = \frac{x(y - 2) + 3(y - 2)}{x(y - 2) - 3(y - 2)} = \frac{(\cancel{y - 2})(x + 3)}{(\cancel{y - 2})(x - 3)} = \frac{x + 3}{x - 3} \quad \blacksquare$$

EXAMPLE 16 Reduce $\dfrac{2x - y}{(4x^2 - y^2) + 2x - y}$ to its lowest terms.

Solution

$$\frac{2x - y}{(4x^2 - y^2) + 2x - y} = \frac{2x - y}{(4x^2 - y^2) + (2x - y)}$$

$$= \frac{(2x - y)}{(2x - y)(2x + y) + (2x - y)}$$

$$= \frac{(\cancel{2x - y})}{(\cancel{2x - y})(2x + y + 1)}$$

$$= \frac{1}{2x + y + 1} \quad \blacksquare$$

Note that when $2x - y$ is cancelled from the numerator, it leaves 1, not 0.

CAUTION

The student should take special caution to remember that it is *not* correct to cancel a quantity from the numerator and denominator unless it is a factor of the *entire* numerator and of the *entire* denominator. $\dfrac{7x}{7 + y} = \dfrac{x}{y}$ is *wrong*, because 7 is not a factor of the denominator. The fraction $\dfrac{8 + 3(x - y)}{5 + 2(x - y)}$ is in its lowest terms. We cannot cancel $(x - y)$ since $(x - y)$ is a factor of only a part of the numerator and part of the denominator.

EXERCISES 6.1

In Exercises 1–12, determine whether the given pairs of fractions are equivalent.

1. $\dfrac{4}{3}, \dfrac{5}{4}$

2. $\dfrac{-3}{17}, \dfrac{3}{-17}$

3. $\dfrac{-5}{-7}, \dfrac{10}{14}$

4. $\dfrac{7x}{3}, \dfrac{21x}{9}$

26. $\dfrac{a^2 - b^2}{(a - b)^2}$

27. $\dfrac{3x + 12}{x^2 - 16}$

5. $\dfrac{8}{x}, \dfrac{16x}{2x^2}$

6. $\dfrac{a + b}{y}, \dfrac{6a + b}{6y}$

28. $\dfrac{x^2 - y^2}{x^3 - y^3}$

29. $\dfrac{4x + 4y}{x^3 + y^3}$

7. $\dfrac{x + y}{1}, \dfrac{ax + ay}{a}$

30. $\dfrac{m^2 + m - 56}{m^2 - m - 42}$

31. $\dfrac{y^2 - 9y + 18}{y^2 + y - 12}$

8. $\dfrac{x^2 + x - 2}{x + 2}, \dfrac{4x - 4}{4}$

32. $\dfrac{3m^2 - 3n^2}{3m^2 + 6mn + 3n^2}$

9. $\dfrac{x^2 - 4}{x - 2}, x + 2$

33. $\dfrac{ax^2 - ax - 12a}{3x^2 + 13x + 12}$

34. $\dfrac{4 - 4x + x^2}{x^2 + x - 6}$

10. $\dfrac{a + b}{1}, \dfrac{a^3 + b^3}{a^2 - ab + b^2}$

35. $\dfrac{x^2 - 4}{6 + x - x^2}$

36. $\dfrac{x^2 - x^3}{x^2 - 1}$

11. $\dfrac{a^2 + b^2}{1}, \dfrac{a^4 - b^4}{a^2 - b^2}$

12. $\dfrac{y - 3}{y + 3}, \dfrac{-3 - y}{3 - y}$

37. $\dfrac{1 - 2x}{8x^2 - 2}$

38. $\dfrac{3 - 2x}{2x^2 - 11x + 12}$

In Exercises 13–21, use the fundamental principle of fractions to determine the missing expressions.

39. $\dfrac{4x^2 - 4ax + a^2}{x^2 + ax - 6x^2}$

40. $\dfrac{x^2 + 2ax + a^2}{x^2 - 2ax + a^2}$

13. $\dfrac{17}{19} = \dfrac{?}{57}$

14. $\dfrac{35}{24} = \dfrac{105}{?}$

41. $\dfrac{x^4 - 16}{x^4 - x^2 - 12}$

15. $\dfrac{2xy^3}{17x^2y} = \dfrac{?}{34x^3y^4}$

16. $\dfrac{36}{m} = \dfrac{?}{m^2np^3}$

42. $\dfrac{a^3 + a - (a^2 + 1)}{a^3 + a^2 - (a + 1)}$

17. $\dfrac{(a - b)}{4} = \dfrac{7(a - b)^3}{?}$

43. $\dfrac{x^2 - 3x - ax + 3a}{3x - x^2 + 3b - bx}$

18. $\dfrac{12}{4x(x - 2)(x + 6)} = \dfrac{?}{16x^2(x - 2)^2(x + 6)}$

44. $\dfrac{x^2 - 12x - y^2 + 12y}{x^2 + 2xy + y^2 - 144}$

19. $\dfrac{x + 4}{x^2 - 16} = \dfrac{x^2 + 4x}{?}$

45. $\dfrac{5x^2 + 8x - 21}{7y - 5xy + 7z - 5xz}$

*20. $\dfrac{-3}{x^2 + 2x + 4} = \dfrac{?}{x^3 - 8}$

46. $\dfrac{(1 + xy)^2 - (x + y)^2}{1 - x^2}$

*21. $\dfrac{ab}{a^2 - ab + b^2} = \dfrac{?}{a^3 + b^3}$

*47. $\dfrac{2mx - my - 12nx + 6ny}{6mx - 3my - 2nx + ny}$

In Exercises 22–50, reduce each fraction to its lowest terms.

*48. $\dfrac{x^2 - a^2 + 2ab - b^2}{x^2 + a^2 + 2ax - b^2}$

22. $\dfrac{5x^3y^4}{3xy^5}$

23. $\dfrac{12m^2n^3}{20m^3n^2}$

*49. $\dfrac{(x^3 - 1)(x^2 - 1)}{(x - 1)^2(x + 1)^3}$

24. $\dfrac{2(a + b)}{a^2 - b^2}$

25. $\dfrac{3x - 3y}{x^2 - y^2}$

*50. $\dfrac{a^2 - a - 2}{a^3 + 1 + (a + 1)^2}$

6.2 MULTIPLICATION OF FRACTIONAL EXPRESSIONS

Recall from arithmetic that when multiplying two fractions, the product is obtained by multiplying numerators together and denominators together. That is, if a/b and c/d are fractions, with $b \neq 0$ and $d \neq 0$, then

$$\frac{a}{b} \cdot \frac{c}{d} = \frac{a \cdot c}{b \cdot d}$$

EXAMPLE 1

$$\frac{4}{5} \cdot \frac{3}{7} = \frac{4 \cdot 3}{5 \cdot 7} = \frac{12}{35}$$

Multiplication of rational expressions is the same as multiplication of rational numbers.

Multiplication of Rational Expressions

If P, Q, R, and S are polynomials, with $Q \neq 0$ and $S \neq 0$, then

$$\frac{P}{Q} \cdot \frac{R}{S} = \frac{P \cdot R}{Q \cdot S}$$

EXAMPLE 2

$$\frac{13x^2}{11y} \cdot \frac{3y^2}{4x^3} = \frac{39x^2y^2}{44x^3y} = \frac{39y}{44x}$$

If the numerators and denominators are written in factored form, and common factors occurring in a numerator and a denominator are cancelled, then the resulting product will be in lowest terms.

EXAMPLE 3 Multiply $\dfrac{x^2 + 3x}{5x} \cdot \dfrac{x - 3}{7x + 21}$.

Solution

$$\frac{x^2 + 3x}{5x} \cdot \frac{x - 3}{7x + 21} = \frac{x(x + 3)}{5x} \cdot \frac{x - 3}{7(x + 3)} = \frac{x - 3}{35} \quad \blacksquare$$

How to Multiply Rational Expressions

1. Completely factor the numerators and denominators of the given fractions.
2. Cancel any factor common to any numerator and any denominator.
3. Multiply the remaining factors of the numerators to find the numerator of the product. Multiply the remaining factors of the denominators to find the denominator of the product. The resulting product will be in lowest terms.

EXAMPLE 4 Multiply $\dfrac{x^2 - 14x - 15}{x^2 - 4x - 45} \cdot \dfrac{x^2 - 6x - 27}{x^2 - 12x - 45}$.

Solution

$$\frac{x^2 - 14x - 15}{x^2 - 4x - 45} \cdot \frac{x^2 - 6x - 27}{x^2 - 12x - 45} = \frac{(x - 15)(x + 1)}{(x + 5)(x - 9)} \cdot \frac{(x - 9)(x + 3)}{(x - 15)(x + 3)}$$

$$= \frac{x + 1}{x + 5} \quad \blacksquare$$

EXAMPLE 5 Multiply $\dfrac{2x^2 - xy - y^2}{x^2 + 2xy + y^2} \cdot \dfrac{x^2 - y^2}{4x^2 + 4xy + y^2}$.

Solution

$$\frac{2x^2 - xy - y^2}{x^2 + 2xy + y^2} \cdot \frac{x^2 - y^2}{4x^2 + 4xy + y^2} = \frac{(2x + y)(x - y)}{(x + y)(x + y)} \cdot \frac{(x + y)(x - y)}{(2x + y)(2x + y)}$$

$$= \frac{(x - y)(x - y)}{(x + y)(2x + y)}$$

$$= \frac{(x - y)^2}{(x + y)(2x + y)} \quad \blacksquare$$

EXAMPLE 6 Multiply $\dfrac{27a^4 - 8a}{6a^2 + 5a - 1} \cdot \dfrac{6a - 1}{9a^2 + 6a + 4} \cdot \dfrac{a + 1}{3a^2 - 2a}$.

Solution

$$\frac{27a^4 - 8a}{6a^2 + 5a - 1} \cdot \frac{6a - 1}{9a^2 + 6a + 4} \cdot \frac{a + 1}{3a^2 - 2a}$$

$$= \frac{a(27a^3 - 8)}{(6a - 1)(a + 1)} \cdot \frac{6a - 1}{(9a^2 + 6a + 4)} \cdot \frac{a + 1}{a(3a - 2)}$$

$$= \frac{a(3a - 2)(9a^2 + 6a + 4)}{(6a - 1)(a + 1)} \cdot \frac{(6a - 1)}{(9a^2 + 6a + 4)} \cdot \frac{(a + 1)}{a(3a - 2)} = 1 \quad \blacksquare$$

EXAMPLE 7 Multiply $\dfrac{ax - bx + ay - by}{x^2 - y^2} \cdot \dfrac{y^2 + 2xy + x^2}{a^2 - b^2}$.

Solution

$$\frac{ax - bx + ay - by}{x^2 - y^2} \cdot \frac{y^2 + 2xy + x^2}{a^2 - b^2}$$

$$= \frac{x(a - b) + y(a - b)}{(x + y)(x - y)} \cdot \frac{(y + x)(y + x)}{(a + b)(a - b)}$$

$$= \frac{(a - b)(x + y)}{(x + y)(x - y)} \cdot \frac{(y + x)(y + x)}{(a + b)(a - b)}$$

$$= \frac{(y + x)(y + x)}{(x - y)(a + b)}$$

$$= \frac{(y + x)^2}{(x - y)(a + b)} \quad \blacksquare$$

EXERCISES 6.2

Find the following products in lowest terms.

1. $\dfrac{5}{2} \cdot \dfrac{3}{4}$

2. $\dfrac{4}{7} \cdot \dfrac{5}{3}$

3. $\dfrac{18}{7} \cdot \dfrac{2}{3}$

4. $\dfrac{3}{8} \cdot \dfrac{24}{5}$

5. $\dfrac{-3}{5} \cdot \dfrac{25}{-12}$

6. $\dfrac{27}{16} \cdot \dfrac{4}{9}$

7. $\dfrac{-5}{-12} \cdot \dfrac{18}{15}$

8. $\dfrac{12}{18} \cdot \dfrac{35}{14}$

9. $\dfrac{ab^2}{x^2y^2} \cdot \dfrac{x^2y}{b^2}$

10. $\dfrac{2x^3y}{c^3d} \cdot \dfrac{c^2d^3}{6xy^3}$

11. $\dfrac{3mn}{2p^4q} \cdot \dfrac{4p^3q^3}{6m^2n^3}$

12. $\dfrac{6x^3y^2}{9a^2b^3} \cdot \dfrac{3a^3b^2}{2x^2y^3}$

13. $\dfrac{4m^3}{10cd} \cdot \dfrac{5c^2}{m^3n}$

14. $\dfrac{6a^3b}{14p^4q} \cdot \dfrac{aq}{a^2b}$

15. $\dfrac{3ax^2}{b^2y} \cdot \dfrac{2ab}{6x^2y}$

16. $\dfrac{12m^4n^2}{9n^3b} \cdot \dfrac{3nb^2}{2bm^4}$

17. $\dfrac{-abc}{axy} \cdot \dfrac{axy}{bcx}$

18. $\dfrac{23a^2c^3}{51x^2y^4} \cdot \dfrac{17x^3y^2}{ac^4} \cdot -cx$

19. $\dfrac{-12m^2c^3x}{y^2} \cdot \dfrac{15y^2}{m^2cx^3} \cdot \dfrac{x^2y^3}{144a^4c^2x}$

20. $21abc \cdot \dfrac{31a^2xy}{7b^2c} \cdot \dfrac{5b}{93xy} \cdot \dfrac{1}{15}$

21. $\dfrac{3x - 6y}{4xy} \cdot \dfrac{16y^2}{9x^3}$

22. $\dfrac{2x^2 + 5x}{6y^3} \cdot \dfrac{18y^5}{11x^3}$

23. $\dfrac{10y}{x^3 - 4x^2} \cdot \dfrac{4x^5}{45x^3z}$

24. $\dfrac{16b}{21a^3} \cdot \dfrac{7ab}{2a + 8b}$

25. $\dfrac{2x - 12}{20xy^2} \cdot \dfrac{8x^2}{3x - 18}$

26. $\dfrac{42x^2y}{5x + 10y} \cdot \dfrac{6y + 3x}{84x^3}$

27. $\dfrac{27a^2b}{a^2 - 4b^2} \cdot \dfrac{3a + 6b}{3ab^3}$

28. $\dfrac{x^2 - 25}{x + 1} \cdot \dfrac{24x^2b}{6x + 30}$

29. $\dfrac{3x + 9}{4x + 12} \cdot \dfrac{5y + 15}{6x - 12}$

30. $\dfrac{6b^2 - 24a^2}{9a - 9b} \cdot \dfrac{ab - a^2}{2a^2 + ab}$

31. $\dfrac{x^2 - x - 12}{3x^2 - 27} \cdot \dfrac{x^2 + 8x + 15}{x^2 - 9x + 20}$

32. $\dfrac{27 - y^3}{6 + 7y - 3y^2} \cdot \dfrac{3y^2 - 4y - 4}{y^2 - 4}$

33. $\dfrac{6x^2 + x - 12}{4 + x - 3x^2} \cdot \dfrac{x^2 + 3x + 2}{2x^2 + 7x + 6}$

34. $\dfrac{y^4 - 125y}{y^2 + 5y + 25} \cdot \dfrac{4y - 2}{2y^2 - 11y + 5}$

35. $\dfrac{5x + 1}{6x^2 + x - 1} \cdot \dfrac{9x^2 - 1}{3x^2 - 14x - 5}$
$\cdot \dfrac{x^2 - 2x - 15}{5x^2 + 16x + 3}$

36. $\dfrac{14x^2 + 19x - 3}{5x^2 - 7x + 2} \cdot \dfrac{25x^2 - 4}{7x - 1}$
$\cdot \dfrac{x - x^2}{(2x + 3)(5x + 2)}$

37. $\dfrac{x^4 + x^2y^2 + y^4}{x^3 + y^3} \cdot \dfrac{x + y}{x^3 - y^3}$

*38. $\dfrac{(y + z)^2 - x^2}{y^2 + yz + yx} \cdot \dfrac{(z + x)^2 - y^2}{(x + y)^2 - z^2}$
$\cdot \dfrac{yx + y^2 - yz}{y + z - x}$

*39. $\dfrac{ab^2 - ad^2}{b^3 + d^3} \cdot \dfrac{ab^2 - abd + ad^2}{a^3 - a}$
$\cdot \dfrac{ab - b + ad - d}{ab}$

*40. $\dfrac{9 - (x + 3)^2}{x^2 - 2xy + 2y^2} \cdot \dfrac{x^4 + 4y^4}{x^3 + 6x^2}$
$\cdot \dfrac{x - xy}{x^2 + 2xy + 2y^2}$

6.3 DIVISION OF FRACTIONAL EXPRESSIONS

Recall from arithmetic that when we divide one fraction by another, we invert the divisor and multiply. That is, if a/b and c/d are fractions, with $b \neq 0$ and $cd \neq 0$, then

$$\frac{a}{b} \div \frac{c}{d} = \frac{a}{b} \cdot \frac{d}{c} = \frac{ad}{bc}$$

EXAMPLE 1

$$\frac{3}{8} \div \frac{15}{16} = \frac{3}{\overset{}{8}} \cdot \frac{\overset{2}{16}}{\underset{5}{15}} = \frac{2}{5}$$

The rule for rational expressions is the same.

Division of Rational Expressions

If P, Q, R, and S are polynomials, with $Q \neq 0$ and $R \cdot S \neq 0$, then

$$\frac{P}{Q} \div \frac{R}{S} = \frac{P}{Q} \cdot \frac{S}{R} = \frac{P \cdot S}{Q \cdot R}$$

EXAMPLE 2

$$\frac{7x^3}{36y^2} \div \frac{x^2}{6y^3} = \frac{7x^{\overset{x}{3}}}{36y^{\underset{6}{2}}} \cdot \frac{6y^{\overset{y}{3}}}{x^{2}} = \frac{7xy}{6}$$

EXAMPLE 3

$$\frac{4x^2 - 8x}{3} \div \frac{x^2 - 4}{18} = \frac{4x^2 - 8x}{3} \cdot \frac{18}{x^2 - 4} = \frac{4x(x-2)}{3} \cdot \frac{\overset{6}{18}}{(x+2)(x-2)}$$

$$= \frac{24x}{x + 2}$$

How to Divide Rational Expressions

1. Invert the divisor.
2. Multiply the dividend and the inverted divisor.

EXAMPLE 4

$$\frac{x^2 - 2x - 15}{x^2 - 25} \div \frac{x^2 + 6x + 9}{x^2 + x - 6} = \frac{x^2 - 2x - 15}{x^2 - 25} \cdot \frac{x^2 + x - 6}{x^2 + 6x + 9}$$

$$= \frac{(x - 5)(x + 3)}{(x + 5)(x - 5)} \cdot \frac{(x + 3)(x - 2)}{(x + 3)(x + 3)}$$

$$= \frac{x - 2}{x + 5}$$

EXAMPLE 5

$$\frac{ax - bx + by - ay}{3a^3 - 3ab^2} \div \frac{ax - ay}{b^2 - 2ab + a^2}$$

Solution

$$\frac{ax - bx + by - ay}{3a^3 - 3ab^2} \div \frac{ax - ay}{b^2 - 2ab + a^2}$$

$$= \frac{ax - bx + by - ay}{3a^3 - 3ab^2} \cdot \frac{b^2 - 2ab + a^2}{ax - ay}$$

$$= \frac{x(a - b) - y(a - b)}{3a(a^2 - b^2)} \cdot \frac{b^2 - 2ab + a^2}{a(x - y)}$$

$$= \frac{(a - b)(x - y)}{3a(a + b)(a - b)} \cdot \frac{(b - a)(b - a)}{a(x - y)}$$

$$= \frac{(b - a)(b - a)}{3a^2(a + b)}$$

$$= \frac{(b - a)^2}{3a^2(a + b)} \quad ∎$$

EXAMPLE 6

$$\frac{a - 3}{a^2 + 2a - 3} \cdot \frac{a^2 - 2a + 1}{a^2 - 2a - 3} \div \frac{a^2 - 9}{a^2 - 1}$$

Solution

$$\frac{a - 3}{a^2 + 2a - 3} \cdot \frac{a^2 - 2a + 1}{a^2 - 2a - 3} \div \frac{a^2 - 9}{a^2 - 1}$$

$$= \frac{a - 3}{a^2 + 2a - 3} \cdot \frac{a^2 - 2a + 1}{a^2 - 2a - 3} \cdot \frac{a^2 - 1}{a^2 - 9}$$

$$= \frac{(a - 3)}{(a + 3)(a - 1)} \cdot \frac{(a - 1)(a - 1)}{(a - 3)(a + 1)} \cdot \frac{(a + 1)(a - 1)}{(a + 3)(a - 3)}$$

$$= \frac{(a - 1)(a - 1)}{(a + 3)(a + 3)(a - 3)}$$

$$= \frac{(a - 1)^2}{(a + 3)^2(a - 3)} \quad ∎$$

EXAMPLE 7

$$\frac{2x + 10}{x^2 + x - 6} \div \left[\frac{x^2 + 4x + 4}{x^3 - 9x} \cdot \frac{x^2 + 2x - 15}{x^2 - 4} \right]$$

Solution   First, we must perform the operation inside the bracket.

$$\frac{2(x + 5)}{(x + 3)(x - 2)} \div \left[\frac{x^2 + 4x + 4}{x^3 - 9x} \cdot \frac{x^2 + 2x - 15}{x^2 - 4} \right]$$

$$= \frac{2(x + 5)}{(x + 3)(x - 2)} \div \left[\frac{(x + 2)(x + 2)}{x(x + 3)(x - 3)} \cdot \frac{(x + 5)(x - 3)}{(x + 2)(x - 2)} \right]$$

$$= \frac{2(x + 5)}{(x + 3)(x - 2)} \div \left[\frac{(x + 2)(x + 5)}{x(x + 3)(x - 2)} \right]$$

$$= \frac{2(\cancel{x + 5})}{(\cancel{x + 3})(\cancel{x - 2})} \cdot \frac{x(\cancel{x + 3})(\cancel{x - 2})}{(x + 2)(\cancel{x + 5})} \quad \text{[invert and multiply]}$$

$$= \frac{2x}{x + 2} \quad \blacksquare$$

EXAMPLE 8 $$\frac{(a + b)^2 - c^2}{a^2 + ab - ac} \cdot \frac{a}{(a + c)^2 - b^2} \div \frac{ab - b^2 - bc}{(a - b)^2 - c^2}$$

Solution $$\frac{(a + b)^2 - c^2}{a^2 + ab - ac} \cdot \frac{a}{(a + c)^2 - b^2} \div \frac{ab - b^2 - bc}{(a - b)^2 - c^2}$$

$$= \frac{(a + b)^2 - c^2}{a^2 + ab - ac} \cdot \frac{a}{(a + c)^2 - b^2} \cdot \frac{(a - b)^2 - c^2}{ab - b^2 - bc}$$

$$= \frac{(\cancel{a + b + c})(a + b - c)}{\cancel{a}(a + b - c)} \cdot \frac{\cancel{a}}{(a + c + b)(a + c - b)}$$

$$\cdot \frac{(a - b + c)(a - b - c)}{b(a - b - c)}$$

$$= \frac{1}{b} \quad \blacksquare$$

EXERCISES 6.3

Find the following quotients in lowest terms.

1. $\dfrac{1}{2} \div \dfrac{3}{4}$

2. $\dfrac{3}{4} \div \dfrac{6}{7}$

3. $\dfrac{4}{5} \div \dfrac{20}{3}$

4. $\dfrac{5}{16} \div \dfrac{-7}{36}$

5. $\dfrac{8}{-9} \div \dfrac{16}{27}$

6. $-36 \div \dfrac{24}{-17}$

7. $\dfrac{1}{a} \div \dfrac{b}{a}$

8. $\dfrac{1}{c} \div \dfrac{2}{c}$

9. $\dfrac{10}{b^7} \div \dfrac{25}{b^3}$

10. $\dfrac{x^2}{y^3} \div \dfrac{x^3}{y^5}$

11. $\dfrac{6b^7}{5x^4} \div \dfrac{9b^3}{10x^5}$

12. $\dfrac{8c^3 r^3}{9d^5 s^2} \div \dfrac{16c^7 r^4}{27d^4 s}$

13. $\dfrac{10a^4 b^7}{9c^3 d^8} \div \dfrac{25ab^3}{36cd^5}$

14. $\dfrac{18r^5 s}{21g^5 h^4} \div \dfrac{8rs^6}{7gh^7}$

15. $\dfrac{a - 2b}{2a - b} \div \dfrac{2a - 4b}{6a - 3b}$

16. $\dfrac{a - b}{a + b} \div \dfrac{5a - 5b}{a + b}$

17. $\dfrac{c + d}{c - d} \div \dfrac{3c + 3d}{4c - 4d}$

18. $\dfrac{x - y}{a^2 + 2ab + b^2} \div \dfrac{x^2 - y^2}{a + b}$

19. $\dfrac{x^2 - y^2}{x^2 + 2xy + y^2} \div \dfrac{x^2 - 3xy + 2y^2}{x^2 + 3xy + 2y^2}$

20. $\dfrac{a^2 + a - 6}{a^2 - a - 12} \div \dfrac{a^2 - 4}{a^2 + 6a + 8}$

21. $\dfrac{y^2 + y - 12}{y^2 + y - 30} \div \dfrac{y^2 + 10y + 24}{y^2 - 2y - 15}$

22. $\dfrac{a^2 + 8a + 15}{a^2 + 3a - 18} \div \dfrac{a^2 + 7a + 10}{a^2 + a - 12}$

23. $\dfrac{x^2 + 5x + 6}{x^2 + 3x - 10} \div \dfrac{x^2 + 7x + 12}{x^2 - 9x + 14}$

24. $\dfrac{a^2 - 6a + 8}{a^2 - 8a + 15} \div \dfrac{a^2 - a - 12}{a^2 - 7a + 10}$

25. $\dfrac{6y^2 - 11y + 3}{8y^2 + 10y - 3} \div \dfrac{8y^2 - 10y - 3}{6y^2 + 11y + 3}$

26. $\dfrac{x^2 - 5x + 6}{x^2 + x - 6} \div (x^2 - 9)$

27. $\dfrac{3a^2 - 10ab + 3b^2}{4a^2 - b^2} \div \dfrac{3a^2 - 7ab + 2b^2}{6a^2 + ab - 2b^2}$

28. $\dfrac{8x^3 + 1}{10x^2 + 3x - 1} \div \dfrac{4x^2 - 2x + 1}{25x^2 - 1}$

29. $\dfrac{2 - 15x + 7x^2}{4 - 8x + 3x^2} \div \dfrac{2x - 14x^2}{4 - 9x^2}$

30. $\dfrac{a^4 - b^4}{2a^2 - ab - b^2} \cdot \dfrac{8a^3 + b^3}{a^2 + b^2} \div \dfrac{a + b}{2}$

31. $\dfrac{x^3 - 8y^3}{x^2 + 2xy - 3y^2} \cdot \dfrac{2x^2 + 5xy - 3y^2}{x^2 + 2xy + 4y^2}$

$\div \dfrac{2x^2 - 5xy + 2y^2}{4x^2 - 3xy - y^2}$

32. $\dfrac{2a^2 - 7ab + 3b^2}{a^2 - 3ab + 9b^2}$

$\div \left[\dfrac{6a^2 + ab - 2b^2}{2a^3 + 54b^3} \div \dfrac{3a^2 + 2ab}{a^2 - 9b^2} \right]$

33. $\dfrac{x^3 + 2x^2y + xy^2}{27x^3 + 64y^3} \cdot \dfrac{6x^2 + 5xy - 4y^2}{3x^2 + 6xy + 3y^2}$

$\div \dfrac{2x^2 - 3xy + y^2}{9x^2 - 12xy + 16y^2}$

34. $\dfrac{4x^2 - 9y^2}{3x^2y^2} \div \left[\dfrac{4x - 6y}{xy} \cdot \dfrac{2x + 3y}{6x^2} \right]$

35. $\dfrac{16a^4 - 1}{2a^2 - a} \div \left[\dfrac{4a^3 + a}{a^2} \cdot \dfrac{2a + 1}{4a} \right]$

*36. $\left[\dfrac{a^2 + 2ab + b^2 - c^2}{a^2 - 2ab + b^2 - c^2} \div \dfrac{a + b + c}{a - b + c} \right]$

$\cdot \dfrac{(a - c)^2 - b^2}{a + b - c}$

*37. $\dfrac{(x + 2y)^2}{4x^2 - 2xy + y^2} \cdot \dfrac{2x^2 - 3xy - 2y^2}{3x^2 - 12y^2}$

$\div \dfrac{(2x + y)}{8x^3 + y^3}$

*38. $\left[\dfrac{3x^2y^2 + 16xy + 5}{27x^3y^3 + 1} \div \dfrac{x^2y + 5x}{x^4} \right]$

$\cdot \dfrac{9x^2y^2 - 3xy + 1}{x^2y}$

*39. $\dfrac{a^2 - 2ab + b^2 - 16}{a^2 - 5ab + 6b^2} \div \dfrac{a - b + 4}{6a - 18b}$

*40. $\dfrac{x^2 - y^2}{x^3 + 3x^2y + 3xy^2 + y^3}$

$\div \dfrac{3y - 3x}{x^2 + 2xy + y^2}$

6.4 ADDITION AND SUBTRACTION OF FRACTIONAL EXPRESSIONS

Addition and Subtraction of Like Fractions

Fractions with the same denominator are called **like fractions**. Two rational numbers having the same denominator can be added or subtracted by adding or subtracting their numerators and writing the result over the common denominator. We then reduce the resulting fraction to lowest terms. That is,

$$\text{For all real numbers } a, b, \text{ and } c, \text{ with } c \neq 0,$$

$$\frac{a}{c} + \frac{b}{c} = \frac{a + b}{c} \quad \text{and} \quad \frac{a}{c} - \frac{b}{c} = \frac{a - b}{c}$$

EXAMPLE 1
$$\frac{9}{11} + \frac{3}{11} = \frac{9 + 3}{11} = \frac{12}{11} \quad \text{and} \quad \frac{17}{21} - \frac{10}{21} = \frac{17 - 10}{21} = \frac{7}{21} = \frac{1}{3}$$

The same rule applies for rational expressions.

Addition of Rational Expressions

If P, Q, and R are polynomials with $R \neq 0$, then

$$\frac{P}{R} + \frac{Q}{R} = \frac{P + Q}{R}$$

EXAMPLE 2
$$\frac{15}{x} + \frac{8}{x} = \frac{15 + 8}{x} = \frac{23}{x}$$

EXAMPLE 3
$$\frac{a + c}{b} + \frac{a - c}{b} + \frac{a + d}{b} = \frac{a + c + a - c + a + d}{b} = \frac{3a + d}{b}$$

EXAMPLE 4
$$\frac{4x - 2y}{z} - \frac{3x - 3y}{z} = \frac{4x - 2y - (3x - 3y)}{z}$$
$$= \frac{4x - 2y - 3x + 3y}{z}$$
$$= \frac{x + y}{z}$$

EXAMPLE 5
$$\frac{x^2}{x + 2} - \frac{4}{x + 2} = \frac{x^2 - 4}{x + 2} = \frac{(x + 2)(x - 2)}{(x + 2)} = x - 2$$

EXAMPLE 6

$$\frac{3x}{2x^2 + 3x} + \frac{5 - x}{2x^2 + 3x} - \frac{2}{2x^2 + 3x} = \frac{3x + 5 - x - 2}{2x^2 + 3x}$$

$$= \frac{2x + 3}{2x^2 + 3x}$$

$$= \frac{(2x + 3)}{x(2x + 3)}$$

$$= \frac{1}{x}$$

Addition and Subtraction of Unlike Fractions

Fractions that have different denominators are called **unlike fractions**. Unlike fractions cannot be added or subtracted directly. Before adding or subtracting unlike fractions, we must express them as fractions with the same denominator, called a **common denominator**. However, when adding or subtracting unlike fractions, we usually use the **least common denominator (LCD)**. The least common denominator is the smallest algebraic expression that is exactly divisible by each of the denominators.

How to Find the LCD

1. Factor each denominator completely. Express repeated factors as powers.
2. Write each different factor that appears in any denominator.
3. Raise each factor in step 2 to the highest power it occurs in *any* denominator.
4. The LCD is the product of all the factors found in step 3.

EXAMPLE 7 Find the LCD for $\dfrac{3x}{16} - \dfrac{5x}{12} + \dfrac{x}{3}$.

Solution
1. $16 = 2 \cdot 2 \cdot 2 \cdot 2 = 2^4$
   $12 = 2 \cdot 2 \cdot 3 = 2^2 \cdot 3$ [denominators in factored form]
   $3 = 3$

2. $2, 3$ [all the different factors]

3. $2^4, 3^1$ [highest powers of factors]

4. $\text{LCD} = 2^4 \cdot 3^1 = 48$ ■

EXAMPLE 8 Find the LCD for $\dfrac{5}{2x} + \dfrac{1}{5x^2} - \dfrac{3}{10x}$.

Solution   1. $2x = 2 \cdot x$

$5x^2 = 5 \cdot x^2$ [denominators in factored form]

$10x = 5 \cdot 2 \cdot x$

2. $2, 5, x$ [all the different factors]

3. $2^1, 5^1, x^2$ [highest powers of factors]

4. LCD $= 2 \cdot 5 \cdot x^2 = 10x^2$ ■

EXAMPLE 9 Find the LCD for $\dfrac{x^2}{x-1} + \dfrac{3x}{4x+8} - \dfrac{1}{2}$.

Solution   1. $x - 1 = (x - 1)$

$4x + 8 = 2 \cdot 2(x + 2) = 2^2(x + 2)$ [denominators in factored form]

$2 = 2$

2. $2, (x - 1), (x + 2)$ [all the different factors]

3. $2^2, (x - 1)^1, (x + 2)^1$ [highest powers of factors]

4. LCD $= 2^2(x - 1)^1(x + 2)^1 = 4(x - 1)(x + 2)$ ■

EXAMPLE 10 Find the LCD for $\dfrac{x+3}{x-4} - \dfrac{x+4}{x-3} - \dfrac{8}{x^2-16}$.

Solution   1. $x - 4 = (x - 4)$

$x - 3 = (x - 3)$ [denominators in factored form]

$x^2 - 16 = (x + 4)(x - 4)$

2. $(x - 3), (x + 4), (x - 4)$ [all the different factors]

3. $(x - 3)^1, (x + 4)^1, (x - 4)^1$ [highest powers of factors]

4. LCD $= (x - 3)(x + 4)(x - 4)$ ■

EXAMPLE 11 Find the LCD for $\dfrac{2a}{a^2-b^2} + \dfrac{3}{a+b} + \dfrac{b}{a^2+2ab+b^2}$.

Solution   1. $a^2 - b^2 = (a + b)(a - b)$

$a + b = (a + b)$ [denominators in factored form]

$a^2 + 2ab + b^2 = (a + b)^2$

2. $(a + b), (a - b)$ [all the different factors]

3. $(a + b)^2, (a - b)$ [highest powers of factors]

4. LCD $= (a + b)^2(a - b)$ ■

We are now ready to add and subtract unlike fractions.

How to Add and Subtract Unlike Fractional Expressions

1. Find the LCD of the terms.
2. Express each fraction as an equivalent fraction whose denominator is the LCD of Step 1.
3. Add or subtract the resulting like fractions.
4. Reduce the resulting fraction to lowest terms, if possible.

EXAMPLE 12 Combine $\dfrac{3x}{16} - \dfrac{5x}{12} + \dfrac{x}{3}$.

Solution 1. LCD = 48 (see Example 7).

2. $\dfrac{(3x) \cdot 3}{16 \cdot 3} - \dfrac{(5x) \cdot 4}{12 \cdot 4} + \dfrac{(x) \cdot 16}{3 \cdot 16}$

$= \dfrac{9x}{48} - \dfrac{20x}{48} + \dfrac{16x}{48}$

3. $= \dfrac{9x - 20x + 16x}{48}$

$= \dfrac{5x}{48}$ ■

EXAMPLE 13 Combine $\dfrac{5}{2x} + \dfrac{1}{5x^2} - \dfrac{3}{10x}$.

Solution 1. LCD = $10x^2$ (see Example 8).

2. $\dfrac{5 \cdot (5x)}{(2x) \cdot (5x)} + \dfrac{1 \cdot 2}{(5x^2) \cdot 2} - \dfrac{3 \cdot x}{(10x) \cdot x}$

$= \dfrac{25x}{10x^2} + \dfrac{2}{10x^2} - \dfrac{3x}{10x^2}$

3. $= \dfrac{25x + 2 - 3x}{10x^2}$

$= \dfrac{22x + 2}{10x^2}$

4. $= \dfrac{2(11x + 1)}{10x^2} = \dfrac{11x + 1}{5x^2}$ ■

EXAMPLE 14   Combine $\dfrac{x^2}{x-1} + \dfrac{3x}{4x+8} - \dfrac{1}{2}$.

Solution   1. LCD $= 4(x-1)(x+2)$ (see Example 9).

2. $\dfrac{x^2 \cdot 4(x+2)}{(x-1)\cdot 4(x+2)} + \dfrac{(3x)\cdot(x-1)}{4(x+2)\cdot(x-1)} - \dfrac{1\cdot 2(x-1)(x+2)}{2\cdot 2(x-1)(x+2)}$

$= \dfrac{4x^2(x+2)}{4(x-1)(x+2)} + \dfrac{3x(x-1)}{4(x-1)(x+2)} - \dfrac{2(x-1)(x+2)}{4(x-1)(x+2)}$

3. $= \dfrac{4x^2(x+2) + 3x(x-1) - 2(x-1)(x+2)}{4(x-1)(x+2)}$

$= \dfrac{4x^3 + 9x^2 - 5x + 4}{4(x-1)(x+2)}$ ∎

EXAMPLE 15   Combine $\dfrac{x+3}{x-4} - \dfrac{x+4}{x-3} - \dfrac{8}{x^2-16}$.

Solution   1. LCD $= (x-3)(x+4)(x-4)$ (see Example 10).

2. $\dfrac{(x+3)(x-3)(x+4)}{(x-4)(x-3)(x+4)} - \dfrac{(x+4)(x-4)(x+4)}{(x-3)(x-4)(x+4)}$

$- \dfrac{8(x-3)}{(x+4)(x-4)(x-3)}$

3. $= \dfrac{(x+3)(x-3)(x+4) - (x+4)(x-4)(x+4) - 8(x-3)}{(x-3)(x+4)(x-4)}$

$= \dfrac{52-x}{(x-3)(x+4)(x-4)}$ ∎

EXAMPLE 16   Combine $\dfrac{2a}{a^2-b^2} + \dfrac{3}{a+b} + \dfrac{b}{a^2+2ab+b^2}$.

Solution   1. LCD $= (a+b)^2(a-b)$ (see Example 11).

2. $\dfrac{2a\cdot(a+b)}{(a+b)(a-b)(a+b)} + \dfrac{3\cdot(a+b)(a-b)}{(a+b)(a+b)(a-b)} + \dfrac{b\cdot(a-b)}{(a+b)(a+b)(a-b)}$

3. $= \dfrac{2a(a+b) + 3(a+b)(a-b) + b(a-b)}{(a+b)^2(a-b)}$

$= \dfrac{2a^2 + 2ab + 3a^2 - 3b^2 + ab - b^2}{(a+b)^2(a-b)}$

$= \dfrac{5a^2 + 3ab - 4b^2}{(a+b)^2(a-b)}$ ∎

EXAMPLE 17 Combine $\dfrac{1-x}{2+2x} - \dfrac{1+x}{2-2x} + \dfrac{2x}{1-x^2}$.

Solution 1.

$$2 + 2x = 2(1+x)$$
$$2 - 2x = 2(1-x)$$
$$1 - x^2 = (1+x)(1-x)$$

[factored denominators]

Thus, LCD $= 2(1+x)(1-x)$.

2. $\dfrac{(1-x)(1-x)}{2(1+x)(1-x)} - \dfrac{(1+x)(1+x)}{2(1-x)(1+x)}$

$+ \dfrac{2x \cdot 2}{(1+x)(1-x) \cdot 2}$

3. $= \dfrac{(1-x)(1-x) - (1+x)(1+x) + 4x}{2(1+x)(1-x)}$

$= \dfrac{1 - 2x + x^2 - 1 - 2x - x^2 + 4x}{2(1+x)(1-x)}$

$= \dfrac{0}{2(1+x)(1-x)}$

$= 0$ ■

EXAMPLE 18 Combine $\dfrac{a}{a-b} + \dfrac{b}{b-a} + a$.

Solution Note that $\dfrac{b}{b-a} = \dfrac{b}{-(a-b)} = -\dfrac{b}{a-b}$. Thus, $\dfrac{a}{a-b} + \dfrac{b}{b-a} + a$

$= \dfrac{a}{a-b} - \dfrac{b}{a-b} + \dfrac{a}{1}$.

1. LCD $= a - b$

2. $\dfrac{a}{a-b} - \dfrac{b}{a-b} + \dfrac{a(a-b)}{a-b}$

3. $= \dfrac{a - b + a(a-b)}{(a-b)}$

$= \dfrac{(a-b) + a(a-b)}{(a-b)}$

4. $= \dfrac{(a-b)(1+a)}{(a-b)}$

$= 1 + a$ ■

EXERCISES 6.4

In Exercises 1–20, perform the indicated operation and reduce to lowest terms.

1. $\dfrac{3}{17} + \dfrac{8}{17}$

2. $\dfrac{2}{7} + \dfrac{8}{7} - \dfrac{4}{7}$

3. $\dfrac{4}{3x} - \dfrac{5}{3x} + \dfrac{2}{3x}$

4. $\dfrac{x + 4}{2} + \dfrac{2x - 1}{2}$

5. $\dfrac{4x}{x + y} + \dfrac{4y}{x + y}$

6. $\dfrac{x^2}{x - y} - \dfrac{y^2}{x - y}$

7. $\dfrac{2a - 3b}{3ab} + \dfrac{4a + 2b}{3ab} + \dfrac{3a + b}{3ab}$

8. $\dfrac{3ab}{a + 2b} + \dfrac{a^2 + 2b^2}{a + 2b}$

9. $\dfrac{k^2 + k}{k^2 - 9} + \dfrac{k - 3}{k^2 - 9}$

10. $\dfrac{3z}{z^2 - 2z - 15} - \dfrac{2z + 5}{z^2 - 2z - 15}$

11. $\dfrac{3x^2 + 2x - y^2}{x^2 - 4y^2} + \dfrac{4x^2 - y^2 + 4y}{x^2 - 4y^2}$

12. $\dfrac{14(x + 1)}{3x^2 - 4x} - \dfrac{2(x + 15)}{3x^2 - 4x}$

13. $\dfrac{3x(3x + 8y)}{9x^2 - 4y^2} - \dfrac{y(3x - 10y)}{9x^2 - 4y^2}$

14. $\dfrac{x^2 - 2y}{x^2 + 4xy + 4y^2} - \dfrac{x + 4y^2}{x^2 + 4xy + 4y^2}$

15. $\dfrac{x^2 - 7ax - 27a^2}{x^2 + 5ax + 6a^2} - \dfrac{x^2 + 2ax}{x^2 + 5ax + 6a^2}$

16. $\dfrac{12x^4 + 40x^2 + 80}{8x^3 - 27} + \dfrac{4x^4 - 4x^2 + 1}{8x^3 - 27}$

17. $\dfrac{10a(a + 3b)}{25a^2 - 9b^2} - \dfrac{b(a + 21b)}{25a^2 - 9b^2}$

18. $\dfrac{4b^2}{6a^2 - 52ab - 18b^2}$
$+ \dfrac{12a(3a + 2b)}{6a^2 - 52ab - 18b^2}$

*19. $\dfrac{x^2(x + 1)}{x^2 - 2ax - 35a^2} - \dfrac{7a^2(7 + 4a)}{x^2 - 2ax - 35a^2}$
$+ \dfrac{ax(4a - 7x)}{x^2 - 2ax - 35a^2}$

*20. $\dfrac{a^2 + ab - b^2}{3(a - b)} - \dfrac{a^3 - b^3 + ab + b}{3(a - b)}$
$+ \dfrac{a}{3(a - b)}$

In Exercises 21–40, combine and reduce to lowest terms.

21. $\dfrac{2}{3} + \dfrac{3}{5} + \dfrac{5}{6} + \dfrac{7}{10}$

22. $\dfrac{11}{35} + \dfrac{13}{42} + \dfrac{17}{30}$

23. $\dfrac{6a}{5b} - \dfrac{5b}{6a}$

24. $\dfrac{11}{5a} + \dfrac{12}{10a} + \dfrac{21}{25a}$

25. $\dfrac{3}{a + 2b} + \dfrac{2}{a - 2b}$

26. $\dfrac{a - 2b}{a^2 - b^2} + \dfrac{1}{a + b}$

27. $\dfrac{x - 2}{x^2 - 16x + 48} - \dfrac{x + 8}{x^2 - 8x - 48}$

28. $\dfrac{9a + 2}{3a^2 - 2a - 8} + \dfrac{7}{3a^2 + a - 4}$

29. $\dfrac{x - 9}{7(x - 2)} + \dfrac{4}{x^2 - 4}$

30. $\dfrac{x + 5}{3(x - 3)} - \dfrac{3x + 7}{x^2 - 9}$

31. $5 + \dfrac{3x}{x - 2}$

32. $x + 7 - \dfrac{x^2}{x - 7}$

33. $\dfrac{2x - 7}{2x + 7} - 1$

34. $\dfrac{12x^2 + 5}{3x + 6} - (4x + 2)$

35. $x^2 + x + 1 + \dfrac{1}{x - 1}$

36. $\dfrac{a + 4}{7(a - 3)} - \dfrac{a - 4}{7(a + 3)}$

37. $\dfrac{x + y}{2(2x - 3y)} + \dfrac{x - y}{2(2x + 3y)} + \dfrac{7xy}{4x^2 - 9y^2}$

38. $\dfrac{2 - x}{x^2 + x - 6} - \dfrac{5}{9 - x^2} - \dfrac{4 - x}{x^2 - 7x + 12}$

*39. $\dfrac{x + y}{(y - z)(z - x)} + \dfrac{y + z}{(z - x)(x - y)}$

$+ \dfrac{z + x}{(x - y)(y - z)}$

*40. $\dfrac{x^2 - (y - z)^2}{(z + x)^2 - y^2} + \dfrac{y^2 - (z - x)^2}{(x + y)^2 - z^2}$

$+ \dfrac{z^2 - (x - y)^2}{(y + z)^2 - x^2}$

6.5 COMPLEX FRACTIONS

A complex fraction is a fraction whose numerator and/or denominator contain one or more fractions.

The following are examples of complex fractions:

$$\dfrac{\dfrac{3}{4}}{\dfrac{8}{5}}, \qquad \dfrac{\dfrac{10x^2y^2}{9z^2}}{\dfrac{5xy^2}{3z}}, \qquad \dfrac{\dfrac{x^2 - 9}{x}}{x + 3}, \qquad \dfrac{\dfrac{a}{b} + 2}{1 - \dfrac{a}{b}}$$

There are two methods used to simplify complex fractions.

How to Simplify Complex Fractions

Method 1. Multiply the numerator and denominator of the complex fraction by the LCD of all the fractions that appear in the numerator and in the denominator. Simplify the results.

Method 2. Simplify the numerator and denominator of the complex fraction. Then divide the simplified numerator by the simplified denominator.

EXAMPLE 1 Simplify $\dfrac{\dfrac{3}{4}}{\dfrac{8}{5}}$.

Solution *Method 1.* LCD = 20

$$\dfrac{\dfrac{3}{4}}{\dfrac{8}{5}} = \dfrac{\dfrac{3}{4} \cdot (20)}{\dfrac{8}{5} \cdot (20)}$$

$$= \dfrac{15}{32}$$

Method 2.

$$\dfrac{\dfrac{3}{4}}{\dfrac{8}{5}} = \dfrac{3}{4} \div \dfrac{8}{5}$$

$$= \dfrac{3}{4} \cdot \dfrac{5}{8}$$

$$= \dfrac{15}{32} \quad \blacksquare$$

EXAMPLE 2 Simplify $\dfrac{\dfrac{x^2 - 9}{x}}{x + 3}$.

Solution *Method 1.* LCD $= x$

$$\frac{\dfrac{x^2 - 9}{x}}{\dfrac{x + 3}{1}} = \frac{\dfrac{x^2 - 9}{x} \cdot (x)}{\dfrac{x + 3}{1} \cdot (x)}$$

$$= \frac{x^2 - 9}{x(x + 3)}$$

$$= \frac{(x + 3)(x - 3)}{x(x + 3)}$$

$$= \frac{x - 3}{x}$$

Method 2.

$$\frac{\dfrac{x^2 - 9}{x}}{x + 3} = \frac{x^2 - 9}{x} \div \frac{(x + 3)}{1}$$

$$= \frac{x^2 - 9}{x} \cdot \frac{1}{(x + 3)}$$

$$= \frac{(x + 3)(x - 3)}{x} \cdot \frac{1}{(x + 3)}$$

$$= \frac{x - 3}{x} \quad ■$$

EXAMPLE 3 Simplify $\dfrac{1 - \dfrac{7}{x} + \dfrac{10}{x^2}}{1 - \dfrac{4}{x^2} - \dfrac{12}{x^2}}$.

Solution *Method 1.*
The LCD of the numerator and de-nominator is x^2.

$$\frac{1 - \dfrac{7}{x} + \dfrac{10}{x^2}}{1 - \dfrac{4}{x} - \dfrac{12}{x^2}} = \frac{\left(1 - \dfrac{7}{x} + \dfrac{10}{x^2}\right) \cdot (x^2)}{\left(1 - \dfrac{4}{x} - \dfrac{12}{x^2}\right) \cdot (x^2)}$$

$$= \frac{x^2 - 7x + 10}{x^2 - 4x - 12}$$

$$= \frac{(x - 5)(x - 2)}{(x - 6)(x + 2)}$$

Method 2.
We first express the numerator and denominator separately as single fractions.

$$\frac{1 - \dfrac{7}{x} + \dfrac{10}{x^2}}{1 - \dfrac{4}{x} - \dfrac{12}{x^2}} = \frac{\dfrac{x^2}{x^2} - \dfrac{7x}{x^2} + \dfrac{10}{x^2}}{\dfrac{x^2}{x^2} - \dfrac{4x}{x^2} - \dfrac{12}{x^2}}$$

$$= \frac{\dfrac{x^2 - 7x + 10}{x^2}}{\dfrac{x^2 - 4x - 12}{x^2}} = \frac{x^2 - 7x + 10}{x^2}$$

$$\div \frac{x^2 - 4x - 12}{x^2}$$

$$= \frac{x^2 - 7x + 10}{x^2} \cdot \frac{x^2}{x^2 - 4x - 12}$$

$$= \frac{(x - 5)(x - 2)}{x^2} \cdot \frac{x^2}{(x - 6)(x + 2)}$$

$$= \frac{(x - 5)(x - 2)}{(x - 6)(x + 2)} \quad ■$$

In the following examples, we use either Method 1 or Method 2.

EXAMPLE 4 Simplify $\dfrac{\dfrac{x-y}{x} - \dfrac{x+y}{y}}{\dfrac{x-y}{y} + \dfrac{x+y}{x}}$.

Solution Using Method 1, we find that the LCD is xy.

$$\frac{\dfrac{x-y}{x} - \dfrac{x+y}{y}}{\dfrac{x-y}{y} + \dfrac{x+y}{x}} = \frac{\left[\dfrac{x-y}{x} - \dfrac{x+y}{y}\right] \cdot (xy)}{\left[\dfrac{x-y}{y} + \dfrac{x+y}{x}\right] \cdot (xy)}$$

$$= \frac{y(x-y) - x(x+y)}{x(x-y) + y(x+y)} = \frac{xy - y^2 - x^2 - xy}{x^2 - xy + xy + y^2}$$

$$= \frac{-x^2 - y^2}{x^2 + y^2} = \frac{-(x^2 + y^2)}{(x^2 + y^2)} = -1 \quad ■$$

EXAMPLE 5 Simplify $\dfrac{a - \dfrac{1}{b}}{1 - \dfrac{a}{\dfrac{1}{b}}}$.

Solution Method 2. We begin by simplifying the term $-\dfrac{a}{1/b}$ in the denominator.

$$-\frac{a}{\dfrac{1}{b}} = -a \div \frac{1}{b} = -a \cdot \frac{b}{1} = -ab$$

Thus,

$$\frac{a - \dfrac{1}{b}}{1 - \dfrac{a}{\dfrac{1}{b}}} = \frac{a - \dfrac{1}{b}}{1 - ab} = \frac{\dfrac{ab-1}{b}}{1 - ab} = \frac{ab-1}{b} \div \frac{1-ab}{1}$$

$$= \frac{(ab-1)^{-1}}{b} \cdot \frac{1}{(1-ab)} = \frac{-1}{b} \quad ■$$

EXAMPLE 6 Simplify $\dfrac{\dfrac{1}{x^2 - y^2}}{\dfrac{1}{x-y} + \dfrac{1}{x+y}}$.

Solution   By Method 2,

$$\frac{\dfrac{1}{x^2-y^2}}{\dfrac{1}{x-y}+\dfrac{1}{x+y}}=\frac{\dfrac{1}{x^2-y^2}}{\dfrac{x+y+x-y}{x^2-y^2}}=\frac{\dfrac{1}{x^2-y^2}}{\dfrac{2x}{x^2-y^2}}=\frac{1}{x^2-y^2}\div\frac{2x}{x^2-y^2}$$

$$=\frac{1}{\cancel{x^2-y^2}}\cdot\frac{\cancel{x^2-y^2}}{2x}=\frac{1}{2x}\quad\blacksquare$$

EXAMPLE 7   Simplify $2-\dfrac{1}{1-\dfrac{3x}{2x-\dfrac{2x}{x+1}}}$.

Solution

$$2-\frac{1}{1-\dfrac{3x}{2x-\dfrac{2x}{x+1}}}=2-\frac{1}{1-\dfrac{3x}{\dfrac{2x(x+1)-2x}{x+1}}}$$

$$=2-\frac{1}{1-\dfrac{3x}{\dfrac{2x^2+2x-2x}{x+1}}}=2-\frac{1}{1-\dfrac{3x}{\dfrac{2x^2}{x+1}}}$$

$$=2-\frac{1}{1-\dfrac{3x(x+1)}{2x^2}}=2-\frac{1}{1-\dfrac{3x^2+3x}{2x^2}}$$

$$=2-\frac{1}{\dfrac{2x^2-(3x^2+3x)}{2x^2}}=2-\frac{1}{\dfrac{2x^2-3x^2-3x}{2x^2}}$$

$$=2-\frac{1}{\dfrac{-x^2-3x}{2x^2}}=2-\frac{2x^2}{-(x^2+3x)}=2+\frac{2x^2}{(x^2+3x)}$$

$$=\frac{2(x^2+3x)+2x^2}{x^2+3x}=\frac{2x^2+6x+2x^2}{x^2+3x}=\frac{4x^2+6x}{x^2+3x}$$

$$=\frac{2\cancel{x}(2x+3)}{\cancel{x}(x+3)}=\frac{2(2x+3)}{x+3}\quad\blacksquare$$

EXERCISES 6.5

Simplify.

1. $\dfrac{\dfrac{3}{4}}{\dfrac{3}{5}}$

2. $\dfrac{\dfrac{5}{8}}{\dfrac{2}{3}}$

3. $\dfrac{\dfrac{x}{2}}{\dfrac{y}{7}}$

4. $\dfrac{\dfrac{x}{4}}{\dfrac{y}{9}}$

5. $\dfrac{\dfrac{a}{x}}{\dfrac{b}{y}}$

6. $\dfrac{\dfrac{c}{x}}{\dfrac{d}{y}}$

27. $\dfrac{\dfrac{x^2}{4} - y^2}{\dfrac{x}{6} + \dfrac{y}{3}}$

28. $\dfrac{\dfrac{a}{b} + \dfrac{a^2}{2b}}{\dfrac{2}{a} + 1}$

7. $\dfrac{\dfrac{3a}{8}}{\dfrac{5b}{3}}$

8. $\dfrac{\dfrac{2x}{5}}{\dfrac{6y}{7}}$

29. $\dfrac{x - \dfrac{2y}{3}}{\dfrac{9x^2}{4} - y^2}$

30. $\dfrac{a^2 + b^2}{\dfrac{a^4}{b^2} - b^2}$

9. $\dfrac{10x^2}{\dfrac{3x}{2}}$

10. $\dfrac{\dfrac{5a}{8}}{6ab}$

31. $\dfrac{3a - \dfrac{b^2}{3a}}{3a - b}$

32. $\dfrac{x^3 - x^2}{x - 2 + \dfrac{1}{x}}$

11. $\dfrac{\dfrac{9y}{4}}{2yz}$

12. $\dfrac{18x^3}{\dfrac{3y}{7}}$

33. $\dfrac{2x + 1}{x + 1 + \dfrac{1}{4x}}$

34. $\dfrac{1 - \dfrac{2}{a} + \dfrac{1}{a^2}}{2(a - 1)}$

13. $\dfrac{2\dfrac{1}{4}}{1 - \dfrac{5}{8}}$

14. $\dfrac{3 - \dfrac{3}{5}}{\dfrac{8}{15}}$

35. $\dfrac{3x - \dfrac{x^2 - 5}{x}}{2x^2 + 5}$

36. $\dfrac{\dfrac{a^2 + 3b^2}{a^2 - b^2} - 1}{\dfrac{4}{a - b}}$

15. $\dfrac{1\dfrac{7}{8}}{2\dfrac{2}{3}}$

16. $\dfrac{5 - \dfrac{1}{2}}{2\dfrac{2}{5}}$

37. $2x - \dfrac{\dfrac{2x - 1}{3x}}{x - \dfrac{2 - x}{3}}$

38. $\dfrac{3x - y}{\dfrac{1 - y}{y} - \dfrac{1 - 3x}{3x}}$

17. $\dfrac{2a}{\dfrac{2a}{3} - 2}$

18. $\dfrac{x^2 - 1}{\dfrac{1}{x} - \dfrac{1}{x^2}}$

39. $\dfrac{\dfrac{a}{b} - \dfrac{b}{a}}{\dfrac{a + b}{2b} + \dfrac{a + b}{2a}}$

40. $\dfrac{\dfrac{x^4 + y^4}{x^2y^2} - 2}{\dfrac{x^2}{y^2} - 1}$

19. $\dfrac{\dfrac{x^2}{3} + \dfrac{x^2}{2}}{5}$

20. $\dfrac{\dfrac{3x}{2} - \dfrac{3}{x}}{3x^2}$

41. $\dfrac{\dfrac{a}{b} - \dfrac{b}{a}}{a^2 - b^2}$

42. $\dfrac{\dfrac{a}{2a - b} - 1}{a - \dfrac{a^2}{2a - b}}$

21. $\dfrac{6ab}{\dfrac{3a}{b} - 3a}$

22. $\dfrac{4a^2x}{\dfrac{4a}{x} + 2a}$

23. $\dfrac{\dfrac{a}{2b} - \dfrac{b}{2a}}{\dfrac{b}{2a} + \dfrac{1}{2}}$

24. $\dfrac{x - \dfrac{1}{x}}{\dfrac{1}{x} + 1}$

43. $\dfrac{\dfrac{x}{x - y} - \dfrac{y}{x}}{\dfrac{x^2 - y^2}{x} + \dfrac{xy + y^2}{x - y}}$

44. $\dfrac{\dfrac{a}{bc} - \dfrac{b}{ac}}{\dfrac{2}{b} + \dfrac{2}{a}}$

25. $\dfrac{3x^2}{x^2 - \dfrac{x^2}{4}}$

26. $\dfrac{2y^2 - \dfrac{2y}{3}}{\dfrac{5y}{3} - y}$

45. $\dfrac{\dfrac{1}{x + 2} - \dfrac{1}{x - 2}}{1 + \dfrac{4}{x^2 - 4}}$

46. $\dfrac{x + y}{\dfrac{2x^2}{x - y} - x}$

47. $\dfrac{1}{x + \dfrac{1}{x + \dfrac{1}{x + 1}}}$ 48. $1 - \dfrac{1}{1 + \dfrac{1}{1 - \dfrac{1}{x}}}$ 49. $1 + \dfrac{1}{1 - \dfrac{1}{1 - \dfrac{1}{2}}}$ *50. $\dfrac{\dfrac{1}{x} + 3}{x + \dfrac{1}{3x + \dfrac{x-1}{x}}}$

6.6 SOLVING FRACTIONAL EQUATIONS

An equation that has a variable in the denominator of one or more terms is called a fractional equation. Thus,

$$\frac{1}{x+1} = 3$$

and

$$\frac{1}{x^2 - x} = \frac{3}{x} - 1$$

are fractional equations.

Recall that if we multiply both sides of an equation by the same nonzero constant, we obtain a new equation that is equivalent to the original equation. We shall use this fact to change the fractional equation to an equation that does not contain fractions and then solve the resulting simple equation.

How to Solve Fractional Equations

1. Multiply each term in the fractional equation by the LCD of all the fractions.
2. Collect terms and solve the resulting simple equation.
3. Check each root in the original fractional equation.

EXAMPLE 1   Solve $\dfrac{3x}{4} + \dfrac{3}{2} = \dfrac{5x}{8}$.

Solution

$$\frac{3x}{4} + \frac{3}{2} = \frac{5x}{8}$$

Multiplying both sides by the LCD = 8, we obtain

$$(8)\frac{3x}{4} + (8)\frac{3}{2} = (8)\frac{5x}{8}$$

$$6x + 12 = 5x$$

$$6x - 5x = -12$$

$$x = -12 \qquad \text{Answer} \quad \blacksquare$$

Check

$$\frac{3(-12)}{4} + \frac{3}{2} \overset{?}{=} \frac{5(-12)}{8}$$

$$-9 + \frac{3}{2} \overset{?}{=} -\frac{60}{8}$$

$$-7\frac{1}{2} \overset{\checkmark}{=} -7\frac{1}{2}$$

Note. Checking is very important in solving fractional equations, since multiplying by an LCD containing the variable can result in an equation that may introduce an extra root. This extra root satisfies the new equation, but is *not* a root of the original equation. Such a root is called an **extraneous root.**

EXAMPLE 2 Solve $\dfrac{-2}{x + 1} = 1 + \dfrac{2x}{x + 1}$.

Solution

$$\frac{-2}{x + 1} = 1 + \frac{2x}{x + 1}$$

Multiplying both sides by the LCD $= x + 1$, we obtain

$$(\cancel{x + 1}) \cdot \frac{-2}{(\cancel{x + 1})} = (x + 1) \cdot 1 + (\cancel{x + 1}) \cdot \frac{2x}{(\cancel{x + 1})}$$

$$-2 = x + 1 + 2x$$

$$-3 = 3x$$

$$-1 = x \quad \blacksquare$$

Check

$$\frac{-2}{-1 + 1} \overset{?}{=} 1 + \frac{2(-1)}{-1 + 1}$$

$$\frac{-2}{0} \overset{?}{=} 1 + \frac{-2}{0}$$

However, division by zero is not permitted. Thus, -1 is an extraneous root and the fractional equation has no solutions.

EXAMPLE 3 Solve $\dfrac{1}{x} + \dfrac{2}{x - 1} = \dfrac{1}{x(x - 1)}$.

Solution

$$\frac{1}{x} + \frac{2}{x-1} = \frac{1}{x(x-1)}$$

Multiplying both sides by the LCD $= x(x-1)$, we get

$$\cancel{x}(x-1) \cdot \frac{1}{\cancel{x}} + x\cancel{(x-1)} \cdot \frac{2}{\cancel{(x-1)}} = \cancel{x}\cancel{(x-1)} \cdot \frac{1}{\cancel{x}\cancel{(x-1)}}$$

$$x - 1 + 2x = 1$$

$$3x = 2$$

$$x = \frac{2}{3} \qquad\qquad \text{Answer} \quad \blacksquare$$

Check

$$\frac{1}{\dfrac{2}{3}} + \frac{2}{\dfrac{2}{3} - 1} \stackrel{?}{=} \frac{1}{\dfrac{2}{3}\left(\dfrac{2}{3} - 1\right)}$$

$$\frac{1}{\dfrac{2}{3}} + \frac{2}{-\dfrac{1}{3}} \stackrel{?}{=} \frac{1}{\dfrac{2}{3}\left(-\dfrac{1}{3}\right)}$$

$$\frac{3}{2} - 6 \stackrel{?}{=} -\frac{9}{2}$$

$$-\frac{9}{2} \stackrel{\checkmark}{=} -\frac{9}{2}$$

EXAMPLE 4 Solve the literal fractional equation for x: $\dfrac{1}{x - 3a} + \dfrac{1}{x + 3a} = \dfrac{4a}{9a^2 - x^2}$.

Solution

$$\frac{1}{x - 3a} + \frac{1}{x + 3a} = \frac{4a}{(3a + x)(3a - x)}$$

Multiplying both sides by the LCD $= (3a + x)(3a - x)$, we get

$$(3a + x)\overset{-1}{\cancel{(3a - x)}} \cdot \frac{1}{\cancel{(x - 3a)}} + (3a + x)(3a - x) \cdot \frac{1}{\cancel{(x + 3a)}}$$

$$= \cancel{(3a + x)(3a - x)} \cdot \frac{4a}{\cancel{(3a + x)(3a - x)}}$$

$$-(3a + x) + (3a - x) = 4a$$

$$-3a - x + 3a - x = 4a$$

$$-2x = 4a$$

$$x = -2a \qquad\qquad \text{Answer} \quad \blacksquare$$

Check

$$\frac{1}{-2a - 3a} + \frac{1}{-2a + 3a} \stackrel{2}{=} \frac{4a}{9a^2 - (-2a)^2}$$

$$\frac{1}{-5a} + \frac{1}{a} \stackrel{2}{=} \frac{4a}{5a^2}$$

$$\frac{4}{5a} \stackrel{\checkmark}{=} \frac{4}{5a}$$

EXAMPLE 5 Solve the following formula for R: $\dfrac{1}{R} = \dfrac{1}{r_1} + \dfrac{1}{r_2} + \dfrac{1}{r_3}$.

Solution

$$\frac{1}{R} = \frac{1}{r_1} + \frac{1}{r_2} + \frac{1}{r_3}$$

Multiplying both sides by the LCD $= Rr_1r_2r_3$, we obtain

$$\cancel{R}r_1r_2r_3 \cdot \frac{1}{\cancel{R}} = R\cancel{r_1}r_2r_3 \cdot \frac{1}{\cancel{r_1}} + Rr_1\cancel{r_2}r_3 \cdot \frac{1}{\cancel{r_2}} + Rr_1r_2\cancel{r_3} \cdot \frac{1}{\cancel{r_3}}$$

$$r_1r_2r_3 = Rr_2r_3 + Rr_1r_3 + Rr_1r_2$$

$$r_1r_2r_3 = R(r_2r_3 + r_1r_3 + r_1r_2)$$

$$R = \frac{r_1r_2r_3}{r_2r_3 + r_1r_3 + r_1r_2}$$ Answer ■

EXAMPLE 6 Solve $I = \dfrac{E}{R + \dfrac{r}{n}}$ for n.

Solution Note that this fractional equation involves a complex fraction. Using Method 1 discussed in Section 6.5, we multiply the numerator and denominator on the right side by n to obtain

$$I = \frac{E \cdot (n)}{\left(R + \dfrac{r}{n}\right) \cdot (n)}$$

$$= \frac{nE}{nR + r}$$

We now multiply both sides by the LCD $= nR + r$, giving

$$(nR + r)I = nE$$

$$nRI + rI = nE$$

$$nRI - nE = -rI$$

$$(RI - E)n = -rI$$

$$n = \frac{-rI}{RI - E} \qquad \text{Answer} \quad \blacksquare$$

EXERCISES 6.6

In Exercises 1–22, solve and check the fractional equations.

1. $\dfrac{x + 5}{x - 3} + \dfrac{4}{x - 3} = 5$

2. $\dfrac{3x + 4}{2x - 3} + 3 = \dfrac{3 - x}{2x - 3}$

3. $\dfrac{15x^2 - 5x - 8}{3x^2 + 6x + 4} = 5$

4. $\dfrac{3}{x - 2} - \dfrac{4}{2x - 1} = \dfrac{1}{x + 4}$

5. $\dfrac{9}{3x - 5} - \dfrac{2}{x - 2} = \dfrac{1}{x - 3}$

6. $\dfrac{4 - x}{1 - x} = \dfrac{12}{3 - x} + 1$

7. $\dfrac{2x}{3x - 4} = \dfrac{4x + 5}{6x - 1} - \dfrac{3}{3x - 4}$

8. $\dfrac{6x + 5}{2x^2 - 2x} - \dfrac{2}{1 - x^2} = \dfrac{3x}{x^2 - 1}$

9. $\dfrac{3x}{2x - 5} - \dfrac{7}{3x + 1} = \dfrac{3}{2}$

10. $\dfrac{4x}{2x - 6} - \dfrac{4}{5x - 15} = \dfrac{1}{2}$

11. $\dfrac{3x}{2x + 3} + \dfrac{2x}{3 - 2x} = \dfrac{2x^2 - 15}{4x^2 - 9}$

12. $\dfrac{2x - 14}{x^2 + 3x - 28} = \dfrac{x + 3}{x + 7} - \dfrac{x - 2}{x - 4}$

13. $\dfrac{x + 2}{x - 4} - \dfrac{2x - 3}{x + 3} = \dfrac{26 - x^2}{x^2 - x - 12}$

14. $\dfrac{2x + 1}{x - 8} - \dfrac{2x - 1}{x + 6} = \dfrac{18x + 34}{x^2 - 2x - 48}$

15. $\dfrac{2x - 1}{2} - \dfrac{x + 2}{2x + 5} = \dfrac{6x - 5}{6}$

16. $\dfrac{x}{3} = \dfrac{9x - 2}{9} - \dfrac{2x^2}{3x - 4}$

17. $\dfrac{2x + 7}{14} - \dfrac{x + 6}{7} = \dfrac{5x - 4}{3x + 1}$

18. $\dfrac{5x + 1}{5} - \dfrac{4x}{5x + 8} = \dfrac{3x - 2}{3} + \dfrac{7}{5x + 8}$

19. $\dfrac{3x - 2}{x + 3} = \dfrac{36 - 4x}{x^2 - 9} - \dfrac{2 + 3x}{3 - x}$

20. $\dfrac{8x + 1}{x - 2} + 4 = \dfrac{7x + 3}{x - 2}$

21. $\dfrac{-2x}{x + 1} = 1 + \dfrac{2}{x + 1}$

22. $\dfrac{x^2 - x + 1}{x - 1} + \dfrac{x^2 + x + 1}{x + 1} = 2x$

In Exercises 23–35, solve the literal fractional equations for x.

23. $\dfrac{ax - b}{bx} - 2 = \dfrac{a - b}{abx} - \dfrac{bx + a}{ax}$

24. $\dfrac{x + n}{x - m} - \dfrac{x - n}{x + m} = \dfrac{2(m + n)^2}{x^2 - m^2}$

25. $\dfrac{x - b}{x - 2a} - \dfrac{x + b}{x + 2a} = \dfrac{4a^2}{x^2 - 4a^2}$

26. $\dfrac{2x}{3x - a} - \dfrac{3a}{x + 2a} = \dfrac{2x^2 + ax}{3x^2 + 5ax - 2a^2}$

27. $\dfrac{1}{x - a} - \dfrac{1}{x - b} = \dfrac{a - b}{x^2 - ab}$

28. $\dfrac{x - a}{a - b} - \dfrac{x + a}{a + b} = \dfrac{2ax}{a^2 - b^2}$

29. $\dfrac{b^3 - 2bx + 1}{a^2 + b^2} = \dfrac{b^2 - x}{b}$

30. $\dfrac{x + 2}{x - m} = \dfrac{b^2m^2 + 4}{(x + 2)(x - m)} + \dfrac{x + m}{x + 2}$

*31. $\dfrac{c^2x - 1}{a - 2b} - \dfrac{3a + 2b}{4b^2} = \dfrac{a}{ab - 2b^2}$

*32. $\dfrac{c^2 + 4x}{2a} - \dfrac{a - 4}{c^2 + 2x} = \dfrac{8x^2 + c^4}{2ac^2 + 4ax}$

*33. $\dfrac{b^2}{d^2 + dx} - \dfrac{b^2 + d^2}{d^2} = \dfrac{x - 3b^2 - d}{d + x}$

*34. $\dfrac{4m + 2x}{4m + 1} - 1 = \dfrac{3n}{8m^2 + 2m}$

*35. $\dfrac{2a^2bx - a + b}{a - b} - 1 = \dfrac{a - 3b}{2b}$

In Exercises 36–45, solve the formulas for the indicated letter.

36. $r = \dfrac{a + b}{s}$ for b 37. $C = \dfrac{E}{R + r}$ for r

38. $s = \dfrac{a - rl}{1 - r}$ for r

39. $\dfrac{1}{D} + \dfrac{1}{d} = \dfrac{1}{F}$ for D

40. $L = \dfrac{Mt - g}{t}$ for t

41. $G = \dfrac{V(P - p)}{F - p}$ for p

42. $P = \dfrac{WL + vl + wg}{Al}$ for l

43. $R = \dfrac{rr_1}{r + r_1}$ for r_1

44. $\dfrac{n}{v} = \dfrac{m}{M + m}$ for m

45. $V = \pi h^2\left(r - \dfrac{h}{3}\right)$ for r

6.7 PROBLEM SOLVING INVOLVING FRACTIONAL EXPRESSIONS

In setting up the equations for word problems, we often obtain equations that are expressed in the form of fractional expressions. In this section, we discuss several types of word problems that lead to fractional equations whose solution will depend on the work completed in the previous section. The reader should also review Chapter 4, where problem-solving techniques were discussed.

Number Problems

EXAMPLE 1 The denominator of a fraction is 2 more than the numerator. If both the numerator and the denominator of the fraction are increased by 1, the resulting fraction equals $\frac{2}{3}$. Find the fraction.

Solution Let x = the numerator of the fraction

Then $x + 2$ = the denominator

Increasing both of these expressions by 1, we obtain the equation

$$\frac{x + 1}{x + 2 + 1} = \frac{2}{3}$$

or

$$\frac{x + 1}{x + 3} = \frac{2}{3}$$

Multiplying both sides by the LCD $= 3(x + 3)$, we obtain

$$3(x + 1) = 2(x + 3)$$
$$3x + 3 = 2x + 6$$
$$x = 3$$

Thus, the fraction is $\dfrac{x}{x + 2} = \dfrac{3}{3 + 2} = \dfrac{3}{5}$. ■

Check

$$\frac{3 + 1}{5 + 1} = \frac{4}{6} = \frac{2}{3}$$

EXAMPLE 2   Write 79 as the sum of two numbers such that the larger number divided by the smaller number yields a quotient of 5 and a remainder of 7.

Solution   Let
$$x = \text{the smaller number}$$

Then
$$79 - x = \text{the larger number}$$

We can set up an equation using the following relationship:

$$\boxed{\dfrac{\text{dividend}}{\text{divisor}} = \text{quotient} + \dfrac{\text{remainder}}{\text{divisor}}}$$

Thus,

$$\frac{79 - x}{x} = 5 + \frac{7}{x}$$

Multiplying both sides by the LCD $= x$, we get

$$79 - x = 5x + 7$$
$$-6x = -72$$
$$x = 12$$

Hence, the numbers are $x = 12$ and $79 - x = 67$. ■

Check

$$\frac{67}{12} = 5 + \frac{7}{12}$$

Work Problems

Work problems often involve fractional expressions. The general approach to such problems is to *represent what part of a job is done in one unit of time.*

EXAMPLE 3 One machine can complete a job in 3 days. Another can do the same piece of work in 2 days. How long will it take if both machines work together?

Solution Note that the number of days required for both machines working together is *not* the sum of 3 and 2. The basic relationship we use is:

If a job can be completed in n days, then $\dfrac{1}{n}$ of the job can be completed in one day.

Let x = the number of days required for both machines working together to complete the job

Then $\dfrac{1}{x}$ = the part of the job both can do in 1 day

$\dfrac{1}{3}$ = the part of the job the first machine can do in 1 day

$\dfrac{1}{2}$ = the part of the job the second machine can do in 1 day.

Thus,

$$\binom{\text{part done by first}}{\text{machine in 1 day}} + \binom{\text{part done by second}}{\text{machine in 1 day}} = \binom{\text{part done together}}{\text{in 1 day}}$$

$$\dfrac{1}{3} \qquad + \qquad \dfrac{1}{2} \qquad = \qquad \dfrac{1}{x}$$

Multiplying both sides by the LCD = $6x$, we obtain

$$2x + 3x = 6$$
$$5x = 6$$
$$x = 1\dfrac{1}{5}$$

Hence, it takes both machines working together $1\frac{1}{5}$ days to complete the job.

■

EXAMPLE 4 An inlet pipe can fill a water tank in 10 hr, while an outlet pipe can empty the tank in 15 hr. If the tank is $\frac{1}{3}$ filled and the outlet pipe is opened, how long from that time will it take to fill the tank?

Solution   Let t = the number of hours it takes to fill the tank with both pipes open

Then

$$\left(\begin{array}{c}\text{part filled by inlet} \\ \text{pipe in 1 hr}\end{array}\right) - \left(\begin{array}{c}\text{part emptied by outlet} \\ \text{pipe in 1 hr}\end{array}\right) = \left(\begin{array}{c}\text{part filled in 1 hr} \\ \text{with both pipes open}\end{array}\right)$$

and

$$\frac{1}{10} \quad - \quad \frac{1}{15} \quad = \quad \frac{1}{t}$$

Multiplying both sides by the LCD = $150t$, we obtain

$$15t - 10t = 150$$
$$5t = 150$$
$$t = 30$$

However, 30 hr is the time it would take if the tank was empty at the beginning. But, only $\frac{2}{3}$ of this time is necessary since the tank is $\frac{1}{3}$ filled: $\frac{2}{3} \cdot 30$ hours = 20 hours. ■

Check   part filled in 20 hours = $(\frac{1}{10} \cdot \text{part/hr}) \cdot 20$ hr = 2 parts

part emptied in 20 hours = $(\frac{1}{15} \cdot \text{part/hr}) \cdot 20$ hr = $1\frac{1}{3}$ parts

(part filled) − (part emptied) = 2 part − $1\frac{1}{3}$ part = $\frac{2}{3}$ part

Thus, it will take 20 hr to fill the remaining $\frac{2}{3}$ of the tank.

Motion Problems

When an object travels at a constant rate (r), the relationship between the distance (d) it travels in time t is given by

$$d = rt \quad \text{or} \quad r = \frac{d}{t} \quad \text{or} \quad t = \frac{d}{r}$$

EXAMPLE 5   Scott Beebe's sailboat can sail at the rate of 18 km per hour in still water. If the boat can go 63 km down the river in the same time it takes to go 45 km up the river, what is the speed of the current in the river?

Solution   Let r = the rate of the current in kilometers per hour.

Since the boat's speed (rate) in still water is 18 km/hr:

$$(18 + r)\frac{\text{km}}{\text{hr}} = \text{speed downstream}$$

$$(18 - r)\frac{\text{km}}{\text{hr}} = \text{speed upstream}$$

	r	d	$t = \dfrac{d}{r}$
Downstream	$(18 + r)\dfrac{\text{km}}{\text{hr}}$	63 km	$t = \left(\dfrac{63}{18 + r}\right)\text{hr}$
Upstream	$(18 - r)\dfrac{\text{km}}{\text{hr}}$	45 km	$t = \left(\dfrac{45}{18 - r}\right)\text{hr}$

Since the time is the same in each direction, we have

$$\text{time downstream} = \text{time upstream}$$

$$\left(\frac{63}{18 + r}\right)\text{hr} = \left(\frac{45}{18 - r}\right)\text{hr}$$

Solving for r, we obtain

$$\frac{63}{18 + r} = \frac{45}{18 - r}$$

Multiplying both sides by the LCD $= (18 + r)(18 - r)$, we get

$$63(18 - r) = 45(18 + r)$$

$$1134 - 63r = 810 + 45r$$

$$324 = 108r$$

$$3 = r$$

Thus, the speed of the current is 3 km per hour. ■

Check
$$\text{speed downstream} = (18 + 3)\frac{\text{km}}{\text{hr}} = 21\frac{\text{km}}{\text{hr}}$$

$$\text{speed upstream} = (18 - 3)\frac{\text{km}}{\text{hr}} = 15\frac{\text{km}}{\text{hr}}$$

Since $t = \dfrac{d}{r}$:
$$\text{time downstream} = \frac{63 \text{ km}}{21 \dfrac{\text{km}}{\text{hr}}} = 3 \text{ hr}$$

$$\text{time upstream} = \frac{45 \text{ km}}{15 \dfrac{\text{km}}{\text{hr}}} = 3 \text{ hr}$$

Hence, time downstream = time upstream = 3 hr.

It is sometimes convenient to solve for an unknown not directly asked for in the solution of the problem.

EXAMPLE 6   The distance from Alameda to Fremont is 21 miles. Scott leaves Alameda at 4 PM, and Russ, who on the average walks 25% faster than Scott, leaves Alameda at 3 minutes past 5 PM. If both arrive at Fremont at the same time, find that time.

Solution   In this problem we shall choose as our unknown the rate at which Scott walks, rather than time.

Notice that both hours and minutes occur in this problem. It is thus necessary to decide whether the unit of time will be 1 hour or 1 minute. It is convenient here to work in hours, so it follows that rate must be expressed in miles per hour.

Let x = Scott's rate in miles per hour. Then $x + 25\%x = x + \frac{1}{4}x = \frac{5}{4}x$ is Russ' rate in miles per hour. Using the fact that time = distance/rate, we obtain

$$\text{Scott's time to walk 21 miles is } \frac{21 \text{ miles}}{x \dfrac{\text{mi}}{\text{hr}}} = \frac{21}{x} \text{hr}$$

$$\text{Russ' time to walk 21 miles is } \frac{21 \text{ mi}}{\dfrac{5}{4}x \dfrac{\text{mi}}{\text{hr}}} = \frac{84}{5x} \text{hr}$$

Now Russ started 1 hr 3 min later than Scott; that is, $1\frac{3}{60} = 1\frac{1}{20}$ hr later than Scott, and arrived at Fremont at the same time. Thus, Scott's time for the walk was $1\frac{1}{20}$ hr more than Russ' time.

$$\text{Scott's time} = \text{Russ' time} + 1\frac{1}{20} \text{ hr}$$

That is,

$$\frac{21}{x} \text{hr} = \frac{84}{5x} \text{hr} + 1\frac{1}{20} \text{hr}$$

Solving for x, we obtain

$$\frac{21}{x} = \frac{84}{5x} + \frac{21}{20}$$

Multiplying both sides by the LCD = $20x$, we get

$$420 = 336 + 21x$$

$$84 = 21x$$

$$4 = x$$

Hence, it takes Scott $\frac{21}{4}$ hr = 5 hr 15 min. Thus, Scott (along with Russ) arrives in Fremont at 4 PM + 5 hr 15 min, or 9:15 PM. ∎

Miscellaneous Fractional Problems

EXAMPLE 7   (Percentage Problem). How much pure alcohol must be added to 50 gal of 50% alcohol to make a 72% alcohol (see Figure 6.1)?

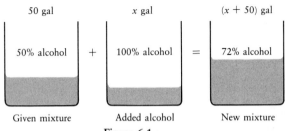

Figure 6.1

Solution   Let x = the number of gallons to be added.

Then   (50%) of 50 gal = 25 gal of pure alcohol in the
                                                given mixture

                    (25 + x) gal = number of gallons of pure alcohol in the
                                                new mixture

                    (50 + x) gal = total number of gallons in the new mixture

Hence,

$$\frac{\text{number of gallons of pure alcohol in new mixture}}{\text{total number of gallons in new mixture}} = 72\%$$

That is,

$$\frac{(25 + x) \text{ gal}}{(50 + x) \text{ gal}} = \frac{72}{100}$$

Solving for x, we obtain

$$\frac{25 + x}{50 + x} = \frac{72}{100}$$

Multiplying both sides by the LCD = $100(50 + x)$, we get

$$100(25 + x) = 72(50 + x)$$

$$2500 + 100x = 3600 + 72x$$

$$28x = 1100$$

$$x = 39\frac{2}{7}.$$

Thus, $39\frac{2}{7}$ gal should be added.   ■

EXAMPLE 8   (Planetary Problem). The planet Venus makes a complete circuit about the sun in $7\frac{1}{2}$ months, and the planet Earth completes a circuit in 12 months. Find the number of months between the two successive times when Venus is between the Earth and the sun. See Figure 6.2.

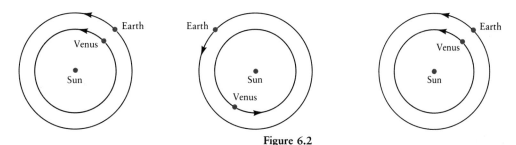

Figure 6.2

Solution   Let x = the number of months *between* the two successive times when Venus is between the Earth and the sun.

Then
$$\frac{1}{7\frac{1}{2}} = \frac{2}{15} = \text{part of circuit Venus makes in one month}$$

and
$$\frac{1}{12} = \text{part of circuit Earth makes in one month}$$

Thus, $\dfrac{2}{15} - \dfrac{1}{12} = \text{part of circuit Venus gains on Earth in one month}$

Also, $\dfrac{1}{x} = \text{part of circuit Venus gains on Earth in one month}$

Hence,

$$\frac{2}{15} - \frac{1}{12} = \frac{1}{x}$$

Multiplying both sides by the LCD = $60x$, we obtain

$$8x - 5x = 60$$

$$3x = 60$$

$$x = 20 \quad \blacksquare$$

Thus, it takes 20 months between the two successive times.

EXERCISES 6.7

In Exercises 1–10, solve the number problems.

1. What number must be subtracted from the numerator and denominator of the fraction $\frac{17}{29}$ to give $\frac{1}{2}$?

2. What number must be added to the numerator and denominator of $\frac{3}{5}$ to give $\frac{2}{3}$?

3. The denominator of a fraction is 5 greater than two times its numerator and the reduced value of the fraction is $\frac{1}{3}$. Find the numerator.

4. The numerator of a certain fraction is 4 less than the denominator. If 5 is added to both the numerator and the denominator, the value of the fraction becomes $\frac{2}{3}$. What is the fraction?

5. In a certain fraction the numerator is 4 less than the denominator. If 1 is added to the numerator and 21 is added to the denominator, the value of the fraction is $\frac{1}{3}$. Find the fraction.

6. Separate 93 into two parts such that the larger number divided by the smaller gives a quotient of 4 and a remainder of 8.

7. Separate 96 into two parts such that the larger divided by the smaller yields a quotient of 18 and a remainder of 1.

8. Find three consecutive positive integers such that the first divided by the sum of the second and third gives a quotient of $\frac{2}{5}$.

9. The sum of the numerator and the denominator of a proper fraction is 70. If each is increased by 7, the value of the fraction becomes $\frac{3}{4}$. Find the fraction.

10. In a given two-place number, the ten's digit is 2 more than the unit's digit. The number divided by the sum of its digits gives 7 as a quotient and 3 as a remainder. What is the number?

Solve the work problems in Exercises 11–16.

11. Claudia can assemble a machine in 4 hr and Daniel can assemble an identical machine in 5 hr. How long will it take the two working together to assemble the machine?

12. Justin can shingle a house in 3 days and Liz can do it in 4 days. How many days are required when both are working?

13. A tank can be filled in $\frac{3}{4}$ hr with two pumps working. The larger pump alone can fill the tank in $1\frac{1}{5}$ hr. How long would it take the smaller pump alone to fill the tank?

14. A, B, and C can wallpaper a house in 8 hr; A and B can do it in 12 hr. How long would it take C alone to wallpaper a house?

15. A water tower can be filled in $2\frac{1}{4}$ hr, but, due to leakage, it requires $2\frac{1}{2}$ hr to fill the tank. With the inlet closed, how long will it take for all the water to leak out?

16. An outlet pipe can empty a swimming pool in 18 hr. With the swimming pool full, this pipe and a second pipe are opened, and together they empty $\frac{3}{5}$ of the pool in 6 hr. If the first pipe is then shut off, how long will it take the second pipe to empty the pool?

Solve the motion problems in Exercises 17–22.

17. A biker travels 100 mi before he gets a flat tire. After fixing the tire, he bikes 300 mi. His average rate before the flat was one-half his average rate after fixing the tire. If he spent a total of 5 hr on his trip, excluding the time spent fixing the tire, what was his average rate on each part of the trip?

18. A Lear jet can travel 200 mi in the same time that a propeller jet travels 150 mi. If the propeller jet travels 20 mph slower than the Lear jet, find the rate of each jet.

19. A motorboat can travel 10 mph in still water. If it can travel 15 mi down a stream in the same time that it can travel 9 mi upstream, what is the rate of the stream?

20. A bus' normal average speed is 42 mph. At

one point in its trip it is delayed for 10 min, and then completes its trip at a speed of 45 mph, arriving 8 min and 40 sec late. How far was it from its destination when the delay occurred?

21. A driver travels 400 mi from her starting point. Returning the same way, she increased her speed by 10 mph and got back in 2 hr less time. What was her speed each way?

22. A man sets out for the park at 4 mph. When he has covered $\frac{2}{3}$ of the distance, he increases his speed to 5 mph, and completes the whole distance in $31\frac{1}{2}$ min. How far was his starting point from the park?

Solve the miscellaneous word problems in Exercises 23–28.

23. How much pure alcohol must be added to 4 gal of water in a radiator to make a 20% solution of alcohol?

24. How much water must be added to 100 lb of a 5% salt solution to make a 4% solution?

25. A woman has equal amounts invested in bonds and stocks, the stocks paying 2% more than the bonds. If the yearly income from the bonds is $360, and from the stocks $480, what is the rate of interest on each? (*Hint:* principal = income/rate.)

26. A man has $820 more invested in real estate than in stocks. The stocks yield $432, and the real estate $641.60, annually. If the rate of interest on the stocks is three-fourths of the rate on the real estate, find the amounts invested.

27. The total resistance R in an electrical circuit consisting of two resistances of r_1 ohms and r_2 ohms connected in parallel is given by the equation $\dfrac{1}{R} = \dfrac{1}{r_1} + \dfrac{1}{r_2}$. Find the larger of the two parallel resistances if it is 2 ohms more than the smaller, and if the total resistance is two-thirds of the smaller resistance.

28. All camera lens have a length f, called the focal length, such that when an object is in focus, the object's distance d_0 from the lens and the distance d_1 from the lens to the film satisfy the equation (see Figure 6.3):

$$\frac{1}{d_0} + \frac{1}{d_1} = \frac{1}{f}$$

If $d_0 = 60$ cm and $d_1 = 15$ cm, find the focal length of the lens.

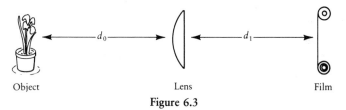

Object   Lens   Film

Figure 6.3

Chapter 6   SUMMARY

TERMS, RULES, AND FORMULAS

1. *Rational Expression.* An expression that can be expressed in the form P/Q, where P and Q are polynomials and $Q \neq 0$.

2. *Equality of Rational Expressions.* If P/Q and R/S are rational expressions, with $Q \neq 0$ and $S \neq 0$, then $P/Q = R/S$ if and only if $P \cdot S = Q \cdot R$.

3. *Fundamental Principle of Fractions.* For all polynomials P, Q, and K, where $Q \neq 0$ and

$K \neq 0$

$$\frac{P}{Q} = \frac{P \cdot K}{Q \cdot K}$$

4. *Multiplication of Rational Expressions.* If P, Q, R, and S are polynomials, with $Q \neq 0$ and $S \neq 0$, then

$$\frac{P}{Q} \cdot \frac{R}{S} = \frac{P \cdot R}{Q \cdot S}$$

5. *Division of Rational Expressions.* If P, Q, R, and S are polynomials with $Q \neq 0$ and $RS \neq 0$, then

$$\frac{P}{Q} \div \frac{R}{S} = \frac{P}{Q} \cdot \frac{S}{R} = \frac{P \cdot S}{Q \cdot R}$$

6. *Addition of LIKE Fractional Expressions.* If P, Q, and R are polynomials with $R \neq 0$, then

$$\frac{P}{R} + \frac{Q}{R} = \frac{P + Q}{R}$$

PROCEDURES

1. *Procedure for Reducing Fractions to Lowest Terms*
   (1) Completely factor the numerator and the denominator.
   (2) Cancel all common factors that exist in both the numerator and the denominator.
2. *Procedure for Multiplying Rational Expressions*
   (1) Completely factor the numerators and denominators of the given fractions.
   (2) Cancel any factor common to any numerator and any denominator.
   (3) Multiply the remaining factors of the numerators to find the numerator of the product. Multiply the remaining factors of the denominators to find the denominator of the product. The resulting product will be in lowest terms.
3. *Procedure for Dividing Rational Expressions*
   (a) Invert the divisor.
   (b) Multiply the dividend and the inverted divisor.

4. *Procedure for Finding the Least Common Denominator* (LCD)
   (1) Factor each denominator completely. Express repeated factors as powers.
   (2) Write each different factor that appears in any denominator.
   (3) Raise each factor in step 2 to the highest power it occurs in *any* denominator.
   (4) The least common denominator (LCD) is the product of all the factors found in step 3.
5. *Procedure for Adding and Subtracting Unlike Fractional Expressions*
   (1) Find the LCD of the terms.
   (2) Express each fraction as an equivalent fraction whose denominator is the LCD of step 1.
   (3) Add or subtract the resulting *like* fractions.
   (4) Reduce the resulting fraction to lowest terms, if possible.
6. *Procedure for Simplifying Complex Fractions*
   Method 1. Multiply the numerator and denominator of the complex fraction by the LCD of all the fractions that appear in the numerator and in the denominator. Simplify the results.
   Method 2. Simplify the numerator and denominator of the complex fraction. Then divide the simplified numerator by the simplified denominator.
7. *Procedure for Solving Fractional Equations*
   (1) Multiply each term in the fractional equation by the LCD of all the fractions.
   (2) Collect terms and solve the resulting simple equation.
   (3) Check each root in the original fractional equation.

Chapter 6 EXERCISES

In Exercises 1–10, find the missing terms.

1. $\dfrac{5x^2}{16} = \dfrac{}{16y^4}$

2. $\dfrac{7c^2 d}{8ef} = \dfrac{}{16e^3 f^2}$

3. $\dfrac{4mn^3}{13r^2 s^6} = \dfrac{8mn^3 r^3}{}$

4. $\dfrac{a}{b - c} = \dfrac{ab - ac}{}$

5. $\dfrac{r}{x + y} = \dfrac{}{x^2 - y^2}$

6. $\dfrac{5n^2}{m - n} = \dfrac{}{2m^2 + mn - 3n^2}$

7. $\dfrac{2c - d}{2c + d} = \dfrac{}{4c^2 - d^2}$

8. $\dfrac{4a + b}{m + r} = \dfrac{}{m^2 + 2mr + r^2}$

9. $\dfrac{c + d}{f - g} = \dfrac{}{f^2 - g^2}$

10. $\dfrac{x^2}{x - 2y} = \dfrac{}{x^3 - 2x^2y}$

In Exercises 11–25, reduce to lowest terms.

11. $\dfrac{6a^2bc^2}{9ab^2c}$

12. $\dfrac{7x^2yz}{28x^3yz^2}$

13. $\dfrac{24a^3c^2x^2}{18a^3x^2 - 12a^2x^3}$

14. $\dfrac{6x^2 - 8xy}{9xy - 12y^2}$

15. $\dfrac{24a^3b^2c^3}{36a^2b^3c^3}$

16. $\dfrac{8a^2bc^2}{12ab^2cd}$

17. $\dfrac{3a^2 - 6ab}{2a^2b - 4ab^2}$

18. $\dfrac{4x^2 - 9y^2}{4x^2 + 6xy}$

19. $\dfrac{20(x^3 - y^3)}{5x^2 + 5xy + 5y^2}$

20. $\dfrac{x^3 - 2xy^2}{x^4 - 4x^2y^2 + 4y^4}$

21. $\dfrac{4(a + b)(b - a)}{16a - 16b}$

22. $\dfrac{4 - (a + b)^2}{4(2 + a + b)(2 - a - b)}$

23. $\dfrac{(a + b) - (a + b)^2}{(a + b)(1 - a - b)}$

24. $\dfrac{3x^2y - 24xy + 36y}{5x^3 - 60x^2 + 180x}$

25. $\dfrac{3a^2 - 3ab + 3bc - 3ac}{5ac - 5bc + 5ad - 5bd}$

In Exercises 26–45, perform the indicated operations and simplify.

26. $\dfrac{7y}{a^3} \cdot \dfrac{5ab^3}{14y^2}$

27. $\dfrac{4x + 12}{7} \cdot \dfrac{14y}{3x + 9}$

28. $\dfrac{abc}{def} \cdot \dfrac{bcd}{efa} \cdot \dfrac{cde}{fab}$

29. $\dfrac{ab^2xy^2}{a^3bx^3y} \cdot \dfrac{a^2bx^2y}{ab^3xy^3}$

30. $\dfrac{8x^2}{x^2 - y^2} \cdot \dfrac{x + y}{2x}$

31. $\dfrac{x^2 - y^2}{xa + ya} \cdot \dfrac{x^2 + y^2}{x - y}$

32. $\dfrac{ax + x^2}{2b - cx} \cdot \dfrac{2bx - cx^2}{(a + x)^2} \cdot \dfrac{a + x}{x^2}$

33. $\dfrac{ac + bc + a^2 + ab}{ab + ac + b^2 + bc} \cdot \dfrac{b + c}{c + a}$

34. $\dfrac{x^2 - (a + b)x + ab}{x^2 - (a + c)x + ac} \cdot \dfrac{x^2 - c^2}{x^2 - b^2}$

35. $\dfrac{x^5y^6}{a^7b^8} \div \dfrac{x^3y^4}{a^5b^6}$

36. $\dfrac{x^2 - y^2}{x + 2y} \div \dfrac{x - y}{3x + 6y}$

37. $\dfrac{x^4 - y^4}{(x - y)^2} \div \dfrac{x^2 + xy}{x - y}$

38. $\dfrac{4(a^2 - ab)}{b(a + b)^2} \div \dfrac{6ab}{a^2 - b^2}$

39. $\dfrac{y^2 + y - 6}{y^2 - 4} \cdot \dfrac{y^2 + y - 12}{y^2 + y - 6}$
$\cdot \dfrac{y^2 - 3y - 10}{y^2 - y - 20}$

40. $\dfrac{2x^3 - 2xy^2}{x + 2y} \div \dfrac{x^2 - y^2}{2x + 4y}$

41. $\dfrac{24ax - 80ay}{14bx - 35by} \div \dfrac{6ax - 20ay}{28cx - 70cy}$

42. $\dfrac{4z - 1}{z - 1} \cdot \dfrac{(z + 1)^2}{16z^2 - 1} \div \dfrac{z^2 - 1}{4z + 1}$

43. $\dfrac{t^2 + 5t + 6}{t^2 + 7t + 12} \cdot \dfrac{t^2 + 9t + 20}{t^2 + 11t + 30}$

44. $\dfrac{4a^2 + 4ab + b^2 - c^2}{4a^2 - b^2 - 2bc - c^2} \div \dfrac{2a + b + c}{2a - b - c}$

45. $\left(\dfrac{x - 1}{x + 1}\right)^4 \div \left(\dfrac{x - 1}{x + 1}\right)^2$

In Exercises 46–55, find the LCD of the expressions.

46. $5ab^3, 7a^2b^4$ 47. $12x^3, 45x^2$

48. $x^2 + x - 6, x^2 + 4x + 4$

49. $x^2 - 9, x^2 + 10x + 21$

50. $x^2 - y^2, x + y, x^2 + xy$

51. $12x - 36, x^2 - 9, x^2 - 5x + 6$

52. $x^{12} + x^{11}, x^{14} + x^{13}$

53. $ax - ay, bx^2 - by^2, cx + cy$

54. $x(a + b)^2, y(a^2 - b^2), z(a - b)^2$

55. $x^2 - 3x - 10, x^2 - 9x + 20$

In Exercises 56–70, perform indicated operations.

56. $\dfrac{5x-1}{8} - \dfrac{3x-2}{7} + \dfrac{x-5}{4}$

57. $\dfrac{2x+5}{x} - \dfrac{x+3}{2x} - \dfrac{27}{8x^2}$

58. $\dfrac{x-4}{x-2} - \dfrac{x-7}{x-5}$

59. $\dfrac{4a^2+b^2}{4a^2-b^2} - \dfrac{2a-b}{2a+b}$

60. $\dfrac{5}{1+2x} - \dfrac{3x}{1-2x} - \dfrac{4-13x}{1-4x^2}$

61. $\dfrac{3}{x-2} + \dfrac{2}{3x+6} + \dfrac{5x}{1-4x^2}$

62. $\dfrac{3x}{1-x^2} - \dfrac{2}{x-1} + \dfrac{2}{x+1}$

63. $\dfrac{1}{(a-b)(a-c)} + \dfrac{1}{(b-c)(b-a)}$
$+ \dfrac{1}{(c-a)(c-b)}$

64. $3x - 4 - \dfrac{9x^2-16}{3x+4}$

65. $\dfrac{1}{1-x} + \dfrac{x}{x^2-1} + \dfrac{x+1}{(x-1)^2}$

66. $\dfrac{1}{4x^2-4x+1} + \dfrac{1-x}{1-2x}$

67. $\dfrac{9}{x^2+7x-18} - \dfrac{8}{x^2+6x-16}$

68. $\dfrac{3x-4}{x-9} - \dfrac{4x-1}{x+2} - \dfrac{x-1}{x^2-7x-18}$

69. $\dfrac{20x^2+7x-3}{9x^2-1} - \dfrac{3x+1}{3x-1}$

70. $\dfrac{x+2y}{x^2-2xy+y^2} - \dfrac{1}{y-x} - \dfrac{2}{x+2y}$

In Exercises 71–75, simplify the complex fraction.

71. $1 - \dfrac{1}{1+\dfrac{1}{x}}$

72. $1 + \dfrac{x}{1+x+\dfrac{2x^2}{1-x}}$

73. $\dfrac{1}{1-\dfrac{1}{1+\dfrac{1}{x}}}$

*74. $\dfrac{1}{1+\dfrac{x}{1+x+\dfrac{2x^2}{1-x}}}$

*75. $\dfrac{x+\dfrac{1}{y}}{x+\dfrac{1}{y+\dfrac{1}{z}}} - \dfrac{1}{y(xyz+x+z)}$

In Exercises 76–90, solve for x.

76. $\dfrac{4x+1}{x+2} = \dfrac{4}{x^2-4} - \dfrac{1+4x}{2-x}$

77. $\dfrac{x+1}{x+5} - \dfrac{1-x}{5-x} = \dfrac{1}{25-x^2}$

78. $\dfrac{1}{x+1} - \dfrac{4}{x^2-1} = \dfrac{1}{1-x}$

79. $\dfrac{x}{x-3} + \dfrac{1}{2(x^2-x-6)} = \dfrac{2}{x+2} + 1$

80. $\dfrac{x^2+x+1}{x+1} = \dfrac{x^2-x+1}{x-1} + \dfrac{x}{1-x^2}$

81. $\dfrac{3}{1+3x} - \dfrac{2}{1-3x} = \dfrac{9x^2}{9x^2-1} - 1$

82. $\dfrac{3}{3x+6} = \dfrac{5}{4x+10} - \dfrac{4}{2x^2+9x+10}$

83. $\dfrac{2x-1}{3x+4} = \dfrac{4x-1}{6x-1} + \dfrac{19}{18x^2+21x-4}$

84. $\dfrac{1}{2}\left(3 + \dfrac{x}{3}\right) - \dfrac{3}{4}\left(\dfrac{3x+7}{6}\right) = 1$

85. $\dfrac{3}{4}\left(2x - \dfrac{8}{x}\right) - \dfrac{2}{3}\left(2x - \dfrac{4}{x}\right) = \dfrac{2x-5}{12}$

86. $\dfrac{c^2}{x} - c = \dfrac{d^2}{x} + d$

87. $\dfrac{2x}{a} + \dfrac{a-4x}{3} + 4 = 3a$

88. $\dfrac{b}{2x} + 3 = \dfrac{a}{3x} - 2c$

*89. $\dfrac{x}{b} - ac + bc + ab = \dfrac{x}{c} + \dfrac{x}{a}$

*90. $\dfrac{1}{a+b} - \dfrac{2ab}{(a+b)^3} = \dfrac{x-b+a}{(a+b)^2}$

91. Separate 135 into two parts such that the quotient is 3 and the remainder is 23, when the larger is divided by the smaller.

92. Find three consecutive positive integers such that the sum of the first and second, divided by one-half of the third, gives a quotient of $3\frac{3}{4}$.

93. Erin can do a piece of work in x hr and Larry can do the same piece of work in y hr. How many hours would be required for them to complete the work together?

94. A tank can be filled in $\frac{3}{4}$ hr with two pumps working. The larger pump alone can fill the tank in $1\frac{1}{5}$ hr. How long would it take the smaller pump alone to fill the tank?

95. A boat, which has a speed of 15 mph in still water, can travel 8 mi up a stream in the same time required to travel 10 mi down the stream. What is the rate of the stream?

96. The area of two rectangular lots are 5000 sq ft and 6000 sq ft, respectively. They have the same width, but the second lot is 20 ft longer than the first lot. Find the length of each.

97. How much pure alcohol must be added to 1 quart of 80% alcohol to make a mixture that will be 90% alcohol?

*98. A certain machine requires an alloy containing half copper and half zinc. How much pure copper must be added to 40 kg of an alloy containing 70% zinc and 30% copper to raise the copper content to 50%?

[*Hint:* percent copper

$$= \frac{\text{final amount of copper}}{\text{final amount of alloy}} \times 100.]$$

*99. In a 100-yard race when Marc gives John half a yard headstart, he wins by a fifth of a second. But, when he gives John $4\frac{1}{2}$ yards headstart, John wins by a fifth of a second. What are their respective times for running 100 yards?

[*Hint:* Suppose Marc runs 100 yards in x seconds. Then John runs $99\frac{1}{2}$ yards in $(x + \frac{1}{5})$ seconds.]

*100. Linda and Joan run two races of half a mile each. In the first heat Linda gives Joan a headstart of 24 yards and beats her by $24\frac{3}{11}$ sec. In the second heat Linda gives Joan a headstart of $\frac{3}{4}$ min and is beaten by 73 yd 1 ft. In what time could Linda and Joan run half a mile? (See *Hint* in Exercise 99.)

7 Exponents, Radicals, and Complex Numbers

In Chapter 2, we worked with exponential notation such as a^n, where n is a positive integer. In this chapter, we expand the exponential notation to include exponents that are positive, zero, negative, or rational numbers. We shall see that the rules discussed in Chapter 2 still hold for these new exponents.

We shall also show that rational exponents may be written using radicals and that the real number system is not adequate to solve all polynomial equations. Thus, it becomes necessary to introduce a new number system called the complex number system.

7.1 POSITIVE INTEGER EXPONENTS

We begin by restating the definition of a^n and the rules given in Chapter 2.

If n is a positive integer and a is any real number, we define a^n by

$$a^n = \underbrace{a \cdot a \cdot \ldots \cdot a}_{n \text{ factors}}$$

We call a^n the nth power of a; n is the exponent, and a is the base. We then showed that if m and n are positive integers and a and b are any real numbers, then

Rule 1
$$a^m \cdot a^n = a^{m+n}$$

EXAMPLE 1   (a) $2^3 \cdot 2^4 = 2^{3+4} = 2^7$ (b) $(-8)^5 \cdot (-8)^6 = (-8)^{5+6} = (-8)^{11}$

(c) $x^5 \cdot x^2 \cdot x^3 = x^{5+2+3} = x^{10}$ (d) $\left(\dfrac{1}{5}\right)^3 \left(\dfrac{1}{5}\right)^{10} = \left(\dfrac{1}{5}\right)^{3+10} = \left(\dfrac{1}{5}\right)^{13}$

Rule 2
$(a^m)^n = a^{mn}$

EXAMPLE 2   (a) $(4^5)^3 = 4^{5 \cdot 3} = 4^{15}$ (b) $(x^4)^3 = x^{4 \cdot 3} = x^{12}$

(c) $[(a+b)^2]^4 = (a+b)^{2 \cdot 4} = (a+b)^8$

Rule 3
(a) $\dfrac{a^m}{a^n} = a^{m-n}$, if m is greater than n, $a \neq 0$.
(b) $\dfrac{a^m}{a^n} = 1$ if m is equal to n, $a \neq 0$.
(c) $\dfrac{a^m}{a^n} = \dfrac{1}{a^{n-m}}$ if m is less than n, $a \neq 0$.

EXAMPLE 3   (a) $\dfrac{x^9}{x^4} = x^{9-4} = x^5$

(b) $\dfrac{x^6}{x^6} = 1$

(c) $\dfrac{x^3}{x^7} = \dfrac{1}{x^{7-3}} = \dfrac{1}{x^4}$

Rule 4
$(ab)^n = a^n b^n$

EXAMPLE 4   (a) $(3x)^4 = 3^4 x^4 = 81x^4$

(b) $(2x^3y)^2 = 2^2(x^3)^2y^2 = 4x^6y^2$

(c) $[4(x + 6)^3]^2 = 4^2[(x + 6)^3]^2 = 16(x + 6)^6$

Rule 5

$$\left(\frac{a}{b}\right)^n = \frac{a^n}{b^n}, \quad \text{provided} \quad b \neq 0$$

EXAMPLE 5   (a) $\left(\dfrac{x^2}{y^4}\right)^3 = \dfrac{(x^2)^3}{(y^4)^3} = \dfrac{x^6}{y^{12}}$

(b) $\left(\dfrac{3x}{4ab^2}\right)^2 = \dfrac{(3x)^2}{(4ab^2)^2} = \dfrac{3^2x^2}{4^2a^2(b^2)^2} = \dfrac{9x^2}{16a^2b^4}$

CAUTION

An exponent indicates the power to be taken of that quantity and only that quantity to which it is attached.

EXAMPLE 6   (a) $2x^3$ means $2 \cdot x \cdot x \cdot x$, whereas $(2x)^3$ means $2 \cdot 2 \cdot 2 \cdot x \cdot x \cdot x$.

(b) $-(y^4)^2$ means $-(y^4)(y^4) = -y^8$, whereas $(-y^4)^2$ means $(-y^4)(-y^4) = y^8$.

EXERCISES   7.1

Simplify, if possible, using the rules of exponents.

1. $2^3 \cdot 2^5$
2. $x^3 \cdot x^6$
3. $(-5)^2(-5)^4$
4. $(a^2)(a^5)$
5. m^3m^{12}
6. $10^4 \cdot 10^7$
7. $x(x^7)$
8. $b^{11}b^{30}$
9. $2x(2^3x^2)$
10. $(3^4x)(3^4x^2)$
11. $(5x^2y^3)(5^2xy^2)$
12. $(-6x^2y)(-6xy)$
13. $(x^3yz^2)(xy^4z^3)(4yz)$
14. $(axy^4)(2b^2x^3y^2)(3abx^2y)$
15. $(x^2)^5$
16. $(y^3)^3$
17. $(-x^4)^3$
18. $(-x^8)^2$
19. $(x^2)^3(y^3)^2$
20. $(x^5)^3(y^4)^4$
21. $(-1)^5$
22. $(-1)^{10}$
23. $(a^2)^2 \cdot (b^3)^3 \cdot (c^4)^4$
24. $(a^2)^4 \cdot (x^3)^3 \cdot (y^4)^2$
25. $\dfrac{x^5}{x^2}$
26. $\dfrac{y^6}{y^4}$
27. $\dfrac{x^3}{x^3}$
28. $\dfrac{(-4)^2}{(-4)^2}$
29. $\dfrac{x^3y^4}{xy^2}$
30. $\dfrac{x^3y}{x^2y^2}$
31. $\dfrac{ab^2}{a^3b^3}$
32. $\dfrac{a^2b^2}{ab^2}$
33. $\dfrac{a^3x^2}{a^3x^2}$
34. $\dfrac{x^3yz^2}{xyz^4}$
35. $\dfrac{-xyz^5}{xy^2z^3}$
36. $\dfrac{a^2b^2c^4}{-a^3bc^2}$
37. $(2x^3)^3$
38. $(-3x)^2$
39. $-(x^2y^3)^4$
40. $(-a^5b^4)^3$
41. $(x^2y^3z)^4$
42. $(x^3y^2z^4)^7$
43. $(2ax^3y^2)^4$
44. $(-3a^2bx)^3$

45. $\left(\dfrac{-x}{4}\right)^3$

46. $\left(\dfrac{a}{x^2}\right)^4$

47. $\left(\dfrac{5x^3}{6y^2}\right)^2$

48. $\left(\dfrac{ab^2}{2x^3}\right)^5$

49. $\left(\dfrac{xy^2z}{a^2b^3c}\right)^{10}$

50. $\left(\dfrac{-a^2b^3}{2x^2y}\right)^8$

51. $(5a^2y^3)^2 \cdot (b^3x^4)^3$

52. $(2ax^2)^2(3bx)^2$

53. $\dfrac{(xy^2)^3}{(xy^3)^2}$

54. $\dfrac{(3x^2y)^3}{(6xy^2)^2}$

55. $\dfrac{(xy^2)^3 \cdot (x^2y^3)^2}{(x^5y)^2}$

56. $\dfrac{(6x^5y^3)^4}{(3xy^2)^3(2x^2y)^5}$

57. $(a^2b^3)^4\left(\dfrac{x^2}{a^2}\right)^2\left(\dfrac{y}{b^3}\right)^3$

58. $\left(\dfrac{ax^2}{y}\right)^2\left(\dfrac{x^2y^3}{a}\right)^3\left(\dfrac{ay}{x^3}\right)^2$

59. $\left(\dfrac{2x}{3x}\right)^2\left(\dfrac{6z}{y^2}\right)^3\left(\dfrac{y^3}{4x}\right)^2$

60. $\left(\dfrac{x^2y}{a}\right)^4\left(\dfrac{x}{a^3y^2}\right)^2\left(\dfrac{a^2}{x^2}\right)^5$

7.2 ZERO AND NEGATIVE EXPONENTS

Now we would like to extend the rules to zero or negative exponents. We begin with a^0, where $a \neq 0$, since 0^0 has no mathematical meaning. If the rule $a^m \cdot a^n = a^{m+n}$ is to hold for a^0, then

$$a^m \cdot a^0 = a^{m+0} = a^m$$

Since the only factor that multiplies with a^m to yield a product of a^m is the number 1, then we must have

---

Rule 6

$$a^0 = 1, \quad \text{provided} \quad a \neq 0$$

EXAMPLE 1 (a) $4^0 = 1$ (b) $7x^0 = 7 \cdot 1 = 7$

(c) $(7x)^0 = 1$ (d) $(4x + 2)^0 = 1$

If the rule $a^m \cdot a^n = a^{m+n}$ is to hold for negative exponents, we must have

$$a^m \cdot a^{-m} = a^{m-m} = a^0 = 1$$

Thus,

$$a^m \cdot a^{-m} = 1$$

If we divide both sides by a^m, we obtain

Rule 7

$$a^{-m} = \frac{1}{a^m}, \quad \text{provided} \quad a \neq 0$$

EXAMPLE 2   (a) $3^{-1} = \frac{1}{3}$ (b) $4^{-2} = \frac{1}{4^2} = \frac{1}{16}$

(c) $x^3 y^{-2} = x^3 \cdot \frac{1}{y^2} = \frac{x^3}{y^2}$

(d) $\frac{1}{a^{-3}} = \frac{1}{\frac{1}{a^3}} = 1 \div \frac{1}{a^3} = 1 \cdot \frac{a^3}{1} = a^3$

(e) $\left(\frac{a}{b}\right)^{-1} = \frac{1}{\left(\frac{a}{b}\right)} = 1 \div \frac{a}{b} = 1 \cdot \frac{b}{a} = \frac{b}{a}$

(f) $\frac{x^{-2}}{y^{-3}} = \frac{\frac{1}{x^2}}{\frac{1}{y^3}} = \frac{1}{x^2} \div \frac{1}{y^3} = \frac{1}{x^2} \cdot \frac{y^3}{1} = \frac{y^3}{x^2}$

From the definition of zero and negative exponents, the following rule holds.

Rule for Negative Exponents

Any factor of the numerator of a fraction may be changed to the denominator, or any factor of the denominator may be changed to the numerator, by changing the sign of the exponent.

EXAMPLE 3   (a) $\frac{x^{-2}}{y^{-3}} = \frac{y^3}{x^2}$ (b) $\frac{5a^{-2}}{b^{-3}c^4} = \frac{5b^3}{a^2 c^4}$

(c) $\frac{3x^2}{5x^{-3}} = \frac{3x^2 x^3}{5} = \frac{3x^5}{5}$ (d) $\frac{7(a+b)^{-2}}{4a^{-3}} = \frac{7a^3}{4(a+b)^2}$

CAUTION
When applying the rule for negative exponents, be careful to distinguish between **factors** and **terms**.

Thus,

$$\frac{1}{x^{-1}y^{-1}} = xy$$

However,

$$\frac{1}{x^{-1} + y^{-1}} = \frac{1}{\dfrac{1}{x} + \dfrac{1}{y}} = \frac{1}{\dfrac{y + x}{xy}} = \frac{xy}{x + y}$$

EXAMPLE 4 Simplify, writing the answer using only positive exponents.

$$\left(\frac{3xy^{-1}}{2x^{-1}y^2}\right)^{-2}$$

Solution

$$\left(\frac{3xy^{-1}}{2x^{-1}y^2}\right)^{-2} = \frac{(3xy^{-1})^{-2}}{(2x^{-1}y^2)^{-2}} \qquad \text{by Rule 5: } \left(\frac{a}{b}\right)^n = \frac{a^n}{b^n}$$

$$= \frac{3^{-2}x^{-2}(y^{-1})^{-2}}{2^{-2}(x^{-1})^{-2}(y^2)^{-2}} \qquad \text{by Rule 4: } (ab)^n = a^n b^n$$

$$= \frac{3^{-2}x^{-2}y^2}{2^{-2}x^2 y^{-4}} \qquad \text{by Rule 2: } \frac{a^m}{a^n} = a^{m-n}$$

$$= \frac{2^2 y^2 y^4}{3^2 x^2 x^2} \qquad \text{by Rule for negative exponents}$$

$$= \frac{4y^6}{9x^4} \qquad \text{by Rule 1: } a^m a^n = a^{m+n} \quad \blacksquare$$

With the introduction of zero and negative exponents, we can summarize the rules of exponents.

Rules of Exponents

1. $a^m a^n = a^{m+n}$ 5. $\left(\dfrac{a}{b}\right)^n = \dfrac{a^n}{b^n}$

2. $(a^m)^n = a^{mn}$ 6. $a^0 = 1$

3. $\dfrac{a^m}{a^n} = a^{m-n}$ 7. $a^{-n} = \dfrac{1}{a^n}$

4. $(ab)^n = a^n b^n$

EXERCISES 7.2

In Exercises 1–96, simplify and express with positive exponents.

1. d^0

2. b^0

3. $(a + b)^0$

4. $(x - y)^0$

5. a^{-3}

6. c^{-2}

7. 4^{-1}

8. 3^{-3}

9. $3y^0$

10. $5x^0$

11. x^5y^0

12. $(13x)^0$

13. $(-8a)^0$

14. a^0y^3

15. $2x^{-1}$

16. ax^{-3}

17. $x^{-2}y^3$

18. $6a^{-5}$

19. $(9a)^{-2}$

20. $(xy)^{-2}$

21. $(ab)^{-5}$

22. $(2c)^{-3}$

23. $-5x^{-3}$

24. $3b^{-2}$

25. $-(4b)^0$

26. $a^{-1}(xy)^0$

27. $(a^2b)^0z^{-2}$

28. $(-19a)^0$

29. $\dfrac{x^{-3}}{y}$

30. $\dfrac{a^{-5}}{c^{-3}}$

31. $\dfrac{a^{-3}}{b}$

32. $\dfrac{x^2}{y^{-1}}$

33. $\left(\dfrac{x}{b}\right)^{-2}$

34. $\dfrac{2^{-4}}{3^{-2}}$

35. $\dfrac{5^{-3}}{2^{-2}}$

36. $\left(\dfrac{3}{x^3}\right)^{-1}$

37. $\dfrac{x^0}{y}$

38. $\dfrac{a^3}{b^0}$

39. $\left(\dfrac{x}{y^3}\right)^0$

40. $(7x - 3a)^0$

41. $(a + 7b)^0$

42. $\left(\dfrac{-3}{x^5}\right)^0$

43. $\dfrac{3c^2}{2d^0}$

44. $\dfrac{5x^0}{3y^2}$

45. $\dfrac{4x^2}{3d^{-2}}$

46. $\dfrac{5a^{-3}}{2b^2}$

47. $(x + y)^{-2}$

48. $(2x + y)^{-1}$

49. $a^{-2}b^{-1}$

50. $2z^0w^0$

51. $r^0 \cdot s^0$

52. $x^{-2}y^{-3}$

53. $x^5 \cdot x^{-3}$

54. $\dfrac{x^n}{x^{-n}}$

55. $\dfrac{a^4}{a^{-3}}$

56. $b^x b^{-y}$

57. $x^2 \cdot x^0$

58. $a^x \cdot a^0$

59. $\dfrac{x^5}{x^0}$

60. $\dfrac{x^n}{x^0}$

61. $5^3 \div 5^0$

62. $3^5 \div 3^{-3}$

63. $2^4 \cdot 2^{-1}$

64. $3^0 \cdot 3^3$

65. $4^n \div 4^{-m}$

66. $5^3 \cdot 5^{-2}$

67. $4^2 \cdot 7^0$

68. $2^8 \div 5^0$

69. $x + x^0$

70. $y^{-1} + y^2$

71. $a + a^{-2}$

72. $x^2 - 3^0$

73. $x^3 - 3x^{-3}$

74. $7c^2 - c^{-2}$

75. $c^2 - 5c^0$

76. $4a^2 + a^0$

77. $(2x^2 - y^{-2})^{-3}$

78. $(3a^2 - b^{-3})^{-2}$

79. $\dfrac{2a - b^0}{(2a - b)^0}$

80. $\dfrac{3x^0 - y^{-2}}{-(3x - y)}$

81. $\dfrac{1}{5a^{-2}}$

82. $\dfrac{3}{2x^{-3}}$

83. $\dfrac{3x^{-2}}{5y^{-3}}$

84. $\dfrac{4a^{-3}}{3b^{-5}}$

85. $\dfrac{xy^{-1}}{x^2y^3}$

86. $\dfrac{b^2c^{-2}}{b^{-4}c}$

87. $xy^{-3}z^{-1}$

88. $\dfrac{ab^{-2}c^3}{a^{-5}b^4c^{-5}}$

89. $\dfrac{5^{-1}x^4y^{-3}}{3^2x^{-2}y^5}$

90. $\dfrac{3x^{-5}y^2}{2^{-2}m^3n^{-3}}$

91. $ab^{-3} - a^{-2}b$

92. $4x^{-2} - 5y^{-3}$

93. $\dfrac{2x}{y^{-2}} + \dfrac{y^2}{2^{-1}x^{-1}}$

94. $\dfrac{3a}{a^{-3}b^2} + \dfrac{7ab^{-2}}{a^{-5}}$

95. $\dfrac{5}{x^{-2} + y^{-2}}$

96. $(5x - y^{-2})^{-5}$

In Exercises 97–100, compute the following:

*97. $\left[6 - 4\left(\dfrac{5}{16}\right)^0\right]^{-2}$

*98. $\dfrac{3\left(\dfrac{2}{3}\right)^{-2} + 4^{-1}}{\left(\dfrac{1}{2}\right)^{-1} + 5}$

*99. $\dfrac{(0.6)^0 - (0.1)^{-1}}{\left(\dfrac{3}{2^3}\right)^{-1}\left(\dfrac{3}{2}\right)^3 + \left(\dfrac{-1}{3}\right)^{-1}}$

*100. $\dfrac{\left(\dfrac{1}{2}\right)^{-2} - 5(-2)^{-2} + \left(\dfrac{2}{3}\right)^{-2}}{2^{-2} + 1^0}$

7.3 SCIENTIFIC NOTATION

In scientific work we often encounter very large or very small numbers. For example,

1. The speed of light is approximately 300,000,000 meters per second.
2. The distance of the earth from the sun is 93,000,000 miles.
3. The thickness of an oil film on water is 0.0000002 inches.

Such numbers can be expressed in a shorter way, using **scientific notation** (or **standard notation**). The number is written as the product of a number between 1 and 10 and an integral power of 10. Thus,

1. $300,000,000 = 3 \times 10^8$
2. $93,000,000 = 9.3 \times 10^7$
3. $0.0000002 = 2 \times 10^{-7}$

DEFINITION

A positive number N is expressed in **scientific notation** when it is written as the product of an integral power of 10 and a number between 1 and 10. That is, a positive number N in scientific notation has the form

$$N = A \times 10^n$$

where A is a number between 1 and 10 and n is an integer.

To determine a method for writing a positive number in scientific notation, consider the change in the exponent of 10 in the following:

$$0.00723 = 7.23 \times 10^{-3}$$
$$0.0723 = 7.23 \times 10^{-2}$$
$$0.723 = 7.23 \times 10^{-1}$$
$$7.23 = 7.23 \times 10^{0}$$
$$72.3 = 7.23 \times 10^{1}$$
$$723 = 7.23 \times 10^{2}$$
$$7230 = 7.23 \times 10^{3}$$

Note that the effect of multiplying or dividing a number in the decimal system by 10 is to shift the position of the decimal point. Thus, writing a number in scientific notation is a matter of counting the number of places we must shift the decimal point.

How to Write a Number in Scientific Notation
1. Move the decimal point to a position such that only one nonzero digit appears to its left, thus obtaining a number between 1 and 10.
2. Multiply this number by 10^n, where $|n|$ equals the number of places the decimal has been moved,
(**a**) n being positive if the decimal point has been moved to the left.
(**b**) n being negative if the decimal point has been moved to the right.

EXAMPLE 1 Write in scientific notation:

	Decimal Notation	*Scientific Notation*
(**a**)	$375 = 3.75.$	3.75×10^2
(**b**)	$.000428$	4.28×10^{-4}
(**c**)	$8 = 8.$	8.0×10^0
(**d**)	$84{,}700{,}000.$	8.47×10^7
(**e**)	0.6735	6.735×10^{-1}

EXERCISES 7.3

Write the numbers in Exercises 1–12 in scientific notation.

1. 56,400
2. 0.0328
3. 0.00075
4. 726
5. 4693
6. 11.82
7. 4
8. 0.007002
9. 5,603.7
10. 0.6×10^{-2}
11. 36×10^4
12. 3.2×10^6

Express each of the quantities in Exercises 13–20 in scientific notation.

13. The diameter of the earth is approximately 7930 mi.
14. The mass of a water molecule is approximately 0.000 000 000 000 000 000 000 03 g.
15. The distance from the earth to the moon is 240,000 mi.
16. A light-year, the distance that light travels in a year, is approximately 5,870,000,000,000 mi.
17. The distance from the earth to the sun is approximately 149,000,000 km.
18. The diameter of the smallest visible particle is about 0.005 cm.
19. The approximate number of seconds in a century is 3,150,000,000.
20. The diameter of the Einstein universe, according to the theory of relativity, is 2,000,000,000 light-years.

Express each of the quantities in Exercises 21–30 in decimal notation.

21. The mass of the earth is approximately 6.6×10^{21} tons.
22. The charge on the electron is about 4.77×10^{-10} electrostatic units.
23. The diameter of an electron is estimated to be 4×10^{-13} cm.
24. The mass of an electron is approximately 9×10^{-28} g.
25. The proton weighs about 1.66×10^{-24} g.
26. The estimated diameter of the average atom is 2×10^{-8} cm.
27. The diameter of the average red blood corpuscle is 8×10^{-5} cm.
28. The speed of the earth in its orbit is approximately 9.769×10^4 feet per second.
29. The greatest rate of plant growth is about 3×10^{-2} mm per second.
30. The radius of the sun is approximately 6.9×10^5 cm.

7.4 PRINCIPAL ROOT

Recall from arithmetic that

$$3 \times 3 = 3^2 = 9 \qquad (3 \text{ squared} = 9)$$

$$3 \times 3 \times 3 = 3^3 = 27 \qquad (3 \text{ cubed} = 27)$$

$$3 \times 3 \times 3 \times 3 = 3^4 = 81 \qquad (3 \text{ to the fourth} = 81)$$

However, one very important question in mathematics is "What number squared is 9?" or "What number cubed is 27?" or, in general, "What number raised to the nth power equals a certain number?" These numbers are called the **roots** of the given number. That is,

3 is the **square root** of 9

3 is the **cube root** of 27

3 is the **fourth root** of 81

The procedure of finding roots is represented by the sign $\sqrt{}$, called the **radical sign**. In order to indicate which type of root is to be found, a number is placed in the crook of the radical sign. This number is called the **index** of the root. The number or expression appearing under the radical sign is called the **radicand**. Thus,

EXAMPLE 1

$$\text{Cube root of 27:} \quad \overset{\text{index}}{\underset{\text{radical}}{\sqrt[3]{27}}} \longleftarrow \text{radicand}$$

Note. If the index of a root is not written explicitly, it is assumed that it is 2.

EXAMPLE 2

$$\sqrt{9} \text{ means } \sqrt[2]{9}$$

Some numbers have more than one real root. For example, the square roots of 9 are the two numbers $+3$ and -3, since $(+3)(+3) = 9$ and $(-3)(-3) = 9$. However, when using the radical sign for expressions that have both a positive and a negative root, we shall agree that only the positive root will be designated. Thus, while the square roots of 9 are both $+3$ and -3, the radical expression $\sqrt{9}$ represents only the root 3. This single value is called the **principal root** of the radicand.

EXAMPLE 3 Compute
(a) $\sqrt{25}$ (b) $\sqrt[3]{-8}$ (c) $\sqrt[3]{64}$ (d) $\sqrt{49}$

Solution (a) $\sqrt{25} = 5$ (b) $\sqrt[3]{-8} = -2$

(c) $\sqrt[3]{64} = 4$ (d) $\sqrt{49} = 7$ ■

EXAMPLE 4 Compute

(a) $\sqrt[5]{32}$ (b) $\sqrt[4]{81}$ (c) $-(\sqrt[3]{125})$ (d) $\sqrt{\dfrac{1}{36}}$

Solution (a) $\sqrt[5]{32} = 2$ (b) $\sqrt[4]{81} = 3$

(c) $-(\sqrt[3]{125}) = -5$ (d) $\sqrt{\dfrac{1}{36}} = \dfrac{1}{6}$ ∎

EXAMPLE 5 Does -9 have a real square root?

Solution No. There is no real square root of (-9) since the square of any real number is nonnegative. ∎

Note. For the present, we shall use only values of the radicand and index for which a real root does exist. In Section 7.11, we shall discuss how to find roots such as $\sqrt{-9}$.

Given the above examples, we now give a formal definition of a principal root.

DEFINITION

Consider $\sqrt[n]{A}$. If A has an nth root that is positive or zero, that root is called the **principal** nth **root** of A. If A does not have a positive nth root but does have a negative nth root, the negative root is the principal nth root of A.

EXAMPLE 6 Evaluate $\sqrt{a^2}$.

Solution $\sqrt{a^2} = a$, if $a > 0$ or $\sqrt{a^2} = -a$, if $a < 0$. Thus, $\sqrt{a^2} = |a|$. ∎

EXAMPLE 7 Evaluate $\sqrt[3]{-8x^3y^6}$.

Solution $\sqrt[3]{-8x^3y^6} = -2xy^2$, because $(-2xy^2)^3 = -8x^3y^6$. ∎

EXAMPLE 8 Evaluate $\sqrt{49a^4} - \sqrt[3]{-27a^6}$.

Solution Since $\sqrt{49a^4} = 7a^2$ and $\sqrt[3]{-27a^6} = -3a^2$, we have

$$\sqrt{49a^4} - \sqrt[3]{-27a^6} = 7a^2 - (-3a^2) = 10a^2$$ ∎

EXAMPLE 9 Evaluate $\sqrt[3]{\dfrac{8}{27}} - \sqrt{\dfrac{16}{9}}$.

Solution

$$\sqrt[3]{\frac{8}{27}} - \sqrt{\frac{16}{9}} = \frac{2}{3} - \frac{4}{3} = \frac{-2}{3} \quad \blacksquare$$

EXERCISES 7.4

In Exercises 1–10, give the square root.
1. 36   2. 100   3. 1
4. 16   5. x^2 6. y^2
7. a^4 8. $a^4 b^2$ 9. $36x^2$
10. $121a^8$

In Exercises 11–20, give the cube root.
11. -27 12. 64   13. -125
14. 8   15. b^3 16. $-a^6$
17. $8a^3 b^3$ 18. $-x^6 y^3$

19. $\dfrac{z^3}{8}$ 20. $-\dfrac{27x^6}{64y^3}$

In Exercises 21–35, give the root indicated.
21. $\sqrt{144}$ 22. $-\sqrt{121}$
23. $\sqrt[3]{-64}$ 24. $\sqrt{400}$
25. $\sqrt[5]{-32}$ 26. $\sqrt[6]{64}$
27. $\sqrt{225}$ 28. $\sqrt{625}$

29. $\sqrt[4]{\dfrac{1}{256}}$ 30. $\sqrt[3]{\dfrac{-1}{27}}$

31. $\sqrt[3]{\dfrac{8}{125}}$ 32. $\sqrt[4]{x^{12}}$

33. $\sqrt[5]{y^{10}}$ 34. $\sqrt[3]{8x^3 y^9}$

35. $\sqrt[5]{-32a^{10}}$

In Exercises 36–44, simplify and combine.
36. $\sqrt{16} + \sqrt[3]{27}$ 37. $\sqrt[3]{-8} + \sqrt[3]{-27}$
38. $\sqrt[5]{32} + \sqrt[4]{256}$ 39. $\sqrt{144} - \sqrt{81}$
40. $\sqrt{9x^4} + \sqrt[3]{8x^6}$
41. $\sqrt[3]{-125x^{12}} - \sqrt{x^8}$
42. $\sqrt[6]{a^6 b^{12}} - \sqrt[7]{a^7 b^{14}}$

43. $\sqrt[3]{\dfrac{125}{64}} - \sqrt{\dfrac{25}{16}}$ 44. $\sqrt{\dfrac{1}{9}} - \sqrt[4]{\dfrac{16}{81}}$

7.5 RATIONAL EXPONENTS

Up to this point, the exponent n of the power a^n has been limited to the integers. We shall now define a^n for other rational values of n; that is, fractional values. We shall present these definitions in such a way that the new exponents satisfy the same rules as integral exponents.

We begin by considering a few easy examples:

1. Consider $(a^{1/2})^2$. By Rule 1, $a^m \cdot a^n = a^{m+n}$, we have

$$a^{1/2} \cdot a^{1/2} = a^{(1/2)+(1/2)} = a^1 = a$$

That is,

$$(a^{1/2})^2 = a$$

Since the result of squaring $a^{1/2}$ is a, it follows that $a^{1/2}$ *is the square root of a.* Thus,

$$a^{1/2} = \sqrt{a}$$

2. Consider $(a^{3/4})^4$. By Rule 1, we have

$$a^{3/4} \cdot a^{3/4} \cdot a^{3/4} \cdot a^{3/4} = a^{(3/4)+(3/4)+(3/4)+(3/4)} = a^3$$

That is,

$$(a^{3/4})^4 = a^3$$

Since the result of raising $a^{3/4}$ to the fourth power is a^3, it follows that $a^{3/4}$ *is the fourth root of a^3*. Thus,

$$a^{3/4} = \sqrt[4]{a^3}$$

3. Consider $(a^{-2/3})^3$. By Rule 1, we have

$$a^{-2/3} \cdot a^{-2/3} \cdot a^{-2/3} = a^{-2/3-2/3-2/3} = a^{-2}$$

That is,

$$(a^{-2/3})^3 = a^{-2}$$

Since the result of cubing $a^{-2/3}$ is a^{-2}, it follows that $a^{-2/3}$ *is the cube root of a^{-2}*. Thus,

$$a^{-2/3} = \sqrt[3]{a^{-2}}$$

In general, by Rule 1, we have

$$\underbrace{a^{m/n} \cdot a^{m/n} \cdot a^{m/n} \cdot \ \cdots \ \cdot a^{m/n}}_{n \text{ factors}} = a^{(m/n)+(m/n)+(m/n)+ \ \cdots \ +(m/n)}$$

$$= (a^{m/n})^n$$

$$= a^m$$

That is,

$$(a^{m/n})^n = a^m$$

Since the result of raising $a^{m/n}$ to the nth power is a^m, it follows that $a^{m/n}$ *is the nth root of a^m*. Thus, we have the following definition.

DEFINITION

If $\dfrac{m}{n}$ is a rational number, where n is a positive integer, and if a is a real number such that $\sqrt[n]{a}$ exists, then

$$a^{m/n} = \sqrt[n]{a^m} = (\sqrt[n]{a})^m$$

The second form, $(\sqrt[n]{a})^m$, follows from Rule 2; that is, $a^{m/n} = (a^m)^{1/n} = (a^{1/n})^m$.

Thus, for any quantity with a fractional exponent, the numerator of the exponent indicates the power to which the quantity is raised, and the denominator of the exponent indicates the root that is to be taken.

$$a^{m/n} = (\sqrt[n]{a})^m$$

root

power

EXAMPLE 1   Evaluate (a) $8^{2/3}$ (b) $\left(2\dfrac{1}{4}\right)^{-1/2}$ (c) $(27)^{-2/3}$

Solution   (a) $8^{2/3} = \sqrt[3]{8^2} = \sqrt[3]{64} = 4$ or $8^{2/3} = (\sqrt[3]{8})^2 = (2)^2 = 4.$

In evaluating numbers raised to fractional powers, it usually is easier to take the root first, then raise the resulting root to the indicated power.

(b) $\left(2\dfrac{1}{4}\right)^{-1/2} = \left(\dfrac{9}{4}\right)^{-1/2} = \left(\sqrt[2]{\dfrac{9}{4}}\right)^{-1} = \left(\dfrac{3}{2}\right)^{-1} = \dfrac{2}{3}$

(c) $(27)^{-2/3} = (\sqrt[3]{27})^{-2} = (3)^{-2} = \dfrac{1}{9}$ ■

EXAMPLE 2   Using the rules of exponents, find

(a) $x^{1/3} \cdot x^{1/2}$ (b) $\dfrac{x^{1/5}}{x^{3/5}}$ (c) $(x^{2/3})^6$ (d) $(x^4 y^{-6})^{-1/2}$

(e) $\left(\dfrac{x^2}{y^{1/5}}\right)^{-5}$

Solution   (a) $x^{1/3} \cdot x^{1/2} = x^{(1/3)+(1/2)} = x^{(2/6)+(3/6)} = x^{5/6}$

(b) $\dfrac{x^{1/5}}{x^{3/5}} = x^{(1/5)-(3/5)} = x^{-2/5} = \dfrac{1}{x^{2/5}}$

(c) $(x^{2/3})^6 = x^{(2/3)6} = x^4$

(d) $(x^4 y^{-6})^{-1/2} = (x^4)^{-1/2}(y^{-6})^{-1/2} = x^{-2}y^3 = \dfrac{y^3}{x^2}$

(e) $\left(\dfrac{x^2}{y^{1/5}}\right)^{-5} = \dfrac{(x^2)^{-5}}{(y^{1/5})^{-5}} = \dfrac{x^{-10}}{y^{-1}} = \dfrac{y}{x^{10}}$ ■

As we have seen, if $m = 1$, then

$$a^{1/n} = \sqrt[n]{a}$$

EXAMPLE 3 (a) $36^{1/2} = \sqrt{36} = 6$

(b) $(-32)^{1/5} = \sqrt[5]{-32} = -2$

(c) $(729)^{1/6} = \sqrt[6]{739} = 3$

EXERCISES 7.5

In Exercises 1–12, express the quantities in their equivalent radical form:

1. $a^{2/3}$ 2. $b^{-1/2}$ 3. $x^{3/4}$
4. $y^{3/2}$ 5. $c^{-2/3}$ 6. $a^{4/5}$
7. $x^{-5/3}$ 8. $x^{-7/2}$ 9. $b^{a/3}$
10. $x^{-a/b}$ 11. $x^{4/3a}$ 12. $y^{-3m/2n}$

Express the quantities in Exercises 13–26 in their equivalent exponent form:

13. $\sqrt{x}$ 14. $\sqrt[3]{a}$ 15. $\sqrt{b^3}$
16. $\sqrt[4]{y^3}$ 17. $\sqrt[3]{a^{-2}}$ 18. $\sqrt[5]{x^3}$
19. $\sqrt[4]{b^{-3}}$ 20. $\sqrt[6]{a^{-5}}$ 21. $\sqrt[m]{x^{-n}}$
22. $\sqrt[p]{a^{-2q}}$ 23. $\dfrac{1}{\sqrt[5]{x^4}}$ 24. $\dfrac{1}{\sqrt[3]{y^m}}$
25. $\dfrac{1}{\sqrt[p]{x^q}}$ 26. $\dfrac{1}{\sqrt[n]{b^{-2a}}}$

Evaluate the quantities in Exercises 27–50.

27. $25^{1/2}$ 28. $49^{1/2}$ 29. $8^{1/3}$
30. $(-27)^{1/3}$ 31. $16^{1/4}$
32. $(-243)^{1/5}$ 33. $64^{1/6}$
34. $81^{1/4}$ 35. $(-64)^{1/3}$
36. $343^{1/3}$ 37. $128^{1/7}$ 38. $(-125)^{1/3}$
39. $4^{3/2}$ 40. $9^{-1/2}$ 41. $8^{2/3}$
42. $16^{-3/4}$ 43. $27^{-2/3}$ 44. $25^{3/2}$

45. $32^{3/5}$ 46. $8^{-4/3}$ 47. $9^{3/2}$
48. $27^{4/3}$ 49. $(-64)^{5/3}$
50. $(-27)^{-5/3}$

In Exercises 51–67, write each of the quantities with a single exponent.

51. $(a^{-3})^{-1/2}$ 52. $(a^{2/3})^{-3/4}$
53. $(x^{-5/2})^{4/3}$ 54. $(x^{3/2})^{-7/3}$
55. $(x^{-5/6})^{-3/2}$ 56. $(\sqrt{a^3})^5$
57. $(\sqrt[3]{y^2})^6$ 58. $(\sqrt{b^{-3}})^4$
59. $(\sqrt[3]{x^{-4}})^2$ 60. $(\sqrt[5]{a^{-3}})^{5/3}$
61. $(\sqrt[3]{p^2})^{-3/4}$ 62. $\left(\sqrt[4]{\dfrac{1}{y^3}}\right)^{-2/5}$

*63. $\left(\sqrt[3]{\dfrac{1}{x^{5/2}}}\right)^{6/5}$ *64. $\left(\sqrt[5]{\dfrac{1}{q^{-2/3}}}\right)^{-3/4}$

65. $\sqrt[3]{(x^{-3})^5}$ *66. $\sqrt[4]{(\sqrt{x^{-4/3}})^{-6/7}}$

*67. $\sqrt[3]{\left(\sqrt[3]{\dfrac{1}{x^{-5/6}}}\right)^{-12/5}}$

Find the value of the quantities in Exercises 68–75.

68. $(2\tfrac{1}{4})^{1/2}$ 69. $(6\tfrac{1}{4})^{-1/2}$
70. $(3\tfrac{3}{8})^{2/3}$ 71. $(-15\tfrac{5}{8})^{-1/3}$
72. $(5\tfrac{1}{16})^{-3/4}$ 73. $(-4\tfrac{17}{27})^{-5/3}$
74. $(4\tfrac{21}{25})^{3/2}$ 75. $(42\tfrac{7}{8})^{-2/3}$

7.6 EVALUATING AND SIMPLIFYING RADICALS

There are three properties of radicals that are useful in evaluating and simplifying radicals.

$$\left.\begin{array}{l}\sqrt[n]{a^n} = a \\ \sqrt[n]{ab} = \sqrt[n]{a}\,\sqrt[n]{b}\end{array}\right\} \qquad a \geq 0,\, b \geq 0$$

$$\sqrt[n]{\dfrac{a}{b}} = \dfrac{\sqrt[n]{a}}{\sqrt[n]{b}}, \qquad a \geq 0,\, b > 0$$

What does it mean to simplify a radical?

To write a radical in its **simplest form** means to

1. Remove any factors from the radicand.
2. Reduce the index to its lowest possible value.
3. Rewrite any fractions so that the denominator does not contain a radical. This is called **rationalizing the denominator**.

We now discuss each of the above procedures.

Removing Factors from the Radicand

1. Factor the radicand so that one or more of its factors has an exponent that is a multiple of the index.
2. Remove any such factor from the radicand and write its root as a factor outside the radical sign.

EXAMPLE 1 Simplify $\sqrt{75}$.

Solution
$$\sqrt{75} = \sqrt{25 \cdot 3}$$
$$= \sqrt{5^2 \cdot 3}$$
$$= 5\sqrt{3} \quad \blacksquare$$

EXAMPLE 2 Simplify $\sqrt{x^3}$.

Solution
$$\sqrt{x^3} = \sqrt{x^2 \cdot x}$$
$$= x\sqrt{x} \quad \blacksquare$$

EXAMPLE 3 Simplify $\sqrt[3]{24x^5y^8}$.

Solution Since the index is 3, we seek factors that have exponents that are a multiple of 3.

$$\sqrt[3]{24x^5y^8} = \sqrt[3]{8 \cdot 3 \cdot x^3 \cdot x^2 \cdot y^6 \cdot y^2}$$
$$= \sqrt[3]{2^3 \cdot 3 \cdot x^3 \cdot x^2 \cdot y^6 \cdot y^2}$$
$$= 2xy^2 \sqrt[3]{3x^2y^2} \quad \blacksquare$$

Reducing the Index

1. Write the radical in exponential form, with each factor of the radicand having a fractional exponent.
2. Reduce each fractional exponent to lowest terms.
3. Rewrite the expression in step 2 in radical form.

EXAMPLE 4 Simplify $\sqrt[4]{36x^2y^2}$.

Solution

$$\sqrt[4]{36x^2y^2} = (6^2x^2y^2)^{1/4}$$
$$= (6^2)^{1/4} \cdot (x^2)^{1/4} \cdot (y^2)^{1/4}$$
$$= 6^{1/2}x^{1/2}y^{1/2}$$
$$= \sqrt{6xy} \quad \blacksquare$$

EXAMPLE 5 Simplify $\sqrt[6]{49a^4b^{12}}$.

Solution

$$\sqrt[6]{49a^4b^{12}} = (7^2a^4b^{12})^{1/6}$$
$$= (7^2)^{1/6} \cdot (a^4)^{1/6} \cdot (b^{12})^{1/6}$$
$$= 7^{1/3}a^{2/3}b^2$$
$$= b^2 \sqrt[3]{7a^2} \quad \blacksquare$$

EXAMPLE 6 Simplify $\sqrt[4]{9x^2y^3}$.

Solution

$$\sqrt[4]{9x^2y^3} = (3^2x^2y^3)^{1/4}$$
$$= (3^2)^{1/4} \cdot (x^2)^{1/4} \cdot (y^3)^{1/4}$$
$$= 3^{1/2} \cdot x^{1/2} \cdot y^{3/4}$$
$$= 3^{2/4}x^{2/4}y^{3/4}$$
$$= \sqrt[4]{9x^2y^3} \quad \blacksquare$$

Thus, the index of $\sqrt[4]{9x^2y^3}$ is already in its lowest form.

Rationalizing the Denominator

If a fraction has as its denominator a radical of index n, rationalize the denominator by multiplying the numerator and denominator of the fraction by that quantity that will make the radicand a perfect nth power.

EXAMPLE 7 Rationalize $\sqrt{\dfrac{3}{5}}$.

Solution

$$\sqrt{\frac{3}{5}} = \frac{\sqrt{3}}{\sqrt{5}}$$

If the denominator of a fraction is a square root, then multiply the numerator and denominator by the given denominator. Thus,

$$\sqrt{\frac{3}{5}} = \frac{\sqrt{3}}{\sqrt{5}} = \frac{\sqrt{3}}{\sqrt{5}} \cdot \frac{\sqrt{5}}{\sqrt{5}} = \frac{\sqrt{15}}{\sqrt{25}} = \frac{\sqrt{15}}{5} \quad \blacksquare$$

EXAMPLE 8 Rationalize $\dfrac{8}{\sqrt[3]{4}}$.

Solution Since the denominator is a cube root, we must multiply the numerator and denominator by that quantity that will make the radicand in the denominator a perfect cube.

$$\frac{8}{\sqrt[3]{4}} = \frac{8}{\sqrt[3]{4}} \cdot \frac{\sqrt[3]{2}}{\sqrt[3]{2}} = \frac{8\sqrt[3]{2}}{\sqrt[3]{8}} = \frac{8\sqrt[3]{2}}{2} = 4\sqrt[3]{2} \quad \blacksquare$$

EXAMPLE 9 Rationalize $\dfrac{6}{\sqrt[3]{x^2 y}}$.

Solution
$$\frac{6}{\sqrt[3]{x^2 y}} = \frac{6}{\sqrt[3]{x^2 y}} \cdot \frac{\sqrt[3]{xy^2}}{\sqrt[3]{xy^2}} = \frac{6\sqrt[3]{xy^2}}{\sqrt[3]{x^3 y^3}} = \frac{6\sqrt[3]{xy^2}}{xy} \quad \blacksquare$$

EXAMPLE 10 Rationalize $\dfrac{9ab^2}{\sqrt{27ab^3}}$.

Solution
$$\frac{9ab^2}{\sqrt{27ab^3}} = \frac{9ab^2}{\sqrt{27ab^3}} \cdot \frac{\sqrt{3ab}}{\sqrt{3ab}} = \frac{9ab^2\sqrt{3ab}}{\sqrt{81a^2 b^4}} = \frac{9ab^2\sqrt{3ab}}{9ab^2} = \sqrt{3ab} \quad \blacksquare$$

EXAMPLE 11 Rationalize $\sqrt{\dfrac{y^2}{y-3}}$.

Solution
$$\sqrt{\frac{y^2}{y-3}} = \frac{\sqrt{y^2}}{\sqrt{y-3}} = \frac{y}{\sqrt{y-3}} = \frac{y}{\sqrt{y-3}} \cdot \frac{\sqrt{y-3}}{\sqrt{y-3}} = \frac{y\sqrt{y-3}}{\sqrt{(y-3)^2}}$$

$$= \frac{y\sqrt{y-3}}{y-3} \quad \blacksquare$$

EXERCISES 7.6

In Exercises 1–40, simplify.

1. $\sqrt{44}$
2. $\sqrt{75}$
3. $\sqrt{333}$
4. $\sqrt[3]{72}$
5. $\sqrt[3]{8000}$
6. $\sqrt[4]{405}$
7. $\sqrt[4]{112}$
8. $2\sqrt{12}$
9. $x\sqrt{45}$
10. $\sqrt{4ab}$
11. $\sqrt[3]{8x^2}$
12. $\sqrt[5]{2x^6}$
13. $\sqrt{12x^3}$
14. $\sqrt[3]{16t^6}$

15. $\sqrt[3]{\dfrac{1}{27}m^2}$
16. $\sqrt[3]{\dfrac{4}{27}x^5 y^2}$

17. $\sqrt[3]{27a^2 b^3}$
18. $\sqrt[3]{27m^4}$
19. $\sqrt{98rs^3}$
20. $\sqrt{32m^4}$
21. $\sqrt[3]{54y^4}$
22. $\sqrt[4]{8c^3 d^5}$

23. $\sqrt[4]{\dfrac{1}{81}x^2 y^7}$
24. $\sqrt[5]{32x^5 y^6}$

25. $\sqrt[6]{64x^7 y}$
26. $\sqrt[4]{225}$
27. $\sqrt[6]{64}$
28. $\sqrt[4]{27}$
29. $\sqrt[6]{144}$
30. $\sqrt[8]{64}$
31. $\sqrt[3]{128m^4}$
32. $\sqrt[3]{108a^7}$

33. $\sqrt[4]{\dfrac{9}{81}ax^8}$
34. $\sqrt[4]{\dfrac{5}{256}x^4}$

35. $\sqrt[5]{243x^2}$
36. $\sqrt[3]{\dfrac{8}{125}xy}$

37. $\sqrt[3]{16x^3m^2}$ 38. $\sqrt[4]{80c^3d^5}$
39. $\sqrt[6]{128a^7}$ 40. $\sqrt[6]{xy^7z}$

Rationalize the denominators in Exercises 41–70.

41. $\dfrac{1}{\sqrt{2}}$ 42. $\dfrac{1}{\sqrt{7}}$ 43. $\dfrac{1}{\sqrt{x}}$

44. $\dfrac{3}{\sqrt{5}}$ 45. $\dfrac{4}{\sqrt{11}}$ 46. $\dfrac{10}{\sqrt{a}}$

47. $\dfrac{x}{\sqrt{3}}$ 48. $\dfrac{a}{\sqrt{b}}$ 49. $\dfrac{\sqrt{3}}{\sqrt{5}}$

50. $\dfrac{\sqrt{5}}{\sqrt{2}}$ 51. $\dfrac{\sqrt{x}}{\sqrt{y}}$ 52. $\dfrac{2\sqrt{6}}{\sqrt{13}}$

53. $\dfrac{6\sqrt{3}}{5\sqrt{2}}$ 54. $\dfrac{a\sqrt{b}}{x\sqrt{y}}$ 55. $\dfrac{6}{\sqrt[3]{2}}$

56. $\dfrac{2x}{\sqrt{6x}}$ 57. $\dfrac{6y}{\sqrt{2y}}$ 58. $\dfrac{4z}{\sqrt{8z}}$

59. $\dfrac{ax}{\sqrt{ax^3}}$ 60. $\dfrac{12xy}{\sqrt{6x}}$

61. $\dfrac{15r^2}{\sqrt{5r}}$ 62. $\dfrac{18c^2d}{\sqrt{4cd}}$

63. $\dfrac{x^2yz}{\sqrt{xy^2z^2}}$ 64. $\dfrac{2c^2}{\sqrt[3]{2c}}$

65. $\dfrac{3x}{\sqrt[3]{3x}}$ 66. $\dfrac{24y}{\sqrt[3]{4y^2}}$ 67. $\dfrac{5z}{\sqrt[3]{4z^2}}$

68. $\dfrac{\sqrt{ax-ay}}{\sqrt{bx-by}}$ 69. $\dfrac{\sqrt{x-1}}{\sqrt{x+1}}$

70. $\dfrac{x+3}{\sqrt{x^2-9}}$

7.7 ADDITION AND SUBTRACTION OF RADICALS

Radical expressions are said to be like radicals when they have *the same index and the same radicand.*

EXAMPLE 1   (a) The expressions $\sqrt{3}$, $6\sqrt{3}$, and $-4\sqrt{3}$ are like radicals because each possesses the same index 2 and the same radicand 3.

(b) The expressions $\sqrt[3]{2}$, $-\sqrt[3]{2}$, and $\frac{8}{9}\sqrt[3]{2}$ are like radicals because each possesses the same index 3 and the same radicand 2.

EXAMPLE 2   (a) The expressions $\sqrt{5}$ and $\sqrt{7}$ are not like expressions because while each possesses the same index 2, they have different radicands, 5 and 7.

(b) The expressions $\sqrt{6}$ and $\sqrt[3]{6}$ are not like expressions because even though they have the same radicand 6, they have different indices, 2 and 3.

To add or subtract like radicals, we use the distributive property as follows:

$$5\sqrt{3} + 4\sqrt{3} = (5+4)\sqrt{3} = 9\sqrt{3}$$

$$6\sqrt[3]{2} - 5\sqrt[3]{2} = (6-5)\sqrt[3]{2} = \sqrt[3]{2}$$

Addition and Subtraction of Like Radicals

1. Add or subtract the coefficients of the radicals.
2. Multiply the obtained sum or difference by the common radical.

EXAMPLE 3 $4\sqrt{2} + 3\sqrt{2} - 5\sqrt{2} = 2\sqrt{2}$

If radicals are not like, it may be possible to add or subtract them by transforming them into like radicals.

EXAMPLE 4 $\sqrt{48} - 5\sqrt{3} + 3\sqrt{12}$

Solution

$$
\begin{aligned}
\sqrt{48} - 5\sqrt{3} + 3\sqrt{12} &= \sqrt{16\cdot3} - 5\sqrt{3} + 3\sqrt{4\cdot3} \\
&= 4\sqrt{3} - 5\sqrt{3} + 6\sqrt{3} \\
&= 5\sqrt{3} \quad \blacksquare
\end{aligned}
$$

EXAMPLE 5 $\sqrt[3]{6} - 5\sqrt[3]{48} - 4\sqrt[3]{162}$

Solution

$$
\begin{aligned}
\sqrt[3]{6} - 5\sqrt[3]{48} - 4\sqrt[3]{162} &= \sqrt[3]{6} - 5\sqrt[3]{8\cdot6} - 4\sqrt[3]{27\cdot6} \\
&= \sqrt[3]{6} - 10\sqrt[3]{6} - 12\sqrt[3]{6} \\
&= -21\sqrt[3]{6} \quad \blacksquare
\end{aligned}
$$

If the radicals cannot be transformed into like radicals, then their addition or subtraction cannot be expressed as a single term.

EXAMPLE 6 $\sqrt{5} + \sqrt[3]{2}$ cannot be expressed as a single term because they are not like terms.

Addition and Subtraction of Radical Expressions

1. Simplify each radical.
2. Combine like radicals using the distributive property.
3. Indicate the sum or difference of the unlike radicals.

EXAMPLE 7   Combine $\sqrt[3]{24} + \sqrt{50} + \sqrt[3]{3} - \sqrt{2} - \sqrt[3]{81} - \sqrt{32}$.

Solution

$$
\begin{aligned}
&\sqrt[3]{24} + \sqrt{50} + \sqrt[3]{3} - \sqrt{2} - \sqrt[3]{81} - \sqrt{32} \\
&= \sqrt[3]{8\cdot3} + \sqrt{25\cdot2} + \sqrt[3]{3} - \sqrt{2} - \sqrt[3]{27\cdot3} - \sqrt{16\cdot2} \\
&= 2\sqrt[3]{3} + 5\sqrt{2} + \sqrt[3]{3} - \sqrt{2} - 3\sqrt[3]{3} - 4\sqrt{2} \\
&= (2\sqrt[3]{3} + \sqrt[3]{3} - 3\sqrt[3]{3}) + (5\sqrt{2} - \sqrt{2} - 4\sqrt{2}) \\
&= 0 + 0 \\
&= 0 \quad \blacksquare
\end{aligned}
$$

EXAMPLE 8   Combine $\sqrt{x^3y^3} - x\sqrt{xy^4} + y\sqrt{x^3y}$.

Solution $\sqrt{x^3y^3} - x\sqrt{xy^4} + y\sqrt{x^3y} = xy\sqrt{xy} - xy^2\sqrt{x} + xy\sqrt{xy}$

$$= (xy\sqrt{xy} + xy\sqrt{xy}) - xy^2\sqrt{x}$$

$$= 2xy\sqrt{xy} - xy^2\sqrt{x} \quad ■$$

EXAMPLE 9   Combine $\sqrt{4 + 4a^2} + \sqrt{9 + 9a^2} - 7\sqrt{1 + a^2}$.

Solution $\sqrt{4 + 4a^2} + \sqrt{9 + 9a^2} - 7\sqrt{1 + a^2}$

$$= \sqrt{4(1 + a^2)} + \sqrt{9(1 + a^2)} - 7\sqrt{1 + a^2}$$

$$= 2\sqrt{1 + a^2} + 3\sqrt{1 + a^2} - 7\sqrt{1 + a^2}$$

$$= -2\sqrt{1 + a^2} \quad ■$$

EXERCISES   7.7

Simplify.

1. $\sqrt{28} + \sqrt{63} - \sqrt{175}$
2. $2\sqrt{18} + \sqrt{32} - \sqrt{128}$
3. $\sqrt{75} - 3\sqrt{48} + \sqrt{147}$
4. $\sqrt{125} - 2\sqrt{180} + \sqrt{245}$
5. $\sqrt{121} + \sqrt{208} + \sqrt{225}$
6. $\sqrt{8} + \sqrt{12} + \sqrt{18} + \sqrt{27}$
7. $\sqrt{44} + \sqrt{50} + \sqrt{98} + \sqrt{99}$
8. $\sqrt[3]{32} + \sqrt[3]{108} - \sqrt[3]{500}$
9. $\sqrt[8]{9} + \sqrt[4]{48} - \sqrt[4]{243}$
10. $a\sqrt{ax^3} + x\sqrt{a^3x} + \sqrt{a^3x^3}$
11. $a\sqrt[3]{ab^4} + b\sqrt[3]{a^4b} - 2\sqrt[3]{a^4b^4}$

12. $\sqrt{\dfrac{1}{2}} + \sqrt[6]{\dfrac{1}{8}} - 2\sqrt{4\dfrac{1}{2}}$

13. $\dfrac{1}{4}\sqrt{2a^3} + a\sqrt{\dfrac{1}{2}a} + \sqrt{\dfrac{1}{8}a^3}$

14. $\sqrt[3]{54} + 3\sqrt[3]{\dfrac{16}{27}} + 2\sqrt[3]{6\dfrac{3}{4}}$

15. $\sqrt{9\sqrt{9}} + \sqrt{\sqrt{729}} + \sqrt[3]{3\sqrt{3}}$
16. $\sqrt{4\sqrt[3]{8}} + \sqrt[3]{2\sqrt{2}} + \sqrt[6]{2\sqrt{8}}$
17. $\sqrt[4]{4} + \sqrt[6]{8} + \sqrt[6]{27} + \sqrt[4]{144}$
18. $\sqrt[3]{40} + \sqrt[3]{24} + \sqrt[3]{81} + \sqrt[3]{135}$

19. $\dfrac{1}{2}\sqrt{\dfrac{1}{2}} + \sqrt{\dfrac{1}{8}} + \sqrt{\dfrac{2}{9}} + \sqrt{\dfrac{1}{18}}$

20. $\sqrt[3]{\dfrac{3}{4}} - 3\sqrt[3]{\dfrac{1}{36}} + \sqrt[3]{\dfrac{2}{9}} - \dfrac{1}{2}\sqrt[3]{\dfrac{16}{9}}$

*21. $a\sqrt[4]{(a - x)^2} + x\sqrt[6]{(a - x)^3}$
     $+ \sqrt{(a + x)(a^2 - x^2)}$

*22. $a\sqrt{a^4 + a^2x^2} + x\sqrt[4]{(a^2x^2 + x^4)^2}$
     $+ \sqrt{(a^2 - x^2)(a^4 - x^4)}$

23. $\sqrt[3]{(a - b)^2} + \sqrt[3]{(a - b)^5}$

24. $\dfrac{3}{4}\sqrt{a^2b} - \dfrac{2}{3}a\sqrt{b} + \dfrac{1}{6}\sqrt{ab}$

25. $\sqrt{a^2b} - \sqrt{b^5} - \sqrt{a^4b^3}$

7.8   MULTIPLICATION OF RADICALS

Recall that for $a, b \geq 0$, $\sqrt[n]{a} \cdot \sqrt[n]{b} = \sqrt[n]{ab}$. This rule, along with the commutative and associative properties of multiplication, is the basis for multiplying radicals.

EXAMPLE 1   (a) $\sqrt[3]{4} \cdot \sqrt[3]{5} = \sqrt[3]{4 \cdot 5} = \sqrt[3]{20}$

(b) $6\sqrt{3} \cdot 2\sqrt{7} = 6 \cdot 2 \cdot \sqrt{3} \cdot \sqrt{7} = 12\sqrt{3 \cdot 7} = 12\sqrt{21}$

Computing the Product of Radical Expressions

1. Express each of the radical factors as an equivalent radical expression having a common index.
2. Multiply the coefficients of the radical factors to find the coefficient of the product.
3. Multiply the radicands to find the radicand of the product, and place this product under the radical of the common order.
4. Simplify if possible.

EXAMPLE 2   Find the product: $\sqrt{5x^2y} \cdot \sqrt{10y}$.

Solution $\sqrt{5x^2y} \cdot \sqrt{10y} = \sqrt{(5x^2y)(10y)} = \sqrt{50x^2y^2} = 5xy\sqrt{2}$ ■

EXAMPLE 3   Find the product: $2\sqrt[3]{4ab^2} \cdot 5\sqrt[3]{3bc}$.

Solution $2\sqrt[3]{4ab^2} \cdot 5\sqrt[3]{3bc} = 10\sqrt[3]{12ab^3c} = 10b\sqrt[3]{12ac}$ ■

In order to multiply radicals with different indices, we first find a common index and express each radical as an equivalent radical with the common index.

EXAMPLE 4   Find the product: $7\sqrt[3]{2x^2y^2} \cdot 2\sqrt{xy}$.

Solution   A common index is 6.

$$\sqrt[3]{2x^2y^2} = (2x^2y^2)^{1/3} = (2x^2y^2)^{2/6} = \sqrt[6]{(2x^2y^2)^2} = \sqrt[6]{4x^4y^4}$$
$$\sqrt{xy} = (xy)^{1/2} = (xy)^{3/6} = \sqrt[6]{(xy)^3} = \sqrt[6]{x^3y^3}$$

Thus,

$$7\sqrt[3]{2x^2y^2} \cdot 2\sqrt{xy} = 7\sqrt[6]{4x^4y^4} \cdot 2\sqrt[6]{x^3y^3}$$
$$= 14\sqrt[6]{4x^7y^7}$$
$$= 14xy\sqrt[6]{4xy}$$ ■

To multiply sums and differences containing radicals, we use the distributive property of multiplication.

EXAMPLE 5   Find the product: $3\sqrt{6}(\sqrt{2} + 2\sqrt{3})$.

Solution   Using the distributive property, we obtain

$$3\sqrt{6}(\sqrt{2} + 2\sqrt{3}) = 3\sqrt{6} \cdot \sqrt{2} + 3\sqrt{6} \cdot 2\sqrt{3}$$

$$= 3\sqrt{12} + 6\sqrt{18}$$
$$= 3\sqrt{4 \cdot 3} + 6\sqrt{9 \cdot 2}$$
$$= 3 \cdot 2\sqrt{3} + 6 \cdot 3\sqrt{2}$$
$$= 6\sqrt{3} + 18\sqrt{2} \quad \blacksquare$$

EXAMPLE 6   Find the product: $(4 + \sqrt{5})(6 - \sqrt{5})$.

Solution   The product can be computed using long multiplication:

$$
\begin{array}{r}
4 + \sqrt{5} \\
\times\, 6 - \sqrt{5} \\
\hline
24 + 6\sqrt{5} \\
- 4\sqrt{5} - \sqrt{5} \cdot \sqrt{5} \\
\hline
24 + 2\sqrt{5} - 5 \qquad = 19 + 2\sqrt{5} \quad \blacksquare
\end{array}
$$

EXAMPLE 7   Find the product: $(3\sqrt{6} - 4\sqrt{10})^2$.

Solution   Using the horizontal method, we obtain

$$(3\sqrt{6} - 4\sqrt{10})^2 = (3\sqrt{6} - 4\sqrt{10})(3\sqrt{6} - 4\sqrt{10})$$
$$= 9\sqrt{36} - 12\sqrt{60} - 12\sqrt{60} + 16\sqrt{100}$$
$$= 9 \cdot 6 - 24\sqrt{60} + 16 \cdot 10$$
$$= 214 - 24\sqrt{4 \cdot 15}$$
$$= 214 - 24 \cdot 2\sqrt{15}$$
$$= 214 - 48\sqrt{15} \quad \blacksquare$$

EXAMPLE 8   Find the product: $(\sqrt{A} + \sqrt{B})(\sqrt{A} - \sqrt{B})$.

Solution $\quad (\sqrt{A} + \sqrt{B})(\sqrt{A} - \sqrt{B}) = \sqrt{A^2} - \sqrt{AB} + \sqrt{AB} - \sqrt{B^2}$
$$= A - B$$

Thus,

$$(\sqrt{A} + \sqrt{B})(\sqrt{A} - \sqrt{B}) = (\sqrt{A})^2 - (\sqrt{B})^2 = A - B \quad \blacksquare$$

EXAMPLE 9   Find the product: $(3\sqrt{2} + 5\sqrt{3})(3\sqrt{2} - 5\sqrt{3})$

Solution Using the results of Example 8, we get

$$(3\sqrt{2} + 5\sqrt{3})(3\sqrt{2} - 5\sqrt{3}) = (3\sqrt{2})^2 - (5\sqrt{3})^2$$

$$= 9 \cdot 2 - 25 \cdot 3 = 18 - 75 = -57 \quad \blacksquare$$

EXERCISES 7.8

Multiply and simplify.

1. $\sqrt{2} \cdot \sqrt{6}$ 2. $\sqrt{7} \cdot \sqrt{6}$
3. $2\sqrt{3} \cdot \sqrt{2}$ 4. $2\sqrt{5} \cdot \sqrt{5}$
5. $4\sqrt{2m} \cdot \sqrt{2m}$ 6. $x\sqrt{2} \cdot \sqrt{2x}$
7. $4\sqrt{11} \cdot \sqrt{11}$ 8. $6\sqrt{ax} \cdot \sqrt{3x}$
9. $3\sqrt{2x} \cdot 3\sqrt{2x}$ 10. $7\sqrt{3y} \cdot 3\sqrt{3y}$
11. $6\sqrt{a+b} \cdot \sqrt{a+b}$
12. $5\sqrt{2x-1} \cdot \sqrt{2x-1}$
13. $2\sqrt{a-b} \cdot \sqrt{a-b}$
14. $\sqrt{5} \cdot \sqrt[3]{3}$ 15. $\sqrt{7} \cdot \sqrt[4]{11}$
16. $\sqrt[4]{4} \cdot \sqrt[6]{5}$ 17. $\sqrt{ab} \cdot \sqrt[5]{a^4b}$
18. $\sqrt[4]{2m^3n} \cdot \sqrt[12]{5m^5n^9}$
19. $\sqrt[5]{a^4b^3} \cdot \sqrt[6]{a^3b^5}$

20. $\sqrt{2}\left(\dfrac{1}{2}\sqrt{6} - 3\sqrt{2}\right)$

21. $\dfrac{\sqrt{3}}{2}\left(2\sqrt{2} + \sqrt{\dfrac{1}{3}}\right)$

22. $(2\sqrt{3} - 1)(\sqrt{3} + 5)$

23. $(5\sqrt{6} + \sqrt{3})(\sqrt{6} - \sqrt{3})$
24. $(\sqrt{3s} - \sqrt{2})(\sqrt{3s} + \sqrt{2})$
25. $(3\sqrt{6} - \sqrt{5})(3\sqrt{6} + \sqrt{5})$
26. $(7 + 9\sqrt{2})(3 - \sqrt{2})$
27. $(2\sqrt{a} - \sqrt{b})(2\sqrt{a} - \sqrt{b})$
28. $(\sqrt{11} - 2\sqrt{2})(\sqrt{11} + 2\sqrt{2})$
29. $(\sqrt{10} - 9)(3\sqrt{10} + 1)$
30. $(m\sqrt{2n} + 3)(m\sqrt{2n} - 3)$
31. $(\sqrt{x+3} - 3)(\sqrt{x+3} + 3)$
32. $(3x + \sqrt{y-2})(3x - \sqrt{y-2})$
33. $(3\sqrt{x} - \sqrt{y})^2$
34. $(2\sqrt{5} + \sqrt{3})^2$
35. $(3\sqrt{7x} - \sqrt{y})(3\sqrt{7x} + \sqrt{y})$
36. $(3\sqrt{5m} + 2\sqrt{n})(3\sqrt{5m} - 2\sqrt{n})$
37. $(2\sqrt{x-1} - 3)^2$
38. $(3 - 4\sqrt{x+2})^2$
39. $(5\sqrt{3} - 2)^2$
40. $(2a\sqrt{2} + 3b)(2a\sqrt{2} - 3b)$

7.9 DIVISION: RATIONALIZATION OF BINOMIAL DENOMINATORS

When the denominator of a radical expression is a binomial containing one or two square root radicals, in order to rationalize such expressions, we take advantage of the results of Example 8 in Section 7.8; that is, $(\sqrt{a} + \sqrt{b})(\sqrt{a} - \sqrt{b}) = a - b$.

EXAMPLE 1 Rationalize the denominator for $\dfrac{7}{5 + \sqrt{3}}$.

Solution Multiplying numerator and denominator by $5 - \sqrt{3}$, we obtain

$$\frac{7}{5 + \sqrt{3}} = \frac{7(5 - \sqrt{3})}{(5 + \sqrt{3})(5 - \sqrt{3})} = \frac{7(5 - \sqrt{3})}{25 - 3} = \frac{7(5 - \sqrt{3})}{22} \quad \blacksquare$$

EXAMPLE 2 Rationalize the denominator for $\dfrac{15}{\sqrt{10} - \sqrt{5}}$.

Solution Multiplying numerator and denominator by $\sqrt{10} + \sqrt{5}$, we obtain

$$\frac{15}{\sqrt{10} - \sqrt{5}} = \frac{15(\sqrt{10} + \sqrt{5})}{(\sqrt{10} - \sqrt{5})(\sqrt{10} + \sqrt{5})}$$

$$= \frac{15(\sqrt{10} + \sqrt{5})}{10 - 5} = 3(\sqrt{10} + \sqrt{5}) \quad \blacksquare$$

The two binomial radicals $(\sqrt{10} - \sqrt{5})$ and $(\sqrt{10} + \sqrt{5})$ are called conjugates of one another. Similarly, the binomial radicals $(5 + \sqrt{3})$ and $(5 - \sqrt{3})$ are conjugates.

DEFINITION

The conjugate of a binomial containing square roots is a binomial that has the same two terms with the sign of the second term changed. That is, the conjugate of $(a\sqrt{b} + c\sqrt{d})$ is $(a\sqrt{b} - c\sqrt{d})$.

EXAMPLE 3 (a) The conjugate of $4 + \sqrt{2}$ is $4 - \sqrt{2}$.

(b) The conjugate of $3\sqrt{5} - 6\sqrt{2}$ is $3\sqrt{5} + 6\sqrt{2}$.

(c) The conjugate of $\sqrt{x} + \sqrt{y}$ is $\sqrt{x} - \sqrt{y}$.

Rationalizing a Binomial Denominator Containing Square Roots

1. Multiply both numerator and denominator of the fraction by the *conjugate* of the denominator.
2. Simplify, if possible.

EXAMPLE 4 Rationalize the denominator for $\dfrac{9}{\sqrt{6} - 2\sqrt{2}}$.

Solution $\dfrac{9}{\sqrt{6} - 2\sqrt{2}} = \dfrac{9(\sqrt{6} + 2\sqrt{2})}{(\sqrt{6} - 2\sqrt{2})(\sqrt{6} + 2\sqrt{2})}$

$$= \frac{9(\sqrt{6} + 2\sqrt{2})}{(\sqrt{6})^2 - (2\sqrt{2})^2} = \frac{9(\sqrt{6} + 2\sqrt{2})}{6 - 8} = \frac{9(\sqrt{6} + 2\sqrt{2})}{-2} \quad \blacksquare$$

EXAMPLE 5 Rationalize the denominator for $\dfrac{\sqrt{x} + \sqrt{y}}{\sqrt{x} - \sqrt{y}}$.

Solution $\dfrac{\sqrt{x} + \sqrt{y}}{\sqrt{x} - \sqrt{y}} = \dfrac{(\sqrt{x} + \sqrt{y})(\sqrt{x} + \sqrt{y})}{(\sqrt{x} - \sqrt{y})(\sqrt{x} + \sqrt{y})}$

$$= \dfrac{(\sqrt{x} + \sqrt{y})^2}{x - y} = \dfrac{x + 2\sqrt{xy} + y}{x - y} \quad \blacksquare$$

EXERCISES 7.9

Rationalize each denominator and simplify, if possible.

1. $\dfrac{6}{3 + \sqrt{5}}$

2. $\dfrac{5}{\sqrt{5} - 2}$

3. $\dfrac{4}{3 - \sqrt{7}}$

4. $\dfrac{6}{\sqrt{3} - 2}$

5. $\dfrac{7}{2\sqrt{2} - 1}$

6. $\dfrac{11}{2\sqrt{3} + 2}$

7. $\dfrac{9}{5 - 2\sqrt{3}}$

8. $\dfrac{10}{6 - 3\sqrt{2}}$

9. $\dfrac{\sqrt{2} + 1}{\sqrt{2} - 1}$

10. $\dfrac{2 - \sqrt{3}}{2 + \sqrt{3}}$

11. $\dfrac{3 - \sqrt{2}}{3 + \sqrt{2}}$

12. $\dfrac{4 + \sqrt{5}}{4 - \sqrt{5}}$

13. $\dfrac{6 + \sqrt{2}}{\sqrt{2} - 1}$

14. $\dfrac{8 - \sqrt{3}}{2 - \sqrt{3}}$

15. $\dfrac{10 + \sqrt{5}}{3 + \sqrt{5}}$

16. $\dfrac{\sqrt{6} - 5}{\sqrt{6} - 2}$

17. $\dfrac{\sqrt{8} + 5}{\sqrt{8} - 5}$

18. $\dfrac{\sqrt{7} - 2}{\sqrt{7} + 2}$

19. $\dfrac{2\sqrt{2} - 1}{2\sqrt{2} + 1}$

20. $\dfrac{2\sqrt{3} - 1}{\sqrt{3} - 1}$

21. $\dfrac{2 - 3\sqrt{2}}{3 - \sqrt{2}}$

22. $\dfrac{2\sqrt{5} - 1}{\sqrt{5} + 2}$

23. $\dfrac{\sqrt{6} - 2}{2\sqrt{6} - 5}$

24. $\dfrac{3\sqrt{6} - 5}{2\sqrt{6} - 1}$

25. $\dfrac{1}{\sqrt{a} - \sqrt{3a + 3}}$

26. $\dfrac{\sqrt{x}}{\sqrt{x + 1} + 1}$

27. $\dfrac{\sqrt{a} + \sqrt{b}}{\sqrt{a} - \sqrt{b}}$

28. $\dfrac{\sqrt{x + 1}}{\sqrt{x + 1} - \sqrt{x}}$

29. $\dfrac{1}{b + \sqrt{b^2 - a^2}}$

30. $\dfrac{\sqrt{x + y} + \sqrt{x - y}}{\sqrt{x + y} - \sqrt{x - y}}$

7.10 EQUATIONS INVOLVING RADICALS

An equation in which the unknown quantity appears under a radical sign is called a **radical equation**. For example,

$$\sqrt{x} = 7 \qquad \sqrt{x - 12} = \sqrt{x} - 2 \quad \text{and} \quad \sqrt[3]{y} - c = d$$

are all radical equations. A radical equation may be solved by raising both sides of the equation to a power equal to the index of the radical.

EXAMPLE 1 Solve $\sqrt{x - 2} = 7$.

Solution

$$\sqrt{x-2} = 7$$

Thus, $(\sqrt{x-2})^2 = (7)^2$

$$x - 2 = 49$$

$$x = 51 \quad \blacksquare$$

Check Letting $x = 51$,

$$\sqrt{51-2} \overset{?}{=} 7$$

$$\sqrt{49} \overset{?}{=} 7$$

$$7 \overset{\checkmark}{=} 7$$

A similar procedure can be used for other radical equations.

How to Solve Radical Equations

1. Transpose the terms of the equation so that one term with a radical stands alone on one side of the equation. If the equation has more than one radical, isolate the more complex radical.
2. Raise each side of the equation to a power equal to the index of the radical.
3. Combine like terms.
4. If radicals still remain, repeat steps 1, 2, and 3.
5. Solve the resulting equation for the unknown.
6. *Check* answers in step 5 in the *original* equation.

EXAMPLE 2 Solve the equation $\sqrt{x-4} + 1 = 0$.

Solution

$$\sqrt{x-4} + 1 = 0$$

$$\sqrt{x-4} = -1 \qquad \text{[isolating radical]}$$

$$(\sqrt{x-4})^2 = (-1)^2 \qquad \text{[squaring both sides]}$$

$$x - 4 = 1$$

$$x = 5 \qquad \text{[solving for } x] \quad \blacksquare$$

Check Substituting $x = 5$ into the original equation, we get

$$\sqrt{5-4} + 1 \overset{?}{=} 0$$

$$\sqrt{1} + 1 \overset{?}{=} 0$$

$$1 + 1 \overset{?}{=} 0$$

$$2 \neq 0$$

Thus, $x = 5$ is not a solution to the original equation. The equation $\sqrt{x-4} + 1 = 0$ has *no* solution.

As Example 2 illustrates, checking is an important step in solving radical equations. The reason for checking is that, in solving radical equations, we raise both sides of the equation to an integral power. This process of raising to an integral power amounts to multiplying both sides by an expression involving the unknown. This procedure may introduce what are called extraneous roots, and these roots must be omitted.

EXAMPLE 3 Solve the equation $\sqrt{x - 12} - \sqrt{x} = -2$.

Solution

$$\sqrt{x - 12} - \sqrt{x} = -2$$

$$\sqrt{x - 12} = -2 + \sqrt{x} \qquad \text{[isolating more complex radical]}$$

$$(\sqrt{x - 12})^2 = (-2 + \sqrt{x})^2 \qquad \text{[squaring both sides]}$$

$$x - 12 = 4 - 4\sqrt{x} + x$$

$$-16 = -4\sqrt{x} \qquad \text{[combining like terms]}$$

$$4 = \sqrt{x} \qquad \text{[dividing by 4]}$$

$$(4)^2 = (\sqrt{x})^2 \qquad \text{[squaring again]}$$

$$16 = x \quad \blacksquare$$

Check Substituting $x = 16$ into the original equation,

$$\sqrt{16 - 12} - \sqrt{16} \overset{?}{=} -2$$

$$\sqrt{4} - \sqrt{16} \overset{?}{=} -2$$

$$2 - 4 \overset{?}{=} -2$$

$$-2 \overset{\checkmark}{=} -2$$

Thus, $x = 16$ is the solution of the given equation.

EXAMPLE 4 Solve the equation $\sqrt[3]{x + 1} = -4$.

Solution

$$\sqrt[3]{x + 1} = -4$$

$$(\sqrt[3]{x + 1})^3 = (-4)^3 \qquad \text{[cubing both sides]}$$

$$x + 1 = -64$$

$$x = -65 \quad \blacksquare$$

Check Substituting $x = -65$ into the original equation,

$$\sqrt[3]{-65 + 1} \overset{?}{=} -4$$

$$\sqrt[3]{-64} \overset{?}{=} -4$$

$$-4 \overset{\checkmark}{=} -4$$

Thus, $x = -65$ is the solution of the given equation.

The method used to solve equations involving radicals can also be used to solve some equations that have rational exponents.

EXAMPLE 5   Solve $x^{-3/5} = 8$.

Solution $x^{-3/5} = 8$

We raise both sides to the $(-5/3)$ power in order to make the exponent of x equal to 1.

$$(x^{-3/5})^{-5/3} = (8)^{-5/3}$$

$$x = (2^3)^{-5/3}$$

$$x = 2^{-5}$$

$$x = \frac{1}{32} \quad \blacksquare$$

Check   Substituting $x = \frac{1}{32}$ into the original equation,

$$\left(\frac{1}{32}\right)^{-3/5} \overset{?}{=} 8$$

$$(2^{-5})^{-3/5} \overset{?}{=} 8$$

$$2^3 \overset{?}{=} 8$$

$$8 \overset{\checkmark}{=} 8$$

Thus, $x = \dfrac{1}{32}$ is the solution.

EXERCISES   7.10

In Exercises 1–30, solve and check.

1. $\sqrt{x} + 3 = 5$
2. $\sqrt{x} - 2 = 4$
3. $\sqrt[3]{x} + 3 = 6$
4. $\sqrt{x + 3} = 4$
5. $\sqrt[3]{x} - 2 = 3$
6. $\sqrt{x + 4} = 5$
7. $\sqrt[3]{x + 6} = 1$
8. $\sqrt{x - 1} = 7$
9. $\sqrt{x + 3} = 9$
10. $(x - 7)^{1/3} = 2$
11. $\sqrt{x} + 2 = 3\sqrt{x} - 4$
12. $\sqrt[3]{x} + 3 = 18 - 2\sqrt[3]{x}$
13. $2\sqrt{x} + 1 = 7 - \sqrt{x}$
14. $4\sqrt[3]{x} - 30 = 5 - 3\sqrt[3]{x}$
15. $20 - 5\sqrt{x} = 3\sqrt{x} - 4$
16. $4 + 2\sqrt{x} = 8 - \sqrt{x}$
17. $5 - 3\sqrt{x} = 2\sqrt{x} - 10$
18. $(x^2 + 3x + 9)^{1/2} = x + 2$

19. $\sqrt{x^2 - 2x - 4} = x - 2$
20. $\sqrt{x^2 - 5x + 9} - x + 2 = 0$
21. $(x^2 - x + 3)^{1/2} + x = 6$
22. $\sqrt{x + 5} - \sqrt{x} = 1$
23. $\sqrt{x + 16} + \sqrt{x} = 8$
24. $\sqrt{x - 2} + \sqrt{6 + x} = 4$
25. $(x - 5)^{1/2} - 1 = \sqrt{x - 14}$
26. $\sqrt{x + 9} = \sqrt{49 + x} - 4$
27. $\dfrac{\sqrt{x} + 3}{\sqrt{x} - 4} = \dfrac{\sqrt{x}}{\sqrt{x} - 3}$
28. $\dfrac{\sqrt{x} - 3}{\sqrt{x} - 4} = \dfrac{\sqrt{x} + 2}{\sqrt{x}}$

29. $\dfrac{\sqrt{x}-1}{\sqrt{x}-2} = \dfrac{\sqrt{x}-4}{\sqrt{x}-3}$

30. $\dfrac{\sqrt{x}+5}{\sqrt{x}-4} = \dfrac{\sqrt{x}-3}{\sqrt{x}+2}$

In Exercises 31–60, solve for x.

31. $\sqrt{x}+a = b$ 32. $\sqrt{x}-c = d$

33. $\sqrt{x}+3 = a$ 34. $x^{1/2}-4 = c$

35. $\sqrt{x}+5 = c+3$

36. $\sqrt{x}-3 = a-4$ 37. $x^{1/2}-a = 3$

38. $\sqrt{x}-b = 4+b$

39. $3\sqrt{x}+\sqrt{a} = 2\sqrt{a}-4\sqrt{x}$

40. $2\sqrt{x}+\sqrt{b} = \sqrt{x}+3\sqrt{b}$

41. $4\sqrt{x}-2\sqrt{c} = 3\sqrt{x}+3\sqrt{c}$

42. $x^{1/2} = a+2^{1/2}$

43. $\sqrt{x}-\sqrt{a} = b^{1/2}$

44. $\sqrt{x}-3\sqrt{a} = \sqrt{c}$

45. $\sqrt{x}-\sqrt{a} = \sqrt{b}-\sqrt{x}$

46. $\sqrt{b}-\sqrt{x} = \sqrt{c}+\sqrt{x}$

47. $(a+x)^{1/2} = \sqrt{2a}$

48. $\sqrt{b-x} = \sqrt{3b}$

*49. $(c-x)^{1/2} = c^{1/2}-d^{1/2}$

*50. $(b+cx)^{1/2} = (2b)^{1/2}-c^{1/2}$

51. $x^{3/5} = -\dfrac{1}{8}$ 52. $x^{3/2} = 27$

53. $27x^{3/2} = 1$ 54. $3x^{2/3} = \dfrac{1}{3}$

55. $3x^{3/2} = -\dfrac{1}{9}$ 56. $27x^{3/2} = -8$

57. $x^{-1/3} = \dfrac{1}{2}$ 58. $x^{-2/3} = 16$

59. $3x^{-1/2} = 9$ 60. $x^{-2} = a^2$

7.11 COMPLEX NUMBERS

Problems involving radicals are often of the form $\sqrt{-b}$, $b > 0$. Such numbers are not real numbers, since the square of either a positive or a negative real number is always positive. However, $\sqrt{-b}$, $b > 0$ means we are looking for a number such that the square of this number is negative.

$$(\sqrt{-b})^2 = [(-b)^{1/2}]^2 = (-b)^1 = -b, \quad b > 0$$

To solve this problem, mathematicians created a new number system based upon an "imaginary unit" i defined by $i = \sqrt{-1}$ and having the property that $i^2 = -1$. That is,

$$\boxed{i = \sqrt{-1} \quad \text{and} \quad i^2 = -1}$$

We shall assume that i satisfies all the algebraic properties we have already developed, with the exception of the rules for inequalities. Thus, we can compute higher powers of i:

$i^1 = i$ $i^5 = i^4 \cdot i = (1) \cdot i = i$

$i^2 = -1$ $i^6 = i^4 \cdot i^2 = (1) \cdot (-1) = -1$

$i^3 = i^2 \cdot i = (-1) \cdot i = -i$ $i^7 = i^4 \cdot i^3 = (1) \cdot (-i) = -i$

$i^4 = i^2 \cdot i^2 = (-1) \cdot (-1) = 1$ $i^8 = i^4 \cdot i^4 = (1) \cdot (1) = 1$

Continuing to compute higher powers of i, we note that i has an important cyclic property. That is,

$$i^1 = i^5 = i^9 = \cdots = i$$
$$i^2 = i^6 = i^{10} = \cdots = -1$$
$$i^3 = i^7 = i^{11} = \cdots = -i$$
$$i^4 = i^8 = i^{12} = \cdots = 1$$

Thus, every integral power of i can be expressed equivalently by either i, -1, $-i$, or 1.

EXAMPLE 1 Compute (a) i^{17} (b) i^{38} (c) i^{100}.

Solution (a) $i^{17} = i^{16} \cdot i^1 = (i^4)^4 \cdot i^1 = (1)^4 \cdot i = i$

(b) $i^{38} = i^{36} \cdot i^2 = (i^4)^9 \cdot i^2 = (1)^9 \cdot (-1) = -1$

(c) $i^{100} = (i^4)^{25} = (1)^{25} = 1$ ▪

The square roots of negative numbers can easily be found in terms of i.

EXAMPLE 2 (a) $\sqrt{-1} = i$, by definition

(b) $\sqrt{-4} = \sqrt{(-1)(4)} = i\sqrt{4} = 2i$

(c) $\sqrt{-9} = \sqrt{(-1)(9)} = i\sqrt{9} = 3i$

(d) $\sqrt{-16} = \sqrt{(-1)(16)} = i\sqrt{16} = 4i$

(e) $\sqrt{-7} = \sqrt{(-1)(7)} = \sqrt{7}\, i$

In general,

$$\boxed{\sqrt{-b} = \sqrt{b}\, i \quad \text{for} \quad b > 0}$$

EXAMPLE 3 (a) $\sqrt{-50} = \sqrt{(-1)(25)(2)} = i\sqrt{(25)(2)} = 5\sqrt{2}\, i$

(b) $(-64)^{1/2} = \sqrt{-64} = \sqrt{(-1)(64)} = i\sqrt{64} = 8i$

Any number of the form bi, where b is a real number, is called a pure imaginary number.

CAUTION
Always convert radicals to imaginary numbers *before* performing other operations, or a contradiction may result.

EXAMPLE 4 Find the product $\sqrt{-9} \cdot \sqrt{-9}$.

Solution $\sqrt{-9} \cdot \sqrt{-9} = (3i)(3i) = 9i^2 = -9$, since $i^2 = -1$. However, the following is *incorrect:* $\sqrt{-9} \cdot \sqrt{-9} = \sqrt{(-9)(-9)} = \sqrt{81} = 9$. ■

If we combine the real numbers and imaginary numbers, we form a larger set called the complex numbers.

DEFINITION

A complex number has the form $a + bi$, where a and b are real numbers. The number a is called the real part of $a + bi$, and b is called the imaginary part of $a + bi$.

$$a + bi$$

$$\uparrow \quad \uparrow$$

real part   imaginary part

EXAMPLE 5 The following are examples of complex numbers and their real and imaginary parts.

(a) $4 + 3i$; $a = 4$, $b = 3$ (b) $1 - 5i$; $a = 1$, $b = -5$

(c) 6; $a = 6$, $b = 0$ (d) $-8i$; $a = 0$, $b = -8$

Note that every real number a can be written as a complex number by letting $b = 0$.

$$a = a + 0i$$

EXAMPLE 6 Write each of the following as a complex number of the form $a + bi$:

(a) 13 (b) $-\dfrac{1}{5}$ (c) $21 + \sqrt{-3}$ (d) $\sqrt{5} - \sqrt{-7}$

Solution (a) $13 = 13 + 0i$ (b) $-\dfrac{1}{5} = -\dfrac{1}{5} + 0i$

(c) $21 + \sqrt{-3} = 21 + \sqrt{3}\, i$ (d) $\sqrt{5} - \sqrt{-7} = \sqrt{5} - \sqrt{7}\, i$ ■

Two complex numbers are said to be equal if their real parts are equal and their imaginary parts are equal. That is,

---

$$a + bi = c + di \quad \text{if and only if} \quad a = c \quad \text{and} \quad b = d$$

EXAMPLE 7 $-8 + xi = y - 2i$ if and only if $x = -2$ and $y = -8$.

EXERCISES   7.11

Express the following quantities in Exercises 1–28
using the imaginary number i.

1. $\sqrt{-36}$ 2. $\sqrt{-25}$ 3. $\sqrt{-49}$

4. $\sqrt{-100}$ 5. $\sqrt{-7}$ 6. $\sqrt{-2}$

7. $\sqrt{-18}$ 8. $\sqrt{-27}$ 9. $\sqrt{-24}$

10. $\sqrt{-48}$ 11. $\sqrt{-72}$ 12. $\sqrt{-54}$

13. $\sqrt{-\dfrac{1}{9}}$ 14. $\sqrt{-\dfrac{1}{4}}$ 15. $\sqrt{-\dfrac{1}{16}}$

16. $\sqrt{-\dfrac{1}{25}}$ 17. $\sqrt{-\dfrac{4}{49}}$ 18. $\sqrt{-\dfrac{16}{81}}$

19. $\sqrt{-\dfrac{7}{10}}$ 20. $\sqrt{-\dfrac{11}{13}}$ 21. $\sqrt{-x^2y}$

22. $\sqrt{-13x}$ 23. $x\sqrt{-x^2y^2}$

24. $2y\sqrt{-xy^3}$ 25. $7 + \sqrt{-45}$

26. $5 - 3\sqrt{-49}$ 27. $-1 + 2\sqrt{-24}$

28. $8 - \sqrt{-96}$

In Exercises 29–36, find the values.

29. i^9 30. i^{10} 31. i^{15} 32. i^{16}

33. $-i^{25}$ 34. $-i^{28}$ 35. i^{50} 36. i^{200}

In Exercises 37 and 38, determine the value of x
and y.

37. $x + \sqrt{-64} = 14 + yi$

38. $x + yi = \sqrt{-81}$

7.12   OPERATIONS WITH COMPLEX NUMBERS

Addition and Subtraction of Complex Numbers

To add complex numbers, we add the real parts and the imaginary parts separately, and express the result in the form $a + bi$. Subtraction is computed in the same manner.

EXAMPLE 1   Add $(2 - 3i) + (3 + 4i)$.

Solution $(2 - 3i) + (3 + 4i) = (2 + 3) + (-3 + 4)i = 5 + i$ ■

EXAMPLE 2   Subtract $(5 - 2i) - (8 - 7i)$.

Solution $(5 - 2i) - (8 - 7i) = (5 - 8) + [-2 - (-7)]i = -3 + 5i$ ■

In general, we have

Addition and Subtraction of Complex Numbers
Addition: $(a + bi) + (c + di) = (a + c) + (b + d)i$ Subtraction: $(a + bi) - (c + di) = (a - c) + (b - d)i$

EXAMPLE 3   Combine $(3 + 5i) + (-1 + 4i) - (2 - 3i)$.

Solution $(3 + 5i) + (-1 + 4i) - (2 - 3i) = (3 - 1 - 2) + (5 + 4 + 3)i = 12i$ ■

EXAMPLE 4 Combine $(-2 - 3i) - (5 - \sqrt{-49}) + (1 - \sqrt{-16})$.

Solution First we express each number in terms of i. Thus,

$$(-2 - 3i) - (5 - \sqrt{-49}) + (1 - \sqrt{-16})$$
$$= (-2 - 3i) - (5 - 7i) + (1 - 4i)$$
$$= (-2 - 5 + 1) + (-3 + 7 - 4)i$$
$$= -6 \quad \blacksquare$$

Multiplication and Division of Complex Numbers

The product of two complex numbers can be obtained by multiplying them the same as any two binomials.

EXAMPLE 5 Multiply $(7 + 2i)(1 - 4i)$.

Solution
$$(7 + 2i)(1 - 4i) = 7 - 28i + 2i - 8i^2$$
$$= 7 - 26i - 8i^2$$
$$= 7 - 26i - 8(-1)$$
$$= 15 - 26i \quad \blacksquare$$

In general, we have

$$(a + bi)(c + di) = ac + adi + bci + bdi^2$$
$$= ac + (ad + bc)i + bd(-1)$$
$$= (ac - bd) + (ad + bc)i$$

Thus,

Multiplication of Complex Numbers

$$(a + bi)(c + di) = (ac - bd) + (ad + bc)i$$

EXAMPLE 6 Multiply $(3 - 2\sqrt{2}i)(-1 + 5\sqrt{2}i)$.

Solution
$$(3 - 2i\sqrt{2})(-1 + 5i\sqrt{2}) = -3 + 15i\sqrt{2} + 2i\sqrt{2} - 10i^2(\sqrt{2})^2$$
$$= -3 + 17i\sqrt{2} - 10(-1)(2)$$
$$= -3 + 17i\sqrt{2} + 20$$
$$= 17 + 17\sqrt{2}i \quad \blacksquare$$

EXAMPLE 7   Simplify $(7 - 3i)^2$.

Solution $(7 - 3i)^2 = (7 - 3i)(7 - 3i) = 49 - 21i - 21i + 9i^2$
$$= 49 - 42i + 9(-1)$$
$$= 40 - 42i \quad \blacksquare$$

EXAMPLE 8   Multiply $(5 + 4i)(5 - 4i)$.

Solution $(5 + 4i)(5 - 4i) = 25 - 20i + 20i - 16i^2$
$$= 25 - 16(-1)$$
$$= 25 + 16$$
$$= 41 \quad \blacksquare$$

DEFINITION

The numbers $(a + bi)$ and $(a - bi)$ are called conjugate complex numbers.

EXAMPLE 9   (a) The conjugate of $7 + 3i$ is $7 - 3i$.

(b) The conjugate of $-6 - 5i$ is $-6 + 5i$.

(c) The conjugate of $2i$ is $-2i$.

(d) The conjugate of 9 is 9, since $9 = 9 + 0i$ and $9 - 0i = 9$.

We now consider the product of conjugate complex numbers.

$$(a + bi)(a - bi) = a^2 - abi + abi - b^2i^2$$
$$= a^2 - b^2(-1)$$
$$= a^2 + b^2$$

which is a real number.

The product of a complex number and its conjugate is a real number:
$$(a + bi)(a - bi) = a^2 + b^2$$

EXAMPLE 10   Find the product of the following complex numbers and their conjugates:

(a) $1 + i$ (b) $(-2 - 6i)$ (c) $-9i$

Solution (a) $(1 + i)(1 - i) = (1)^2 + (1)^2 = 2$

(b) $(-2 - 6i)(-2 + 6i) = (-2)^2 + (6)^2 = 4 + 36 = 40$

(c) Since $-9i = 0 - 9i$, its conjugate is $0 + 9i$ and

$(0 - 9i)(0 + 9i) = 0^2 + (9)^2 = 81.$ ■

The division of one complex number by another is accomplished by transforming the fraction into a fraction with a *real* denominator. Since the product of a complex number and its conjugate is a real number, the division of two complex numbers can be achieved by multiplying both numerator and denominator by the conjugate of the denominator.

EXAMPLE 11 Divide $7 - 6i$ by $4 + i$. Express the result in the form $a + bi$.

Solution The conjugate of $4 + i$ is $4 - i$.

$$\frac{7 - 6i}{4 + i} = \frac{7 - 6i}{4 + i} \cdot \frac{4 - i}{4 - i} = \frac{28 - 7i - 24i + 6i^2}{(4)^2 + (1)^2} = \frac{22 - 31i}{17}$$

Thus,

$$\frac{7 - 6i}{4 + i} = \frac{22}{17} - \frac{31i}{17}$$ ■

EXAMPLE 12 Divide $\dfrac{5 - \sqrt{2}i}{3 + \sqrt{2}i}$

Solution The conjugate of $3 + i\sqrt{2}$ is $3 - i\sqrt{2}$.

$$\frac{5 - i\sqrt{2}}{3 + i\sqrt{2}} = \frac{5 - i\sqrt{2}}{3 + i\sqrt{2}} \cdot \frac{3 - i\sqrt{2}}{3 - i\sqrt{2}} = \frac{15 - 5i\sqrt{2} - 3i\sqrt{2} + i^2(\sqrt{2})^2}{(3)^2 + (\sqrt{2})^2}$$

$$= \frac{15 - 8\sqrt{2}i - 2}{9 + 2}$$

$$= \frac{13 - 8\sqrt{2}i}{11}$$

$$= \frac{13}{11} - \frac{8\sqrt{2}}{11}i$$ ■

EXERCISES 7.12

In Exercises 1–16, perform the indicated operations and express the results in the form a + bi.

1. $(4 - 3i) + (2 - 4i)$
2. $(12 - 5i) + (12 + 5i)$
3. $(4 + 3i) + (3 - 4i)$
4. $(5 - 4i) - (-1 + 3i)$
5. $(6 - 3i) - (7 + i)$
6. $(3 - 2i) - (5 - 3i)$
7. $\left(\dfrac{1}{3} - \dfrac{1}{6}i\right) - \left(\dfrac{5}{6} + \dfrac{2}{3}i\right)$

8. $\left(\frac{1}{2} + \frac{1}{3}i\right) + \left(-\frac{1}{3} + \frac{1}{2}i\right)$

9. $(0.4 + 0.3i) - (0.1 - 0.5i)$
10. $(0.25 - 0.25i) + (0.75 + 0.75i)$
11. $7 - \sqrt{-4} - 5 + \sqrt{-1}$
12. $6 - \sqrt{-9} - 2 + \sqrt{-1}$
13. $8 + \sqrt{-12} + 3 - \sqrt{-75}$
14. $5 - \sqrt{-98} + 3 - \sqrt{-32}$
15. $9 + 2\sqrt{-72} - 4 - 5\sqrt{-18}$
16. $8 - 5\sqrt{-12} + 7 - 2\sqrt{-48}$

*17. For what condition is the sum of two complex numbers $(a + bi)$ and $(c + di)$ a real number?

*18. For what condition is the sum of two complex numbers an imaginary number?

*19. For what condition is the difference of two complex numbers a real number?

*20. For what condition is the difference of two complex numbers an imaginary number?

In Exercises 21–35, multiply and write in simplest form.

21. $(3 + 4i)(5 + 2i)$ 22. $(3 + 2i)(1 - 3i)$
23. $(3 + 2i)(3 - 2i)$ 24. $(2 + i)(1 + 5i)$
25. $(7 + 4i)(3 + 6i)$
26. $(3 - 2i)(-3 + i)$
27. $(-7 + 3i)(-3 + 2i)$
28. $\left(\frac{1}{2} + \frac{1}{3}i\right)\left(\frac{1}{2} - \frac{1}{3}i\right)$
29. $(4 + 2i)^2$ 30. $(6 - 3i)(3 + i)$

31. $(1 + i)^2$ 32. $(2 - 3i)^2$
33. $(4 - 2i)^2$ 34. $(1 - i)^2$
35. $(5 + 3i)^2$

In Exercises 36–50, divide and write in simplest form.

36. $\frac{6}{1 - 2i}$ 37. $\frac{4}{1 + 2i}$

38. $\frac{2i}{1 + i}$ 39. $\frac{4i}{2 - 3i}$

40. $\frac{1 + i}{1 - i}$ 41. $\frac{1 - i}{1 + i}$

42. $\frac{1 + \sqrt{3}i}{1 - \sqrt{3}i}$ 43. $\frac{4 - \sqrt{2}i}{1 + \sqrt{2}i}$

44. $\frac{5 - 2i}{1 - 2i}$ 45. $\frac{4 - 3i}{1 + 3i}$

46. $\frac{\sqrt{5} - i}{\sqrt{5} - 2i}$ 47. $\frac{-\sqrt{3} + \sqrt{6}i}{-1 + \sqrt{3}i}$

*48. $\frac{-3\sqrt{2} + i}{1 + 3\sqrt{2}i}$ *49. $\frac{a + \sqrt{b}i}{a - \sqrt{b}i}$

*50. $\frac{b + ai}{a - bi}$

*51. For what condition will the product of two complex numbers, $a + bi$ and $c + di$, be a real number?

*52. For what condition will the product of two complex numbers, $a + bi$ and $c + di$, be an imaginary number?

Chapter 7   SUMMARY

TERMS, RULES, AND FORMULAS

1. *Rules of Exponents.*   The rules of exponents hold for all rational exponents. Let m and n be rational numbers, and a and b be any real number. Then

   (1) $a^m \cdot a^n = a^{m+n}$ (2) $(a^m)^n = a^{mn}$

   (3) $\frac{a^m}{a^n} = a^{m-n},\ a \neq 0$

   (4) $(ab)^n = a^n b^n$

   (5) $\left(\frac{a}{b}\right)^n = \frac{a^n}{b^n},\ b \neq 0$

   (6) $a^0 = 1,\ a \neq 0$

   (7) $a^{-n} = \frac{1}{a^n},\ a \neq 0$ (8) $a^{1/n} = \sqrt[n]{a}$

   (9) $a^{m/n} = \sqrt[n]{a^m} = (\sqrt[n]{a})^m$

2. *Scientific Notation.*   Any positive number N can be written in the form
   $$N = A \times 10^n,\ 1 \leq A < 10,\ n \text{ an integer}$$

3. *Radical*

   $$\text{index} \to \sqrt[n]{A}$$

   radical $\overset{\_\!\!\uparrow\uparrow}{}$ radicand

   The symbol $\sqrt[n]{A}$ always represents the principal nth root of A.

4. *Rational Exponent*

5. *Simplifying Radicals*

$$\sqrt[n]{a^n} = a, \quad \sqrt[n]{ab} = \sqrt[n]{a}\,\sqrt[n]{b}, \quad \sqrt[n]{\frac{a}{b}} = \frac{\sqrt[n]{a}}{\sqrt[n]{b}}$$

PROCEDURES

1. *Radical in Its Simplest Forms.* To write a radical in its simplest form means to
  (a) Remove any factors from the radicand.
  (b) Reduce the index to its lowest possible value.
  (c) Rewrite any fractions so that the denominator does not contain a radical. This is called *rationalizing the denominator.*
2. *Addition and Subtraction of Radicals.* *Like radicals* have the same index and radicand.
  (a) Addition and subtraction of *like* radicals: Add their coefficients
  (b) Addition and subtraction of *unlike* radicals:
    (i) Simplify each radical.
    (ii) Combine like radicals using the distributive property.
    (iii) Indicate the sum or difference of the unlike radicals.
3. *Multiplication of Radicals.* If factors have common index, use the formula
$$\sqrt[n]{a}\,\sqrt[n]{b} = \sqrt[n]{ab}, \quad a,b \geq 0$$
then simplify.
4. *Division of Radicals.* The conjugate of $a\sqrt{b} + c\sqrt{d}$ is $a\sqrt{b} - c\sqrt{d}$.
  (a) *Monomial denominator.* Multiply the numerator and denominator by that quantity which will make the radicand in the denominator a perfect *n*th power.
  (b) *Binomial denominator.* Multiply the numerator and denominator by the conjugate of the denominator.
5. *Solving Radical Equations.*
  (1) Transpose the terms of the equation so that one term with a radical stands alone on one side of the equation. If the equation has more than one radical, isolate the more complex radical.
  (2) Raise each side of the equation to a power equal to the index of the radical.
  (3) Combine like terms.
  (4) If radicals still remain, repeat Steps 1, 2, and 3.
  (5) Solve the resulting equation for the unknown.
  (6) *Check* answers in Step 5 in the *original* equation.
6. *Complex Numbers.*
  (a) A complex number (see Figure 7.1) is a number of the form $a + bi$, where a and b are real numbers, and $i = \sqrt{-1}$.
  $$a + bi$$
  real part $\overset{\curvearrowleft}{}$ $\underset{}{\curvearrowright}$ imaginary part
  (b) *Addition of complex numbers:*
  $$(a + bi) + (c + di)$$
  $$= (a + c) + (b + d)i$$
  (c) *Subtraction of complex numbers:*
  $$(a + b) - (c + di)$$
  $$= (a - c) + (b - d)i$$

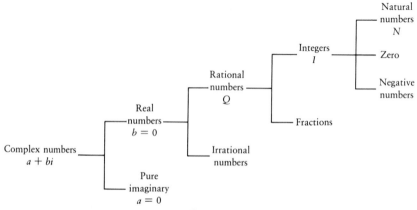

Figure 7.1

(d) *Multiplication of complex numbers:*
$$(a + bi)(c + di)$$
$$= (ac - bd) + (ad + bc)i$$
The *conjugate* of $a + bi$ is $a - bi$, and
$$(a + bi)(a - bi) = a^2 + b^2$$

(e) *Division of complex numbers:*
Multiply numerator and denominator by

the conjugate of the denominator:
$$\frac{c + di}{a + bi} = \frac{(c + di)}{(a + bi)} \cdot \frac{(a - bi)}{(a - bi)}$$
$$= \frac{(ac + bd) + (ad - bc)i}{a^2 + b^2}$$

Chapter 7 EXERCISES

In Exercises 1–20, perform the indicated operations using laws of exponents. Simplify the results when possible.

1. $x^2 \cdot x^4$

2. $(3x^2)^2$

3. $3(x^2)^2$

4. $3(x^2 + 2)^0$

5. $[4(y^2 + 3y)]^0$

6. $(x^2 \cdot x^3)^2$

7. $\left(\dfrac{x^3}{x^{-2}}\right)^2$

8. $\left(\dfrac{-4x}{x^2}\right)^3$

9. $(a^3 \cdot a^2 \cdot a^{-5})^5$

10. $\dfrac{b^2}{ac} \cdot \dfrac{a^2 c}{b} \cdot \dfrac{c}{2b}$

11. $a^{-x} \cdot a^{1+x}$

12. $y^{2a+5} \cdot y^{a-5}$

13. $\dfrac{(ab)^{2x-3}}{(ab)^{3x-4}}$

14. $\dfrac{(x^2 y^n)^{-5}}{5x^3 y^{3n}}$

15. $(x^x)^x$

16. $(3)^{2a+4}(9)^{1-a}$

17. $\dfrac{(36)^{a-3}(6)^{a+7}}{(216)^a}$

18. $(125)^{a-b}(25)^{a+b}(5)^2$

19. $\left(\dfrac{b^{3a+2}}{b^{2a-2}}\right)^3$

20. $\left(\dfrac{x^{2m-n}}{x^{m+n}}\right) \div \left(\dfrac{x^{3m+n}}{x^{2m-3n}}\right)$

Express the quantities in Exercises 21–23 in radical form.

21. $x^{3/2}$

22. $(x^2 y)^{2/5}$

23. $\left(\dfrac{1 + x^2}{x}\right)^{2/3}$

Express the quantities in Exercises 24–26 in exponent form.

24. $\sqrt[4]{a^3 b}$

25. $\sqrt[4]{\dfrac{b}{(c + d)^3}}$

26. $\sqrt[5]{x^2 y^{-3} z}$

Simplify the quantities in Exercises 27–40.

27. $\sqrt{12}$

28. $\sqrt{24}$

29. $\sqrt{75}$

30. $\sqrt{27}$

31. $\sqrt{125}$

32. $\sqrt[3]{54}$

33. $\sqrt[3]{24}$

34. $\sqrt[4]{32}$

35. $\sqrt[4]{162}$

36. $\sqrt[6]{576}$

37. $\sqrt[3]{a^5 b^4 c^3}$

38. $\sqrt[4]{243 a^{13} b^{14}}$

39. $\sqrt[n]{2^{n+1} x^{4n+5}}$

40. $\sqrt[2n-1]{x^{2n} y^{4n-1}}$

Rationalize the denominators in Exercises 41–55.

41. $\dfrac{1}{\sqrt{3}}$

42. $\sqrt{\dfrac{1}{2}}$

43. $\sqrt{\dfrac{3}{2}}$

44. $\sqrt{\dfrac{5}{4}}$

45. $\sqrt{\dfrac{5}{32}}$

46. $\sqrt{\dfrac{7}{8}}$

47. $\sqrt{\dfrac{a^2}{bc}}$

48. $\sqrt{\dfrac{9ab}{98c^3}}$

49. $\sqrt[4]{\dfrac{b^{4n}}{c^{3n}}}$

50. $\sqrt[n+1]{\dfrac{1}{x^n}}$

51. $\dfrac{1}{\sqrt[3]{3}}$

52. $\dfrac{\sqrt[4]{3}}{\sqrt[4]{4}}$

53. $\sqrt{\dfrac{x - y}{x + y}}$

54. $\sqrt{\dfrac{5}{72}}$

55. $\sqrt{\dfrac{(x + y)^3}{x}}$

Perform the indicated operations in Exercises 56–80 and simplify when possible.

56. $\sqrt{5} - 3\sqrt{5} + 4\sqrt{5}$

57. $a\sqrt[n]{y} - b\sqrt[n]{y} + c\sqrt[n]{y}$

58. $\sqrt{8} - \sqrt{2}$

59. $\sqrt{3} - \sqrt{12}$

60. $\sqrt{\dfrac{1}{3}} + \sqrt{27}$

61. $3\sqrt{50} - 2\sqrt{8}$

62. $3\sqrt{72} - 5\sqrt{50} + 2\sqrt{98}$

63. $-2\sqrt{27} + 2\sqrt{48} + \sqrt{75}$

64. $\sqrt{28} + \dfrac{1}{\sqrt{7}} + \sqrt{\dfrac{7}{9}}$

65. $\sqrt[3]{32} + \sqrt[3]{500}$

66. $3\sqrt[3]{\dfrac{1}{4}} - 2\sqrt[3]{\dfrac{2}{27}} + \sqrt[3]{\dfrac{4}{27}}$

67. $\sqrt{x^3y} + \sqrt{xy} - \sqrt{xy^3}$

68. $\sqrt[3]{a^4b^5} - 3\sqrt[3]{a^7b^5} + 2\sqrt[3]{a^4b^2}$

69. $\sqrt{x+y} - 2\dfrac{1}{\sqrt{x+y}}$

70. $\dfrac{x^2}{2}\sqrt{\dfrac{2y^3}{x^5}} - \sqrt{\dfrac{y^3}{2x}} + x\sqrt{\dfrac{y^3}{2x^3}}$

71. $2\sqrt{\dfrac{x^2}{4}} + 3\sqrt{\dfrac{x}{2}} - \dfrac{1}{2}\sqrt{2x}$

72. $8\sqrt{36} - 3\sqrt{12} + 7\sqrt{3}$

73. $\sqrt{\dfrac{x}{3}} + \dfrac{2}{3}\sqrt{3x} - 3\sqrt{8x^2}$

74. $12\sqrt{\dfrac{x}{2}} - 2\sqrt{8x} + \sqrt{3x^2}$

75. $\sqrt{\dfrac{x}{4}} + \dfrac{3}{4}\sqrt{x} + 2x\sqrt{2x^2}$

76. $\sqrt[3]{24x^2} - \sqrt[3]{4x^2} + \sqrt[3]{81x^2}$

77. $\sqrt{\dfrac{1}{3}xy} + \dfrac{2}{3}\sqrt{3xy} - \sqrt{xy}$

78. $8\sqrt{6x} - 2\sqrt{54x} + 6\sqrt{x}$

79. $16\sqrt{\dfrac{3}{4}} + 3\sqrt{48} - \sqrt{18}$

80. $\sqrt[3]{\dfrac{3}{4}} - 3\sqrt[3]{\dfrac{1}{36}} + \sqrt[3]{\dfrac{2}{9}} - \dfrac{1}{2}\sqrt[3]{\dfrac{16}{9}}$

Perform the indicated operations in Exercises 81–100 and simplify.

81. $\sqrt[3]{4} \cdot \sqrt[3]{12}$ 82. $\sqrt{3x^3y} \cdot \sqrt{6xy^5}$

83. $2\sqrt{6} \cdot 5\sqrt{18}$ 84. $7\sqrt[3]{9} \cdot 9\sqrt[3]{21}$

85. $2\sqrt{3\dfrac{1}{3}} \cdot 3\sqrt{5\dfrac{5}{6}}$

86. $\sqrt{\dfrac{2}{3}} \cdot \sqrt{\dfrac{5}{6}} \cdot \sqrt{\dfrac{3}{10}}$

87. $\sqrt{6}(3\sqrt{2} + 2\sqrt{3} + \sqrt{6})$

88. $\sqrt{30}(\sqrt{6} + \sqrt{10} + \sqrt{15})$

89. $(7 + \sqrt{13})^{1/2}(7 - \sqrt{13})^{1/2}$

90. $(8 - \sqrt{15})^{1/2}(8 + \sqrt{15})^{1/2}$

91. $(2\sqrt[3]{3} + 3\sqrt[3]{2})(\sqrt[3]{9} - 2\sqrt[3]{4})$

92. $(3 + \sqrt{x-9})^2$

93. $(a - \sqrt{x-a^2})^2$

94. $(\sqrt{x} - \sqrt{a-x})^2$

95. $\dfrac{1}{2}(\sqrt{x} + \sqrt{x-2})^2$

96. $\dfrac{1}{2}(\sqrt{a+x} - \sqrt{a-x})^2$

*97. $\dfrac{1}{2}(\sqrt{x-b} + \sqrt{x+b})^2$

*98. $(\sqrt{x-1} - \sqrt{x^2+x+1})^2$

*99. $\dfrac{1}{2}(\sqrt{a+\sqrt{b}} - \sqrt{a-\sqrt{b}})^2$

*100. $\dfrac{1}{2}(\sqrt{a+b+c} - \sqrt{a-b-c})^2$

Rationalize the denominators in Exercises 101–105.

101. $\dfrac{1}{2 + \sqrt{5}}$ 102. $\dfrac{4}{4 - \sqrt{2}}$

103. $\dfrac{3}{\sqrt{3} + \sqrt{2}}$ 104. $\dfrac{5}{5\sqrt{3} - 2\sqrt{5}}$

105. $\dfrac{\sqrt{5} - \sqrt{3}}{\sqrt{5} + 2\sqrt{3}}$

Solve and check the equations in Exercises 106–109.

106. $1 + \sqrt{x} = \sqrt{x+2}$

107. $\sqrt{x-4} = \sqrt{x+1} - 1$

108. $\sqrt{x-5} - \sqrt{x+7} = 6$

109. $1 - \sqrt{5x-4} = \sqrt{5x+9}$

*110. $\dfrac{\sqrt{x+1} - \sqrt{x}}{\sqrt{x+1} + \sqrt{x}} = \dfrac{1}{3}$

In Exercises 111–130, perform the indicated operations.

111. $(5 - 2i) + (6 + 3i)$

112. $(20 - 7i) - (15 + 8i)$

113. $(9 + 5i) - (11 + 18i)$

114. $(6 + 2i) - (3 - i)$

115. $-(-7 + 3i) - (5 + 2i)$

116. $5 - (3 - 4i)$

117. $(4 + 8i) - 5i$

118. $(2 + 5i)(-3 + 6i)$

119. $(9 + i)(6 - 4i)$

120. $(\sqrt{2} + i)(\sqrt{2} - i)$

121. $(5 + 8i)(5 - 8i)$

122. $5(6 - 3i)$

123. $(3i)(2 + 6i)$

124. $(2i)(8i)$

125. i^3

126. $\dfrac{4 + 3i}{2 + 5i}$

127. $\dfrac{7 + 6i}{3 - 4i}$

128. $\dfrac{-6 + 3i}{-1 + 2i}$

129. $\dfrac{3i}{4 + 7i}$

130. $\dfrac{2 + 6i}{4i}$

In Exercises 131–133, solve for x and y.

131. $(x + yi)(2 - 5i) = 1 + 5i$

132. $(x + yi)(-4 + i) = -9 + 7i$

133. $(x + yi)(3i) = 6$

134. Show that $2 + 3i$ satisfies the equation $x^2 - 4x + 13 = 0$. What can be said about $2 - 3i$?

135. Show that $5 + 3i$ satisfies $x^2 - (6 + 4i)x + (2 + 8i) = 0$.

8 Quadratic Equations and Inequalities

In this chapter, we discuss several methods for solving quadratic equations and inequalities. In solving quadratic equations in the form

$$ax^2 + bx + c = 0$$

any one of the following methods may be used:

1. factoring
2. completing the square
3. quadratic formula

Of these methods, factoring is the simplest method when the solutions are rational and the factors are easily found. In other cases, the quadratic formula is used. However, the quadratic formula applies to any quadratic equation whether the solutions are real or complex, rational or irrational.

While the method of completing the square may be used to solve quadratic equations, it is more involved than the quadratic formula and is seldom used in practical work. However, its importance will become clear in deriving the quadratic formula.

We begin by defining a quadratic equation and an incomplete quadratic equation.

DEFINITION

A quadratic equation in one variable is an equation of the form

$$ax^2 + bx + c = 0$$

where a, b, and c are real numbers and $a \neq 0$.

Note. While the coefficients a, b, and c of the quadratic equation are restricted to the set of real numbers, we shall see that the solutions of such equations may be complex numbers.

EXAMPLE 1 The following are quadratic equations in one variable:

(a) $2x^2 + 5x - 3 = 0$ $(a = 2, b = 5, c = -3)$

(b) $-x^2 + 8x = 0$ $(a = -1, b = 8, c = 0)$

(c) $3x^2 - 27 = 0$ $(a = 3, b = 0, c = -27)$

An incomplete quadratic equation is a quadratic equation in which $b = 0$ or $c = 0$, or both $b = 0$ and $c = 0$.

EXAMPLE 2 The following are incomplete quadratic equations:

(a) $x^2 - 6 = 0$ $(b = 0)$

(b) $x^2 - 7x = 0$ $(c = 0)$

(c) $3x^2 = 0$ (both $b = 0$ and $c = 0$)

8.1 SOLVING QUADRATIC EQUATIONS OF THE FORM $ax^2 + c = 0$

The simplest type of quadratic equation is the incomplete quadratic equation of the form $ax^2 + c = 0$, $a \neq 0$. The solution is found by a direct use of the definition of square root. The following examples illustrate the method known as the square root method.

EXAMPLE 1 Solve by the square root method: $x^2 - 7 = 0$.

Solution
$$x^2 - 7 = 0$$
$$x^2 = 7$$
$$x = \pm\sqrt{7} \quad \blacksquare$$

Note. $\pm\sqrt{7}$ is a short way to signify $+\sqrt{7}$ or $-\sqrt{7}$.

EXAMPLE 2 Solve $2x^2 - 5 = 0$.

Solution
$$2x^2 - 5 = 0$$
$$2x^2 = 5$$
$$x^2 = \frac{5}{2}$$
$$x = \pm\sqrt{\frac{5}{2}} \quad \text{or} \quad \pm\frac{\sqrt{10}}{2} \quad \blacksquare$$

EXAMPLE 3   Solve $4x^2 + 64 = 0$.

Solution
$$4x^2 + 64 = 0$$
$$4x^2 = -64$$
$$x^2 = -16$$
$$x = \pm\sqrt{-16} = \pm 4i \quad \blacksquare$$

EXAMPLE 4 $\left(x - \dfrac{7}{2}\right)^2 - \dfrac{23}{4} = 0$

Solution
$$\left(x - \frac{7}{2}\right)^2 - \frac{23}{4} = 0$$
$$\left(x - \frac{7}{2}\right)^2 = \frac{23}{4}$$
$$\left(x - \frac{7}{2}\right) = \pm\sqrt{\frac{23}{4}}$$
$$x - \frac{7}{2} = \pm\frac{\sqrt{23}}{2}$$
$$x = \frac{7}{2} \pm \frac{\sqrt{23}}{2}$$
$$x = \frac{7 \pm \sqrt{23}}{2} \quad \blacksquare$$

Note. $x = \dfrac{7 \pm \sqrt{23}}{2}$ is short for $x = \dfrac{7 + \sqrt{23}}{2}$ or $x = \dfrac{7 - \sqrt{23}}{2}$.

EXAMPLE 5   Solve $\dfrac{2}{x^2 + 1} = \dfrac{3}{2x^2 - 11}$.

Solution
$$\frac{2}{x^2 + 1} = \frac{3}{2x^2 - 11}$$

Multiplying both sides by $(x^2 + 1)(2x^2 - 11)$,

$$2(2x^2 - 11) = 3(x^2 + 1)$$
$$4x^2 - 22 = 3x^2 + 3$$
$$x^2 = 25$$
$$x = \pm 5 \quad \blacksquare$$

EXAMPLE 6   Solve for x: $x^2 + a^2x^2 = (a^2 - 1)^2 - 2ax^2$.

Solution
$$x^2 + a^2x^2 = (a^2 - 1)^2 - 2ax^2$$
$$x^2 + a^2x^2 + 2ax^2 = (a^2 - 1)^2$$

$$(a^2 + 2a + 1)x^2 = (a^2 - 1)^2$$

$$(a + 1)^2 x^2 = (a^2 - 1)^2$$

$$x^2 = \frac{(a^2 - 1)^2}{(a + 1)^2}$$

$$x^2 = \left(\frac{a^2 - 1}{a + 1}\right)^2$$

$$x = \pm\left[\frac{a^2 - 1}{a + 1}\right] = \pm\left[\frac{(a + 1)(a - 1)}{(a + 1)}\right]$$

$$x = \pm(a - 1) \quad \blacksquare$$

EXERCISES 8.1

Solve the equations in Exercises 1–44.

1. $x^2 - 16 = 0$ 2. $x^2 - 25 = 0$
3. $x^2 = 9$ 4. $-2x^2 = -4$
5. $4x^2 = 1$ 6. $25x^2 = 121$
7. $x^2 = a^2$ 8. $x^2 = 99 - 10x^2$
9. $80 - 2x^2 = 3x^2$
10. $-5 + 0.75x^2 - 22 = 0$
11. $11x^2 - 44 = 5x^2 + 10$
12. $2x^2 - 7 = 25$ 13. $2(x^2 + 7) = 112$
14. $3(11 - 3x^2) - 15 = 0$
15. $2(3x^2 - 4) = 3x^2 + 19$
16. $a^2 - x^2 = b^2$ 17. $2x^2 + 8 = 0$
18. $9x^2 + 64 = 0$ 19. $2x^2 + 12 = 0$
20. $4x^2 + 11 = 0$ 21. $-5x^2 - 7 = 0$
22. $(x + 1)^2 - 36 = 0$
23. $(x + 2)^2 = 49$ 24. $(x - 5)^2 - 64 = 0$
25. $(x - 2)^2 - 1 = 0$
26. $(x - 3)^2 - 9 = 0$
27. $(x + 3)^2 - b^2 = 0$
28. $(x - 5)^2 + 9 = 0$
29. $(2x + 4)^2 + 3 = 0$
30. $(x - 3)^2 + 2 = 0$
31. $\dfrac{11}{x^2 + 6} - \dfrac{6}{x^2 - 4} = 0$
32. $\dfrac{3}{x^2 - 3} = \dfrac{5}{x^2 + 1}$
33. $\dfrac{x + 4}{x - 2} + \dfrac{x - 4}{x + 2} = 5$

34. $\dfrac{x + 6}{7} = \dfrac{4}{x - 6}$

35. $8 = \dfrac{3}{1 + x} + \dfrac{3}{1 - x}$

36. $\dfrac{3x + 4}{x} = \dfrac{x}{3x - 4}$

37. $(x + 3)(x + 3) + (x - 3)^2 = 4(x^2 - 9)$
38. $(x - 4)^2 + 4 = -8x + 29$
39. $(a - x)^2 - (x - a)(3x + a) = 0$

40. $\dfrac{x^2 + x + 1}{x - 1} - 6 = \dfrac{x^2 - x + 1}{x + 1}$

41. $\sqrt{(x + 2)(x + 3)} = \sqrt{5x + 31}$
42. $\sqrt{(x + 4)(x - 6)} = \sqrt{6 - 2x}$
43. $\sqrt{(x + 5)(x + 1)} = \sqrt{14 + 6x}$
44. $\sqrt{x^2 + 21} = 2x + \sqrt{x^2 - 3}$
45. Solve for r: $A = \pi r^2$ (area of a circle)

46. Solve for r: $V = \dfrac{1}{3}\pi r^2 h$ (volume of a cone)

47. Solve for t: $s = \dfrac{gt^2}{2}$

48. Solve for d: $F = k\dfrac{m_1 m_2}{d^2}$ (force of attraction)

49. Solve for i: $p = i^2 r$ (p = power, i = intensity, r = resistance)

50. Solve for c: $a^2 + b^2 = c^2$ (Pythagorean theorem)

8.2 SOLVING QUADRATIC EQUATIONS BY FACTORING

The following theorem provides a means for solving quadratic equations when the equation can be factored into linear equations.

The Zero-Product Property of Real Numbers

If a and b are real numbers and $ab = 0$, then $a = 0$, or $b = 0$, or both $a = 0$ and $b = 0$.

Proof Assume $ab = 0$ and $a \neq 0$. Then we can divide both sides by a.

$$ab = 0$$

$$\frac{\cancel{a}b}{\cancel{a}} = \frac{0}{a}$$

$$b = 0$$

Similarly, if $b \neq 0$, we can show that $a = 0$. ▫

The following examples illustrate the use of this theorem in solving factored quadratic equations.

EXAMPLE 1 Solve $x^2 + 4x - 12 = 0$.

Solution Factoring the terms on the left side of the equation, we have

$$x^2 + 4x - 12 = 0$$

$$(x + 6)(x - 2) = 0$$

However, as stated in the above theorem, this equation implies that either

$$x + 6 = 0 \quad \text{or} \quad x - 2 = 0$$

Solving each of these linear equations gives

$$x = -6 \quad \text{or} \quad x = 2. \quad ■$$

Check

$$x = -6 \qquad\qquad\qquad x = 2$$

$$(-6)^2 + 4(-6) - 12 \stackrel{?}{=} 0 \qquad (2)^2 + 4(2) - 12 \stackrel{?}{=} 0$$

$$36 - 24 - 12 \stackrel{?}{=} 0 \qquad\qquad 4 + 8 - 12 \stackrel{?}{=} 0$$

$$12 - 12 \stackrel{?}{=} 0 \qquad\qquad\qquad 12 - 12 \stackrel{?}{=} 0$$

$$0 \stackrel{\checkmark}{=} 0 \qquad\qquad\qquad\qquad 0 \stackrel{\checkmark}{=} 0$$

EXAMPLE 2 Solve $x^2 - 11x = 0$.

Solution

$$x^2 - 11x = 0$$

$$x(x - 11) = 0$$

$$x = 0 \quad \text{or} \quad x - 11 = 0$$
$$x = 11$$

Thus, $x = 0$ or $x = 11$. ■

Note. It is important to emphasize that only when the product of factors is *zero* can it be deduced that one or more of them is zero. If the product of two factors is not zero, say 10, we cannot deduce that either of them is 10. One factor may be 5 and the other 2. Thus it is important to transpose all terms to one side before factoring.

EXAMPLE 3 Solve $-3x - 9 = -2x^2$.

Solution Transposing all terms to the left side (making the coefficient of x^2 positive), we obtain

$$2x^2 - 3x - 9 = 0$$
$$(2x + 3)(x - 3) = 0$$

Thus,
$$2x + 3 = 0 \quad \text{or} \quad x - 3 = 0$$
$$x = -\frac{3}{2} \quad \text{or} \quad x = 3 \quad ■$$

EXAMPLE 4 Solve $30x^2 - 85x + 25 = 0$.

Solution Before factoring, we reduce the coefficients to their lowest terms by dividing both sides by 5:

$$30x^2 - 85x + 25 = 0$$
$$6x^2 - 17x + 5 = 0$$
$$(3x - 1)(2x - 5) = 0$$

Thus,
$$3x - 1 = 0 \quad \text{or} \quad 2x - 5 = 0$$
$$x = \frac{1}{3} \quad \text{or} \quad x = \frac{5}{2} \quad ■$$

EXAMPLE 5 Solve $x^2 - 6ax + 9a^2 = 0$ for x.

Solution
$$x^2 - 6ax + 9a^2 = 0$$
$$(x - 3a)(x - 3a) = 0$$

Thus,
$$x - 3a = 0 \quad \text{or} \quad x - 3a = 0$$
$$x = 3a \quad \text{or} \quad x = 3a \quad ■$$

The solution $x = 3a$ is called a double root of the equation.
Some quadratic equations involve group factoring.

EXAMPLE 6   Solve $x^2 - 2ax + 8x = 16a$.

Solution

$$x^2 - 2ax + 8x - 16a = 0$$
$$x(x - 2a) + 8(x - 2a) = 0$$
$$(x - 2a)(x + 8) = 0$$

Thus,
$$x - 2a = 0 \quad \text{or} \quad x + 8 = 0$$
$$x = 2a \quad \text{or} \quad x = -8 \quad \blacksquare$$

EXAMPLE 7   Solve $(x + a)(x + 1) = 2(x + a)$.

Solution

$$(x + a)(x + 1) = 2(x + a)$$
$$(x + a)(x + 1) - 2(x + a) = 0$$
$$(x + a)(x + 1 - 2) = 0$$
$$(x + a)(x - 1) = 0$$

Thus,
$$x + a = 0 \quad \text{or} \quad x - 1 = 0$$
$$x = -a \quad \text{or} \quad x = 1 \quad \blacksquare$$

EXAMPLE 8   Solve $\dfrac{7}{x + 4} - \dfrac{4}{x + 1} = \dfrac{5}{x + 2} - \dfrac{2}{x - 1}$.

Solution

$$\frac{7}{x + 4} - \frac{4}{x + 1} = \frac{5}{x + 2} - \frac{2}{x - 1}$$

Multiply both sides by $(x + 4)(x + 1)(x + 2)(x - 1)$:

$$7(x + 1)(x + 2)(x - 1) - 4(x + 4)(x + 2)(x - 1)$$
$$= 5(x + 4)(x + 1)(x - 1) - 2(x + 4)(x + 1)(x + 2)$$

$$7(x^3 + 2x^2 - x - 2) - 4(x^3 + 5x^2 + 2x - 8)$$
$$= 5(x^3 + 4x^2 - x - 4) - 2(x^3 + 7x^2 + 14x + 8)$$

$$7x^3 + 14x^2 - 7x - 14 - 4x^3 - 20x^2 - 8x + 32$$
$$= 5x^3 + 20x^2 - 5x - 20 - 2x^3 - 14x^2 - 28x - 16$$

$$3x^3 - 6x^2 - 15x + 18 = 3x^3 + 6x^2 - 33x - 36$$

$$-12x^2 + 18x + 54 = 0$$

$$2x^2 - 3x - 9 = 0$$

$$(2x + 3)(x - 3) = 0$$

Thus, $\qquad 2x + 3 = 0 \quad \text{or} \quad x - 3 = 0$

$$x = -\frac{3}{2} \quad \text{or} \qquad x = 3 \quad \blacksquare$$

EXERCISES 8.2

Solve by factoring.

1. $x^2 - 5x + 6 = 0$
2. $x^2 - 4x + 3 = 0$
3. $x^2 - 14x + 13 = 0$
4. $x^2 - 7x + 12 = 0$
5. $x^2 - 19x + 88 = 0$
6. $x^2 + 4x + 3 = 0$
7. $x^2 + 10x + 21 = 0$
8. $x^2 + 9x + 20 = 0$
9. $x^2 + 6x + 9 = 0$
10. $-x^2 + x + 6 = 0$
11. $x^2 - 3x - 10 = 0$
12. $x^2 - 12x - 45 = 0$
13. $-x^2 - 12x + 28 = 0$
14. $x^2 + 10x - 39 = 0$
15. $x^2 + 2x - 15 = 0$
16. $-x^2 - x + 30 = 0$
17. $x^2 - 4x = 0$ 18. $8x = 20 - x^2$
19. $x^2 = x + 12$
20. $x^2 - 1 = x + 1$
21. $x^2 + 143 = 24x$
22. $x^2 + 50 = 15x - 6$
23. $x^2 = 7ax$ 24. $x^2 - 2bx = 0$
25. $x^2 + 3x = 0$ 26. $x^2 = -ax$
27. $6x^2 - 11x + 4 = 0$
28. $10x^2 - 21x + 9 = 0$
29. $2x^2 + 5x - 3 = 0$
30. $18x = 16 - 13x^2$
31. $9x^2 - 13x - 10 = 0$
32. $10x^2 = 9x + 9$
33. $11x^2 + 24x = -4$
34. $12x^2 + 32x + 21 = 0$
35. $11x^2 = 4x + 7$ 36. $8x - 3x^2 = 5$
37. $x^2 = 2bx - b^2$ 38. $7x^2 - 21x = 0$
39. $7a^2 = 12ax - 5x^2$
40. $18x^2 = 13bx + 11b^2$
41. $3a^2 - 4ax - 4x^2 = 0$

42. $4x^2 - (x + 1)^2 = 0$
43. $x - 10 = 6x(2 - x)$
44. $6x^2 + 7x + 2 = 0$
45. $12x^2 + 5 = 19x$
46. $10x^2 = 13x + 30$
47. $2ax^2 - 6a^2x = 15a - 5x$
48. $x^2 - ax - bx + ab = 0$
49. $x^2 - 2ax + 4ab = 2bx$
50. $3cx^2 + 2x = 4c + 6c^2x$
51. $(1 - x)(x + 2) = (1 - x)$
52. $2x(4x - 3) = x$
53. $(x - a)(x - 1) = (x - a)$
54. $(x - 4)^2 = (x - 4)$
55. $x^2 - 25 = (x + 5)$
56. $17x^2 + 11x - 5 = 5x^2 - 6x + 2$
57. $(3x - 1)(2x + 7) = (5x + 2)(1 - 2x)$
58. $(2x + 1)(x + 3) = x^2 + 4x + 1$
59. $x + \dfrac{1}{x} = 2$ 60. $x + 2 = \dfrac{1}{x + 2}$
61. $\dfrac{x}{x - 1} + \dfrac{2}{x - 1} = 2x$
62. $\dfrac{3x - 2}{x + 2} - \dfrac{x - 2}{x + 2} = \dfrac{2}{3}$
63. $\dfrac{1}{x - 1} - \dfrac{2}{x} = \dfrac{3}{x + 1} - \dfrac{4}{x + 2}$
64. $\dfrac{3}{x - 3} - \dfrac{8}{x + 2} = \dfrac{1}{x - 4} - \dfrac{6}{x + 6}$
*65. $\dfrac{3}{3x - 2a} - \dfrac{2}{x} = \dfrac{9}{3x + 2a} - \dfrac{12}{3x + 4a}$
*66. $\dfrac{x}{b} + \dfrac{b}{10x} = \dfrac{-(2a^2 + 5)}{10a}$
*67. $\dfrac{1}{x + a} + \dfrac{3}{4x + a} = \dfrac{2}{2x - a}$
*68. $\dfrac{a}{3x - a} - \dfrac{2a}{5x + a} = \dfrac{1}{6}$

8.3 SOLVING QUADRATIC EQUATIONS BY COMPLETING THE SQUARE

The simplest and fastest method of solving a quadratic equation is by factoring. However, not all quadratic equations are factorable. For example, the equation $x^2 - 6x + 7 = 0$ is not easily factorable. In this section, we discuss another method for solving quadratic equations that solves quadratic equations, regardless of whether the quadratic equation is factorable or not.

We begin by considering two special products from Chapter 2:

$$(x + a)^2 = x^2 + 2ax + a^2$$
$$(x - a)^2 = x^2 - 2ax + a^2$$

Note that in either case the constant term a^2 is the square of one-half of the coefficient of x; that is, the square of $\frac{1}{2}(2a)$ or of $\frac{1}{2}(-2a)$.

EXAMPLE 1   What must be added to each of the following expressions to make the sum a perfect square?

(a) $x^2 + 10x$ (b) $x^2 - 6x$ (c) $x^2 + 3x$

Solution   (a) The square of $\frac{1}{2}$ of 10 is $(5)^2 = 25$. Thus, adding 25, we obtain

$$x^2 + 10x + 25 = (x + 5)^2$$

(b) The square of $\frac{1}{2}$ of -6 is $(-3)^2 = 9$, and

$$x^2 - 6x + 9 = (x - 3)^2$$

(c) The square of $\frac{1}{2}$ of 3 is $\left(\frac{3}{2}\right)^2 = \frac{9}{4}$, and

$$x^2 + 3x + \frac{9}{4} = \left(x + \frac{3}{2}\right)^2 \quad \blacksquare$$

We now give a method, known as completing the square, for solving *any* quadratic equation.

How to Complete the Square

1. Simplify and write the quadratic so that the terms in x^2 and x are on the left side of the equation and the constant term is on the right side of the equation.
2. Make the coefficient of x^2 equal to 1 by dividing both sides of the equation by the coefficient of x^2.
3. Add to each side of the equation ($\frac{1}{2}$ coefficient of x)^2, and simplify.
4. Take the square root of each side, prefixing $\pm$ to the right side.
5. Solve for x.

EXAMPLE 2 Solve $x^2 + 6x + 7 = 0$.

Solution Note that the coefficient of x^2 is 1, so step 2 can be omitted.

1.
$$x^2 + 6x = -7$$

3.
$$x^2 + 6x + \left\{\frac{1}{2}\cdot 6\right\}^2 = -7 + \left\{\frac{1}{2}\cdot 6\right\}^2$$

$$x^2 + 6x + (3)^2 = -7 + 9$$

$$(x + 3)^2 = 2$$

4.
$$x + 3 = \pm\sqrt{2}$$

5.
$$x = -3 \pm \sqrt{2}$$

Thus, $x = -3 + \sqrt{2}$ or $x = -3 - \sqrt{2}$. ■

EXAMPLE 3 Solve $2x^2 - 4x - 9 = 0$.

Solution
$$2x^2 - 4x - 9 = 0$$

1.
$$2x^2 - 4x = 9$$

2.
$$x^2 - 2x = \frac{9}{2}$$

3.
$$x^2 - 2x + \left\{\frac{1}{2}\cdot(-2)\right\}^2 = \frac{9}{2} + \left\{\frac{1}{2}\cdot(-2)\right\}^2$$

$$x^2 - 2x + (-1)^2 = \frac{9}{2} + 1$$

$$(x - 1)^2 = \frac{11}{2}$$

4.
$$x - 1 = \pm\sqrt{\frac{11}{2}} = \pm\frac{\sqrt{22}}{2}$$

5.
$$x = 1 \pm \frac{\sqrt{22}}{2}$$

Thus, $x = 1 + \dfrac{\sqrt{22}}{2}$ or $x = 1 - \dfrac{\sqrt{22}}{2}$. ■

EXAMPLE 4 Solve $3x^2 - 4x + 2 = 0$.

Solution
$$3x^2 - 4x + 2 = 0$$

1.
$$3x^2 - 4x = -2$$

2. $$x^2 - \frac{4}{3}x = -\frac{2}{3}$$

3. $$x^2 - \frac{4}{3}x + \left\{\frac{1}{2}\cdot\left(-\frac{4}{3}\right)\right\}^2 = -\frac{2}{3} + \left\{\frac{1}{2}\cdot\left(-\frac{4}{3}\right)\right\}^2$$

$$x^2 - \frac{4}{3}x + \left\{-\frac{2}{3}\right\}^2 = -\frac{2}{3} + \left\{-\frac{2}{3}\right\}^2$$

$$\left(x - \frac{2}{3}\right)^2 = -\frac{2}{9}$$

4. $$x - \frac{2}{3} = \pm\sqrt{-\frac{2}{9}} = \pm\frac{\sqrt{2}}{3}i$$

5. $$x = \frac{2}{3} \pm \frac{\sqrt{2}}{3}i$$

Thus, $x = \dfrac{2 + \sqrt{2}\,i}{3}$ or $x = \dfrac{2 - \sqrt{2}\,i}{3}$. ■

EXAMPLE 5 Solve $4x^2 - 8ax + 3a^2 = 0$ for x.

Solution $$4x^2 - 8ax + 3a^2 = 0$$

1. $$4x^2 - 8ax = -3a^2$$

2. $$x^2 - 2ax = -\frac{3}{4}a^2$$

3. $$x^2 - 2ax + \left\{\frac{1}{2}\cdot(-2a)\right\}^2 = -\frac{3}{4}a^2 + \left\{\frac{1}{2}\cdot(-2a)\right\}^2$$

$$x^2 - 2ax + (-a)^2 = -\frac{3}{4}a^2 + (-a)^2$$

$$(x - a)^2 = \frac{a^2}{4}$$

4. $$x - a = \pm\frac{a}{2}$$

$$x = a \pm \frac{a}{2}$$

Thus, $x = a + \dfrac{a}{2} = \dfrac{3a}{2}$ or $x = a - \dfrac{a}{2} = \dfrac{a}{2}$. ■

EXERCISES 8.3

Solve the following equations by completing the square.

1. $x^2 + 8x + 15 = 0$
2. $x^2 - 8x + 12 = 0$

3. $x^2 - 2x - 35 = 0$
4. $x^2 + 2x - 3 = 0$
5. $x^2 - 4x - 5 = 0$ 6. $x^2 = 6x + 7$
7. $x^2 - 10x + 9 = 0$
8. $x^2 + 8x - 9 = 0$
9. $x^2 - 5x + 5 = 0$
10. $x^2 - 3x + 2 = 0$
11. $x^2 + 3x - 4 = 0$ 12. $x^2 - 5x = 14$
13. $x^2 + x - 2 = 0$ 14. $x^2 + 5x = 6$
15. $x^2 - x = 12$
16. $x^2 - 7x + 10 = 0$
17. $x^2 + 9x = 10$
18. $x^2 + 4x - 4 = 0$
19. $x^2 + 2x - 4 = 0$
20. $x^2 - 2x - 1 = 0$
21. $x^2 + 4x - 3 = 0$ 22. $x^2 - 4x = 6$
23. $x^2 + 6x - 3 = 0$ 24. $x^2 - 6x = 1$
25. $x^2 + 3x - 1 = 0$
26. $x^2 - 3x - 2 = 0$ 27. $x^2 + x = 5$
28. $x^2 - x = 3$
29. $x^2 - 5x - 7 = 0$
30. $x^2 + 5x - 3 = 0$

31. $2x^2 - 5x + 2 = 0$
32. $3x^2 + 10x + 3 = 0$
33. $6x^2 + x - 2 = 0$
34. $4x^2 - 2x = 1$
35. $5x^2 + 2x = 7$
36. $3x^2 + 5x - 2 = 0$
37. $3x^2 - 2x - 3 = 0$
38. $8x^2 + 4x - 1 = 0$
39. $4x^2 - 3x - 2 = 0$
40. $6x^2 + 8x - 5 = 0$
41. $2x^2 - 10x - 9 = 0$
42. $10x^2 + 4x - 3 = 0$
43. $x^2 - 8x + 25 = 0$
44. $x^2 - 10x + 29 = 0$
45. $2x^2 - 6x + 17 = 0$
46. $5x^2 - 4x + 4 = 0$
47. $2x^2 - 10x + 44 = 0$
48. $18 = 9x - 2x^2$
49. $3x^2 - 2ax = 5a^2$
50. $2x^2 - 13ax = 7a^2$
*51. $4x^2 + 17ax + 15a^2 = 0$
*52. $6x^2 - 19ax = -15a^2$

8.4 THE QUADRATIC FORMULA

The general quadratic equation, $ax^2 + bx + c = 0$, can be solved by the method of completing the square.

$$ax^2 + bx + c = 0$$

1. $ax^2 + bx = -c$ — [taking the constant to right side]

2. $x^2 + \dfrac{b}{a}x = -\dfrac{c}{a}$ — [dividing by coefficient of x^2]

3. $x^2 + \dfrac{b}{a}x + \left(\dfrac{b}{2a}\right)^2 = -\dfrac{c}{a} + \left(\dfrac{b}{2a}\right)^2$ — $\left[\text{adding } \left(\dfrac{1}{2} \text{ coefficient of } x\right)^2 \text{ to both sides}\right]$

$$\left(x + \dfrac{b}{2a}\right)^2 = -\dfrac{c}{a} + \dfrac{b^2}{4a^2}$$ — [simplifying]

$$\left(x + \dfrac{b}{2a}\right)^2 = \dfrac{b^2 - 4ac}{4a^2}$$

4. $x + \dfrac{b}{2a} = \pm\sqrt{\dfrac{b^2 - 4ac}{4a^2}}$ — [taking the square root of both sides]

$$x + \dfrac{b}{2a} = \pm\dfrac{\sqrt{b^2 - 4ac}}{2a}$$

5.
$$x = \frac{-b \pm \sqrt{b^2 - 4ac}}{2a}$$
   [solving for x]

Since we have applied the method of completing the square to the standard form of the quadratic equation $ax^2 + bx + c = 0$, we have derived a formula that gives the solutions of *any* quadratic equation in one variable. This formula is called the quadratic formula.

Quadratic Formula

$$x = \frac{-b \pm \sqrt{b^2 - 4ac}}{2a}, \, a \neq 0$$

Thus, the solutions of the quadratic equation $ax^2 + bx + c = 0$ are

$$x = \frac{-b + \sqrt{b^2 - 4ac}}{2a} \quad \text{and} \quad x = \frac{-b - \sqrt{b^2 - 4ac}}{2a}$$

EXAMPLE 1   Solve $8x^2 - 10x - 3 = 0$ by the quadratic formula.

Solution   In this example, $a = 8$, $b = -10$, and $c = -3$. Substituting these values into the quadratic formula, we get

$$x = \frac{-(-10) \pm \sqrt{(-10)^2 - 4(8)(-3)}}{2(8)}$$

$$= \frac{10 \pm \sqrt{100 + 96}}{16} = \frac{10 \pm \sqrt{196}}{16}$$

$$= \frac{10 \pm 14}{16}$$

Thus, $x = \dfrac{10 + 14}{16} = \dfrac{24}{16} = \dfrac{3}{2}$ and $x = \dfrac{10 - 14}{16} = \dfrac{-4}{16} = -\dfrac{1}{4}$. ■

EXAMPLE 2   Solve $x^2 - 6x = 11$ by the quadratic formula.

Solution   In order to use the quadratic formula, we must put the quadratic equation in standard form

$$x^2 - 6x - 11 = 0$$

Now we can determine a, b, and c:

$$a = 1 \qquad b = -6 \qquad c = -11$$

Substituting into the quadratic formula, we obtain

$$x = \frac{-(-6) \pm \sqrt{(-6)^2 - 4(1)(-11)}}{2(1)}$$

$$= \frac{6 \pm \sqrt{36 + 44}}{2} = \frac{6 \pm \sqrt{80}}{2}$$

$$= \frac{6 \pm 4\sqrt{5}}{2} = 3 \pm 2\sqrt{5}$$

Thus, $x = 3 + 2\sqrt{5}$ and $x = 3 - 2\sqrt{5}$. ■

EXAMPLE 3 Solve $2x^2 - 3x + 5 = 0$ by the quadratic formula.

Solution In this example, $a = 2$, $b = -3$, and $c = 5$. Therefore,

$$x = \frac{-(-3) \pm \sqrt{(-3)^2 - 4(2)(5)}}{2(2)}$$

$$= \frac{3 \pm \sqrt{9 - 40}}{4} = \frac{3 \pm \sqrt{-31}}{4}$$

$$= \frac{3 \pm \sqrt{31}\, i}{4}$$

Thus, $x = \dfrac{3 + \sqrt{31}\, i}{4}$ and $x = \dfrac{3 - \sqrt{31}\, i}{4}$. ■

EXAMPLE 4 Solve for x using the quadratic formula: $mx^2 - nx - p = 0$.

Solution Here we have $a = m$, $b = -n$, and $c = -p$. Substituting into the quadratic formula gives

$$x = \frac{-(-n) \pm \sqrt{(-n)^2 - 4(m)(-p)}}{2(m)}$$

$$= \frac{n \pm \sqrt{n^2 + 4mp}}{2m}$$ ■

CAUTION

When using the quadratic formula, we must be careful to use it properly. The following are common mistakes that should be avoided.

1. The equation should be put in the *standard form*
$$ax^2 + bx + c = 0$$
before determining a, b, and c. For example, the equation

$x^2 = -7x + 4$ should be written in the form $x^2 + 7x - 4 = 0$. Then $a = 1$, $b = 7$, and $c = -4$.

2. When determining a, b, and c, be sure to *include their sign*. If $-2x^2 + 5x - 2 = 0$, then $a = -2$, $b = 5$, and $c = -2$.

3. Be sure to write the quadratic formula correctly:

$$x = \frac{-b \pm \sqrt{b^2 - 4ac}}{2a} \quad not \quad x = -b \pm \frac{\sqrt{b^2 - 4ac}}{2a}$$

EXERCISES 8.4

Solve the equations in Exercises 1–39 by the quadratic formula.

1. $3x^2 - 2x - 5 = 0$
2. $21x^2 = -x + 2$
3. $6x^2 - 5x - 4 = 0$
4. $x^2 - 5x + 4 = 0$
5. $x^2 + 4x = 5$
6. $2x^2 + 5x = -2$
7. $3x^2 - 10x = -3$
8. $2x^2 - 5x + 3 = 0$
9. $30x^2 = 17x - 2$
10. $15x^2 = 75x - 90$
11. $6x^2 - 13x + 6 = 0$
12. $7x^2 = 23x - 6$
13. $9x^2 - 2x = 0$
14. $12x^2 - 13x - 35 = 0$
15. $5x^2 - 7x - 6 = 0$
16. $x^2 - 2x = 1$
17. $-6x - 3 = -x^2$
18. $9x^2 - 12x - 1 = 0$
19. $2x^2 + x - 5 = 0$
20. $x^2 + 2x - 5 = 0$
21. $7x^2 = 2x - 5$
22. $x^2 - 6x + 10 = 0$
23. $x^2 - x + 1 = 0$
24. $2x^2 - 6x + 5 = 0$
25. $2x^2 + 10x + 15 = 0$
26. $3x^2 - 8x + 7 = 0$
27. $6x^2 + ax - 2a^2 = 0$
28. $x^2 - kx - 6 = 0$
29. $x^2 + 2 = 2\sqrt{3}x$
30. $9x^2 - 6kx = 4k + 4$
31. $mx^2 - m^2x = x - m$
32. $2x^2 + ax + 2bx = -ab$
33. $ax^2 - bx + c = x^2$
*34. $25x^2 - 10px = 6 - 2p - 25x$
*35. $x^2 + ax + bx = (a + b)^2$
*36. $2a^2x^2 + 5b^2 = 6abx$
*37. $a(x^2 + bc) = x(b + a^2c)$
*38. $2x^2 - (a + 2b)x + ab = 0$
*39. $a^2x^2 - x^2 - 2a^2x + a^2 = 0$
40. Solve $s = s_0 + v_0t - 16t^2$ for t.
41. The sum S of the first n counting numbers $1, 2, 3, \ldots, n$ is given by the formula

$$S = \frac{1}{2}n(n + 1).$$ Find n for $S = 105$.

42. The number of straight lines l determined by n points, no three of which lie on the same line, is given by $l = \dfrac{n(n - 1)}{2}$. How many such points determine 21 lines?

*43. Solve for y in terms of x: $2x^2 + 5xy + y^2 + 4x - y = 7$.

*44. Solve for x in terms of y: $3x^2 + 4xy + y^2 - 7x - 5y + 4 = 0$.

*45. Solve for y in terms of x: $x^2 + 3xy + y^2 + 2x - y = 4$.

8.5 THE NATURE OF THE ROOTS OF A QUADRATIC EQUATION

The Quadratic Discriminant: $b^2 - 4ac$

Quite often it is helpful to obtain certain information about the roots of a quadratic equation without completely solving the equation. For instance, are the roots real or imaginary, equal or unequal, rational or irrational?

The quadratic formula states that the roots of the equation

$$ax^2 + bx + c = 0$$

are

$$r_1 = \frac{-b + \sqrt{b^2 - 4ac}}{2a} \quad \text{and} \quad r_2 = \frac{-b - \sqrt{b^2 - 4ac}}{2a}$$

The quantity $b^2 - 4ac$, which appears under the radical sign, is called the discriminant of the equation. If a, b, and c are real numbers, the nature of the roots can be determined by the value of the discriminant. Examining the roots r_1 and r_2, we find that

1. If $b^2 - 4ac > 0$, then $\sqrt{b^2 - 4ac}$ will be positive and r_1 and r_2 will be real and unequal.
2. If $b^2 - 4ac = 0$, then $\sqrt{b^2 - 4ac} = \sqrt{0} = 0$ and r_1 and r_2 will be real and equal, $r_1 = r_2 = \frac{-b}{2a}$.
3. If $b^2 - 4ac < 0$, then $\sqrt{b^2 - 4ac}$ will be imaginary (square root of a negative number) and r_1 and r_2 will be imaginary and unequal, that is, a complex conjugate pair.

The following table summarizes our results.

The Nature of Quadratic Roots

Given the quadratic equation $ax^2 + bx + c = 0$, where a, b, and c are rational, and $a \neq 0$,

1. (a) If $b^2 - 4ac > 0$ and is *not* a perfect square, the roots are real, unequal, and irrational conjugates.
   (b) If $b^2 - 4ac > 0$ and is a perfect square, the roots are real, unequal, and rational.
2. If $b^2 - 4ac = 0$, the roots are real, equal, and rational.
3. If $b^2 - 4ac < 0$, the roots are two complex conjugates.

EXAMPLE 1   Determine the nature of the roots of $3x^2 + 2x - 6 = 0$.

Solution   Using $a = 3$, $b = 2$, and $c = -6$, we evaluate the discriminant $b^2 - 4ac$:

$$b^2 - 4ac = (2)^2 - 4(3)(-6) = 4 + 72 = 76 > 0$$

Since the discriminant is greater than zero, the roots are real and unequal. Also, because $b^2 - 4ac = 76$ is not a perfect square, the roots are irrational numbers. ■

EXAMPLE 2   Determine the nature of the roots of $x^2 + 22x + 120 = 0$.

Solution   Substituting $a = 1$, $b = 22$, and $c = 120$ into $b^2 - 4ac$, we obtain

$$b^2 - 4ac = (22)^2 - 4(1)(120) = 484 - 480 = 4 > 0$$

The discriminant is positive and a perfect square; thus, the roots are real, unequal, and rational.   ■

EXAMPLE 3   Determine the nature of the roots of $x^2 - 14x + 49 = 0$.

Solution   Setting $a = 1$, $b = -14$, and $c = 49$, we evaluate the discriminant

$$b^2 - 4ac = (-14)^2 - 4(1)(49) = 196 - 196 = 0$$

The discriminant is 0 and the roots are real and equal; that is, a double real root.   ■

EXAMPLE 4   Determine the nature of the roots of $x^2 - 12x + 40 = 0$.

Solution   Since $a = 1$, $b = -12$, and $c = 40$, the discriminant is

$$b^2 - 4ac = (-12)^2 - 4(1)(40) = 144 - 160 = -16 < 0$$

Since the discriminant is negative, the roots are complex conjugates.   ■

The discriminant can be used to find other information about equations.

EXAMPLE 5   Determine the values of k for which the equation

$$5x^2 - 4x + 2 + k(4x^2 - 2x - 1) = 0$$

will have equal roots.

Solution   Putting the equation in the standard form:

$$(5 + 4k)x^2 + (-4 - 2k)x + (2 - k) = 0$$

and

$$a = 5 + 4k \qquad b = -4 - 2k \qquad c = 2 - k$$

If the roots are to be equal, the discriminant $b^2 - 4ac$ must be zero. Thus,

$$(-4 - 2k)^2 - 4(5 + 4k)(2 - k) = 0$$
$$16 + 16k + 4k^2 - 4(10 + 3k - 4k^2) = 0$$

$$20k^2 + 4k - 24 = 0$$
$$5k^2 + k - 6 = 0$$
$$(5k + 6)(k - 1) = 0$$

Thus,

$$5k + 6 = 0 \quad \text{or} \quad k - 1 = 0$$
$$k = -\frac{6}{5} \quad \text{or} \quad k = 1$$

Hence, for $k = -\frac{6}{5}$ or $k = 1$, the equation will have equal roots. ■

The Sum and Product of the Roots

The roots of the quadratic equation $ax^2 + bx + c = 0$ may be written as

$$r_1 = \frac{-b}{2a} + \frac{\sqrt{b^2 - 4ac}}{2a} \quad \text{and} \quad r_2 = \frac{-b}{2a} - \frac{\sqrt{b^2 - 4ac}}{2a}$$

If we add their values, the terms involving the radicals cancel, and we obtain

$$r_1 + r_2 = -\frac{b}{a}$$

If we multiply their values, we obtain

$$r_1 r_2 = \frac{b^2 - (b^2 - 4ac)}{4a^2} = \frac{4ac}{4a^2} = \frac{c}{a}$$

Therefore, we have the following theorem.

Theorem 1

If r_1 and r_2 are the roots of $ax^2 + bx + c = 0$, then

$$r_1 + r_2 = -\frac{b}{a} \quad \text{and} \quad r_1 r_2 = \frac{c}{a}$$

This theorem is useful in checking the roots of a quadratic equation.

EXAMPLE 6 Given the equation $6x^2 + (2k - 8)x - 3k = 0$, determine the value of k so that:

(a) The product of the roots will be 10.

(b) One root will be the negative of the other.

(c) One root will be 3.

Solution   For this equation, we have $a = 6$, $b = (2k - 8)$, and $c = -3k$. Then

(a) The product $= \dfrac{c}{a} = \dfrac{-3k}{6} = -\dfrac{k}{2}$. If the product is to be 10, we have

$-\dfrac{k}{2} = 10$, or $k = -20$.

(b) If one root is the negative of the other, then their sum is zero. Thus,

$$\text{sum} = -\frac{b}{a} = \frac{-(2k - 8)}{6} = 0 \quad \text{or} \quad \frac{-k - 4}{3} = 0 \quad \text{or} \quad k = -4$$

(c) If one root is 3, then $x = 3$ must satisfy the equation:

$$6(3)^2 + (2k - 8)(3) - 3k = 0$$
$$54 + 6k - 24 - 3k = 0$$
$$3k = -30$$
$$k = -10 \quad \blacksquare$$

Finding a Quadratic Equation from Known Roots

The formulas for the sum and product of roots of a quadratic equation are also useful in deriving the factored form of any quadratic expression $ax^2 + bx + c$. If r_1 and r_2 are the roots of the quadratic equation $ax^2 + bx + c = 0$, we may write

$$ax^2 + bx + c = a\left(x^2 + \frac{b}{a}x + \frac{c}{a}\right)$$
$$= a[x^2 - (r_1 + r_2)x + r_1 r_2]$$
$$= a(x - r_1)(x - r_2)$$

Theorem 2

If r_1 and r_2 are roots of a quadratic equation, then the equation can be written in the forms

$$a(x - r_1)(x - r_2) = 0, \qquad a \neq 0 \tag{1}$$

or

$$a[x^2 - (r_1 + r_2)x + r_1 r_2] = 0, \qquad a \neq 0 \tag{2}$$

EXAMPLE 7 Find a quadratic equation with integral coefficients having roots 1 and 3.

Solution Using form (2) of Theorem 2, we have $r_1 + r_2 = 1 + 3 = 4$ and $r_1 r_2 = 3$, and one such quadratic equation is

$$a(x^2 - 4x + 3) = 0 \qquad a \neq 0 \quad \blacksquare$$

EXAMPLE 8 Find a quadratic equation having roots $2 \pm \sqrt{3}$.

Solution Using form (1) of Theorem 2, we obtain

$$a[x - (2 + \sqrt{3})] \cdot [x - (2 - \sqrt{3})] = 0$$
$$a(x^2 - 4x + 1) = 0 \quad \blacksquare$$

EXAMPLE 9 Find a quadratic equation having roots $1 \pm i$.

Solution Using form (2) of Theorem 2, we have $r_1 + r_2 = (1 + i) + (1 - i) = 2$ and $r_1 r_2 = (1 + i)(1 - i) = 1^2 + 1^2 = 2$. Thus, $x^2 - 2x + 2 = 0$ is such an equation. $\blacksquare$

EXERCISES 8.5

For each of the equations in Exercises 1–15, determine the nature of the roots.

1. $x^2 - 7x + 8 = 0$
2. $x^2 + 8x + 17 = 0$
3. $9x^2 - 12x + 4 = 0$
4. $x^2 - 5x + 6 = 0$
5. $3x^2 + 2x = -4$
6. $3x^2 = 15$
7. $2x^2 + 3x = 0$
8. $x^2 + x + 1 = 0$
9. $10x + 1 = -25x^2$
10. $2x^2 + 9x - 4 = 0$
11. $16x^2 + 25 = 40x$
12. $x^2 + 23 = 17x$
13. $4x^2 - 5x - 9 = 0$
14. $2x^2 + 4x + 6 = 0$
15. $x^2 - 6x + 9 = 0$

In Exercises 16–24, determine the value or values of the constant k for which the equations will have equal roots.

16. $k(x^2 + x + 1) = x + 1$
17. $x^2 + 2x + 1 = k(2x + 3)$
18. $x^2 + 9 = k(x + 4)$
19. $7x^2 - 1 = 2kx(4x - 1)$

20. $(kx + 1)(4 - 3x) = \dfrac{9}{2}$
21. $4x^2 - 12x + k = 0$
*22. $(k + 1)x^2 - 20x + 9k = 2$
*23. $(5k + 1)x^2 - 60x + 3k = -4$
*24. $2akx^2 - 2bkx = -b^2$
25. Prove that if a, b, and c are real numbers and $x^2 + 2ax + a^2 - b^2 - c^2 = 0$, then the roots are real.
*26. Assume that a, b, and c are real numbers. Prove that zero is a root of $ax^2 + bx + c = 0$, if and only if $c = 0$.
*27. Use the results of Exercise 26 to determine the value of k such that $4x^2 - 7kx + k + 4 = 0$ has zero as a root.
*28. Assume a, b, and c are real numbers. Prove that zero is the *only* root of $ax^2 + bx + c = 0$, if and only if $b = 0$ and $c = 0$.
*29. Use the results of Exercise 28 to determine the values of h and k such that $5x^2 + 7hx - 14x + 3k - 12 = 0$ has only the zero root.
*30. Prove that if a, b, and c are real numbers and a and c are of opposite sign, then the roots of $ax^2 + bx + c = 0$ are real and unequal.

In Exercises 31–45, without solving the equations, find the sum and the product of their roots.

31. $3x^2 - 15x + 4 = 0$

32. $x^2 + 4x - 9 = 0$

33. $2x^2 - 3x + 7 = 0$

34. $7 - 5x = 3x^2$ **35.** $x^2 + 3x = 4$

36. $2x + 3 = 6x^2$ **37.** $49 - 16x^2 = 0$

38. $4x^2 - 4x - 1 = 0$

39. $3x^2 - 7x - 20 = 0$

40. $x^2 + 4x = -2$ **41.** $x^2 = 5x + 3$

42. $2x^2 = 7x$ **43.** $3x^2 + x - 4 = 0$

44. $4x^2 - 3x = -9$

45. $8x^2 - 7 = 0$

In Exercises 46–65, form a quadratic equation, with integral coefficients, having the given numbers as roots.

46. $-2, 4$

47. $\dfrac{1}{3}, \dfrac{3}{4}$

48. $\dfrac{4}{5}, -\dfrac{1}{2}$

49. $-\dfrac{3}{8}, -\dfrac{1}{3}$

50. $6, \dfrac{2}{3}$

51. $-7, 0$

52. $-3, 4$

53. $2, -3$

54. $-2, \dfrac{5}{2}$

55. $\pm\sqrt{5}$

56. $2\pm\sqrt{3}$

57. $\pm 3i$

58. $\pm\dfrac{3}{2}i$

59. $1\pm i$

60. $4\pm 3i$

61. $\dfrac{-3\pm\sqrt{5}}{2}$

62. $\dfrac{-2 \pm 4i\sqrt{3}}{5}$

63. $\dfrac{6 \pm 3i\sqrt{7}}{2}$

64. $-\dfrac{a}{b}, \dfrac{b}{a}$

65. $\dfrac{\sqrt{a}}{\sqrt{a} \pm \sqrt{a - b}}$

In Exercises 66–75, determine k so that the stated condition is satisfied:

66. $5x^2 + kx - 4 = 0$ has roots whose sum equals 4.

67. $x^2 + 5x - 3k = 0$ has roots whose product equals -12.

68. $4x^2 - 3x - 2k = 0$ has roots whose product equals 8.

69. $4x^2 - 13x = k$ has roots whose product equals $\dfrac{3}{4}$.

70. $x^2 - kx = 14$ has roots whose sum equals -5.

71. $x^2 - kx - 30 = 0$ has roots whose sum equals 7.

*72. $2x^2 - kx^2 + 4x + 5k = 0$, one root is the reciprocal of the other.

*73. $kx^2 + \sqrt{2}x - 1 = 0$, one root is the reciprocal of the other.

74. $3kx^2 + 5x + 7kx + 9 = 0$, the roots are numerically equal but opposite in sign.

75. $x^2 + 3kx + h^2 = x + 5$, the roots are numerically equal but opposite in sign.

8.6 EQUATIONS IN QUADRATIC FORM

An equation that contains only two powers of the unknown number or expression, the exponent of one term being twice the exponent in the other, or an equation that can be reduced to this form, is an **equation in quadratic form**.

EXAMPLE 1   The following are equations in quadratic form:

(a) $x^4 - x^2 - 12 = 0$ is quadratic in x^2, since $(x^2)^2 = x^4$.

(b) $x^{-1} - 2x^{-1/2} + 1 = 0$ is quadratic in $x^{-1/2}$ since $(x^{-1/2})^2 = x^{-1}$.

(c) $(x + 3)^4 + 3(x + 3)^2 = 4$ is   quadratic   in $(x + 3)^2$ since $[(x + 3)^2]^2 = (x + 3)^4$.

Using a technique called **substitution of variables**, an equation in quadratic form can be transformed into a quadratic equation of a single variable and solved as usual.

EXAMPLE 2 Solve $x^4 - x^2 - 12 = 0$.

Solution The equation $x^4 - x^2 - 12 = 0$ may be written

$$(x^2)^2 - x^2 - 12 = 0$$

Let $u = x^2$ and substituting gives a quadratic equation in the variable u:

$$u^2 - u - 12 = 0$$

Solving for u, we obtain

$$(u - 4)(u + 3) = 0$$
$$u - 4 = 0 \quad \text{or} \quad u + 3 = 0$$
$$u = 4 \quad \text{or} \quad u = -3$$

Since $u = x^2$, we must solve the equations

$$
\begin{array}{c|c}
x^2 = 4 & x^2 = -3 \\
x = \pm\sqrt{4} & x = \pm\sqrt{-3} \\
x = \pm 2 & x = \pm\sqrt{3}\,i
\end{array}
$$

The original equation $x^4 - x^2 - 12 = 0$ has four solutions: 2, -2, $\sqrt{3}\,i$, and $-\sqrt{3}\,i$. ■

EXAMPLE 3 Solve $x^{-1} - 2x^{-1/2} + 1 = 0$.

Solution The equation $x^{-1} - 2x^{-1/2} + 1 = 0$ may be written

$$(x^{-1/2})^2 - 2(x^{-1/2}) + 1 = 0$$

Let $u = x^{-1/2}$; then the equation becomes

$$u^2 - 2u + 1 = 0$$
$$(u - 1)(u - 1) = 0$$
$$u - 1 = 0 \quad \text{or} \quad u - 1 = 0$$
$$u = 1 \text{ is a double root}$$

Since $u = x^{-1/2}$, we have

$$x^{-1/2} = 1$$

Raising both sides to the -2 power, we obtain

$$x = (1)^{-2}$$
$$x = 1 \quad ■$$

EXAMPLE 4   Solve $(x + 3)^4 + 3(x + 3)^2 = 4$.

Solution   The equation $(x + 3)^4 + 3(x + 3)^2 = 4$ may be written

$$[(x + 3)^2]^2 + 3(x + 3)^2 - 4 = 0$$

Letting $u = (x + 3)^2$, we get

$$u^2 + 3u - 4 = 0$$
$$(u + 4)(u - 1) = 0$$
$$u + 4 = 0 \quad \text{or} \quad u - 1 = 0$$
$$u = -4 \quad \text{or} \quad u = 1$$

Since $u = (x + 3)^2$, we solve the equations

$$
\begin{array}{c|c}
(x + 3)^2 = -4 & (x + 3)^2 = 1 \\
x + 3 = \pm 2i & x + 3 = \pm 1 \\
x = -3 \pm 2i & x = -3 \pm 1 \\
& x = -2 \quad \text{or} \quad x = -4
\end{array}
$$

The equation $(x + 3)^4 + 3(x + 3)^2 = 4$ has four solutions: $-3 + 2i$, $-3 - 2i$, -2, and -4.   ■

EXAMPLE 5   Solve $\left(x + \dfrac{1}{x} \right)^2 - 5\left(x + \dfrac{1}{x} \right) - 6 = 0$.

Solution   Let $u = \left(x + \dfrac{1}{x} \right)$. Then the original equation becomes

$$u^2 - 5u - 6 = 0$$
$$(u - 6)(u + 1) = 0$$
$$u = 6 \quad \text{or} \quad u = -1$$

Since $u = \left(x + \dfrac{1}{x} \right)$, we solve the equations

$$
\begin{array}{c|c}
x + \dfrac{1}{x} = 6 & x + \dfrac{1}{x} = -1 \\
x^2 + 1 = 6x & x^2 + 1 = -x \\
x^2 - 6x + 1 = 0 & x^2 + x + 1 = 0
\end{array}
$$

Using quadratic formulas, we get

$$x = \frac{-(-6) \pm \sqrt{(-6)^2 - 4(1)(1)}}{2(1)} \qquad\qquad x = \frac{-1 \pm \sqrt{1 - 4(1)(1)}}{2(1)}$$

$$= 3 \pm 2\sqrt{2} \qquad\qquad = \frac{-1 \pm \sqrt{3}\,i}{2}$$

The four roots of the given equation are $3 + 2\sqrt{2}, 3 - 2\sqrt{2}, \dfrac{-1 + \sqrt{3}\,i}{2}$, and $\dfrac{-1 - \sqrt{3}\,i}{2}$. ◼

EXAMPLE 6 Give a substitution of variable that will yield a quadratic equation.

(a) $\dfrac{5}{x^4} + \dfrac{2}{x^2} - 3 = 0$ (b) $x^{-2/3} + x^{-1/3} = 0$

(c) $\dfrac{5}{(3x - 4)^2} - \dfrac{13}{(3x - 4)} = 6$ (d) $(4x^4 - 4x^3 + x^2) - 2(2x^2 - x) = 3$

Solution (a) $\dfrac{5}{x^4} + \dfrac{2}{x^2} - 3 = 0$

The substitution $u = \dfrac{1}{x^2}$ results in the quadratic equation $5u^2 + 2u - 3 = 0$.

(b) $x^{-2/3} + x^{-1/3} = 0$

The substitution $u = x^{-1/3}$ results in the quadratic equation $u^2 + u = 0$.

(c) $\dfrac{5}{(3x - 4)^2} - \dfrac{13}{(3x - 4)} = 6$

The substitution $u = \dfrac{1}{3x - 4}$ results in the quadratic equation $5u^2 - 13u = 6$.

(d) $(4x^4 - 4x^3 + x^2) - 2(2x^2 - x) = 3$ can be written

$$x^2(4x^2 - 4x + 1) - 2x(2x - 1) = 3$$
$$x^2(2x - 1)^2 - 2x(2x - 1) = 3$$

The substitution $u = x(2x - 1)$ results in the quadratic equation $u^2 - 2u = 3$. ◼

EXAMPLE 7 Solve $12\left(\dfrac{x^2 + 1}{x^2 - 1}\right) + 25\left(\dfrac{x^2 - 1}{x^2 + 1}\right) - 35 = 0$.

Solution Let $u = \dfrac{x^2 + 1}{x^2 - 1}$; then $\dfrac{x^2 - 1}{x^2 + 1} = \dfrac{1}{u}$, and the given equation becomes

$$12u + 25 \cdot \dfrac{1}{u} - 35 = 0$$

or

$$12u^2 + 25 - 35u = 0$$

or

$$12u^2 - 35u + 25 = 0$$

$$(3u - 5)(4u - 5) = 0$$

$$u = \frac{5}{3} \quad \text{or} \quad u = \frac{5}{4}$$

Since $u = \dfrac{x^2 + 1}{x^2 - 1}$, we must solve the equations

$\dfrac{x^2 + 1}{x^2 - 1} = \dfrac{5}{3}$	$\dfrac{x^2 + 1}{x^2 - 1} = \dfrac{5}{4}$
$3(x^2 + 1) = 5(x^2 - 1)$	$4(x^2 + 1) = 5(x^2 - 1)$
$3x^2 + 3 = 5x^2 - 5$	$4x^2 + 4 = 5x^2 - 5$
$8 = 2x^2$	$9 = x^2$
$4 = x^2$	$\pm 3 = x$
$\pm 2 = x$	

The solutions of the original equation are 2, -2, 3, and -3. ■

CAUTION

Quite often radical equations lead to quadratic equations. Therefore, remember that extraneous roots may occur. Thus, the solutions to the derived quadratic equation must be checked in the *original* equation.

EXAMPLE 8 Solve $\sqrt{3x - 5} + \sqrt{x - 9} = 2\sqrt{x - 1}$.

Solution Squaring both sides, we obtain

$$(\sqrt{3x - 5} + \sqrt{x - 9})^2 = (2\sqrt{x - 1})^2$$

$$(\sqrt{3x - 5})^2 + 2\sqrt{3x - 5} \cdot \sqrt{x - 9} + (\sqrt{x - 9})^2 = 4(x - 1)$$

$$3x - 5 + 2\sqrt{(3x - 5)(x - 9)} + x - 9 = 4(x - 1)$$

$$2\sqrt{(3x - 5)(x - 9)} = 10$$

$$\sqrt{(3x - 5)(x - 9)} = 5$$

Now, square both sides again:

$$(3x - 5)(x - 9) = 25$$

$$3x^2 - 32x + 45 = 25$$

$$3x^2 - 32x + 20 = 0$$

$$(3x - 2)(x - 10) = 0$$

$$x = \frac{2}{3} \quad \text{or} \quad x = 10 \quad \blacksquare$$

Check $x = \frac{2}{3}$:

$$\sqrt{3\left(\frac{2}{3}\right) - 5} + \sqrt{\frac{2}{3} - 9} \overset{?}{=} 2\sqrt{\frac{2}{3} - 1}$$

$$\sqrt{-3} + \sqrt{-\frac{25}{3}} \overset{?}{=} 2\sqrt{-\frac{1}{3}}$$

$$\sqrt{3}\,i + \frac{5\sqrt{3}\,i}{3} \neq \frac{2\sqrt{3}\,i}{3}$$

Thus, $x = \frac{2}{3}$ is *not* a solution.

$x = 10$:

$$\sqrt{3(10) - 5} + \sqrt{10 - 9} \overset{?}{=} 2\sqrt{10 - 1}$$

$$\sqrt{25} + \sqrt{1} \overset{?}{=} 2\sqrt{9}$$

$$5 + 1 \overset{?}{=} 2 \cdot 3$$

$$6 \overset{\checkmark}{=} 6$$

Thus, $x = 10$ is a solution.

EXERCISES 8.6

Solve the following equations:

1. $x^4 - 7x^2 + 12 = 0$
2. $x^4 - 2x^2 = 3$
3. $x^4 + 7x^2 + 10 = 0$
4. $2x^4 - 5x^2 + 2 = 0$
5. $2x^4 - x^2 - 3 = 0$
6. $x^4 - 4x^2 + 3 = 0$
7. $x^4 - 13x^2 + 36 = 0$
8. $x^4 - 10x^2 + 9 = 0$
9. $x^4 - 8x^2 - 9 = 0$
10. $2x^4 + 5x^2 - 12 = 0$
11. $x^4 - 81 = 0$
12. $x^4 - 14x^2 + 45 = 0$
13. $x^6 - 9x^3 + 8 = 0$
14. $x^6 - 10x^3 + 16 = 0$
15. $8x^6 - 9x^3 + 1 = 0$
16. $8x^6 + 7x^3 - 1 = 0$
17. $x^6 + 9x^3 + 8 = 0$
18. $x^6 - 9x^3 + 8 = 0$
19. $x^8 - 17x^4 + 16 = 0$
20. $x - 3x^{1/2} + 2 = 0$
21. $x^{2/3} + 2x^{1/3} = 8$
22. $7x^{1/4} + 12 = x^{1/2}$

23. $x^{2/3} + 9x^{1/3} + 18 = 0$
24. $x^{2/5} + x^{1/5} = 2$
25. $x^{1/5} + x^{2/5} - 6 = 0$
26. $x^{-2} - x^{-1} - 20 = 0$
27. $9x^{-4} - 37x^{-2} + 4 = 0$
28. $4x^{-4} - 5x^{-2} + 1 = 0$
29. $12x^{-1} - 7x^{-1/2} + 1 = 0$
30. $3x^{-2/3} - 5x^{-1/3} = 2$
31. $(x + 1)^2 + 4(x + 1) - 45 = 0$
32. $(x + 5)^2 - 22 = -9(x + 5)$
33. $(x - 3)^2 + 6(x - 3) + 8 = 0$
34. $(x + 2)^2 + 7(x + 2) + 10 = 0$
35. $(x + 3)^2 + 20 = 9(x + 3)$
36. $(x + 2)^2 - 12(x + 2) = 13$
37. $(x^2 + x)^2 - 5(x^2 + x) = -6$
38. $(x^2 + 2x)^2 - 2(x^2 + 2x) = 3$
39. $(2x^2 - x)^2 - 4(2x^2 - x) = -3$
40. $(2x^2 + 3x)^2 = 8(2x^2 + 3x) + 9$
41. $(3x^2 - 4x)^2 = 3(3x^2 - 4x) + 4$
42. $3(x^2 + 3x)^2 = 2(x^2 + 3x) + 5$
43. $\left(2x - \dfrac{5}{x}\right)^2 - 6\left(2x - \dfrac{5}{x}\right) = 27$

44. $\left(x - \dfrac{6}{x}\right)^2 + \left(x - \dfrac{6}{x}\right) = 2$

45. $2\left(2x - \dfrac{1}{x}\right)^2 - 5\left(2x - \dfrac{1}{x}\right) = 7$

46. $\left(2x + \dfrac{1}{x}\right)^2 - 4\left(2x + \dfrac{1}{x}\right) + 3 = 0$

47. $\dfrac{x + 1}{2x^2} + \dfrac{36x^2}{x + 1} = 9$

48. $\dfrac{x^2 + 6}{x + 3} - 12\left(\dfrac{x + 3}{x^2 + 6}\right) + 4 = 0$

*49. $x - \sqrt{x} = 8 - \dfrac{12}{x - \sqrt{x}}$

*50. $\dfrac{a + \sqrt{a^2 + x^2}}{x} + \dfrac{x}{a + \sqrt{x^2 + a^2}} = 2\sqrt{2}$

51. $x - 2\sqrt{2x - 9} = 3$

52. $x + 5 = 3\sqrt{x + 5}$

53. $2x - 5\sqrt{x - 1} = 5$

54. $6x = 4\sqrt{6x - 4} + 1$

55. $2\sqrt{25 - x^2} - x = 5$

56. $\sqrt{32x - 47} - 1 = 2\sqrt{6x - 9}$

57. $\sqrt{27x + 55} = 3 + \sqrt{9x + 22}$

58. $\sqrt{7x - 6} - \sqrt{2x - 11} = 5$

59. $3\sqrt{9 - 2x} - 2 = \sqrt{57 - 14x}$

60. $\sqrt{96x - 47} = 4\sqrt{4x - 2} + 1$

8.7 QUADRATIC INEQUALITIES

A quadratic inequality is an inequality that may be expressed equivalently as

$$ax^2 + bx + c > 0 \qquad ax^2 + bx + c < 0$$
$$\text{or}$$
$$ax^2 + bx + c \geq 0 \qquad ax^2 + bx + c \leq 0$$

where x represents the variable, and a, b, and c represent real numbers, $a \neq 0$. Unfortunately, the quadratic formula cannot be used to solve quadratic inequalities. The procedure we present here for solving a quadratic inequality is based on the facts that:

$$\text{(positive number)} \cdot \text{(positive number)} = \text{positive number}$$

$$\text{(negative number)} \cdot \text{(negative number)} = \text{positive number}$$

and

$$\text{(positive number)} \cdot \text{(negative number)} = \text{negative number.}$$

If we let A and B represent two factors, we can restate the above facts.

> If A and B are two real factors and
>
> $AB > 0$, then $\begin{cases} (1)\ A > 0 \text{ and } B > 0 \\ \text{or} \\ (2)\ A < 0 \text{ and } B < 0 \end{cases}$
>
> $AB < 0$, then $\begin{cases} (1)\ A > 0 \text{ and } B < 0 \\ \text{or} \\ (2)\ A < 0 \text{ and } B > 0 \end{cases}$

EXAMPLE 1 Solve $x^2 - 3x > 4$.

Solution First, we express the inequality in terms of an equivalent inequality having zero on the right side.

$$x^2 - 3x > 4$$
$$x^2 - 3x - 4 > 0$$

Factor the left side, obtaining

$$(x + 1)(x - 4) > 0$$

If the product $(x + 1)(x - 4)$ is to be greater than zero, then

both factors are positive	*or*	*both factors are negative*
$x + 1 > 0$ and $x - 4 > 0$		$x + 1 < 0$ and $x - 4 < 0$
$x > -1$ and $\qquad x > 4$		$x < -1$ and $\qquad x < 4$
The only numbers which satisfy *both* conditions satisfy $x > 4$, whose solution set is the intersection of their solution sets.		The only numbers which satisfy *both* conditions satisfy $x < -1$, whose solution set is the intersection of their solution sets.

The complete solution set may be graphically displayed on the real number line.

Thus, the solution set for $x^2 - 3x > 4$ is $\{x \,|\, x < -1 \text{ or } x > 4\}$. ■

Note. The numbers -1 and 4 are the roots of the quadratic equation $x^2 - 3x - 4 = 0$ and these points divide the real number line into three parts. With this in mind, the following is a procedure for solving quadratic inequalities.

How to Solve Quadratic Inequalities

1. Express the given inequality (if necessary) in terms of an equivalent inequality having zero on the right side.
2. Transform the inequality in step 1 into an equation by replacing the inequality sign with an $=$ sign.
3. Solve the resulting quadratic equation.
4. Mark the solutions obtained in step 3 on a real number line, thus dividing the number line into three parts, each corresponding to a set of numbers. If the equation has only one double root, then there will be only two sets of numbers.
5. Test each set by choosing a number in the set and substituting it for the variable in the given inequality. If the inequality is true for the selected number, then all the numbers in that set satisfy the inequality.

EXAMPLE 2   Solve $x^2 + x \le 6$.

Solution
$$x^2 + x \le 6$$

1. $\qquad x^2 + x - 6 \le 0$

2. $\qquad x^2 + x - 6 = 0$

3. $\qquad (x + 3)(x - 2) = 0$

$$x = -3 \quad \text{or} \quad x = 2$$

4.

$\longleftarrow \{x \mid x < -3\} \longrightarrow \longleftarrow \{x \mid -3 < x < 2\} \longrightarrow \longleftarrow \{x \mid x > 2\} \longrightarrow$

5.

Set	Chosen Number	$x^2 + x \le 6$	
$\{x \mid x < -3\}$	-4	$(-4)^2 + (-4) \le 6$	False
$x = -3$	-3	$(-3)^2 + (-3) \le 6$	True
$\{x \mid -3 < x < 2\}$	0	$(0)^2 + (0) \le 6$	True
$x = 2$	2	$(2)^2 + (2) \le 6$	True
$\{x \mid x > 2\}$	3	$(3)^2 + (3) \le 6$	False

Thus, the solution set for $x^2 + x \le 6$ is $\{x \mid -3 \le x \le 2\}$.   ◼

EXAMPLE 3   Solve $x^2 - 4x - 13 > 0$.

Solution
$$x^2 - 4x - 13 > 0$$

2. $\qquad x^2 - 4x - 13 = 0$

3. Using the quadratic formula, we obtain

$$x = 2 - \sqrt{17} \quad \text{or} \quad x = 2 + \sqrt{17}$$

4.

$\longleftarrow \{x \mid x < 2 - \sqrt{17}\} \longrightarrow \longleftarrow \{x \mid 2 - \sqrt{17} < x < 2 + \sqrt{17}\} \longrightarrow \longleftarrow \{x \mid x > 2 + \sqrt{17}\} \longrightarrow$

5.

Set	Chosen Number	$x^2 - 4x - 13 > 0$	
$\{x \mid x < 2 - \sqrt{17}\}$	-4	$(-4)^2 - 4(-4) - 13 > 0$	True
$\{x \mid 2 - \sqrt{17} < x < 2 + \sqrt{17}\}$	0	$(0)^2 - 4(0) - 13 > 0$	False
$\{x \mid x > 2 + \sqrt{17}\}$	8	$(8)^2 - 4(8) - 13 > 0$	True

Thus, the solution set for $x^2 - 4x - 13 > 0$ is $\{x \mid x < 2 - \sqrt{17}$ or $x > 2 + \sqrt{17}\}$.   ◼

The method given above (with slight modifications) can be used to solve certain inequalities even though they are not quadratic inequalities. Although

$$\frac{ax + b}{cx + d} > 0$$

is not a quadratic inequality, its solution is the same as the quadratic inequality $(ax + b)(cx + d) > 0$ because both expressions are positive when $(ax + b)$ and $(cx + d)$ have the same sign.

Similarly,

$$\frac{ax + b}{cx + d} < 0$$

has the same solution as the quadratic inequality $(ax + b)(cx + d) < 0$ since both expressions are negative when $(ax + b)$ and $(cx + d)$ have opposite signs.

EXAMPLE 4 Solve $\dfrac{3x + 2}{2x - 1} \geq 0$.

Solution Solving the corresponding quadratic inequality $(3x + 2)(2x - 1) \geq 0$, we get

$\left\{x \mid x \leq -\dfrac{2}{3} \text{ or } x \geq \dfrac{1}{2}\right\}$. However, for $x = \dfrac{1}{2}$, the denominator $(2x - 1)$ in the given inequality is equal to zero, and division by zero is undefined. Thus, we must modify the set by eliminating the value $\dfrac{1}{2}$. Hence, the solution to the original inequality $\dfrac{3x + 2}{2x - 1} \geq 0$ is the set $\left\{x \mid x \leq -\dfrac{2}{3} \text{ or } x > \dfrac{1}{2}\right\}$. ∎

EXAMPLE 5 Solve the inequality $(x + 3)(x - 2)(x - 4) > 0$.

Solution Even though this is a cubic (third-degree) inequality, we use the same procedure.

1. $$(x + 3)(x - 2)(x - 4) > 0$$

2. $$(x + 3)(x - 2)(x - 4) = 0$$

3. $$x = -3 \quad \text{or} \quad x = 2 \quad \text{or} \quad x = 4$$

4.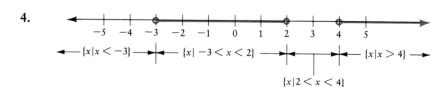

5.

Set	Chosen Number	$(x + 3)(x - 2)(x - 4) > 0$	
$\{x \mid x < -3\}$	-4	$(-1)(-6)(-8) > 0$	False
$\{x \mid -3 < x < 2\}$	0	$(3)(-2)(-4) > 0$	True
$\{x \mid 2 < x < 4\}$	3	$(6)(1)(-1) > 0$	False
$\{x \mid x > 4\}$	5	$(8)(3)(1) > 0$	True

Thus, the solution set for $(x + 3)(x - 2)(x - 4) > 0$ is the set $\{x \mid -3 < x < 2$ or $x > 4\}$. ∎

EXERCISES 8.7

In Exercises 1–30, solve the quadratic inequalities.

1. $x^2 - 49 < 0$
2. $x^2 + 16 > 0$
3. $x^2 - 6x + 8 > 0$
4. $x^2 + 7x + 6 < 0$
5. $4x^2 + 12x + 9 < 0$
6. $x^2 - 2x + 4 > 0$
7. $3x^2 + 4 > -8x$
8. $7x - 6x^2 > -3$
9. $4x^2 + 3x < 0$
10. $4 + 2x - x^2 < 0$
11. $x^2 - 5x + 4 > 0$
12. $x^2 + 4x < 0$
13. $2x^2 - 7x - 15 \geq 0$
14. $12x^2 - 17x - 105 < 0$
15. $x^2 - 4x > 0$
16. $x^2 + 6x + 9 \leq 0$
17. $12x^2 - 4x + 3 < 0$
18. $3x^2 + 2x + 1 > 0$
19. $x^2 + 13x + 36 \leq 0$
20. $-2 + x - 3x^2 > 0$
21. $-5 + 4x - 3x^2 < 0$
22. $x^2 - 3x > 10$
23. $4x^2 + 9 > 12x$
24. $4x - x^2 < 5$
25. $(x - 5)x + 4x > 2$
26. $(x + 5)x \leq 2(x^2 + 2)$

27. $(x + 4)(x + 5) - 5 \geq 5$
28. $\dfrac{1}{3}x^2 - 3x + 6 < 0$
29. $2(x + 2)^2 - \dfrac{7}{2} \geq 2x$
30. $2x > \dfrac{x^2}{2} - 5x + \dfrac{11}{2}$

In Exercises 31–40, solve the inequalities.

31. $\dfrac{x - 5}{x - 2} > 0$
32. $\dfrac{2x + 3}{5x + 4} < 0$
33. $\dfrac{2x + 3}{2x - 1} < 0$
34. $\dfrac{x - 1}{x + 1} > 0$
35. $\dfrac{2x + 2}{x + 1} \geq 0$
36. $\dfrac{5x - 2}{3x - 2} \leq 0$
37. $(x + 2)(x - 1)(2x - 3) > 0$
38. $(3x - 2)(x - 1)(x + 2) > 0$
39. $x^3 + 10x^2 + 24x < 0$
40. $(x^2 + 2x - 3)(x^2 - 25) < 0$

Chapter 8 SUMMARY

TERMS, RULES, AND FORMULAS

1. *Square Root Method.* A quadratic equation of the form $ax^2 + c = 0$ has the solution

$$x = \pm\sqrt{-\dfrac{c}{a}}$$

2. *The Zero-Product Property of Real Numbers.* If a and b are real numbers and $ab = 0$, then $a = 0$, or $b = 0$, or both $a = 0$ and $b = 0$.

3. *Factoring Quadratic Equations.* If a quadratic equation can be factored into the product of linear factors

$$(px + q)(rx + s) = 0$$

then

$$x = -\dfrac{q}{p} \quad \text{or} \quad x = -\dfrac{s}{r}$$

4. *The Quadratic Formula.* The solution of

$ax^2 + bx + c = 0$, $a \neq 0$, is given by

$$x = \dfrac{-b \pm \sqrt{b^2 - 4ac}}{2a}$$

5. *The Nature of the Roots of a Quadratic Equation.* If a, b, c are rational and $a \neq 0$, then
   (a) If $b^2 - 4ac > 0$ and a perfect square, the roots are real, unequal, and rational.
   (b) If $b^2 - 4ac > 0$ and not a perfect square, the roots are real, unequal, and irrational conjugates.
   (c) If $b^2 - 4ac = 0$, there is one real, rational root (double root).
   (d) If $b^2 - 4ac < 0$, the roots are complex conjugates.

6. *The Sum and Product of the Roots.* If r_1 and r_2 are the roots of $ax^2 + bx + c = 0$, then

$$r_1 + r_2 = -\dfrac{b}{a} \quad \text{and} \quad r_1 r_2 = \dfrac{c}{a}$$

7. *Finding a Quadratic Equation from Known Roots.* If r_1 and r_2 are roots of a quadratic equation, then the equation can be written in the forms

   (a) $a(x - r_1)(x - r_2) = 0$ $(a \neq 0)$ or

   (b) $a[x^2 - (r_1 + r_2)x + r_1 r_2] = 0$ $(a \neq 0)$

8. If A and B are two factors and $AB > 0$, then

   (a) $A > 0$ and $B > 0$ or

   (b) $A < 0$ and $B < 0$

   If A and B are two factors and $AB < 0$, then

   (c) $A > 0$ and $B < 0$ or

   (d) $A < 0$ and $B > 0$

PROCEDURES

1. *Completing the Square*

   (1) Simplify and write the quadratic so that the terms in x^2 and x are on the left side of the equation and the constant term is on the right side of the equation.

   (2) Make the coefficient of x^2 equal to 1 by dividing both sides of the equation by the coefficient of x^2.

   (3) Add to each side of the equation ($\frac{1}{2}$ coefficient of x)^2 and simplify.

   (4) Take the square root of each side, prefixing $\pm$ to the right side.

   (5) Solve for x.

2. *Solving Quadratic Inequalities*

   (1) Express the given inequality (if necessary) in terms of an equivalent inequality having zero on the right side.

   (2) Transform the inequality in step 1 into an equation by replacing the inequality sign with an $=$ sign.

   (3) Solve the resulting quadratic equation.

   (4) Mark the solutions obtained in step 3 on a real number line, thus dividing the number line into three parts, each corresponding to a set of numbers. If the equation has only one solution, then there will be only two sets of numbers.

   (5) Test each set by choosing a number in the set and substituting it for the variable in the given inequality. If the inequality is true for the selected number, then all the numbers in that set satisfy the inequality.

Chapter 8   EXERCISES

In Exercises 1–12, solve using the square root method.

1. $2x^2 - 1 = 5$ 2. $5x^2 + 1 = 11$

3. $16 - 3x^2 = 1$ 4. $4x^2 - 5 = 7$

5. $20 - 3x^2 = 2$ 6. $2x^2 + 5 = 21$

7. $x^2 + a = b$ 8. $x^2 - a = b$

9. $\dfrac{a}{b}x^2 + c = d$ 10. $ax^2 + b = c$

11. $cx^2 - a = b$ 12. $\dfrac{a}{b} - \dfrac{1}{c}x^2 = \dfrac{b}{c}$

In Exercises 13–24, solve by factoring.

13. $x^2 - 14x + 45 = 0$

14. $x^2 - 9x = 52$

15. $x^2 - 54 = -3x$

16. $x^2 + 4ax - 21a^2 = 0$

17. $x^2 + x = bx + b$

18. $px^2 - p^2 x = x - p$

19. $2x^2 + x - 3 = 0$

20. $5x + 4 = 6x^2$

21. $7x^2 - 5 = 2x$

22. $x + 2 = 15x^2$

23. $2x^2 - 9x - 161 = 0$

24. $31x + 14 = 10x^2$

In Exercises 25–36, solve by completing the square.

25. $x^2 + 2x - 6 = 0$

26. $x^2 + x + 1 = 0$

27. $x^2 + 8x + 2 = 0$

28. $3x^2 + 8x + 1 = 0$

29. $x^2 - 2x + 2 = 0$

30. $5x^2 + 12x - 4 = 0$

31. $2x^2 - 5x + 4 = 0$

32. $4x^2 - 8x + 5 = 0$

33. $2x^2 + 6x + 5 = 0$

34. $3x^2 - 6x + 2 = 0$

*35. $x^2 - 2x - m = 0$

*36. $ax^2 = 5 - x^2 + 7x$

In Exercises 37–66, solve using the quadratic formula.

37. $2x^2 + 3x - 2 = 0$

38. $3x^2 + 5x - 2 = 0$

39. $5x^2 - 3x = 2$ 40. $18x^2 + 9x = 2$

41. $5x^2 = 4x + 12$ 42. $3x^2 = 7x + 6$

43. $3x^2 = 25x + 50$

44. $2x^2 = 8 + 3x + 12$

45. $4x^2 - 12x + 1 = 0$

46. $x^2 + 2x - 1 = 0$

47. $(x - 2)^2 = 2x + 1$

48. $3x^2 = 6x - 2$ 49. $3x(3x - 1) = 1$

50. $7x^2 + 1 = 6x$

51. $4x(2x - 1) - 1 = 0$

52. $3(x + 1)^2 = -2x$

53. $x(2 - 3x) = 1$

54. $2x(x - 1) + 3 = 0$

55. $2x(x - 3) + 5 = 0$

56. $5x^2 = x - 1$ 57. $\sqrt{2}\,x^2 = x + \sqrt{2}$

58. $x^2 = \sqrt{3}\,x + 1$

59. $3x^2 + (3\sqrt{2})x = 4$

60. $3x^2 - \sqrt{15}\,x = -2$

61. $\sqrt{7}\,x^2 + 5x + \sqrt{7} = 0$

*62. $x^2 = 6ax - a^2$

*63. $x^2 = 2abx - a^2x$

*64. $ax^2 - 2ax + x = 2$

*65. $bx^2 = (b + 2)x - 2$

*66. $x(x + a - b) = ab$

Without solving, determine the nature of the roots in Exercises 67–75.

67. $x^2 - 5x = -6$

68. $3x^2 + 2x = -4$

69. $3x^2 - 15 = 0$

70. $2x^2 = -3x$

71. $x^2 + x + 1 = 0$

72. $10x + 1 = -25x^2$

73. $2x^2 - 4 = -9x$

74. $16x^2 - 40x = -25$

75. $x^2 + 23 = 17x$

In Exercises 76–81, for what value of k does the equation

76. $2x^2 + kx + 2 = 0$ have equal roots?

77. $3x^2 - kx + 1 = 0$ have equal roots?

78. $x^2 + 2kx + 3 = 0$ have unequal real roots?

79. $2x^2 - kx + 5 = 0$ have unequal real roots?

80. $3x^2 + kx - 3 = 0$ have real roots?

81. $5x^2 - kx + 2 = 0$ have no real roots?

Find the sum and product of the roots of each equation in Exercises 82–84.

82. $3x^2 + 5x - 7 = 0$

83. $(3x + 1)^2 = 3x - 9$

84. $x(x - 2) = 3x - 9$

In Exercises 85–90, find an equation whose roots are:

85. $2, 5$ 86. $\dfrac{1}{2}, \dfrac{3}{4}$ 87. $\sqrt{5}, -\sqrt{5}$

88. $\sqrt{3} \pm 2$ 89. $\dfrac{1 \pm 2i}{2}$ 90. $\dfrac{3 \pm i}{3}$

In Exercises 91–100, solve.

91. $x^4 - 11x^2 + 28 = 0$

92. $x^{-4} - 9x^{-2} = 0$

93. $x^{-1} - 2x^{-1/2} + 1 = 0$

94. $x^{-2/3} + 2x^{-1/3} + 1 = 0$

95. $\sqrt{8x + 25} = 2 - x$

96. $\sqrt{8x + 9} = x + 2$

97. $\sqrt{3x + 4} - \sqrt{2x - 4} = 2$

98. $\sqrt{4x + 2} - \sqrt{4x - 2} = 2$

99. $\left(x + \dfrac{1}{x}\right)^2 + 4\left(x + \dfrac{1}{x}\right) = 12$

100. $(x^2 - x) - 8(x^2 - x) + 12 = 0$

In Exercises 101–110, solve the quadratic inequalities.

101. $x^2 - 8x + 7 < 0$

102. $2x^2 - 9x - 5 > 0$

103. $x^2 - 6x + 9 > 0$

104. $4x^2 - 12x + 9 < 0$

105. $6x^2 + 13x - 15 < 0$

106. $2x^2 - 6x + 3 > 0$

107. $2x^2 + x - 2 < 0$

108. $\dfrac{4x + 8}{x + 2} > 0$

109. $\dfrac{3x - 5}{x + 3} \leq 0$

110. $\dfrac{3x - 1}{2x - 5} > 0$

9 Problem Solving Involving Quadratic Equations and Quadratic Inequalities

Many physical, biological, and economical problems can be expressed as a quadratic equation. In this chapter, we shall discuss several types of problems that can be solved using quadratic equations or quadratic inequalities.

9.1 NUMBER PROBLEMS

EXAMPLE 1 Determine two numbers that differ by 4 and whose product is 480.

Solution Let x be the smaller number; then $x + 4$ must be the larger number. Thus,

$$x(x + 4) = 480$$

or

$$x^2 + 4x - 480 = 0$$

Factoring, we obtain

$$(x - 20)(x + 24) = 0$$

so

$$x = 20 \quad \text{or} \quad x = -24$$

If $x = 20$, then $x + 4 = 24$. If $x = -24$, then $x + 4 = -20$. Thus, the two numbers are 20 and 24, or -24 and -20. ■

Check $$(20)(24) = 480 \quad \text{and} \quad (-24)(-20) = 480$$

EXAMPLE 2 Find two numbers whose sum is 20 and such that the sum of their squares is 250.

Solution Let x be one number; then the second number is given by $(20 - x)$. Since the sum of their squares is 250, we have

$$x^2 + (20 - x)^2 = 250$$
$$x^2 + 400 - 40x + x^2 = 250$$
$$2x^2 - 40x + 150 = 0$$

Dividing both sides by 2, we get

$$x^2 - 20x + 75 = 0$$

Factoring, $$(x - 5)(x - 15) = 0$$

Thus, $$x = 5 \quad \text{or} \quad x = 15$$

If $x = 5$, then $20 - x = 15$. If $x = 15$, then $20 - x = 5$. Hence, the two parts are 5 and 15. ■

Check $$5^2 + 15^2 = 25 + 225 = 250$$

EXAMPLE 3 Find the number which, increased by its reciprocal, is equal to $\frac{50}{7}$.

Solution Let $$x = \text{the number}$$

Then $$\frac{1}{x} = \text{the reciprocal}$$

Thus, $$x + \frac{1}{x} = \frac{50}{7}$$

Multiplying both sides by $7x$ yields

$$7x^2 + 7 = 50x$$
$$7x^2 - 50x + 7 = 0$$
$$(7x - 1)(x - 7) = 0$$
$$7x - 1 = 0 \quad \text{or} \quad x - 7 = 0$$
$$x = \frac{1}{7} \quad \text{or} \quad x = 7$$

If $x = \frac{1}{7}$, then $\frac{1}{x} = 7$. If $x = 7$, then $\frac{1}{x} = \frac{1}{7}$. Hence, the number is 7 or $\frac{1}{7}$. ■

Check   If $x = \frac{1}{7}$, then $x + \frac{1}{x} = \frac{1}{7} + \frac{1}{1/7} = \frac{1}{7} + 7 = 7\frac{1}{7} = \frac{50}{7}$.

If $x = 7$, then $x + \frac{1}{x} = 7 + \frac{1}{7} = 7\frac{1}{7} = \frac{50}{7}$.

EXAMPLE 4   A number consists of three digits, the tens digit being 1 more than the hundreds digit and 1 less than the units digit. The number is equal to the product of the units digit and a two-digit number whose tens digit is twice the original hundreds digit and whose units digit is the sum of the original tens and units digits. Find the number.

Solution   Let x be the hundreds digit. Then $(x + 1)$ is the tens digit and $(x + 2)$ is the units digit:

$$\underline{\quad x \quad} \qquad (x + 1) \qquad (x + 2)$$

hundreds   tens   units
digit   digit   digit

Thus, the *value* of the number is

$$100x + 10(x + 1) + (x + 2) = 111x + 12 \qquad \text{(A)}$$

This number is equal to the product of the units digit, $(x + 2)$, and a two-digit number whose tens digit is $2x$ and whose units digit is $(x + 1) + (x + 2)$; that is, $(2x + 3)$. Thus, this *value* is

$$(x + 2)[10(2x) + (2x + 3)] \qquad \text{(B)}$$

Setting value (A) = value (B), we get

$$111x + 12 = (x + 2)[10(2x) + (2x + 3)]$$
$$111x + 12 = (x + 2)(22x + 3)$$
$$111x + 12 = 22x^2 + 47x + 6$$

Transposing, $$22x^2 - 64x - 6 = 0$$
$$11x^2 - 32x - 3 = 0$$
$$(11x + 1)(x - 3) = 0$$
$$11x + 1 = 0 \quad \text{or} \quad x - 3 = 0$$
$$x = -\frac{1}{11} \quad \text{or} \quad x = 3$$

Since $-\frac{1}{11}$ is *not* a digit, we have $x = 3$. Hence, the number is

$$x \quad (x + 1) \quad (x + 2) = 345 \quad \blacksquare$$

EXERCISES 9.1

1. The sum of two numbers is 8 and their product is 15. Find the numbers.

2. If the square of a certain number is subtracted from six times that number, the result is 9.

3. If 10 is added to three times a number, the result equals the square of the number. Find the number.

4. Divide 50 into two parts such that their product is 429.

5. Separate 20 into two parts whose product is 75.

6. The sum of two numbers is 18 and the sum of their squares is 170. Find the numbers.

7. When the square of a number is added to 63, the result is equal to sixteen times the number. Find the number.

8. When a number is increased by 20, the result is equal to nine times the square root of the number. Find the number.

9. Divide 32 into two parts such that the sum of the squares of the parts is 514.

10. The product of two consecutive even integers is 168. Find the numbers.

11. Find two consecutive integers such that the sum of their squares is equal to 613.

12. The sum of a number and its reciprocal is 1. Find the number.

13. The sum of a number and its reciprocal is $\frac{5}{2}$. Find the number.

14. A number minus its reciprocal is $-\frac{5}{6}$. Find the number.

15. Find two numbers that differ by 4, and are such that the sum of their reciprocals is $\frac{5}{24}$.

16. Find two numbers that differ by 3, and are such that the difference of their reciprocals is $\frac{1}{18}$.

17. One number is four times another. If each is increased by 3, the sum of their reciprocals is $\frac{4}{27}$. Find the numbers.

18. The numerator of a fraction is 4 less than the denominator. One divided by the fraction is $\frac{56}{45}$ greater than the fraction. Find the fraction.

19. The denominator of a certain fraction exceeds its numerator by 5. The sum of the fraction and its reciprocal is $\frac{97}{36}$. What is the fraction?

20. Find two consecutive integers such that the sum of their squares minus five times the larger equals 100.

21. Two numbers differ by 24. The square root of the larger number differs from the square root of the smaller number by 2. What are the numbers?

22. A number consists of two digits whose average is 5. If 13 is subtracted from three times the number, the result will be twice the square of the units digit. Find the number.

23. The tens digit of a number consisting of two digits is $1\frac{1}{3}$ times the units digit. If 30 is subtracted from the number, the result is equal to the sum of the digits multiplied by half the tens digit. Find the number.

24. A number of two digits is such that twice the tens digit together with three times the units digit is equal to 20. If the square of the sum of the digits is decreased by the sum of the digits, the result is equal to the original number. Find the number.

*25. A number consists of three digits. It is equal to the sum of the cubes of its digits, decreased by three times the product of its digits, and increased by twice the product of its middle digit and the difference of the squares of its tens and units digits. If the hundreds digit is 1 more, and the tens digit is 3 more than the units digit, find the number.

9.2 GEOMETRIC PROBLEMS

In this section, we discuss several geometric problems that can be solved using quadratic equations. We begin by stating an important theorem from geometry.

Pythagorean Theorem

In any right triangle, the square of the length of the hypotenuse equals the sum of the squares of the lengths of the other two sides.

The hypotenuse of a right triangle is the longest side and is opposite the right angle. Figure 9.1 illustrates the Pythagorean theorem:

$$c^2 = a^2 + b^2$$

where c is the length of the hypotenuse, and a and b are the lengths of the other two sides.

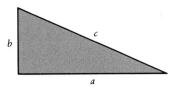

Figure 9.1

EXAMPLE 1   The diagonal of a square is 6 cm longer than a side. Find the length of a side.

Solution   Let x = the number of centimeters in the length of a side

Then $x + 6$ = the number of centimeters in the length of the diagonal

See Figure 9.2. Using the Pythagorean theorem, we have

$$[(x + 6) \text{ cm}]^2 = (x \text{ cm})^2 + (x \text{ cm})^2$$
$$(x + 6)^2 \text{ cm}^2 = x^2 \text{ cm}^2 + x^2 \text{ cm}^2$$
$$(x + 6)^2 \text{ cm}^2 = (2x^2) \text{ cm}^2$$
$$(x + 6)^2 = 2x^2$$
$$x^2 + 12x + 36 = 2x^2$$
$$0 = x^2 - 12x - 36$$

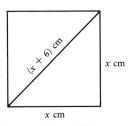

x cm

Figure 9.2

From the quadratic formula, we obtain

$$x = \frac{12 \pm \sqrt{(12)^2 - 4(1)(-36)}}{2}$$

$$x = \frac{12 \pm \sqrt{288}}{2} = \frac{12 \pm 12\sqrt{2}}{2}$$

$$x = 6 \pm 6\sqrt{2}$$

Since we assume distance is nonnegative, we reject $6 - 6\sqrt{2}$. Thus, the side is $(6 + 6\sqrt{2})$ cm. ■

EXAMPLE 2 A wood frame of uniform thickness has outside dimensions of 46 in. and 32 in. Find the thickness of the frame, if it encloses an area of 1040 sq in.

Solution Let $x =$ the thickness, in inches, of the frame. Then $(46 - 2x)$ in. is the inside width, and $(32 - 2x)$ in. is the inside height. See Figure 9.3. Thus,

$$[(46 - 2x) \text{ in.}] \cdot [(32 - 2x) \text{ in.}] = 1040 \text{ in.}^2$$

$$(46 - 2x)(32 - 2x) \text{ in.}^2 = 1040 \text{ in.}^2$$

$$(46 - 2x)(32 - 2x) = 1040$$

$$1472 - 156x + 4x^2 = 1040$$

$$4x^2 - 156x + 432 = 0$$

$$x^2 - 39x + 108 = 0$$

$$(x - 36)(x - 3) = 0$$

$$x = 36 \quad \text{or} \quad x = 3$$

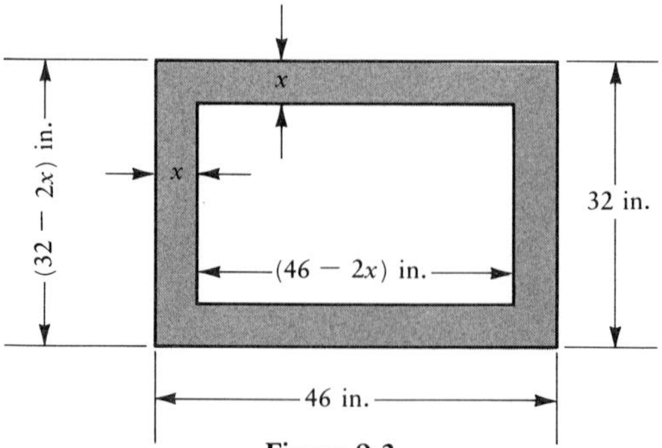

Figure 9.3

If $x = 36$, then $32 - 2x = 32 - 72 = -40$. However, we assume that the height cannot be negative, and we reject this value. Hence, the thickness of the frame is 3 in. ■

EXAMPLE 3 A rectangular sheet of cardboard is three times as long as it is wide. From each corner, a 2-cm square is cut out and the ends are turned up so as to form a box.

If the volume of the box is 512 cu cm, find the dimensions of the sheet of cardboard.

Solution   Let x = the width, in cm, of the cardboard.

Then $3x$ = the length, in cm, of the cardboard.

When the squares are cut out and the ends turned up,

the width of the box is $(x - 4)$ cm

the length of the box is $(3x - 4)$ cm

and the height of the box is 2 cm. See Figure 9.4. Thus,

$$(\text{height}) \cdot (\text{width}) \cdot (\text{length}) = \text{volume of box}$$
$$(2 \text{ cm}) \cdot [(x - 4) \text{ cm}] \cdot [(3x - 4) \text{ cm}] = 512 \text{ cm}^3$$
$$[2(x - 4)(3x - 4)] \text{ cm}^3 = 512 \text{ cm}^3$$
$$2(x - 4)(3x - 4) = 512$$
$$(x - 4)(3x - 4) = 256$$
$$3x^2 - 16x - 240 = 0$$
$$(3x + 20)(x - 12) = 0$$
$$x = -\frac{20}{3} \quad \text{or} \quad x = 12$$

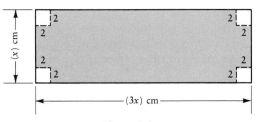

Figure 9.4

Rejecting the negative value, we have

width = 12 cm   length = 36 cm   height = 2 cm ■

EXAMPLE 4   A theorem in geometry states that if a tangent line, $\overline{PT}$, and a secant line, $\overline{PAB}$, are drawn to a circle from an external point, P, then the square of the distance $\overline{PT}$ equals the product of the distances $\overline{PB}$ and $\overline{PA}$. See Figure 9.5. How far must a 5-in. chord be extended in order that a tangent from its end point will be 6 in.?

Solution   Let x = the extension, in inches, of the chord. Then $PA = x$ in., $PB = (x + 5)$ in., and $PT = 6$ in. Using the results of the theorem, we get

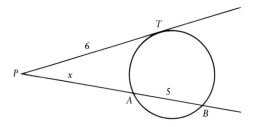

$$\overline{PT}^2 = PB{\cdot}PA$$
Figure 9.5

$$(6\ \text{in.})^2 = (x\ \text{in.})[(x\ +\ 5)\ \text{in.}]$$
$$36\ \text{in.}^2 = [x(x\ +\ 5)]\ \text{in.}^2$$
$$36 = x(x\ +\ 5)$$
$$36 = x^2\ +\ 5x$$
$$0 = x^2\ +\ 5x\ -\ 36$$
$$0 = (x\ +\ 9)(x\ -\ 4)$$
$$x = -9\quad\text{or}\quad x = 4$$

Rejecting the negative value, we have an extension of 4 in. ■

EXERCISES 9.2

1. The altitude of a certain right triangle is 6 ft more than its base. The hypotenuse is 30 ft. What are its base and altitude?
2. A piece of wire 56-cm long is bent into the form of a right triangle whose hypotenuse is 25 cm. Find the other two sides of the triangle.
3. The length of a certain rectangle is 3 in. more than the width. The area is 108 sq in. Find the dimensions.
4. A field is 11 yd longer than it is wide, and has an area of 6 acres. Find its length and width.
5. The area of a certain rectangle is 405 sq ft. The sum of its base and altitude is 42 ft. What are its dimensions?
6. The total area of two certain squares is 765 sq in. The side of one exceeds the side of the other by 3 in. How long is the side of each square?
7. The perimeter of a tennis court is 288 ft and the area is 312 sq yd. Find the dimensions of the court.
8. The length and width of a rectangular lot are 6 ft and 4 ft. Both have to be increased by the

same amount so the area is doubled. What are the dimensions of the new rectangle?
9. A picture that is 9 in. wide by 12 in. long is surrounded by a uniform frame. The area of the frame alone is 162 sq in. Find the width of the frame.
10. A rectangle of sides 13 in. and 9 in. has a strip of uniform width cut off all around it. If this reduces the area by 57 sq in., find the width of the strip.
11. A lawn is 30 ft by 40 ft. How wide a strip must be cut around it when mowing the grass to cut half of the lawn?
12. An open box containing 480 cm³ is made by cutting out a 5-cm square from each corner of a square piece of metal and turning up the sides. Find the dimensions of the original piece of metal.
13. From a square piece of cardboard, 8 in. on a side, a box is made by cutting a square from each corner and folding the remaining cardboard to form a box with an open top. How large must the squares be in order that the box will hold nine times as much as a cubical

box, one face of which is the same as one of the squares cut out?

14. How far from a circle whose radius is 6 cm must a point be taken in order that the whole secant from that point through the center of the circle will be twice the tangent from the point?

15. A theorem in geometry states that if two chords intersect within a circle, then the product of the segments of one equals the product of the segments of the other. If a 14-in. chord divides a second chord into segments of 3 in. and 8 in., into what two segments is the 14-in. chord divided?

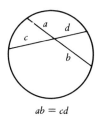

$$ab = cd$$

16. There are two cubes whose edges differ by 3 in. and whose volumes differ by 657 cu in. Find the length of the side of the smaller cube.

17. Given a right circular cylinder of radius r and altitude h, then the *total surface area, T,* is given by

$$T = 2\pi r^2 + 2\pi rh$$

Suppose a right circular cylinder has altitude h and a diameter of the same length. By what length can the altitude be decreased and the diameter increased so that the total surface area will be one-fourth larger?

18. A right circular cylinder has altitude h and a diameter of the same length. By what length can the altitude be decreased and the diameter increased so that the volume will remain the same?

19. The corners of a square the length of whose side is 4 in. are cut off in such a way that a regular octagon remains. What is the length of a side of this octagon?

*20. From a string whose length equals the perimeter of a square, an 8-cm piece is cut off. The remaining piece is equal in length to the perimeter of another square whose area is nine-sixteenths of the first square. How long was the original string?

9.3 MOTION PROBLEMS

EXAMPLE 1 A biker travels 300 mi at a uniform rate. If the rate had been 5 mph more, the trip would have taken 2 hr less. Find the rate of the biker.

Solution Let $\qquad x =$ the rate of the biker in miles per hour

Then $\qquad \dfrac{300 \text{ mi}}{x \text{ mi/hr}} = \dfrac{300}{x}$ hr is time of the biker for the trip, and

$\dfrac{300 \text{ mi}}{(x + 5) \text{ mi/hr}} = \dfrac{300}{x + 5}$ hr is time of biker for trip at faster rate.

Thus,

$$(\text{time at faster rate}) + 2 \text{ hr} = (\text{time at slower rate})$$

that is, $\qquad \dfrac{300}{x + 5} \text{ hr} + 2 \text{ hr} = \dfrac{300}{x} \text{ hr}$

Solving for x, we have

$$\frac{300}{x + 5} + 2 = \frac{300}{x}$$

$$300x + 2x(x + 5) = 300(x + 5)$$

$$300x + 2x^2 + 10x = 300x + 1500$$

$$2x^2 + 10x - 1500 = 0$$

$$x^2 + 5x - 750 = 0$$

$$(x + 30)(x - 25) = 0$$

$$x = -30 \quad \text{or} \quad x = 25$$

Rejecting the negative value, the biker travels at a rate of 25 mph. ■

EXAMPLE 2 Mark and John each hike 24 mi. The sum of their speeds is 7 mph, and the sum of the times taken is 14 hr. Find their speeds.

Solution Let $\qquad\qquad\qquad x = $ Mark's speed in miles per hour

Then $\qquad\qquad\qquad 7 - x = $ John's speed in miles per hour

The time Mark takes to hike 24 mi is $\dfrac{24}{x}$ hr

The time John takes to hike 24 mi is $\dfrac{24}{7 - x}$ hr

Using the fact that

$$(\text{Mark's time}) + (\text{John's time}) = 14 \text{ hr}$$

we get $\qquad\qquad \left(\dfrac{24}{x}\right) \text{hr} + \left(\dfrac{24}{7 - x}\right) \text{hr} = 14 \text{ hr}$

Solving for x, we get

$$\frac{24}{x} + \frac{24}{7 - x} = 14$$

$$24(7 - x) + 24x = 14x(7 - x)$$

$$168 - 24x + 24x = 98x - 14x^2$$

$$14x^2 - 98x + 168 = 0$$

$$x^2 - 7x + 12 = 0$$

$$(x - 4)(x - 3) = 0$$

$$x = 4 \quad \text{or} \quad x = 3$$

If $x = 4$, then $7 - x = 3$. If $x = 3$, then $7 - x = 4$. Hence, their speeds are 3 mph and 4 mph. ■

Note. The problem does not differentiate between Mark and John; thus, the dual answers should be expected.

EXAMPLE 3 Two cars on separate roads were 144 mi apart and started at the same time towards a point where the two roads met at right angles. They both arrived at the intersection at the same time. The speed of car A was 30 mph and car B 40 mph. How long did it take them to meet?

Solution Let

$$x = \text{the time, in hours, taken for both cars to meet at the intersection.}$$

Then $30\dfrac{\text{mi}}{\text{hr}} \cdot x \text{ hr} = (30x) \text{ mi}$ is distance traveled by car A

and $40\dfrac{\text{mi}}{\text{hr}} \cdot x \text{ hr} = (40x) \text{ mi}$ is distance traveled by car B

See Figure 9.6. By the Pythagorean theorem, we have

$$[(30x) \text{ mi}]^2 + [(40x) \text{ mi}]^2 = (144 \text{ mi})^2$$

$$900x^2 \text{ sq mi} + 1600x^2 \text{ sq mi} = 20736 \text{ sq mi}$$

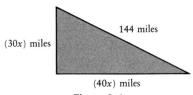

(30x) miles

144 miles

(40x) miles

Figure 9.6

Solving for x,

$$900x^2 + 1600x^2 = 20736$$

$$2500x^2 = 20736$$

$$x^2 = 8.29$$

$$x = \pm\sqrt{8.29} \approx \pm 2.88$$

Thus, both cars travel for 2.88 hr. ■

Free Falling Body

In physics it has been shown that if a body is *thrown vertically upward* with a velocity v_0 units per second, then at the end of t sec, its *height, S*, above the starting point is given by the formula

$$S = v_0 t - \frac{1}{2}gt^2$$

and its velocity is

$$V = v_0 - gt$$

If the distances are measured in feet, g is approximately equal to 32 ft/sec^2. These formulas are only approximate, since they ignore the effect of air resistance.

If the body is *thrown vertically downward* with a velocity v_0 units per second, then at the end of t seconds the distance of the body *below* the starting point is

$$S = v_0 t + \frac{1}{2} g t^2$$

and its velocity is

$$V = v_0 + gt$$

EXAMPLE 4   If a ball is thrown upward with a velocity of 64 ft/sec, what will be:

(a) its height above the starting point at the end of 2 seconds

(b) its velocity at the end of 2 seconds

(c) its position at the end of 6 seconds

(d) its velocity at the end of 6 seconds

Solution   (a) Using Formula (1) with $g = 32$, $v_0 = 64$, and $t = 2$, we get

$$S = 64 \, \frac{\text{ft}}{\text{sec}} \cdot 2 \text{ sec} - \frac{1}{2} \cdot 32 \, \frac{\text{ft}}{\text{sec}^2} \cdot 4 \text{ sec}^2 = 64 \text{ ft}$$

(b) Using Formula (1) with $g = 32$, $v_0 = 64$, and $t = 2$, we get

$$V = 64 \, \frac{\text{ft}}{\text{sec}} - 32 \, \frac{\text{ft}}{\text{sec}^2} \cdot 2 \text{ sec} = 32 \, \frac{\text{ft}}{\text{sec}}$$

(c) For $t = 6$ sec, we obtain

$$S = 64 \, \frac{\text{ft}}{\text{sec}} \cdot 6 \text{ sec} - \frac{1}{2} \cdot 32 \, \frac{\text{ft}}{\text{sec}^2} \cdot 36 \text{ sec}^2 = -192 \text{ ft}$$

The minus sign means 192 ft *below* the starting point.

(d) For $t = 6$ sec, we have

$$V = 64 \frac{\text{ft}}{\text{sec}} - 32 \frac{\text{ft}}{\text{sec}^2} \cdot 6 \text{ sec} = -128 \frac{\text{ft}}{\text{sec}}$$

The minus sign means $128 \frac{\text{ft}}{\text{sec}}$ *downward*. ■

EXAMPLE 5 If a ball is thrown upward with a velocity of 112 ft/sec, how long will it be before it is 160 ft above the starting point? Explain the two answers.

Solution Using Formula (1) with $S = 160$ ft, $v_0 = 112 \frac{\text{ft}}{\text{sec}}$, and $g = 32 \frac{\text{ft}}{\text{sec}^2}$, we have

$$160 \text{ ft} = 112 \frac{\text{ft}}{\text{sec}} \cdot t \text{ sec} - \frac{1}{2} \cdot 32 \frac{\text{ft}}{\text{sec}^2} \cdot t^2 \text{ sec}^2$$

$$160 \text{ ft} = (112t) \text{ ft} - (16t^2) \text{ ft}$$

Solving for t, we get

$$160 = 112t - 16t^2$$
$$16t^2 - 112t + 160 = 0$$
$$t^2 - 7t + 10 = 0$$
$$(t - 5)(t - 2) = 0$$
$$t = 5 \quad \text{or} \quad t = 2$$

Thus, the ball reaches a height of 160 ft above the starting point 2 sec and 5 sec after being thrown upward. The first time, 2 sec, is on the way up and the second time, 5 sec, is on the way down. See Figure 9.7. ■

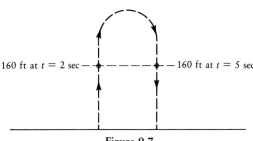

160 ft at $t = 2$ sec   160 ft at $t = 5$ sec

Figure 9.7

EXERCISES 9.3

1. A truck traveled 202.5 mi and returned. By increasing the speed 4.5 mph, the driver saved $\frac{1}{2}$ hour on the return trip. Find the speed on the first trip.

2. After traveling 72 mi at x mph, a boat increased its speed 10 mph for 98 mi. If the entire trip was made in $7\frac{1}{2}$ hr, what was the speed at first?

3. A sailboat made a run of 50 mi and returned. Due to the change of wind and tide, the speed on the return trip was 4 mph less than going out. If the entire trip was made in 10 hr and 25 min, what was the rate of speed going out?

4. Three hours after leaving the terminal, a bus was pursued by a car with a speed of 21 mph more than that of the bus. If the car overtook the bus 196 mi from the terminal, find the speed of each.

5. A train traveled 180 miles and returned. By increasing the speed 8 mph, the engineer saved $\frac{3}{4}$ hr on the return trip. Find the rate of speed at first.

6. A woman walks to the post office, $\frac{1}{2}$ mi away, and back in 25 min, her rate of walking each way being uniform. If the sum of these two rates is 5 mph, find what each rate is.

7. Two people each walk 21 mi. If the sum of their speeds is $7\frac{1}{4}$ mph, and the sum of their times is 11 hr 36 min, find what these times are.

8. A man walks for 9 mi. After a rest he walks for 5 mi at a rate $\frac{1}{2}$ mph slower than before, and takes 1 hr less than before he rested. What were his two rates of walking?

9. A woman drives to a city 21 mi away. She returns by a route that is 4 mi shorter, but since her speed is 2 mph less, she saves only 5 min on the return trip. At what speed did she drive to the city?

10. A man walks to his studio and back in $3\frac{1}{4}$ hr, his rate on the return trip being $\frac{1}{2}$ mph slower than on the outward trip. If he lives $5\frac{1}{4}$ mi from his studio, find the rates of walking.

11. If an object is thrown upward with a velocity of 32 ft/sec, (a) how long will it rise ($v = 0$),

and (b) after how many seconds will it be back at the starting point ($s = 0$)?

12. A rock is dropped from a bridge. The sound of its splash is heard 3.5 sec later. Neglecting the time required for sound to travel, find the height of the bridge above the water.

13. If a package is thrown vertically downward from a plane with a velocity of 100 ft/sec, how long will it be before it strikes the ground 900 ft below? What will its velocity be when it strikes?

14. If the package in Exercise 13 were dropped instead of thrown, what would its velocity be on striking the ground?

15. If a ball is thrown upward with a velocity of 48 ft/sec from a building that is 120 ft high, how far above the ground will it rise? What will be its velocity when it strikes the ground?

16. Two formula-one cars race over a 330-mi course. One car, running at a speed 5 mph greater than the other, wins by 55 min. Find the two speeds.

17. Two trucks each have a distance of 200 mi to cover. Their speeds differ by 10 mph, and their times by 1 hr. How long does the faster truck take?

*18. Two men walk towards one another starting from two towns 34 mi apart, and meet in 4 hr. If one takes 1 min 40 sec longer to walk a mile than the other, find their speeds.

*19. If a person whose stride is $2\frac{1}{5}$ ft were to take $\frac{1}{8}$ sec longer over each step, his speed would be reduced by 1 mph. At what rate is he walking?

*20. Two women are cycling along a road in opposite directions. Their speeds differ by 3 mph, and they started from points 54 mi apart. After the moment of crossing, the slower cyclist takes $2\frac{1}{2}$ hr to reach the other one's starting point. Find the speed of each.

9.4 WORK PROBLEMS

EXAMPLE 1   Construction team A takes 3 hr longer to build an oil rig than construction team B. After team A has been working on an oil rig for 5 hr, they are joined by team B and together complete the job in 3 additional hr. How long would it take each team to complete the oil rig, working alone?

Solution   Let $x =$ the number of hours it takes team B to complete the job alone

Then $x + 3 =$ the number of hours it takes team A to complete the job alone

Thus, $\dfrac{1}{x} =$ amount of work done by team B in 1 hr working alone, measured in $\dfrac{\text{rigs}}{\text{hr}}$

and $\dfrac{1}{x + 3} =$ amount of work done by team A in 1 hr working alone, measured in $\dfrac{\text{rigs}}{\text{hr}}$

Since team B works for 3 hr and team A for all 8 hr, and using the relationship

$$\begin{pmatrix} \text{work done by} \\ \text{team } B \end{pmatrix} + \begin{pmatrix} \text{work done by} \\ \text{team } A \end{pmatrix} = 1 \text{ whole job}$$

$$\left(\frac{1}{x}\frac{\text{rigs}}{\text{hr}}\right)3\,\text{hr} + \left(\frac{1}{x + 3}\frac{\text{rigs}}{\text{hr}}\right)8\,\text{hr} = 1 \text{ rig}$$

$$\frac{3}{x}\,\text{rig} + \frac{8}{x + 3}\,\text{rig} = 1 \text{ rig}$$

Solving for x,

$$\frac{3}{x} + \frac{8}{x + 3} = 1$$

$$3(x + 3) + 8x = x(x + 3)$$

$$3x + 9 + 8x = x^2 + 3x$$

$$0 = x^2 - 8x - 9$$

$$0 = (x - 9)(x + 1)$$

$$x = 9 \qquad \text{or} \qquad x = -1$$

Rejecting a negative value for time, we have

$$\text{time for team } B = 9 \text{ hr}$$

$$\text{time for team } A = (9 + 3) \text{ hr} = 12 \text{ hr} \quad \blacksquare$$

EXAMPLE 2   A pipe can fill a tank in 10 minutes less than another pipe, but takes 10 min 25 sec longer than the two pipes together. Determine the time it takes each pipe to fill the tank, working alone.

Solution   Let x = the time it takes the first pipe to fill the tank, in minutes

Then $x + 10$ = the time it takes the second pipe to fill the tank, in minutes

Thus, the first pipe fills $\dfrac{1}{x}$ of the tank in 1 min. The second pipe fills $\dfrac{1}{x + 10}$ of the tank in 1 min. Hence, the two pipes working together fill $\left(\dfrac{1}{x} + \dfrac{1}{x + 10}\right)$ of the tank in 1 min. But the problem states that the two pipes together take $\left(x - 10\,\dfrac{5}{12}\right)$ min (25 sec = $\dfrac{25}{60} = \dfrac{5}{12}$ min). Therefore, together the two pipes fill $\dfrac{1}{x - 10\,\dfrac{5}{12}}$ of the tank in 1 min. We have two expressions indicating the amount of the tank filled when both pipes are working together. Thus,

$$\frac{1}{x} + \frac{1}{x + 10} = \frac{1}{x - 10\,\dfrac{5}{12}}$$

$$\frac{x + 10 + x}{x(x + 10)} = \frac{12}{12x - 125}$$

Multiplying both sides by the common denominator, we get

$$(2x + 10)(12x - 125) = 12x(x + 10)$$

$$24x^2 - 130x - 1250 = 12x^2 + 120x$$

$$12x^2 - 250x - 1250 = 0$$

$$6x^2 - 125x - 625 = 0$$

$$(x - 25)(6x + 25) = 0$$

$$x = 25 \quad \text{or} \quad x = -\frac{25}{6}$$

Rejecting the negative value for time, we have

$$\text{time for first pipe} = 25 \text{ min}$$

$$\text{time for second pipe} = (25 + 10) \text{ min} = 35 \text{ min} \quad \blacksquare$$

EXERCISES 9.4

1. John Bonner working alone can do a piece of work in 5 hr less than the time required by Mark Willard to do the job alone. Working together, they complete the job in 6 hr. How long does it take each to do the job alone?

2. Working together, two carpenters can com-

plete a kitchen in 4 days. The slower carpenter, working alone, requires 6 days more than the faster carpenter to do the job. How long does it take each carpenter to do the job working alone?

3. It takes a boy 5 hr longer to paint a room than it takes Barbara. If Barbara and two boys can paint a room in 6 hr, how long would it take Barbara to paint the room alone?

4. A tank can be filled by two pipes in 3 hr. The larger pipe alone would take $1\frac{3}{4}$ hr less time than the smaller pipe. In what time can each pipe fill the tank alone?

5. A swimming pool can be filled with one pipe in 1 hr less time than another pipe. When both pipes are open, the tank can be filled in 1 hr and 12 min. In what time can each pipe alone fill the pool?

6. A water tank can be filled by two pipes in $33\frac{1}{3}$ min. If the larger pipe takes 15 min less than the smaller to fill the tank, find in what time the tank will be filled by each pipe alone.

7. A and B together can do a piece of work in $14\frac{2}{3}$ days. A can do the job in 12 days less than B alone. Find the time in which A alone can do the work.

8. A and B working together can do a piece of work in 3 days. If they were working separately, A would take 8 days longer than B. How long would A take alone?

9. The hot pipe takes 3 min longer to fill a bath than the cold pipe. The two together take 6 min 40 sec. How long does the cold pipe take?

10. A tank can be filled in $\frac{1}{2}$ hr less time than it can be emptied. When both inlet and outlet pipes are open, it can be filled in 28 hr. In what time can it be filled with the outlet closed?

*11. How long will it take each of two pipes separately to fill a tank if one of them alone takes $9\frac{3}{5}$ min longer to fill the tank than the other, and 15 min longer than the two together?

*12. It took a number of men as many hours as there were men to do a job. If there had been 6 more men, the work would have been done in 8 hr. How many men were working on the job?

*13. A and B separately can do a piece of work in the same time. C takes 4 weeks longer. If the three work together, the time taken is 9 weeks less than that of C. Find the time C takes.

9.5 MISCELLANEOUS PROBLEMS

EXAMPLE 1   There are as many square feet in the surface of a certain sphere as there are cubic feet in its volume. Find its radius.

Solution   Let

$$r = \text{the number of feet in the radius}$$

Then

$$4\pi r^2 = \text{the number of square feet in the surface}$$

and

$$\frac{4\pi r^3}{3} = \text{the number of cubic feet in the volume}$$

Thus,

$$\frac{4\pi r^3}{3} = 4\pi r^2$$

$$4\pi r^3 = 12\pi r^2$$

$$4\pi r^3 - 12\pi r^2 = 0$$

$$4\pi r^2(r - 3) = 0$$

$$r = 0 \quad \text{or} \quad r = 3$$

Hence, the radius is 3 ft. ■

EXAMPLE 2 A stereo dealer paid $1800 for a group of radios. She sold all but 5 of the radios at $20 more per radio than she paid, thereby making a profit of $200 on the transaction. How many radios did the dealer buy?

Solution Let $\qquad\qquad x =$ the number of radios the merchant bought

Then $\qquad \dfrac{1800}{x} =$ the cost of each radio

The relationship used to set up the equation is:

(the number of radios sold) · (the selling price of each radio)

$$= \text{(dealer cost)} + \text{(profit)}$$

Thus,

$$(x - 5) \text{ radios} \cdot \left(\frac{1800}{x} + 20\right) \frac{\text{dollars}}{\text{radio}} = \$1800 + \$200$$

$$\left[(x - 5)\left(\frac{1800}{x} + 20\right)\right] \text{ dollars} = \$2000$$

Solving for x, we get

$$(x - 5)\left(\frac{1800}{x} + 20\right) = 2000$$

$$1800 + 20x - \frac{9000}{x} - 100 = 2000$$

$$20x - \frac{9000}{x} = 300$$

$$20x^2 - 9000 = 300x$$

$$20x^2 - 300x - 9000 = 0$$

$$x^2 - 15x - 450 = 0$$

$$(x - 30)(x + 15) = 0$$

$$x = 30 \quad \text{or} \quad x = -15$$

Rejecting -15, we have $x = 30$. Hence, the dealer bought 30 radios. ■

Check The dealer bought 30 radios at $60 each. She sold 25 radios at $80 each, thus taking in $2000.

EXAMPLE 3   A mother is three times as old as her daughter. The product of the mother's age in 3 years' time and the daughter's age 3 years ago, is twenty times the present age of the daughter. Find their ages.

Solution   Let x denote the daughter's present age, in years

Then $3x$ denotes the mother's present age, in years

In 3 years' time, the mother's age will be $(3x + 3)$ years, while 3 years ago, the daughter's age was $(x - 3)$ years. Thus, the problem states that

$$[(3x + 3)(x - 3)] \text{ years} = (20x) \text{ years}$$

Solving for x, we get

$$(3x + 3)(x - 3) = 20x$$
$$3x^2 - 6x - 9 = 20x$$
$$3x^2 - 26x - 9 = 0$$
$$(3x + 1)(x - 9) = 0$$
$$x = -\frac{1}{3} \quad \text{or} \quad x = 9$$

Rejecting the negative value, we have $x = 9$. Hence, the daughter's age is 9 years, and the mother's age is 27.   ◼

EXAMPLE 4   A gardener knows that it will take 720 small square tiles to cover a patio, but if she uses tiles 4 in. longer on each side, it will take only 80 tiles. Find the length of a side of each tile.

Solution   Let x = the length of a side of the smaller tile, in inches

Then $x + 4$ = the length of a side of the larger tile, in inches

Using the relationship

area of patio using small tiles = area of patio using large tiles

we get

$$(720x^2) \text{ sq in.} = [80(x + 4)^2] \text{ sq in.}$$

Solving for x,

$$720x^2 = 80(x + 4)^2$$

Dividing both sides by 80, we obtain

$$9x^2 = (x + 4)^2$$

$$9x^2 = x^2 + 8x + 16$$
$$8x^2 - 8x - 16 = 0$$
$$x^2 - x - 2 = 0$$
$$(x - 2)(x + 1) = 0$$
$$x = 2 \quad \text{or} \quad x = -1$$

Rejecting $x = -1$, we find that the smaller tile is 2 in. by 2 in. and the larger tile is 6 in. by 6 in. ◼

EXERCISES 9.5

1. A circular swimming pool is surrounded by a walk 8 ft wide. The area of the walk is $\frac{11}{25}$ of the area of the pool. Find the radius of the pool.

2. If the radius of a circle were increased by 1 yd, its area would be increased by 15π sq yd. Find the radius of the circle.

3. A farmer bought a certain number of cows for $1260. Having lost 4 of them, he sold the remaining cows for $10 a head more than they cost him, and made $260 on the entire transaction. How many cows did he buy?

4. A group had a dinner that cost $60. If there had been 5 more people, each would have had to pay $1 less. How many people were at the dinner?

5. A manufacturer was offered $90 for his boxes, but he held them for a higher price. He finally sold all but 5 boxes for $90, thereby receiving 20 cents more per box than he was first offered. How many boxes did he have?

6. A store made a net profit of $20 on apples in one day. The next day a profit of $21 was made by selling 5 bushels more than on the preceding day, but the profit was 10 cents less per bushel. How many bushels were sold on the first day?

7. A certain number of people divided 99 bundles of firewood. If each had received 2 bundles more, he would have received as many bundles as there were people. How many people were there?

8. In a small theater, 600 people are seated on benches of equal length. If there were 10 fewer benches, it would be necessary that 2 people more should sit on each bench. How many benches are there?

9. The sum of the ages of a father and his son is 80 years. One-fourth of the product of their ages, in years, exceeds the father's age by 240. How old are they?

10. Erin is 4 years older than Larry, and the product of their present ages is three times what the product of their ages was 4 years ago. Find their present ages.

11. A woman is six times as old as her son, and next year her age will be equal to the square of her son's age. Find their present ages.

12. A certain rectangle is inscribed in a circle of radius r. Find the dimensions of the rectangle, if its perimeter is $\frac{16}{3} r$.

*13. A certain cylinder is inscribed in a sphere of radius r. Find the radius of the cylinder, if its lateral area is equal to half the area of the sphere.

*14. The circumference of a back wheel of a wagon exceeds that of a front wheel by 8 in. In traveling 1 mi, the back wheel makes 88 less revolutions than the front wheel. Find the circumference of each wheel.

*15. The circumference of the back wheel of a wagon is 5 ft more than that of the front wheel. If the back wheel makes 150 fewer revolutions than the front wheel in going 1 mi, find the circumference of each wheel.

9.6 APPLICATIONS INVOLVING QUADRATIC INEQUALITIES

EXAMPLE 1 An object shot upwards with a speed of 400 ft/sec will be, after 4 sec, at a height of approximately $s = 400t - 16t^2$. For what time interval is the object more than 2000 ft above ground level?

Solution The condition that the object be more than 2000 ft above ground level can be expressed by the inequality

$$400t - 16t^2 > 2000$$

Thus,

$$-16t^2 + 400t - 2000 > 0 \text{ [dividing by } -16]$$

$$t^2 - 25t + 125 < 0 \text{ [note change in sense of inequality]}$$

Solving the associated equation, we get

$$t^2 - 25t + 125 = 0$$

$$t = \frac{25 \pm \sqrt{(25)^2 - 4(1)(125)}}{2}$$

$$t = \frac{25 \pm \sqrt{125}}{2}$$

$$t = \frac{25 \pm 5\sqrt{5}}{2}$$

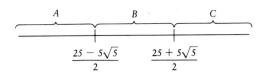

	Set	Chosen Number	$400t - 16t^2 > 2000$
A:	$t < \dfrac{25 - 5\sqrt{5}}{2}$	-5	$-2000 - 400 \not> 2000$
B:	$\dfrac{25 - 5\sqrt{5}}{2} < t < \dfrac{25 + 5\sqrt{5}}{2}$	10	$4000 - 1600 > 2000 \ \checkmark$
C:	$t > \dfrac{25 + 5\sqrt{5}}{2}$	20	$8000 - 6400 \not> 2000$

Thus, the object will be more than 2000 ft above ground level between the time interval

$$\frac{25 - 5\sqrt{5}}{2} < t < \frac{25 + 5\sqrt{5}}{2}$$

that is, $6.91 \sec < t < 18.09 \sec$ ▨

EXAMPLE 2   A manufacturer of video tape recorders finds that when x units are made and sold per week, its profits, in thousands of dollars, is given by $x^2 - 40x - 6000$.

(a) At least how many units must be manufactured and sold each week to make a profit?

(b) When is the manufacturer losing money?

Solution   (a) The manufacturer makes a profit when

$$x^2 - 40x - 6000 > 0$$

Once again, to solve this inequality, we begin by solving the associated equation

$$x^2 - 40x - 6000 = 0$$
$$(x - 100)(x + 60) = 0$$
$$x = 100 \quad \text{or} \quad x = -60$$

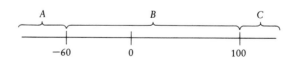

Set	Chosen Number	$x^2 - 40x - 6000$
A: $x < -60$	-100	$10{,}000 + 4000 - 6000 > 0$
B: $-60 < x < 100$	0	$0 - 0 - 6000 < 0$
C: $x > 100$	200	$40{,}000 + 8000 - 6000 > 0$ ✓

Thus, the inequality is satisfied by $x < -60$ or $x > 100$. However, negative values for x make no sense, since x represents units manufactured. Hence, if the manufacturer makes *at least* 100 units per week, she will make a profit.

(b) A loss will occur if $x^2 - 40x - 6000 < 0$. The values of x, manufactured units, for which this occurs is $0 < x < 100$.   ▨

EXERCISES   9.6

1. A particle moves vertically (up and down) under the following law of motion: $s = 8t - t^2$, where t is in seconds and s is in feet. For what time interval is the object (a) above ground? (b) above 12 ft?

2. A manufacturer of calculators finds that when x units are made and sold per week, its profit is given by $x^2 - 40x$ (in thousands of dollars).

 (a) At least how many units must be manufactured and sold each week to make a profit?

 (b) For what values of x is the manufacturer losing money?

3. Repeat Exercise 2 if the profit is given by $x^2 - 20x - 1500$.

4. If the path of an object moves according to $s = 960t - 16t^2$, where t is in seconds and s is in feet, then for what time interval is the object (a) above ground? (b) above 1280 ft?

10 Graphing Equations in Two Variables

In this chapter we introduce the concept of linear equations in two variables and their graphs. In particular, we shall discuss the ideas of the distance between two points in the plane; vertical, horizontal, parallel, and perpendicular lines; the slope of a line; and equations of lines. Then we briefly discuss a class of graphs called conic sections: circles, parabolas, ellipses, and hyperbolas.

We begin our study of linear equations by introducing the Cartesian coordinate system.

10.1 CARTESIAN COORDINATES AND THE DISTANCE FORMULA

Cartesian Coordinates

In Chapter 1, we saw that a point P on a number line can be represented by exactly one real number, and each real number can be represented by exactly one point. And just as the set of real numbers is an algebraic representation of the real number line, in this section we shall introduce the set of "ordered pairs" as an algebraic representation of the plane.

We start by taking two number lines; one horizontal and one vertical, that intersect at right angles such that both number lines have the same zero point, called the origin (see Figure 10.1). The two number lines are called axes. The horizontal line, which extends indefinitely to the right and left, is called the x-axis and the vertical line, which extends indefinitely up and down, is called the y-axis. This system is often referred to as the Cartesian coordinate system, named in honor of the French mathematician René Descartes (1596–1650).

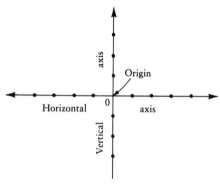

Figure 10.1

The position of any point in the plane can be described by giving its distances from the axes. We choose some unit of length (inches, centimeters, etc.) in which to express distances, and we agree that distances measured along any line that is parallel to the x-axis are positive if measured to the right, and negative if measured to the left. Distances along any line parallel to the y-axis are positive if measured upward and negative if measured downward.

Let P be any point of the plane and let its distance from the y-axis be x and its distance from the x-axis be y. Then x is called the abscissa of the point P and y is called the ordinate of P. The point P is said to have the coordinates (x, y). Thus, the position of a point can be located if we know its coordinates. Conversely, if a point is located in the plane, we can find its coordinates by measuring its distance from the coordinate axes (see Figure 10.2). In Figure 10.2, the points $(3, 1)$, $(1, 3)$, $(-2, 2)$, $(-4, -2)$, and $(2, -1)$ are plotted. Note that the points $(3, 1)$ and $(1, 3)$ are different. Therefore, the order in which a pair of real numbers is listed is very important in locating a point; thus, we use the idea of ordered pair. With this representation, we see that every point P in the Cartesian plane uniquely determines an ordered pair of real numbers (x, y) and, conversely, every ordered pair of real numbers (x, y) is represented uniquely by a point P in the plane.

When writing the coordinates of a point, we must pay attention to their algebraic signs as well as the numerical values. The signs are used to indicate

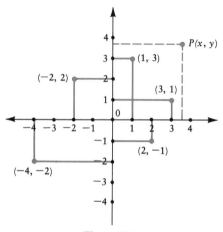

Figure 10.2

the directions in which the distances are measured from the axes. Looking at the Cartesian system we see that the coordinate axes divide the plane into four regions called quadrants (see Figure 10.3). Quadrant I consists of all points (x, y) for which both x and y are positive, Quadrant II consists of all points (x, y) for which x is negative and y is positive, Quadrant III consists of all points (x, y) for which both x and y are negative, and Quadrant IV consists of all points (x, y) for which x is positive and y is negative.

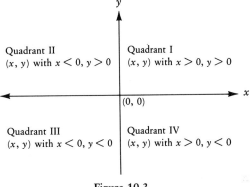

Figure 10.3

Note that a point on a coordinate axis belongs to no quadrant. Points that lie on an axis have zero for one or both coordinates. If the first coordinate is zero, $(0, y)$, then the point lies on the y-axis. Those points with their second coordinate being zero, $(x, 0)$, lie on the x-axis. The point $(0, 0)$ represents the origin.

EXAMPLE 1 Plot each point and indicate which quadrant or coordinate axis contains the point.

(a) $(4, -3)$

(b) $(2, 3)$

(c) $(-1, -3)$

(d) $(0, 2)$

(e) $\left(\dfrac{1}{2}, 0\right)$

(f) $\left(-\dfrac{3}{2}, 1\right)$

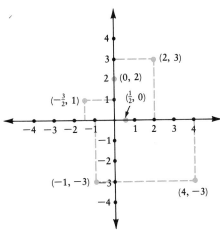

Figure 10.4

Solution   (a) $(4, -3)$ lies in Quadrant IV   (b) $(2, 3)$ lies in Quadrant I

(c) $(-1, -3)$ lies in Quadrant III   (d) $(0, 2)$ lies on y-axis

(e) $\left(\dfrac{1}{2}, 0\right)$ lies on x-axis   (f) $\left(-\dfrac{3}{2}, 1\right)$ lies in Quadrant II

See Figure 10.4. ■

When are two ordered pairs equal?

DEFINITION

Two ordered pairs (x_1, y_1) and (x_2, y_2) are considered to be equal if and only if $x_1 = x_2$ and $y_1 = y_2$.

EXAMPLE 2 $(3, 7) \neq (7, 3)$ and $(-2, 1) \neq (2, 1)$, whereas $(7, -5) = (7, -5)$.

EXAMPLE 3   Find the values of x and y such that $(x, -5) = (3, y - 1)$.

Solution $(x, -5) = (3, y - 1)$ if and only if $x = 3$ and $-5 = y - 1$; that is, $x = 3$ and $y = -4$. ■

The Distance Formula

One of the important features of a Cartesian coordinate system is the existence of a simple formula that gives the distance between two points in terms of their coordinates. To derive such a formula, we first consider two special cases: 1. two points on a horizontal line, and 2. two points on a vertical line.

Distance Between Two Points Along a Horizontal Line

The distance between two points lying on a horizontal line is given by the absolute value of the difference of the x-coordinates. That is, the distance between the points (x_1, y) and (x_2, y) is $|x_2 - x_1|$. See Figure 10.5.

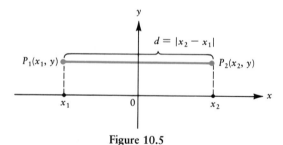

Figure 10.5

EXAMPLE 4   The distance between the points $(-4, 3)$ and $(5, 3)$ is $|5 - (-4)| = |5 + 4| = |9| = 9$.

Note. Since $|x_2 - x_1| = |x_1 - x_2|$, it makes no difference which point is taken to be (x_1, y) or (x_2, y).

Distance Between Two Points Along a Vertical Line

The distance between two points lying on a vertical line is given by the absolute value of the difference of the y-coordinates. That is, the distance between the points (x, y_1) and (x, y_2) is $|y_2 - y_1|$. See Figure 10.6.

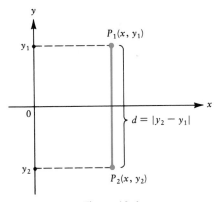

Figure 10.6

EXAMPLE 5   The distance between the points $(-5, 6)$ and $(-5, 1)$ is $|1 - 6| = |-5| = 5$.

We now consider two points that do not lie on a horizontal or vertical line. See Figure 10.7. Let $P_1(x_1, y_1)$ and $P_2(x_2, y_2)$ represent two points that do not lie on a horizontal or vertical line. The line segment joining P_1 and P_2 is the hypotenuse of a right triangle with vertices (x_1, y_1), (x_2, y_2), and (x_2, y_1). The lengths of the legs are $|x_2 - x_1|$ and $|y_2 - y_1|$. By the Pythagorean theorem, it follows that

$$d^2 = |x_2 - x_1|^2 + |y_2 - y_1|^2$$

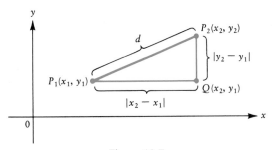

Figure 10.7

However, for any real number a, $|a|^2 = a^2$. Thus,

$$d^2 = (x_2 - x_1)^2 + (y_2 - y_1)^2$$

Since distance is taken to be positive, we take the positive square root:

$$d = \sqrt{(x_2 - x_1)^2 + (y_2 - y_1)^2}$$

Note. The formulas given for distances along a horizontal line $[(y_2 - y_1) = 0]$ and a vertical line $[(x_2 - x_1) = 0]$ are special cases of this formula.

Distance Formula in the Plane

For any points $P_1(x_1, y_1)$ and $P_2(x_2, y_2)$, the distance, d, between them is given by

$$d = \sqrt{(x_2 - x_1)^2 + (y_2 - y_1)^2}$$

EXAMPLE 6 Find the distance between the points $(4, 5)$ and $(6, -7)$.

Solution Let $(x_1, y_1) = (4, 5)$ and $(x_2, y_2) = (6, -7)$. Then

$$d = \sqrt{(6 - 4)^2 + (-7 - 5)^2} = \sqrt{(2)^2 + (-12)^2} = \sqrt{148} = 2\sqrt{37} \quad \blacksquare$$

EXAMPLE 7 Show that the points $A(-2, -5)$, $B(1, -4)$, and $C(0, -1)$ are the vertices of a right triangle.

Solution The distance $AB = \sqrt{[1 - (-2)]^2 + [-4 - (-5)]^2} = \sqrt{(3)^2 + (1)^2} = \sqrt{10}$.
The distance $BC = \sqrt{(0 - 1)^2 + [-1 - (-4)]^2} = \sqrt{(-1)^2 + (3)^2} = \sqrt{10}$.
The distance $CA = \sqrt{[0 - (-2)]^2 + [-1 - (-5)]^2} = \sqrt{(2)^2 + (4)^2} = 2\sqrt{5}$.
Note that (distance AB)^2 + (distance BC)^2 = (distance CA)^2

$$(\sqrt{10})^2 + (\sqrt{10})^2 = (2\sqrt{5})^2$$

$$10 + 10 = 20$$

Thus, by the Pythagorean theorem, the triangle with vertices $A(-2, -5)$, $B(1, -4)$, and $C(0, -1)$ is a right triangle. $\quad \blacksquare$

EXAMPLE 8 Find the radius of a circle with center at $(1, -3)$ and which passes through the point $(2, 5)$.

Solution The radius of a circle is the distance from the center to any point on the circle. Thus, we must find the distance between $(1, -3)$ and $(2, 5)$.

$$r = \sqrt{(2 - 1)^2 + [5 - (-3)]^2} = \sqrt{(1)^2 + (8)^2} = \sqrt{65} \quad \blacksquare$$

EXERCISES   10.1

In Exercises 1–12, determine in which quadrant, if any, each of the points lies.

1. $(3, 4)$ 2. $(6, 2)$
3. $(-2, 3)$ 4. $(-4, 5)$
5. $(-3, -4)$ 6. $(-5, -1)$
7. $(2, -3)$ 8. $(4, -1)$
9. $(3, 0)$ 10. $(-2, 0)$
11. $(0, 4)$ 12. $(0, -1)$

Find the distance between each of the pairs of points in Exercises 13–24.

13. $(2, 3), (6, 3)$ 14. $(-1, 3), (-1, 8)$
15. $(-4, 2), (3, 2)$ 16. $(2, 4), (1, -6)$
17. $(-3, 5), (-6, 2)$ 18. $(6, 3), (-5, 3)$
19. $(-4, -3), (-4, 5)$ 20. $(2, 6), (2, 2)$
21. $(4, 1), (-3, 1)$ 22. $(5, 3), (-4, 2)$
23. $(3, 5), (-6, -3)$ 24. $(a, b), (b, c)$

Prove by using the distance formula that the three points given in Exercises 25–30 are the vertices of a right triangle.

25. $(-1, 4), (2, -1), (-3, -4)$
26. $(0, 0), (2, 6), (4, 2)$
27. $(0, 9), (-8, 1), (6, 3)$
28. $(-10, 6), (-6, -1), (8, 7)$
29. $(6, 5), (0, 3), (2, 1)$
30. $(-1, -4), (-5, -6), (-3, 0)$

Prove by using the distance formula that the three points given in Exercises 31–36 are the vertices of an isoceles triangle.

31. $(-5, 1), (-6, 5), (-2, 4)$
32. $(4, 4), (2, 1), (1, 2)$
33. $(0, 0), (6, 9), (-9, 6)$
34. $(5, 6), (2, -2), (8, -2)$

35. $(-1, 2), (-2, -2), (2, -1)$
36. $(2, -2), (-3, 3), (3, 5)$

Prove by using the distance formula that the four points given in Exercises 37–42 are the vertices of a parallelogram.

37. $(-2, 2), (-4, 6), (0, 6), (-2, 10)$
38. $(2, 2), (-2, 0), (2, 0), (-2, -2)$
39. $(1, 2), (1, -11), (-5, -3), (7, -6)$
40. $(5, 2), (-2, 1), (2, 5), (1, -2)$
41. $(-2, -3), (3, -1), (1, 4), (-4, 2)$
42. $(10, 5), (15, 10), (5, 10), (10, 15)$

43. Find all values of x so that the distance between the points $(-2, 3)$ and (x, x) is 5.
44. Show that the points $(0, 5)$, $(4, -3)$, and $(4, 1)$ are equidistant from $(-2, -1)$.
45. Given points $A(x_1, y_1)$ and $B(x_2, y_2)$, prove that the point $B\left(\dfrac{x_1 + x_2}{2}, \dfrac{y_1 + y_2}{2}\right)$ is the midpoint of the line segment joining A and B.
46. Using the result of Exercise 45, find the midpoint of $A(2, 1)$ and $B(11, 6)$.
47. A line segment CD has its midpoint at $(2, 4)$ and point C at $(-3, 7)$. Find the coordinates of point D.
48. Use the distance formula to determine whether the points $A(0, -3)$, $B(8, 3)$, and $C(11, 7)$ are collinear.
49. Find a point on the y-axis equidistant from both points $(3, 2)$ and $(-1, 4)$.
50. Find a point on the x-axis equidistant from both points $(4, 1)$ and $(1, 2)$.

10.2   LINEAR EQUATIONS AND THEIR GRAPHS

The equation $3x + 2 = 8$ is an equation in one variable. Using the technique discussed in Chapter 3, we know that its solution is $x = 2$.

$$3x + 2 = 8$$
$$3(2) + 2 \overset{?}{=} 8$$
$$8 \overset{\checkmark}{=} 8$$

An equation such as $y = 3x + 1$ is an equation in two variables, x and y. A solution of this equation is an ordered pair that satisfies the equation. For example, the ordered pair $(0, 1)$ is a solution of the equation $y = 3x + 1$ since it satisfies the equation

$$y = 3x + 1$$
$$1 \overset{?}{=} 3(0) + 1$$
$$1 \overset{?}{=} 0 + 1$$
$$1 \overset{\checkmark}{=} 1$$

Also, the ordered pair $(-2, -5)$ is a solution of the equation $y = 3x + 1$. We can assign x any real value and there will be one corresponding real value for y; and thus we can find as many ordered pairs of real numbers which satisfy the given equation. Thus, the equation $y = 3x + 1$ has infinitely many solutions.

Equations such as $y = 3x + 1$, $-2x + y = 1$, and $-x + y = 0$ are called linear equations or first-degree equations in two variables. As mentioned above, the solution of a linear equation in two variables x and y is an ordered pair (x, y) that satisfies the equation. The solution set of a linear equation is the set of ordered pairs that satisfy the equation.

EXAMPLE 1   Find three solutions of $-2x + y = 1$.

Solution   Sometimes it is easier to deal with an equation in a different but equivalent form. Thus, the equation $-2x + y = 1$ can be written in the equivalent form $y = 2x + 1$. We now choose any three values for x. For example, let $x = 0, 1$, and -1.

Substitute $x = 0$: $y = 2(0) + 1 = 1$ $(0, 1)$ is a solution
Substitute $x = 1$: $y = 2(1) + 1 = 3$ $(1, 3)$ is a solution
Substitute $x = -1$: $y = 2(-1) + 1 = -1)$ $(-1, -1)$ is a solution

Thus, $(0, 1)$, $(1, 3)$, and $(-1, -1)$ are three solutions of the equation $-2x + y = 1$. ■

Graph of Linear Equations in Two Variables

Since the Cartesian coordinate system can be used to plot ordered pairs of real numbers, we can use it to plot the solutions of a linear equation in two variables, and thus obtain the graph of the linear equation.

EXAMPLE 2   Plot the three solutions obtained in Example 1 for the equation $-2x + y = 1$.

Solution   The three solutions obtained were $(0, 1)$, $(1, 3)$, and $(-1, -1)$.

If we study Figure 10.8, we notice that these points lie on the same line; that is, we can draw a straight line that passes through all three points. Note also that the point $(-2, -3)$ lies on this line and it too satisfies the equation $-2x + y = 1$; that is, $-2(-2) + (-3) = 1$. In fact, it can be shown that every point on the line in Figure 10.8 satisfies the linear equation $-2x + y = 1$, and, conversely, every solution of $-2x + y = 1$ lies on that line. ■

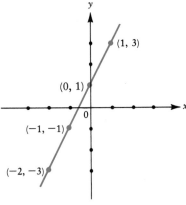

Figure 10.8

DEFINITION

The graph of a linear equation in two variables consists of *all* points whose coordinates satisfy the equation.

In general, the graph of a linear equation in two variables is a straight line in the Cartesian plane.

When graphing an equation in two variables, the first step is to solve the given equation for one variable (usually y) in terms of the other variable (usually x). In Example 1, we solved for y in terms of x. In this form the value of y depends on the value assigned to x. When the value of one unknown depends on that of the other, the unknown to which any value may be given is called the independent variable and the other unknown is called the dependent variable. In Example 1, x is the independent variable and y is the dependent variable.

How to Graph a Linear Equation in Two Variables

1. Solve the equation for y.
2. Prepare a table showing at least three values of x and the corresponding values for y.
3. Join two of these points with a straight line and check its accuracy by noting whether or not the third point lies on the line.

EXAMPLE 3 Graph $x + y = 2$.

Solution Solving for y, we get $y = -x + 2$.

if $x =$	then $y =$
0	2
1	1
2	0

See Figure 10.9. ■

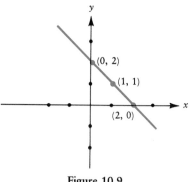

Figure 10.9

EXAMPLE 4 Graph $y = \dfrac{1}{2}x$.

Solution Since the equation is already written in terms of y, we begin by assigning values to x and finding their corresponding y values.

if $x =$	then $y =$
0	0
2	1
−2	−1

See Figure 10.10. ■

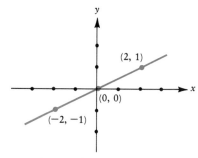

Figure 10.10

> The general form of a linear equation in two variables x and y is
>
> $$ax + by = c$$
>
> where a, b, and c are real numbers with a and b not both zero.

So far we have only considered the graphs of linear equations where a and b are nonzero. We now consider linear equations where $a = 0$ or $b = 0$.

Case 1

If $a = 0$ and $b \neq 0$, then the general linear equation becomes $0x + by = c$ or $by = c$. Solving for y, we get $y = c/b$. Since b and c are constants, c/b is also a constant and the linear equation takes the form $y = k$, where k is a real number.

EXAMPLE 5 Graph $y = 3$.

Solution In order to graph $y = 3$, we must determine ordered pairs that satisfy $y = 3$. It is easy to see that any such ordered pair must have 3 as its y-coordinate. However, the x-coordinate may take any value. Thus $(1, 3),(-1, 3)$, and $(2, 3)$ all satisfy the equation $y = 3$. Plotting these points we see that they all lie on the horizontal line that crosses the y-axis at $(0, 3)$. See Figure 10.11. ■

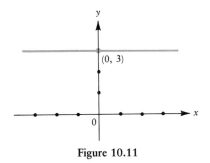

Figure 10.11

In general, the graph of $y = k$, where k is any real number, is the horizontal line that passes through the point $(0, k)$ on the y-axis.

EXAMPLE 6 Graph the equations $y = 1$, $y = -2$, and $y = \frac{5}{2}$.

Solution See Figure 10.12. ■

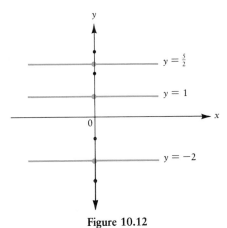

Figure 10.12

Case 2

If $a \neq 0$ and $b = 0$, then the general linear equation becomes $ax + 0y = c$, or $ax = c$. Solving for the variable x, we get $x = c/a$. Thus, the linear equation takes the form $x = k$, where k is a real number.

EXAMPLE 7 Graph $x = -2$.

Solution Any ordered pair satisfying the equation $x = -2$ has -2 as its x-coordinate. Thus, the ordered pairs $(-2, -1)$, $(-2, 1)$, and $(-2, 2)$ all satisfy the equation. Plotting these points we see that they lie on the vertical line that crosses the x-axis at $(-2, 0)$. See Figure 10.13. ▪

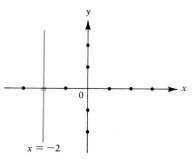

$x = -2$

Figure 10.13

In general, the graph of $x = k$, where k is any real number, is the vertical line that passes through the point $(k, 0)$ on the x-axis.

EXAMPLE 8 Graph the equations $x = -1$, $x = \dfrac{3}{2}$, and $x = 3$.

Solution See Figure 10.14. ▪

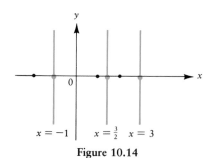

$x = -1$ $x = \frac{3}{2}$ $x = 3$

Figure 10.14

EXERCISES 10.2

Find three pairs of values that will satisfy each of the equations in Exercises 1–12.

1. $x + y = 4$
2. $2x + y = 8$
3. $x - 3y = 3$
4. $2x + 1 = 3$
5. $3x + 2y = 5$
6. $3x - 5y = 10$
7. $3x + 4y = 12$
8. $2x + 3y = 9$
9. $7x - y = 7$
10. $-x - y = 13$
11. $7x + 5y = 0$
12. $-8x + 13y = 0$

Graph each equation in Exercises 13–30 on its own Cartesian plane.

13. $y = 2x + 3$
14. $y = -3x + 2$
15. $y = \dfrac{1}{2}x + 1$
16. $y = -\dfrac{1}{2}x + 5$
17. $y - x = 0$
18. $y + x = 0$

19. $y = \dfrac{1}{3}x$ **20.** $x = 3y$ **25.** $y = -1$ **26.** $y = \dfrac{7}{2}$

21. $3x - y = 6$ **22.** $3x - 2y = 12$

23. $5x + 2y = 20$ **24.** $3x = 5 - 2y$ **27.** $x = 4$ **28.** $x = -\dfrac{1}{2}$

29. $x = 0$ **30.** $y = 0$

10.3 THE SLOPE OF A LINE

Consider the line given in Figure 10.15. If we start at point $R(-3, -2)$ and move to point $Q(-1, 0)$, how far have we moved? We move 2 units up and 2 units to the right. Thus, we had a vertical change of 2 units and a horizontal change of 2 units. If we move from point $Q(-1, 0)$ to point $M(3, 4)$, there is a vertical change of 4 units and a horizontal change of 4 units. As we move from point $M(3, 4)$ to point $P(0, 1)$, the vertical change is -3, since we move down, and the horizontal change is -3, since we move to the left.

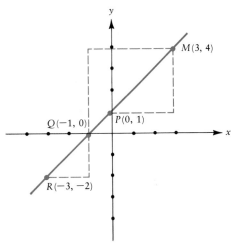

Figure 10.15

We now consider the *ratio* of the vertical change to the horizontal change in each case of these three cases.

1. Moving from $R(-3, -2)$ to $Q(-1, 0)$:

$$\frac{\text{vertical change}}{\text{horizontal change}} = \frac{2}{2} = 1$$

2. Moving from $Q(-1, 0)$ to $M(3, 4)$:

$$\frac{\text{vertical change}}{\text{horizontal change}} = \frac{4}{4} = 1$$

3. Moving from $M(3, 4)$ to $P(0, 1)$:

$$\frac{\text{vertical change}}{\text{horizontal change}} = \frac{-3}{-3} = 1$$

In each case, if we label one point (x_1, y_1) and the second point (x_2, y_2), we see that the ratio of the vertical change $y_2 - y_1$ to the horizontal change $x_2 - x_1$ is the same. The ratio is called the slope of a line. It can be shown that this ratio is the same for any two points lying on the line.

The Slope Formula

If a line passes through points $P_1(x_1, y_1)$ and $P_2(x_2, y_2)$, then its slope is given by the formula

$$m = \frac{y_2 - y_1}{x_2 - x_1} = \frac{\text{vertical change}}{\text{horizontal change}} = \frac{\text{rise}}{\text{run}} \quad (x_1 \neq x_2)$$

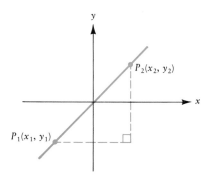

The slope can be taken as a measure of the steepness or inclination of the line passing through the points P_1 and P_2.

EXAMPLE 1 Determine the slope of the line passing through the given pair of points. Graph the lines.

(a) $(1, 3); (4, 5)$ (b) $(-2, 3), (0, 2)$

Solution (a) Let $(x_1, y_1) = (1, 3)$ and $(x_2, y_2) = (4, 5)$. Then the slope

$$m = \frac{y_2 - y_1}{x_2 - x_1} = \frac{5 - 3}{4 - 1} = \frac{2}{3}$$

Also, if we had let $(x_1, y_1) = (4, 5)$ and $(x_2, y_2) = (1, 3)$, we would have

$$m = \frac{y_2 - y_1}{x_2 - x_1} = \frac{3 - 5}{1 - 4} = \frac{-2}{-3} = \frac{2}{3}$$

(b) Let $(x_1, y_1) = (-2, 3)$ and $(x_2, y_2) = (0, 2)$. Then

$$m = \frac{2 - 3}{0 - (-2)} = \frac{-1}{2}$$

Similarly, if we let $(x_1, y_1) = (0, 2)$ and $(x_2, y_2) = (-2, 3)$, then

$$m = \frac{3 - 2}{-2 - 0} = \frac{1}{-2} = -\frac{1}{2}$$

See Figure 10.16.   ■

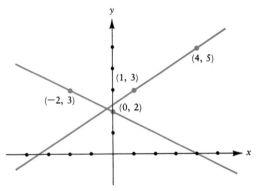

Figure 10.16

Note. Since $(y_2 - y_1) = -(y_1 - y_2)$ and $(x_2 - x_1) = -(x_1 - x_2)$, we have $\frac{y_2 - y_1}{x_2 - x_1} = \frac{-(y_1 - y_2)}{-(x_1 - x_2)} = \frac{(y_1 - y_2)}{(x_1 - x_2)}$. So the slope of a line is the same regardless which point is called (x_1, y_1) and which is called (x_2, y_2).

Horizontal and vertical lines are special cases. Consider the following example.

EXAMPLE 2   Determine the slope of (a) the horizontal line $y = 2$ and (b) the vertical line $x = 1$.

Solution   See Figure 10.17.

(a) On a horizontal line, every point has the same y-coordinate. Thus, the points $(-1, 2)$ and $(3, 2)$ lie on the line $y = 2$, and the slope $= \frac{2 - 2}{3 - (-1)} = \frac{0}{4} = 0$.

(b) On a vertical line, every point has the same x-coordinate. Thus, the points $(1, 4)$ and $(1, 3)$ lie on the line $x = 1$, and $x_2 - x_1 = 1 - 1 = 0$. Hence the denominator in the slope formula is 0. Because we cannot divide by zero, the vertical line $x = 1$ is said to have no slope.   ■

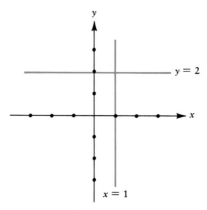

Figure 10.17

In general, for any horizontal line the expression $(y_2 - y_1) = 0$, and

every horizontal line has a slope of zero

For any vertical line, the expression $(x_2 - x_1) = 0$ and

every vertical line has no slope

As Examples 1 and 2 illustrate, the slope of a line may be positive, negative, zero, or no slope. Each of these cases can be interpreted geometrically as follows:

Slope	Geometric Interpretation	Graph
Positive	Line rises upward to right	
Negative	Line falls downward to right	
Zero	Horizontal line	

Slope	Geometric Interpretation	Graph
No Slope	Vertical line	

Parallel and Perpendicular Lines

We now state two important theorems about lines.

Parallel Line Theorem

If two nonvertical lines are parallel, then their slopes are equal. Conversely, if the slopes of two nonvertical lines are equal, then the lines are parallel. That is, two nonvertical lines with slopes m_1 and m_2 are parallel if and only if $m_1 = m_2$.

EXAMPLE 3 Show that the line passing through $(-2, 0)$ and $(0, 1)$ is parallel to the line passing through $(1, -1)$ and $(5, 1)$.

Solution The slope of the line passing through $(-2, 0)$ and $(0, 1)$ is

$$m_1 = \frac{1 - 0}{0 - (-2)} = \frac{1}{2}$$

The slope of the line passing through $(1, -1)$ and $(5, 1)$ is

$$m_2 = \frac{1 - (-1)}{5 - 1} = \frac{2}{4} = \frac{1}{2}$$

Since $m_1 = m_2$, the lines are parallel. ■

Perpendicular Line Theorem

If two nonvertical lines are perpendicular, then their slopes are negative reciprocals. Conversely, if two nonvertical lines have slopes that are negative reciprocals, the lines are perpendicular. That is, two nonvertical lines with slopes m_1 and m_2 are perpendicular if and only if $m_2 = \dfrac{-1}{m_1}$.

EXAMPLE 4 Show that the line passing through $(-1, -2)$ and $(3, 0)$ is perpendicular to the line passing through $(2, 0)$ and $(0, 4)$.

Solution The slope of the line passing through $(-1, -2)$ and $(3, 0)$ is

$$m_1 = \frac{0 - (-2)}{3 - (-1)} = \frac{2}{4} = \frac{1}{2}$$

The slope of the line passing through $(2, 0)$ and $(0, 4)$ is

$$m_2 = \frac{4 - 0}{0 - 2} = \frac{4}{-2} = -2$$

Since $m_1 = \dfrac{-1}{m_2}, \dfrac{1}{2} = \dfrac{-1}{-2}$, the lines are perpendicular. ■

EXAMPLE 5 Using the slope, show that the following four points form the vertices of a parallelogram: $A(-3, 1)$, $B(3, 3)$, $C(4, 0)$, and $D(-2, -2)$.

Solution Slope of line through AB: $\quad m_1 = \dfrac{3 - 1}{3 - (-3)} = \dfrac{2}{6} = \dfrac{1}{3}$

Slope of line through BC: $\quad m_2 = \dfrac{0 - 3}{4 - 3} = \dfrac{-3}{1} = -3$

Slope of line through CD: $\quad m_3 = \dfrac{-2 - 0}{-2 - 4} = \dfrac{-2}{-6} = \dfrac{1}{3}$

Slope of line through DA: $\quad m_4 = \dfrac{1 - (-2)}{-3 - (-2)} = \dfrac{3}{-1} = -3$

Since $m_1 = m_3$, the line through AB is parallel to the line through CD. Since $m_2 = m_4$, the line through BC is parallel to the line through DA. Thus, the points form the vertices of a parallelogram. ■

EXERCISES 10.3

In Exercises 1–12, determine the slope of the line passing through each pair of points.

1. $(-2, -3)$; $(4, 6)$
2. $(-2, 8)$; $(12, -4)$
3. $(-3, 2)$; $(-5, 8)$
4. $(3, 6)$; $(6, 3)$
5. $(0, 0)$; $(3, 4)$
6. $(-5, 12)$; $(0, 0)$
7. $(6, -14)$; $(-3, 4)$
8. $(-6, 4)$; $(3, -9)$
9. $(12, -3)$; $(15, -3)$
10. $(6, -3)$; $(6, 7)$
11. (a, b), (b, a)
12. $(a + b, b)$; $(a, a + b)$

In Exercises 13–16, the three points A, B, and C are collinear; that is, they lie on the same line, if the slope of AB is equal to the slope of BC. Use this fact to show that the points in each of the following sets are collinear.

13. $(0, -4)$, $(-2, -2)$, $(-4, 0)$
14. $(2, 1)$, $(0, -1)$, $(2, 2)$
15. $(6, 8)$, $(0, 0)$, $(3, 4)$
16. $(-3, 6)$, $(-1, 2)$, $(0, 0)$

In Exercises 17–20, determine the value of a so that the three points are collinear.

17. $(2, 3)$, $(4, 6)$, $(a, 9)$
18. $(0, 0)$, $(-1, -2)$, $(3, a)$
19. $(1, -1)$, $(-2, 5)$, $(2, a)$
20. $(1, -1)$, $(2, 1)$, (a, a)
21. Show that the line through $(1, -2)$ and $(6, 5)$ is parallel to the line through $(-2, -4)$ and $(3, 3)$.
22. Show that the line through $(2, 0)$ and $(0, 3)$ is parallel to the line through $(4, 0)$ and $(0, 6)$.
23. Show that the line through $(2, -2)$ and $(-2, -3)$ is perpendicular to the line through $(0, 4)$ and $(1, 0)$.
24. Show that the line through $(5, -1)$ and

$(-2, -3)$ is perpendicular to the line through $(-4, 8)$ and $(3, 2)$.
25. Using slopes, show that the points $A(0, 3)$, $B(3, 4)$, $C(5, -2)$, and $D(2, -3)$ form the vertices of a rectangle.
26. Using slopes and the distance formula, show that the points $A(3, 2)$, $B(0, -1)$, $C(0, 5)$, and $D(-3, 2)$ form the vertices of a square.
27. Using slopes, prove that the triangle $A(0, 0)$, $B(10, -4)$, and $C(2, 5)$ is a right triangle.
28. Find the slopes of the medians of the triangle $A(0, 0)$, $B(2, -3)$, and $C(1, -5)$.
*29. Show that the diagonals of a square intersect at right angles. (*Hint:* Assume that one corner is at $(0, 0)$ and two sides are on the axes.)
*30. Prove that the diagonals of a rhombus are perpendicular bisectors of one another.

10.4 EQUATIONS OF A LINE

Up to now we have been given an equation that represents the line or given two points that graphically determine the unique line passing through the two points. We now reverse the situation and see what information is needed to determine an equation of a line. We shall discuss three ways in which an equation for a line can be derived:

1. two-point equation of a line
2. point-slope equation of a line
3. slope-intercept equation of a line

The Two-Point Equation of a Line

Suppose we wish to find an equation of a line that passes through two points $P_1(x_1, y_1)$ and $P_2(x_2, y_2)$. See Figure 10.18. Let $P(x, y)$ be any other point on the

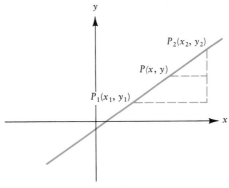

Figure 10.18

line passing through $P_1(x_1, y_1)$ and $P_2(x_2, y_2)$. Using the given points P_1 and P_2, we get

$$m = \frac{(y_2 - y_1)}{(x_2 - x_1)}$$

Using the general point $P(x, y)$ and either of the given points, choosing $P_1(x_1, y_1)$, we also get

$$m = \frac{(y - y_1)}{(x - x_1)}$$

Equating the two expressions for m, we obtain

$$\frac{y - y_1}{x - x_1} = \frac{y_2 - y_1}{x_2 - x_1} \quad \text{or} \quad y - y_1 = \frac{y_2 - y_1}{x_2 - x_1}(x - x_1)$$

Two-Point Equation Theorem

An equation of a nonvertical line passing through points $P_1(x_1, y_1)$ and $P_2(x_2, y_2)$ is given by

$$y - y_1 = \frac{y_2 - y_1}{x_2 - x_1}(x - x_1)$$

EXAMPLE 1   Find the equation of the line passing through $(0, 3)$ and $(2, 7)$.

Solution   Let $P_1(x_1, y_1) = (0, 3)$ and $P_2(x_2, y_2) = (2, 7)$. Thus, $x_1 = 0$, $y_1 = 3$, $x_2 = 2$, and $y_2 = 7$. Substituting into the two-point formula, we have

$$y - 3 = \frac{7 - 3}{2 - 0}(x - 0)$$

or $y - 3 = 2x$. Simplifying, we obtain $y = 2x + 3$.   ■

The Point-Slope Equation of a Line

Let us suppose that we know the slope m of a line and a point $P_1(x_1, y_1)$ through which it passes. We can obtain the equation of the line by referring to the two-point formula,

$$\frac{y - y_1}{x - x_1} = \frac{y_2 - y_1}{x_2 - x_1}$$

The right side of this formula is an expression for the slope. However, we are given the slope as m. Replacing the right side of the two-point formula by m, we obtain

$$\frac{y - y_1}{x - x_1} = m \quad \text{or} \quad y - y_1 = m(x - x_1)$$

Point-Slope Equation Theorem

An equation of a nonvertical line passing through the point $P_1(x_1, y_1)$ and having slope m is given by

$$y - y_1 = m(x - x_1)$$

EXAMPLE 2   Find an equation of the line having slope $-\frac{2}{3}$ and passing through the point $(9, -5)$.

Solution   We have $m = -\frac{2}{3}$ and $(x_1, y_1) = (9, -5)$. Using the point-slope equation, we get

$$y - (-5) = -\frac{2}{3}(x - 9)$$

or

$$y + 5 = -\frac{2}{3}x + 6$$

Solving for y, we get $y = -\frac{2}{3}x + 1$.   ■

EXAMPLE 3   Find an equation of the horizontal line passing through $(3, -7)$.

Solution   Any horizontal line has zero slope. Thus, substituting $m = 0$, $x_1 = 3$, and $y_1 = -7$ into the point-slope equation gives

$$y - (-7) = 0(x - 3) \quad \text{or} \quad y = -7 \quad ■$$

The Slope-Intercept Equation of a Line

Any nonvertical line crosses the y-axis. The point where this happens is called the y-intercept of the line. Since the y-intercept is a point that lies on the y-axis, it has coordinates $(0, b)$. See Figure 10.19.

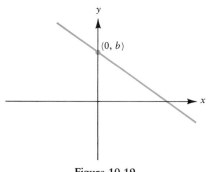

Figure 10.19

EXAMPLE 4   Given the linear equation $y = 2x + 5$, the y-intercept can be found by setting $x = 0$. Thus, $y = 2(0) + 5 = 5$, and the y-intercept is $(0, 5)$.

Slope-Intercept Equation Theorem

An equation of a nonvertical line with slope m and y-intercept b is given by

$$y = mx + b$$

Proof   The point-slope theorem says that given the slope m and point (x_1, y_1), an equation of the line satisfying these conditions is

$$y - y_1 = m(x - x_1)$$

If we use the y-intercept as the point, we have $(x_1, y_1) = (0, b)$. Substituting $(0, b)$ into the point-slope equation, we get

$$y - b = m(x - 0)$$

or

$$y = mx + b \quad \blacksquare$$

EXAMPLE 5   Find an equation of the line with slope -3 and y-intercept 7.

Solution   We have $m = -3$ and $b = 7$. Thus, an equation is

$$y = -3x + 7 \quad \blacksquare$$

Note. One of the most important uses of the slope-intercept form is to find the slope and y-intercept directly from the equation.

EXAMPLE 6   Find the slope and y-intercept of the line whose equation is $2x - 3y - 4 = 0$.

Solution   Our goal is to write the equation $2x - 3y - 4 = 0$ in the form $y = mx + b$. This can be achieved by solving the equation for y in terms of x:

$$2x - 3y - 4 = 0$$
$$-3y = -2x + 4$$
$$y = \frac{2}{3}x - \frac{4}{3}$$

Comparing $y = mx + b$ and $y = \frac{2}{3}x - \frac{4}{3}$, we have a slope of $m = \frac{2}{3}$ and a y-intercept of $b = -\frac{4}{3}$. ■

CAUTION

1. If we are given the equation $5y + 2x - 6 = 0$, do *not* write $m = 2$ and $b = -6$. We must first solve for y before determining the slope m and y-intercept b,

$$y = -\frac{2}{5}x + \frac{6}{5} \quad \text{and} \quad m = -\frac{2}{5} \quad \text{and} \quad b = \frac{6}{5}.$$

2. If we are given the equation $y = 7x - 3$, do *not* write $m = 7$ and $b = 3$. The sign is also part of each answer; that is, $m = 7$ and $b = -3$.

In this section we have developed three forms for linear equations in two variables. The general linear equation in two variables is an equation of the form $Ax + By + C = 0$, where A, B, and C are constants and not both A and B are zero. If $B \neq 0$, the equation can be written in the slope-intercept form as

$$y = \left(-\frac{A}{B}\right)x + \left(-\frac{C}{B}\right)$$

with slope $m = -\frac{A}{B}$ and y-intercept $b = -\frac{C}{B}$. If $B = 0$, then $A \neq 0$ and the equation can be written as

$$x = -\frac{C}{A}$$

the equation of a vertical line.

EXAMPLE 7   Find an equation of the line passing through the point $(-2, 3)$ and parallel to the line $3x - 6y = 14$.

Solution The line we are looking for is parallel to the line whose equation is $3x - 6y = 14$. Therefore, the two lines must have the same slope. Putting $3x - 6y = 14$ into slope-intercept form, we get $y = \frac{1}{2}x - \frac{7}{3}$, and the slope is $\frac{1}{2}$. Thus the line we seek must also have slope $\frac{1}{2}$ and contain the point $(-2, 3)$. Using the point-slope equation with $m = \frac{1}{2}$ and $(x_1, y_1) = (-2, 3)$, we have the equation

$$y - 3 = \frac{1}{2}[x - (-2)]$$

or

$$y = \frac{1}{2}x + 4 \quad \blacksquare$$

EXAMPLE 8 Write an equation of the line passing through $(2, 9)$ and perpendicular to $y = 3x + 4$.

Solution The slope of the given line is $m = 3$. Therefore, the slope of the perpendicular line is $-\frac{1}{3}$. Using the point-slope equation with $m = -\frac{1}{3}$ and $(x_1, y_1) = (2, 9)$, we have the equation

$$y - 9 = -\frac{1}{3}(x - 2)$$

or

$$y = -\frac{1}{3}x + \frac{29}{3} \quad \blacksquare$$

EXAMPLE 9 Find the conditions on the coefficients such that the lines $Ax + By + C = 0$ and $Dx + Ey + F = 0$ are parallel.

Solution The slope of the line $Ax + By + C = 0$ and the slope of the line $Dx + Ey + F = 0$ can be found by putting each equation in slope-intercept form:

$$y = -\frac{A}{B}x - \frac{C}{B} \quad \text{and} \quad y = -\frac{D}{E}x - \frac{F}{E}$$

If the lines are to be parallel, their slopes must be equal. That is,

$$-\frac{A}{B} = -\frac{D}{E}$$

or, equivalently,

$$AE - BD = 0 \quad \blacksquare$$

EXERCISES   10.4

In Exercises 1–9, find an equation of the line passing through each of the pairs of points.

1. $(1, 2), (1, 5)$ **2.** $(0, 0), (-1, 3)$
3. $(6, 2), (-5, 3)$ **4.** $(-3, -2), (5, 7)$
5. $(-3, -2), (1, 1)$
6. $(0, 0), (-1, 4)$ **7.** $(-2, 3), (4, -1)$
8. $(0, 0), (4, 5)$ **9.** $(-4, 3), (2, -5)$

Find the equations of the medians of the triangles whose vertices are as given in Exercises 10 and 11.

10. $(2, 3), (0, -2), (6, -3)$
11. $(-3, 5), (-5, 3), (6, 10)$
12. Find an equation of the line passing through $(1, 1)$ and parallel to the line passing through $(3, 4)$ and $(5, -8)$.
13. Find an equation of the line passing through $(1, 1)$ and perpendicular to the line passing through $(3, 4)$ and $(5, -8)$.

In Exercises 14–19, find an equation of the line passing through the given point with the given slope.

14. $m = 0, (-5, 8)$

15. $m = -\dfrac{1}{2}, (5, 4)$

16. $m = -3, (-1, -2)$

17. $m = -\dfrac{1}{3}, (1, 4)$

18. $m = \dfrac{1}{2}, (5, 6)$

19. $m = -1, (-4, 3)$

In Exercises 20–25, find an equation of the line having slope m and y-intercept b.

20. $m = -3, b = 5$

21. $m = 4, b = 0$ **22.** $m = \dfrac{3}{2}, b = -1$

23. $m = 6, b = \dfrac{1}{2}$ **24.** $m = -1, b = 7$

25. $m = -\dfrac{1}{5}, b = 3$

In Exercises 26–31, find the slope and y-intercept of each of the lines.

26. $3x - y = 6$ **27.** $5x - 2y - 10 = 0$
28. $2x + y + 4 = 0$
29. $2x - 3y - 6 = 0$
30. $3x + 5y - 20 = 0$
31. $3x - 4y - 12 = 0$

In Exercises 32–35, find an equation of the line through the given point which is (a) parallel and (b) perpendicular to the given line.

32. $(1, 2), x - y - 3 = 0$
33. $(4, 6), x - 2y - 4 = 0$
34. $(0, 0), 3x - 5y - 15 = 0$
35. $(6, 8), x - 5y - 6 = 0$
36. Find an equation of the x-axis.
37. Find an equation of the y-axis.
38. Find an equation of the line through $(4, -3)$ parallel to the x-axis.
39. Find an equation of the line through $(4, -3)$ parallel to the y-axis.
40. Find an equation of the line through $(-6, 5)$ perpendicular to the x-axis.
41. Find an equation of the line through $(-6, 5)$ perpendicular to the y-axis.
42. Find an equation of the line through the origin whose slope is three times that of the line $y - 3x + 5 = 0$.
43. Prove that the two points $(5, 2)$ and $(6, -15)$ subtend a right angle at the origin.
44. Find the value of k so that the line $3x + ky = 9$ and the line passing through $(7, -2)$ and $(5, -1)$ are parallel.
45. Find the value of k so that the lines $-2x + ky = 1$ and $3x - 8y = 7$ are perpendicular.
46. Find the conditions on the coefficients such that the lines $Ax + By + C = 0$ and $Dx + Ey + F = 0$ are perpendicular.
47. Find an equation of the line that makes equal intercepts on the axes and passes through the point $(3, 1)$.
48. Find an equation of the line that makes equal intercepts on the axes and passes through the point $(2, 5)$.
49. Show that the line having x-intercept a $(a \neq 0)$ and y-intercept b $(b \neq 0)$ has an equation

$$\frac{x}{a} + \frac{y}{b} = 1$$

*50. Find an equation of the line passing through $(3, 3)$ and forming with the axes a triangle of area of 24.

10.5 SYMMETRY

The work involved in graphing an equation can be greatly reduced if the graph has certain symmetry properties.

Symmetry with Respect to the y-axis

EXAMPLE 1 Consider the graph of the equation $y = x^2$ (see Figure 10.20). Note that the value of y is the same for $x = 1$ and $x = -1$, similarly for $x = 2$ and $x = -2$. Such curves are said to be symmetric with respect to the y-axis.

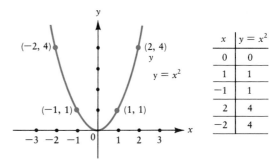

x	$y = x^2$
0	0
1	1
-1	1
2	4
-2	4

Figure 10.20

DEFINITION

If the equation of a curve is unchanged when x is replaced by $-x$, then the curve is symmetric with respect to the y-axis. That is, a curve is called symmetric with respect to the y-axis if for each point (x, y) on the curve, the point $(-x, y)$ is also on the curve.

EXAMPLE 2 The three curves in Figure 10.21 are symmetric with respect to the y-axis.

Note. Geometrically, for curves that are symmetric with respect to the y-axis, that portion of the curve to the left of the y-axis is the mirror image of the portion to the right of the y-axis.

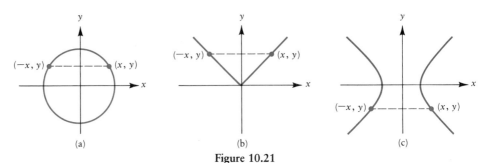

(a) (b) (c)

Figure 10.21

EXAMPLE 3   Without sketching the graph, determine symmetry with respect to the y-axis for: (a) $3x^2 - 2y^2 + y = 4$ (b) $xy = 7$

Solution   (a) $3x^2 - 2y^2 + y = 4$

Replacing x by $-x$, we get

$$3(-x)^2 - 2y^2 + y = 4$$
$$3x^2 - 2y^2 + y = 4$$

Since the equation is unchanged, the curve is symmetric with respect to the y-axis.

(b) $xy = 7$

Replacing x by $-x$, we get

$$(-x)y = 7, \quad \text{or,} \quad -xy = 7$$

The equation is *not* the same as the original equation, $xy = 7$. Thus, the curve is not symmetric with respect to the y-axis.   ■

Symmetry with Respect to the x-axis

EXAMPLE 4   Consider the graph of $x = y^2$ (see Figure 10.22). Note that the value of x is the same for $y = 1$ and $y = -1$, similarly for $y = 2$ and $y = -2$. Such curves are said to be symmetric with respect to the x-axis.

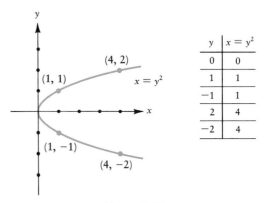

y	$x = y^2$
0	0
1	1
-1	1
2	4
-2	4

Figure 10.22

DEFINITION

If the equation of a curve is unchanged when y is replaced by $-y$, then the curve is symmetric with respect to the x-axis. That is, a curve is symmetric with respect to the x-axis if for each point (x, y) on the curve, the point $(x, -y)$ is also on the curve.

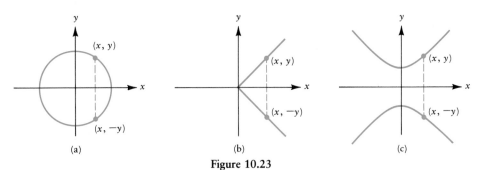

Figure 10.23

EXAMPLE 5 The three curves in Figure 10.23 are symmetric with respect to the x-axis.

Note. Geometrically, for curves that are symmetric with respect to the x-axis, that portion of the curve below the x-axis is the mirror image of the portion above the x-axis.

EXAMPLE 6 Without sketching the graph, determine symmetry with respect to the x-axis for:

(a) $x = |y|$ (b) $5x^2 - y^2 + 2y = 0$

Solution (a) $x = |y|$

Replacing y by $-y$, we get $x = |-y|$. Since $|-y| = |y|$, we once again obtain $x = |y|$. Thus, the curve is symmetric with respect to the x-axis. See Figure 10.23(b).

(b) $5x^2 - y^2 + 2y = 0$

Replacing y by $-y$, we obtain

$$5x^2 - (-y)^2 + 2(-y) = 0$$
$$5x^2 - y^2 - 2y = 0$$

The resulting equation is *not* the same as the original equation, $5x^2 - y^2 + 2y = 0$. Thus, the curve is not symmetric with respect to the x-axis. ∎

Symmetry with Respect to the Origin

EXAMPLE 7 Consider the graph of $y = x^3$ (see Figure 10.24). Note that the point $(1, 1)$ and the point $(-1, -1)$ lie on the curve.

DEFINITION

If the equation of a curve is unchanged when x is replaced by $-x$ and y is replaced by $-y$, then the curve is symmetric with respect to the origin. That is, a curve is symmetric with respect to the origin if for each point (x, y) on the curve, the point $(-x, -y)$ is also on the curve.

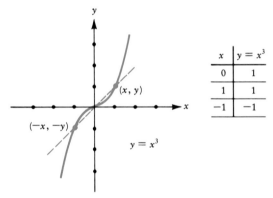

Figure 10.24

EXAMPLE 8 The three curves in Figure 10.25 are symmetric with respect to the origin.

> **Note.** Geometrically, for curves that are symmetric with respect to the origin, for each point P on the curve there is a corresponding point Q on the curve such that the origin bisects the line segment joining P and Q.

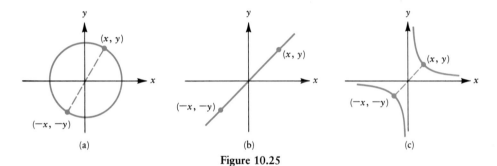

Figure 10.25

EXAMPLE 9 Without sketching the graph, determine symmetry with respect to the origin for:

(a) $2x^2 + 5y^2 = 10$ (b) $y = x^2 + 4$

Solution (a) $2x^2 + 5y^2 = 10$

Replacing x by $-x$ and y by $-y$, we get

$$2(-x)^2 + 5(-y)^2 = 10$$
$$2x^2 + 5y^2 = 10$$

Since the equation is unchanged, the curve is symmetric with respect to the origin.

(b) $y = x^2 + 4$

Replacing x by $-x$ and y by $-y$, we have

$$-y = (-x)^2 + 4$$

$$-y = x^2 + 4$$
$$y = -x^2 - 4$$

Since the equation is changed, it is not symmetric with respect to the origin. ■

In Examples 2(a), 5(a), and 8(a), we see that the circle with its center at the origin is symmetric with respect to the y-axis and the origin. In general, it can be shown that

---

Symmetry Property of Curves

A curve that satisfies any two of the symmetry properties also satisfies the third symmetry property.

EXERCISES 10.5

In Exercises 1–8, determine the symmetry, if any, of the given curves.

1.

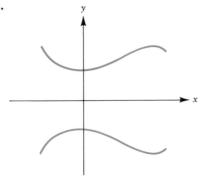

3.

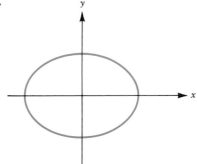

2.

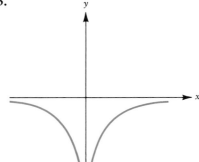

4.

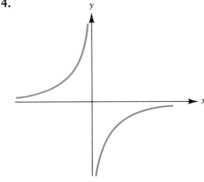

5.

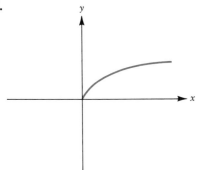

6.

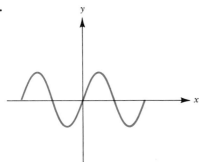

7.

8.

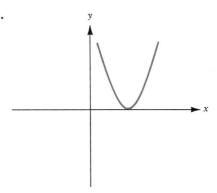

In Exercises 9–29, determine, without graphing, whether the graph is symmetric with respect to the x-axis, y-axis, the origin, or none.

9. $y = x - 2$ 10. $7x = 4y$

11. $y = 3x^2$ 12. $y^2 = x + 3$

13. $y = 5 - 3x^2$ 14. $y = x^3 + 2$

15. $y = (x + 4)^2$ 16. $x^2 = y^2 + 10$

17. $y = -yx^2 + x$ 18. $y^3 = x^2 - 1$

19. $y = \dfrac{1}{x^2}$ 20. $y = \dfrac{1}{x^2 - 1}$

21. $y^2 = 2x^2 + 3$ 22. $5x^2 - 6y^2 = 49$

23. $y^2 = \dfrac{1}{x^2}$ 24. $y = x - \dfrac{1}{x^2}$

25. $xy = 3$ 26. $x^4 - 2y^3 = y$

27. $y = \dfrac{x}{5 + x^2}$ 28. $y^2 = |x| - 2$

29. $y = |x| + 3$

10.6 INTRODUCTION TO THE CONIC SECTIONS

In the following sections we introduce some fundamental concepts of analytic geometry. This area of mathematics deals with the relationships between the geometry of graphs and the algebraic representations of the graphs. In particular, we shall study a **conic (conic section)**, which is a curve of intersection of a plane with a right-circular cone of two nappes; the four types of curves that result are the **circle**, the **parabola**, the **ellipse**, and the **hyperbola**.

Geometrically, a cone can be thought of as having two nappes, extending indefinitely in both directions. A **generator** of a cone is any line lying in the cone and all the generators of a cone intersect in a common point, V, called the **vertex** of the cone. The line OV is called the **axis** (see Figure 10.26). If the plane cutting

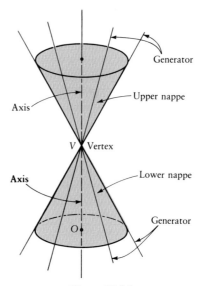

Figure 10.26

a right-circular cone is not parallel to any generator of the cone, then the conic obtained is an *ellipse*. A special case of the ellipse is a *circle*, which results if the cutting plane is also perpendicular to the axis of the cone. If the cutting plane is parallel to one and only one generator of the cone, then the conic obtained is a *parabola*. If the cutting plane is parallel to two generators, it intersects both nappes of the cone and a *hyperbola* is obtained. The graphs of these conics are shown in Figure 10.27.

Later in this chapter, we shall see that all conic sections can be algebraically represented by the equation

$$Ax^2 + Cy^2 + Dx + Ey + F = 0 \tag{1}$$

where not both A and C are zero. Also, we shall give conditions on the constants in equation (1) that determine whether the graph of that equation is a circle, parabola, ellipse, or hyperbola.

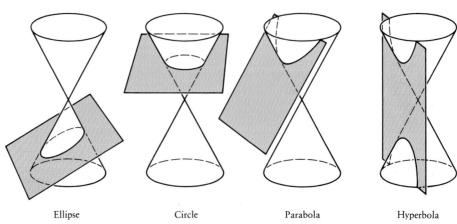

| Ellipse | Circle | Parabola | Hyperbola |

Figure 10.27

10.7 THE CIRCLE

As a conic section, a circle is the curve obtained from the intersection of a right-circular cone by a plane perpendicular to the axis of the cone (see Figure 10.28).Geometrically, the definition of a circle is as follows.

> **DEFINITION**
>
> A circle is the set of all points in a plane at a given distance from a fixed point. The fixed point is called the center of the circle and the fixed distance is called the radius of the circle.

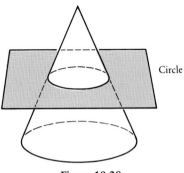

Figure 10.28

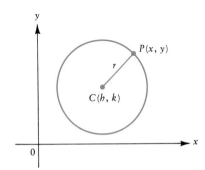

Figure 10.29

If the center C is taken at the point (h, k) and the radius is r (see Figure 10.29), then the equation of the circle can be obtained by using the distance formula. Let $P(x, y)$ be any point on the circle. The distance from the center C to the point P must equal r. That is,

$$\sqrt{(x - h)^2 + (y - k)^2} = r$$

Squaring both sides gives

$$(x - h)^2 + (y - k)^2 = r^2$$

and we can state the following theorem.

> **Theorem 1**
>
> The equation in standard form of the circle whose center is (h, k) and whose radius equals r is:
>
> $$(x - h)^2 + (y - k)^2 = r^2$$

EXAMPLE 1   Find an equation of the circle with center $(-3, 4)$ and radius equal to 7.

Solution   We have $h = -3$, $k = 4$, and $r = 7$. Substituting into the equation $(x - h)^2 + (y - k)^2 = r^2$, we have

$$(x - (-3))^2 + (y - 4)^2 = (7)^2$$

or

$$(x + 3)^2 + (y - 4)^2 = 49$$

Squaring and then combining terms, we obtain

$$x^2 + y^2 + 6x - 8y - 24 = 0 \quad \blacksquare$$

EXAMPLE 2   Find an equation of the circle with center at the origin and radius r.

Solution   If the center is at the origin, then we have $h = 0$ and $k = 0$. Thus, we have

$$x^2 + y^2 = r^2 \quad \blacksquare$$

EXAMPLE 3   Given the standard form of a circle, $(x + 5)^2 + y^2 = 36$, find the center and radius.

Solution   The standard form of a circle is $(x - h)^2 + (y - k)^2 = r^2$. Therefore, for $(x + 5)^2 + y^2 = 36$, we have $x - h = x + 5$, or $h = -5$, and $y - k = y$, or $k = 0$. Thus, the center $(h, k) = (-5, 0)$ and the radius $r = 6$. $\quad \blacksquare$

EXAMPLE 4   Express $4x^2 + 4y^2 - 8x + 24y + 4 = 0$ in standard form.

Solution   The standard form can be obtained by completing the square on the terms involving x and the terms involving y. To complete the square for the x-terms and y-terms, the coefficients of x^2 and y^2 terms must be 1. Thus, dividing each side of the equation by 4, we obtain

$$x^2 + y^2 - 2x + 6y + 1 = 0$$

Completing the square gives

$$(x^2 - 2x \quad) + (y^2 + 6y \quad) = -1$$
$$(x^2 - 2x + 1) + (y^2 + 6y + 9) = -1 + 1 + 9$$
$$(x - 1)^2 + (y + 3)^2 = 9$$

Thus, we have a circle with center $(1, -3)$ and radius 3. $\quad \blacksquare$

EXAMPLE 5 Express $x^2 + y^2 + 2x - 4y + 5 = 0$ in standard form.

Solution Completing the square yields

$$(x^2 + 2x \quad) + (y^2 - 4y \quad) = -5$$
$$(x^2 + 2x + 1) + (y^2 - 4y + 4) = -5 + 1 + 4$$
$$(x + 1)^2 + (y - 2)^2 = 0$$

Therefore, we have a circle with center $(-1, 2)$ and radius 0. This represents a point and such a circle is called a degenerate circle (or point circle). ■

EXAMPLE 6 Express $x^2 + y^2 - 6x + 4y + 14 = 0$ in standard form.

Solution Completing the square we obtain

$$(x - 3)^2 + (y + 2)^2 = -1$$

There are no real numbers (x, y) that satisfy this equation. Thus, because of the form of this equation, we say that this equation represents an imaginary circle. ■

In all examples of circles, whether real, a point, or imaginary, we saw that the equation of a circle could be put in standard form by completing the square. However, to achieve the standard form, the coefficients of x^2 and y^2 must be equal. Thus, we have the following theorem.

Theorem 2

The graph of the equation $Ax^2 + Cy^2 + Dx + Ey + F = 0$ is a circle if $A = C \neq 0$.

EXERCISES 10.7

In Exercises 1–10, find the equations of the circles.

1. Center at $(4, 6)$; radius $= 3$
2. Center at $(3, -1)$; radius $= 4$
3. Center at $(-2, -3)$; radius $= 6$
4. Center at $(a, -a)$; radius $= 2a$
5. Center at $(-a, -a)$; radius $= a$
6. Center at (b, b); radius $= b\sqrt{3}$
7. Center at $(6, 0)$ and passing through the origin
8. Center at $(2, 4)$ and passing through the point $(5, 0)$

*9. Diameter, the segment from $(6, 2)$ to $(8, 4)$
*10. Diameter, the segment from $(0, 0)$ to $(4a, 4a)$

In Exercises 11–16, determine the center and radius of each of the circles.

11. $(x - 1)^2 + (y + 6)^2 = 36$
12. $(x + 3)^2 + y^2 = 9$
13. $x^2 + \left(y + \dfrac{1}{3}\right)^2 = \dfrac{1}{4}$ **14.** $x^2 + y^2 = 5$
15. $(x + \pi)^2 + (y - 2)^2 = 16$
16. $(x - b)^2 + (y + a)^2 = c^2$

In Exercises 17–26, reduce each equation to standard form, and determine whether the equation represents a real circle, a point, or an imaginary circle. If it represents a real circle, find the center and radius.

17. $x^2 + y^2 - 6x + 10y - 2 = 0$
18. $x^2 + y^2 - 6x + 8y = 0$
19. $x^2 + y^2 + 5 = 0$
20. $x^2 + y^2 - 2x + 10y + 19 = 0$
21. $2x^2 + 2y^2 + 8x - 4y + 10 = 0$
22. $4x^2 + 4y^2 + 24x - 4y + 1 = 0$

23. $3x^2 + 3y^2 - 4x + 2y + 6 = 0$
24. $x^2 + y^2 - 2x - 6y + 10 = 0$
25. $x^2 + y^2 - 4x - 12 = 0$
26. $x^2 + y^2 + 10x - 24y + 175 = 0$
*27. Write an equation of the circle circumscribing the triangle formed by the coordinate axes and the line $2x + 3y = 6$.
*28. Find an equation of the circle circumscribing the triangle whose sides are $x = 0$, $y = 0$, and the line $2x - y + 4 = 0$.

10.8 THE PARABOLA

As a conic section, a parabola is the curve obtained by the intersection of a right-circular cone by a plane that is parallel to one and only one generator of the cone (see Figure 10.30). The geometrical definition of a parabola is as follows.

DEFINITION

A parabola is the set of all points in a plane equidistant from a fixed point and a fixed line. The fixed point is called the focus and the fixed line is called the directrix.

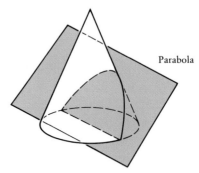

Parabola

Figure 10.30

The equation of a parabola is particularly simple if we choose as focus the point $F(p, 0)$, where p is positive, and as directrix the line $x = -p$ (see Figure 10.31). From Figure 10.31, the definition requires that

$$\text{distance } \overline{FP} = \text{distance } \overline{RP}$$

$$\sqrt{(x - p)^2 + y^2} = \sqrt{(x + p)^2 + (y - y)^2}$$

$$\sqrt{(x - p)^2 + y^2} = \sqrt{(x + p)^2}$$

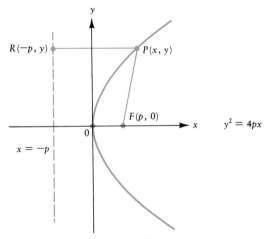

Figure 10.31

Squaring both sides of the equation, we obtain

$$(x - p)^2 + y^2 = (x + p)^2$$
$$x^2 - 2px + p^2 + y^2 = x^2 + 2px + p^2$$

or

$$y^2 = 4px$$

Thus, we have the following theorem.

Theorem 3

An equation of the parabola having its focus at $(p, 0)$, $p > 0$, and its directrix the line $x = -p$, is

$$y^2 = 4px$$

The line through the focal point drawn perpendicular to the directrix is called the axis of symmetry of the parabola. The point of intersection of this line and the curve is called the vertex. The parabola in Figure 10.31 has the x-axis as its axis of symmetry and the origin $(0, 0)$ as its vertex.

If $p > 0$, then the graph of the parabola with focus at $(-p, 0)$ and the line $x = p$ as the directrix is given in Figure 10.32. The following theorem gives the form of the equation of such parabolas.

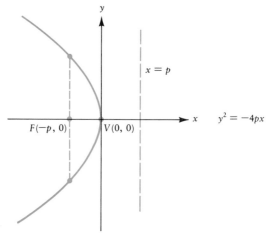

Figure 10.32

Theorem 4

An equation of the parabola having its focus at $(-p, 0)$, $p > 0$, and its directrix the line $x = p$, is

$$y^2 = -4px$$

EXAMPLE 1 Find an equation of the parabola with focus at $(3, 0)$ and directrix $x = -3$.

Solution In this case $p = 3$, and the equation is

$$y^2 = 12x \quad \blacksquare$$

EXAMPLE 2 Find the coordinates of the focus and an equation of the directrix of the parabola whose equation is $y^2 = -7x$.

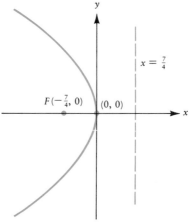

Figure 10.33

Solution   The equation $y^2 = -7x$ is of the form $y^2 = 4px$. Thus, $4p = -7$ or $p = -\frac{7}{4}$. Hence, the focus has coordinates $(-\frac{7}{4}, 0)$ and an equation of the directrix is $x = \frac{7}{4}$ (see Figure 10.33).

If we take the directrix parallel to the x-axis and the focus on the y-axis (see Figure 10.34), we obtain the equations $x^2 = 4py$ and $x^2 = -4py$, depending on whether the focus lies above or below the origin.   ■

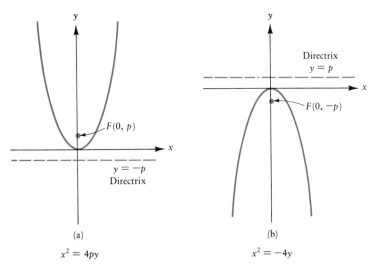

(a)
$x^2 = 4py$

(b)
$x^2 = -4y$

Figure 10.34

EXAMPLE 3   Find the coordinates of the focus and an equation of the directrix of the parabola $x^2 = 2y$.

Solution   The equation $x^2 = 2y$ is of the form $x^2 = 4py$. Thus, $4p = 2$, or $p = \frac{1}{2}$. Hence the focus is $F(0, \frac{1}{2})$ and the directrix is $y = -\frac{1}{2}$ (see Figure 10.35).   ■

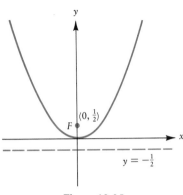

Figure 10.35

Table 10.1 summarizes the graphs, descriptions, and equations of the four parabolas having a vertex at the origin and a vertical or horizontal directrix.

Table 10.1

Graph	Description	Equation
	Vertex at the origin. Parabola opens in the positive x-direction. Symmetric with respect to x-axis.	$y^2 = 4px$ (p positive)
	Vertex at the origin. Parabola opens in the negative x-direction. Symmetric with respect to x-axis	$y^2 = 4px$ (p negative)
	Vertex at the origin. Parabola opens in the positive y-direction Symmetric with respect to y-axis	$x^2 = 4py$ (p positive)
	Vertex at the origin. Parabola opens in the negative y-direction Symmetric with respect to y-axis.	$x^2 = 4py$ (p negative)

But what if the vertex of the parabola is at some point other than the origin, say the point (h, k)? The result will depend on whether the axis of symmetry is parallel to the x-axis or the y-axis. The following theorem gives the standard form of the parabola with vertex at (h, k).

Theorem 5

An equation of a parabola with vertex at $V(h, k)$ and axis of symmetry parallel to the x-axis is

$$(y - k)^2 = 4p(x - h)$$

If $p > 0$, then the parabola opens to the right.
If $p < 0$, then the parabola opens to the left.
The axis of symmetry is the horizontal line $y = k$.

The graphs of these parabolas are given in Figure 10.36.

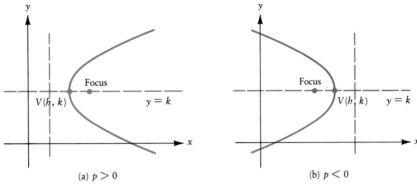

(a) $p > 0$ (b) $p < 0$

Figure 10.36

EXAMPLE 4   Sketch the equation $(y + 1)^2 = -12(x - 4)$.

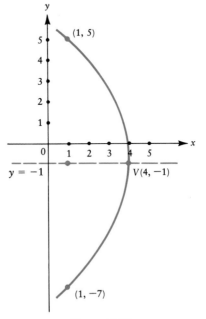

Figure 10.37

Solution The equation $(y + 1)^2 = -12(x - 4)$ can be written as $[y - (-1)]^2 = -12(x - 4)$. Since the equation is of the form $(y - k)^2 = 4p(x - h)$, we have $h = 4$, $k = -1$, and $4p = -12$, or $p = -3$. Therefore, the vertex has coordinates $(h, k) = (4, -1)$, and since $p = -3 < 0$, the parabola opens to the left. See Figure 10.37. ■

For parabolas with the axis of symmetry parallel to the y-axis, we have the following theorem.

Theorem 6

An equation of a parabola with vertex at $V(h, k)$ and axis of symmetry parallel to the y-axis is

$$(x - h)^2 = 4p(y - k)$$

If $p > 0$, then the parabola opens upward.
If $p < 0$, then the parabola opens downward.
The axis of symmetry is the vertical line $x = h$.

The graphs of these parabolas are given in Figure 10.38.

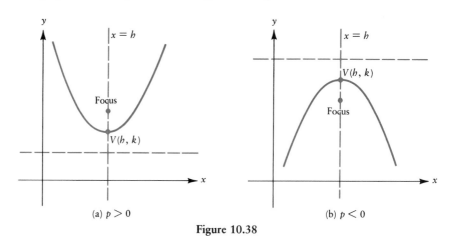

(a) $p > 0$ (b) $p < 0$

Figure 10.38

EXAMPLE 5 Sketch the equation $(x - 1)^2 = 8(y + 3)$.

Solution We rewrite the equation as

$$(x - 1)^2 = 8[y - (-3)]$$

Since the equation is of the form $(x - h)^2 = 4p(y - k)$, we have $h = 1$, $k = -3$, and $4p = 8$, or $p = 2$. Thus, the vertex is $(h, k) = (1, -3)$ and since $p = 2 > 0$, the parabola opens upward. See Figure 10.39. ■

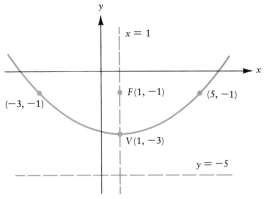

Figure 10.39

EXAMPLE 6 Discuss the graph of the equation $y = x^2 + 2x - 2$.

Solution We complete the square on the right side of the equation $y = x^2 + 2x - 2$. First we take the constant to the left side to obtain

$$y + 2 = x^2 + 2x$$

Now, add ($\frac{1}{2}$ coefficient of x)$^2 = (\frac{1}{2} \cdot 2)^2 = (1)^2 = 1$ to both sides:

$$y + 2 + 1 = x^2 + 2x + 1 = (x + 1)^2$$

Thus, we have

$$y + 3 = (x + 1)^2$$

This is a parabola of the form $(x - h)^2 = 4p(y - k)$. Therefore, $h = -1$, $k = -3$, and $4p = 1$, or $p = \frac{1}{4}$. Thus, the curve $y = x^2 + 2x - 2$ is a parabola with vertex $(-1, -3)$. Also, since $p = \frac{1}{4} > 0$, the parabola opens upward. The graph is given in Figure 10.40. ■

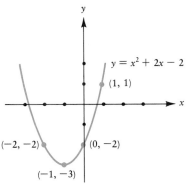

Figure 10.40

Note. The graph of the equation $y = ax^2 + bx + c$, $a \neq 0$, is always a parabola.

If we remove the parentheses from the standard form $(x - h)^2 = 4p(y - k)$, then we get an equation of the form

$$Ax^2 + Dx + Ey + F = 0 \qquad (1)$$

Conversely, by completing the square on x in equation (1), we can show that equation (1) can be written in the standard form $(x - h)^2 = 4p(y - k)$.

Similarly, removing the parentheses from the standard form $(y - k)^2 = 4p(x - h)$, we see that the equation

$$Cy^2 + Dx + Ey + F = 0 \qquad (2)$$

represents a parabola.

We can now state these results as a theorem.

Theorem 7

The graph of the equation $Ax^2 + Cy^2 + Dx + Ey + F = 0$ is a parabola if either A or C equals zero, but not both.

Focusing Property of a Parabola

One very useful property of a parabolic surface is its ability to reflect to its focus any ray of energy that comes in parallel to its axis (see Figure 10.41(a)). The mirror in an astronomical reflecting telescope has a parabolic shape. Conversely, if a small source of energy is placed at the focus of a parabolic surface, the energy will reflect as a parallel beam. An example of this is the parabolic reflectors in car headlights (see Figure 10.41(b)).

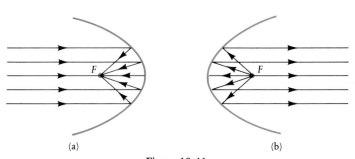

(a)   (b)

Figure 10.41

EXERCISES 10.8

In Exercises 1–16, sketch the curve and locate the vertex.

1. $y^2 = 4x$

2. $x^2 = 4y$

3. $x^2 = 16y$

4. $y^2 = -16x$

5. $y^2 = 64x$

6. $5x^2 = 18y$

7. $x^2 = -10y$

8. $y^2 = -8x$

9. $y^2 = x$

10. $x^2 = -3y$

11. $y^2 = \dfrac{9}{4}x$

12. $y^2 = -7x$

13. $(x - 2)^2 = 4(y - 1)$

14. $(y + 1)^2 = 8(x + 3)$

15. $(y + 1)^2 = 6(x + 3)$

16. $(x - 4)^2 = -10(y + 3)$

In Exercises 17–22, write the equation in standard form and sketch the graph.

17. $y^2 - 2y - 8x + 25 = 0$

18. $x^2 - 4x - 2y - 8 = 0$

19. $x^2 - 6x - \dfrac{1}{2}y + 7 = 0$

20. $y^2 - 6y - x = 0$

21. $4x - y^2 + 6y - 1 = 0$

22. $2y + x^2 - 8x + 18 = 0$

23. Find the set of all points equidistant from the line $x + 2 = 0$ and the point $(4, 0)$.

*24. Let a, b, and c be constants with $a > 0$. Show that $y = ax^2 + bx + c$ is the equation of a parabola opening upwards. Find the coordinates of the vertex V.

10.9 THE ELLIPSE

If we cut a right-circular cone with a plane that is parallel to no generator, then the curve we obtain is an **ellipse** (see Figure 10.42). Geometrically we have the following definition of an ellipse.

DEFINITION

An **ellipse** is the collection of all points in the plane, the sum of whose distances from two fixed points is a constant. The two fixed points are called the **foci** and the midpoint of the segment joining the foci is called the **center** of the ellipse.

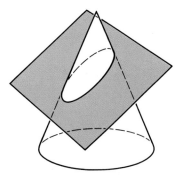

Figure 10.42

To derive an equation of an ellipse, we choose our coordinate system such that the foci are the points $F_1(c, 0)$ and $F_2(-c, 0)$, where $c > 0$. Thus, the line

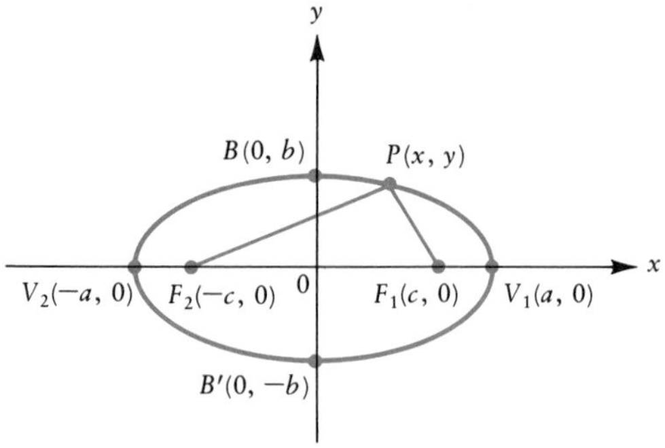

Figure 10.43

F_1F_2 lies along the x-axis and the center is located at the origin (see Figure 10.43). Let the constant sum in the definition be $2a$, $a > c$, and let $P(x, y)$ be any point on the ellipse. Then from the definition of an ellipse and the distance formula we have

$$\overline{F_2P} + \overline{PF_1} = 2a$$

or

$$\sqrt{(x + c)^2 + (y - 0)^2} + \sqrt{(x - c)^2 + (y - 0)^2} = 2a$$

or

$$\sqrt{(x + c)^2 + y^2} = 2a - \sqrt{(x - c)^2 + y^2}$$

Squaring both sides, we get

$$x^2 + 2cx + c^2 + y^2 = 4a^2 - 4a\sqrt{(x - c)^2 + y^2} + x^2 - 2cx + c^2 + y^2$$

or $\qquad -4a^2 + 4cx = -4a\sqrt{(x - c)^2 + y^2}$

or $\qquad -a^2 + cx = -a\sqrt{(x - c)^2 + y^2}$

Squaring both sides again,

$$a^4 - 2a^2cx + c^2x^2 = a^2x^2 - 2ca^2x + a^2c^2 + a^2y^2$$

or $\qquad (a^2 - c^2)x^2 + a^2y^2 = a^2(a^2 - c^2)$

Since $a > c$, then we have $a^2 > c^2$ or $a^2 - c^2 > 0$. Dividing through by $a^2(a^2 - c^2)$, the equation becomes

$$\frac{x^2}{a^2} + \frac{y^2}{a^2 - c^2} = 1$$

Since $a^2 - c^2$ is positive, we can write $a^2 - c^2 = b^2$ and we have the standard form of the equation of an ellipse centered at the origin:

> **Ellipse with Center (0, 0) and Major Axis along x-axis**
>
> $$\frac{x^2}{a^2} + \frac{y^2}{b^2} = 1, \, a > b \qquad (1)$$

Because the equation contains only even powers of x and y, the graph is symmetric with respect to the x-axis, y-axis, and the origin (see Figure 10.43). The longer axis of symmetry, $\overline{V_1 V_2}$, is called the major axis and the shorter axis of symmetry, $\overline{BB'}$, is called the minor axis. The points V_1 and V_2 are called the vertices of the ellipse.

If we take as the foci the points $F_1(0, c)$ and $F_2(0, -c)$, and proceed as above, we obtain

> **Ellipse with Center (0, 0) and Major Axis along y-axis**
>
> $$\frac{x^2}{b^2} + \frac{y^2}{a^2} = 1, \, a > b \qquad (2)$$

In this case, the major axis and the vertices are along the y-axis (see the following example).

EXAMPLE 1   Discuss the graph of $4x^2 + y^2 = 16$ and find its vertices.

Solution   Dividing both sides of the equation by 16, we obtain

$$\frac{x^2}{4} + \frac{y^2}{16} = 1$$

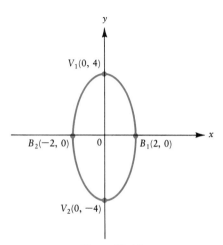

Figure 10.44

which is the standard form (2) for an ellipse whose major axis lies along the y-axis; $a^2 = 16$ so $a = 4$, $b^2 = 4$, or $b = 2$. The vertices are located at $V_1(0, 4)$, $V_2(0, -4)$, $B_1(2, 0)$, and $B_2(-2, 0)$. The graph is given in Figure 10.44.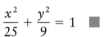

EXAMPLE 2 Find an equation of the ellipse with vertices $(\pm 5, 0)$ and $(0, \pm 3)$.

Solution The graph of the ellipse is given in Figure 10.45. The major axis lies along the x-axis, with $a = 5$ and $b = 3$. Hence, the standard form (1) gives the equation

$$\frac{x^2}{25} + \frac{y^2}{9} = 1$$

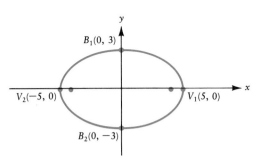

Figure 10.45

Center Not at Origin

If the center of the ellipse is at a point (h, k) and its major axis is parallel to either the x-axis or y-axis, then an equation for the ellipse is given by one of the following formulas.

Suppose the center of an ellipse is at (h, k) and the major axis is parallel to the x-axis. Then, the equation of such an ellipse is

Ellipse with Center (h, k) and Major Axis Parallel to x-axis
$$\frac{(x - h)^2}{a^2} + \frac{(y - k)^2}{b^2} = 1, \quad a > b \qquad (3)$$

See Figure 10.46.

Similarly, if the center of an ellipse is at (h, k) and the major axis is parallel to the y-axis, then the equation of such an ellipse is

Ellipse with Center (h, k) and Major Axis Parallel to y-axis
$$\frac{(x - h)^2}{b^2} + \frac{(y - k)^2}{a^2} = 1, \quad a > b \qquad (4)$$

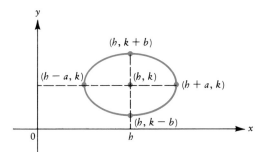

Figure 10.46

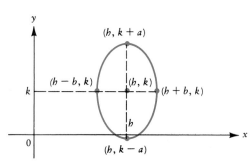

Figure 10.47

See Figure 10.47. Equations (3) or (4) are referred to as the standard form for an ellipse.

EXAMPLE 3 Determine the center, major axis, and vertices for the ellipse $\dfrac{(x-6)^2}{36} + \dfrac{(y+4)^2}{16} = 1$.

Solution The equation is of the form (3). Thus, the center is $(h, k) = (6, -4)$ and the major axis is parallel to the x-axis. Since $a^2 = 36$ and $b^2 = 16$, then $a = 6$ and $b = 4$. From Figure 10.46 we see that the vertices have the coordinates: $(h, k + b) = (6, -4 + 4) = (6, 0)$; $(h, k - b) = (6, -4 - 4) = (6, -8)$; $(h - a, k) = (6 - 6, -4) = (0, -4)$; $(h + a, k) = (6 + 6, -4) = (12, -4)$. The graph of

$$\frac{(x-6)^2}{36} + \frac{(y+4)^2}{16} = 1$$

is given in Figure 10.48. ■

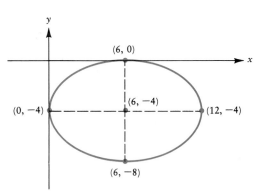

Figure 10.48

EXAMPLE 4 Discuss the graph of $9y^2 + 16x^2 - 54y + 128x + 193 = 0$.

Solution The equation can be put in standard form by completing the square. We begin by grouping the terms in x and those in y, and factoring out the coefficients of the square terms.

$$9y^2 + 16x^2 - 54y + 128x + 193 = 0$$
$$16x^2 + 128x + 9y^2 - 54y = -193$$
$$16(x^2 + 8x) + 9(y^2 - 6y) = -193$$

Completing the square for x and y, we obtain

$$16(x^2 + 8x + 16) + 9(y^2 - 6y + 9) = -193 + 81 + 256$$
$$16(x + 4)^2 + 9(y - 3)^2 = 144$$

Dividing by 144, we get

$$\frac{(x + 4)^2}{9} + \frac{(y - 3)^2}{16} = 1$$

This equation shows that the graph is an ellipse of form (4) with center $(-4, 3)$ and major axis parallel to the y-axis. The graph of the ellipse $\frac{(x + 4)^2}{9} + \frac{(y - 3)^2}{16} = 1$ is given in Figure 10.49. ■

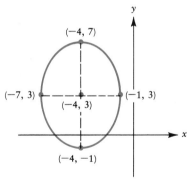

Figure 10.49

If we expand equations (3) or (4), we see that the **general form** of the equation of an ellipse is given by the following theorem.

Theorem 8

The equation

$$Ax^2 + Cy^2 + Dx + Ey + F = 0$$

represents an ellipse if A and C are of the same sign (that is, $AC > 0$).

Note. If $A = C$, then we have the equation of a circle. Thus, a circle is a special case of an ellipse.

Among the many applications of the ellipse are architecture, map projections that are designed to preserve relative areas, and astronomy. For example, Kepler's first law states that all the planets have elliptical orbits, with the sun at one of the foci.

According to Kepler's second law, a line joining a planet to the sun will sweep out equal areas in equal time. As illustrated in Figure 10.50, let a planet move in its elliptical orbit from P to Q in a certain time and from R to S in the same length of time. Then, area A = area B.

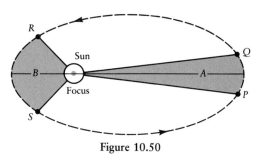

Figure 10.50

EXERCISES 10.9

In Exercises 1–12, for each ellipse, determine the vertices. Sketch the ellipse.

1. $9x^2 + 16y^2 = 576$
2. $4x^2 + 9y^2 = 36$
3. $25x^2 + 4y^2 = 100$
4. $x^2 + 4y^2 = 16$
5. $4x^2 + y^2 = 36$
6. $9x^2 + 25y^2 = 225$
7. $9x^2 + 16y^2 = 144$
8. $4x^2 + 3y^2 = 11$
9. $12x^2 + 5y^2 = 23$
10. $25x^2 + 36y^2 = 900$
11. $9x^2 + 25y^2 = 900$
12. $4x^2 + 25y^2 = 625$

In Exercises 13–24, determine the center and vertices of each ellipse. Sketch the ellipse.

13. $4(x + 2)^2 + 9(y - 1)^2 = 36$
14. $25(x - 2)^2 + 9(y - 3)^2 = 225$
15. $(x - 2)^2 + 4(y + 2)^2 = 16$
16. $4(x - 2)^2 + (y + 2)^2 = 16$
17. $9(x - 1)^2 + 16(y + 2)^2 = 144$
18. $(x + 5)^2 + 4(y - 2)^2 = 36$
19. $4(x + 2)^2 + 16(y - 1)^2 = 64$
20. $25(x + 1)^2 + 16(y - 2)^2 = 400$
21. $9x^2 + 4y^2 - 18x + 16y - 11 = 0$
22. $16x^2 + 4y^2 + 96x - 8y + 84 = 0$
23. $x^2 + 2y^2 + 6x + 7 = 0$
24. $2x^2 + 5y^2 + 20x - 30y + 75 = 0$

10.10 THE HYPERBOLA

A **hyperbola** is a curve formed if the cutting plane is parallel to two generators and intersects both nappes of a right-circular cone (see Figure 10.51). The two parts of a hyperbola are called **branches**. A more formal geometric definition of a hyperbola is given as follows.

> **DEFINITION**
>
> A **hyperbola** is the set of all points $P(x, y)$ in the plane such that the absolute value of the difference between the distances from P to two fixed points is constant. The two fixed points are called the **foci** of the hyperbola.

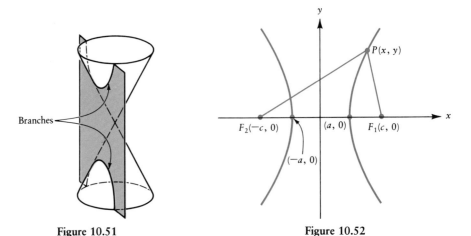

Figure 10.51   Figure 10.52

To derive an equation of a hyperbola, we consider a hyperbola such that the distance between the foci is $2c$, and the constant difference is $2a, a < c$. The graph of such a hyperbola is given in Figure 10.52. We choose the line through the foci as the x-axis and the midpoint between the foci as the origin. Thus, the coordinates of the foci are $(c, 0)$ and $(-c, 0)$. By the definition of a hyperbola, we have

$$\overline{F_2P} - \overline{F_1P} = \pm 2a$$

where the positive sign is used for a point P on the right of the y-axis and the negative sign is used for a point of the left of the y-axis. In either case, the distance formula gives $\overline{F_2P} = \sqrt{(x + c)^2 + y^2}$ and $\overline{F_1P} = \sqrt{(x - c)^2 + y^2}$. Therefore,

$$\sqrt{(x + c)^2 + y^2} - \sqrt{(x - c)^2 + y^2} = \pm 2a$$

or
$$\sqrt{(x + c)^2 + y^2} = \pm 2a + \sqrt{(x - c)^2 + y^2}$$

Squaring both sides, we get

$$x^2 + 2cx + c^2 + y^2 = 4a^2 \pm 4a\sqrt{(x - c)^2 + y^2} + x^2 - 2cx + c^2 + y^2$$

Gathering like terms,

$$4cx - 4a^2 = \pm 4a\sqrt{(x - c)^2 + y^2}$$

or
$$cx - a^2 = \pm a\sqrt{(x - c)^2 + y^2}$$

Squaring both sides again, we obtain

$$c^2x^2 - 2a^2cx + a^4 = a^2x^2 - 2a^2cx + a^2c^2 + a^2y^2$$

or
$$(c^2 - a^2)x^2 - a^2y^2 = a^2(c^2 - a^2)$$

Since $c > a$, $c^2 > a^2$, and $c^2 - a^2 > 0$, we can let $b^2 = c^2 - a^2$, and the equation becomes

$$b^2x^2 - a^2y^2 = a^2b^2$$

or

$$\frac{x^2}{a^2} - \frac{y^2}{b^2} = 1 \tag{1}$$

Therefore, we have the following theorem.

Theorem 9

An equation of a hyperbola with center $C(0, 0)$, foci $(c, 0)$ and $(-c, 0)$, and vertices $(a, 0)$ and $(-a, 0)$, is given by

$$\frac{x^2}{a^2} - \frac{y^2}{b^2} = 1, \text{ where } c^2 = a^2 + b^2$$

Conversely, any equation of the form $\dfrac{x^2}{a^2} - \dfrac{y^2}{b^2} = 1$ represents a hyperbola. The proof consists in showing that the steps taken in obtaining equation (1) can be reversed.

EXAMPLE 1   Discuss the graph of $\dfrac{x^2}{16} - \dfrac{y^2}{9} = 1$.

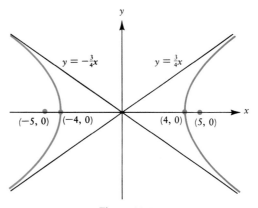

Figure 10.53

Solution The equation takes the form of equation (1) with $a^2 = 16$, $a = 4$, and $b^2 = 9$, $b = 3$. Since $c^2 = a^2 + b^2$, we have $c^2 = 16 + 9 = 25$ and $c = 5$. Thus, the equation $\dfrac{x^2}{16} - \dfrac{y^2}{9} = 1$ is a hyperbola with center $C(0, 0)$, foci $(5, 0)$ and $(-5, 0)$, and vertices $(4, 0)$ and $(-4, 0)$. The graph is given in Figure 10.53. Note that each branch of the hyperbola gets closer to the lines $y = \dfrac{b}{a}x = \dfrac{3}{4}x$ and $y = -\dfrac{b}{a}x = -\dfrac{3}{4}x$. These lines are called the asymptotes of the hyperbola. ■

DEFINITION

The asymptotes of the hyperbola $\dfrac{x^2}{a^2} - \dfrac{y^2}{b^2} = 1$ are the lines given by $y = \dfrac{b}{a}x$ and $y = -\dfrac{b}{a}x$.

If the foci of the hyperbola lie on the y-axis, say at points $(0, c)$ and $(0, -c)$, then an equation of the hyperbola may be derived as for the equation (1). Such an equation is

$$\frac{y^2}{a^2} - \frac{x^2}{b^2} = 1 \tag{2}$$

and we have the following theorem:

Theorem 10

An equation of the hyperbola with center $C(0, 0)$, foci $(0, c)$ and $(0, -c)$, and vertices $(0, a)$ and $(0, -a)$, is

$$\frac{y^2}{a^2} - \frac{x^2}{b^2} = 1, \text{ where } c^2 = a^2 + b^2$$

and equations of the asymptotes are

$$y = \frac{a}{b}x \quad \text{and} \quad y = -\frac{a}{b}x$$

EXAMPLE 2 Analyze the equation $4x^2 - 16y^2 + 25 = 0$.

Solution
$$4x^2 - 16y^2 = -25$$

Divide both sides by -25:

$$\frac{16y^2}{25} - \frac{4x^2}{25} = 1$$

This equation can be written

$$\frac{y^2}{\left(\frac{25}{16}\right)} - \frac{x^2}{\left(\frac{25}{4}\right)} = 1$$

and this is a hyperbola of the form

$$\frac{y^2}{a^2} - \frac{x^2}{b^2} = 1$$

where $a^2 = \left(\frac{25}{16}\right)$ or $a = \frac{5}{4}$, and $b^2 = \left(\frac{25}{4}\right)$ or $b = \frac{5}{2}$. Using the fact that $c^2 = a^2 + b^2 = \frac{25}{16} + \frac{25}{4} = \frac{125}{16}$, we get $c = \sqrt{\frac{125}{16}} = \frac{5\sqrt{5}}{4}$. Thus, the equation $4x^2 - 16y^2 + 25 = 0$ is a hyperbola with center at $(0, 0)$, foci $F_1\left(0, \frac{5\sqrt{5}}{4}\right)$ and $F_2\left(0, \frac{-5\sqrt{5}}{4}\right)$, vertices $V_1\left(0, \frac{5}{4}\right)$ and $V_2\left(0, -\frac{5}{4}\right)$, and asymptotes $y = \frac{a}{b}x = \frac{1}{2}x$ and $y = -\frac{a}{b}x = -\frac{1}{2}x$. See Figure 10.54. ■

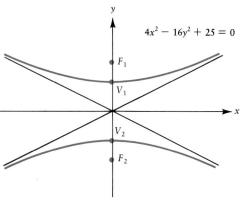

$$4x^2 - 16y^2 + 25 = 0$$

Figure 10.54

Table 10.2 summarizes our results concerning the graphs, description, and equations of hyperbolas having their center at the origin and foci on the x-axis or y-axis.

Table 10.2

Graph	Description	Equation
	Foci on the x-axis. Center at the origin.	$\dfrac{x^2}{a^2} - \dfrac{y^2}{b^2} = 1$
	Foci on the y-axis. Center at the origin.	$\dfrac{y^2}{a^2} - \dfrac{x^2}{b^2} = 1$

Note. Rather than memorizing the equations of the asymptotes of a hyperbola, they can be obtained by substituting 0 for the 1 on the right side of the hyperbola equation and then solving for y in terms of x.

Center Not at Origin

Just as with the other conic sections, we can describe hyperbolas with center at (h, k) instead of at the origin.

Theorem 11

An equation of a hyperbola with center (h, k) and foci $F_1(h + c, k)$ and $F_2(h - c, k)$ is

$$\frac{(x - h)^2}{a^2} - \frac{(y - k)^2}{b^2} = 1, \text{ where } c^2 = a^2 + b^2$$

The vertices are $V_1(h + a, k)$ and $V_2(h - a, k)$ and equations of the asymptotes are

$$y = \frac{b}{a}(x - h) + k \quad \text{and} \quad y = -\frac{b}{a}(x - h) + k$$

EXAMPLE 3   Find the center, vertices, foci, and equations of the asymptotes for the hyperbola $4(x - 3)^2 - 9(y + 5)^2 = 36$.

Solution   Putting $4(x - 3)^2 - 9(y + 5)^2 = 36$ in standard form, we get

$$\frac{(x - 3)^2}{9} - \frac{(y + 5)^2}{4} = 1$$

Thus, $h = 3, k = -5, a^2 = 9$ or $a = 3, b^2 = 4$ or $b = 2$, and $c^2 = a^2 + b^2 = 9 + 4 = 13$ or $c = \sqrt{13}$. Hence, we obtain:

Center: $C(h, k) = (3, -5)$

Vertices: $V_1(h + a, k) = (6, -5)$ and $V_2(h - a, k) = (0, -5)$

Foci: $F_1(h + c, k) = (3 + \sqrt{13}, -5)$ and $F_2(h - c, k) = (3 - \sqrt{13}, -5)$

Asymptotes: $y = \dfrac{b}{a}(x - h) + k = \dfrac{2}{3}(x - 3) - 5$ and

$$y = -\frac{b}{a}(x - h) + k = -\frac{2}{3}(x - 3) - 5$$

See Figure 10.55.   ■

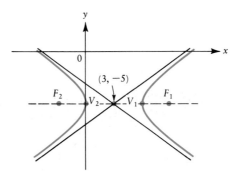

Figure 10.55

Theorem 12

An equation of a hyperbola with center (h, k) and foci $F_1(h, k + c)$ and $F_2(h, k - c)$ is

$$\frac{(y - k)^2}{a^2} - \frac{(x - h)^2}{b^2} = 1, \text{ where } c^2 = a^2 + b^2$$

The vertices are $V_1(h, k + a)$ and $V_2(h, k - a)$ and equations of the asymptotes are

$$y = \frac{a}{b}(x - h) + k \quad \text{and} \quad y = -\frac{a}{b}(x - h) + k$$

EXAMPLE 4   Find the center, vertices, foci, and equations of the asymptotes for the hyperbola $y^2 - 4x^2 - 8x - 4y - 4 = 0$.

Solution   To put the equation in standard form, we complete the square on x and y:

$$y^2 - 4x^2 - 8x - 4y - 4 = 0$$
$$(y^2 - 4y) - 4(x^2 + 2x) = 4$$
$$(y^2 - 4y + 4) - 4(x^2 + 2x + 1) = 4 + 4 - 4$$
$$(y - 2)^2 - 4(x + 1)^2 = 4$$
$$\frac{(y - 2)^2}{4} - \frac{(x + 1)^2}{1} = 1 \quad \blacksquare$$

Thus, $h = -1$, $k = 2$, $a^2 = 4$ or $a = 2$, $b^2 = 1$ or $b = 1$, and $c^2 = a^2 + b^2 = 4 + 1 = 5$ or $c = \sqrt{5}$. Hence, we obtain:

Center: $C(h, k) = (-1, 2)$
Vertices: $V_1(h, k + a) = (-1, 4)$ and $V_2(h, k - a) = V_2(-1, 0)$
Foci: $F_1(h, k + c) = (-1, 2 + \sqrt{5})$ and $F_2(h, k - c) = F_2(-1, 2 - \sqrt{5})$
Asymptotes: $y = \dfrac{a}{b}(x - h) + k = \dfrac{2}{1}(x + 1) + 2 = 2x + 4$ and

$$y = -\frac{a}{b}(x - h) + k = -\frac{2}{1}(x + 1) + 2 = -2x$$

See Figure 10.56.

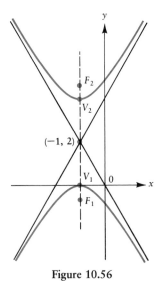

Figure 10.56

*EXAMPLE 5   Find an equation of the hyperbola having foci $(8, 17)$ and $(8, -9)$ and vertices at $(8, 9)$ and $(8, -1)$.

Solution   Since the center is the midpoint of the line connecting the foci, we have the center at $(8, 4)$. Thus, $h = 8$ and $k = 4$. Since the foci, vertices, and center lie on a line parallel to the y-axis, Theorem 12 applies. Using Theorem 12 and the fact that $h = 8$ and $k = 4$, we have $F_1(8, 4 + c) = (8, 17)$ or $c = 13$. Similarly, $V_1(8, 4 + a) = (8, 9)$ or $a = 5$. The relationship $b^2 = c^2 - a^2$ gives $b = 12$. Hence, we obtain the equation

$$\frac{(y - 4)^2}{25} - \frac{(x - 8)^2}{144} = 1 \quad \blacksquare$$

We now give a theorem that states those conditions on the general quadratic equation that gives a hyperbola.

Theorem 13

The equation $Ax^2 + Cy^2 + Dx + Ey + F = 0$ represents a hyperbola if A and C are opposite in sign (that is, if $AC < 0$).

EXERCISES   10.10

In Exercises 1–15, for each hyperbola with center at the origin, find the vertices, foci, and equations of the asymptotes. Sketch each hyperbola.

1. $x^2 - y^2 = 25$
2. $4x^2 - 45y^2 = 180$
3. $y^2 - x^2 = 9$
4. $9x^2 - 16y^2 = 144$
5. $x^2 - y^2 = 16$
6. $y^2 - x^2 = 64$
7. $y^2 - x^2 = 4$
8. $25x^2 - 144y^2 = 3600$
9. $9y^2 - 25x^2 = 225$
10. $y^2 - x^2 = 36$
11. $9x^2 - 4y^2 = 36$
12. $16y^2 - x^2 = 16$
13. $4y^2 - 9x^2 = 25$
14. $x^2 - 25y^2 = -100$
15. $25x^2 - y^2 = 100$

In Exercises 16–19, find an equation of the hyperbola satisfying the given conditions.

16. Vertices at $(-4, 0)$ and $(4, 0)$, foci at $(-6, 0)$ and $(6, 0)$
17. Vertices at $(3, 0)$ and $(-3, 0)$, foci at $(5, 0)$ and $(-5, 0)$
18. Foci at $(0, 1)$ and $(0, -1)$, vertices at $(0, \frac{1}{2})$ and $(0, -\frac{1}{2})$

19. Foci at $(0, 13)$ and $(0, -13)$, vertices at $(0, 5)$ and $(0, -5)$

In Exercises 20–31, find the center, vertices, foci, and equations of the asymptotes for each hyperbola.

20. $16(x - 1)^2 - 9(y - 2)^2 = 144$
21. $(y + 1)^2 - 4(x - 1)^2 = 4$
22. $9(x - 1)^2 - 16(y + 2)^2 = 144$
23. $4(x - 3)^2 - 9(y - 1)^2 = 36$
24. $25(x + 2)^2 - 144(y - 3)^2 = 3600$
25. $16(y + 1)^2 - x^2 = 16$
26. $y^2 - 4(x + 4)^2 = 36$
27. $(y + 7)^2 - (x - 3)^2 = 16$
28. $x^2 - 3y^2 - 4x + 18y = 50$
29. $9x^2 - 4y^2 + 36x + 24y = -36$
30. $16x^2 - 9y^2 + 90y - 81 = 0$
31. $4x^2 - y^2 - 8x - 12 = 0$
*32. Find an equation of a hyperbola with center $(5, -4)$, vertex $(5, 2)$, and passing through the point $(4, 8)$.
*33. Find an equation of the hyperbola with asymptotes $y = 3x$ and $y = -3x$ and foci at $(0, 6)$ and $(0, -6)$.

Chapter 10 SUMMARY

ORDERED PAIRS

1. *Equality of Ordered Pairs.*

$(x_1, y_1) = (x_2, y_2)$ if and only if $x_1 = x_2$ and $y_1 = y_2$

2. *Distance Formula.* The distance between (x_1, y_1) and (x_2, y_2) is given by

$$d = \sqrt{(x_2 - x_1)^2 + (y_2 - y_1)^2}$$

LINES

1. *Slope of a Line.*

$$m = \frac{y_2 - y_1}{x_2 - x_1} = \frac{\text{vertical change}}{\text{horizontal change}} = \frac{\text{rise}}{\text{run}}$$

2. *Parallel Lines.* Slopes are equal: $m_1 = m_2$

3. *Perpendicular Lines.* Slopes are negative reciprocals: $m_2 = \dfrac{-1}{m_1}$

4. *Horizontal Lines.* $y = k$, k constant; slope $= 0$

5. *Vertical Lines.* $x = h$, h constant; undefined slope

6. *Two-Point Formula.* $\dfrac{y - y_1}{x - x_1} = \dfrac{y_2 - y_1}{x_2 - x_1}$

7. *Point-Slope Formula.* $y - y_1 = m(x - x_1)$

8. *Slope-Intercept Formula.* $y = mx + b$

9. *General Line Equation.* $Ax + By + C = 0$

CONIC SECTIONS

1. *Circle:* A second-degree equation of the form

$$Ax^2 + Cy^2 + Dx + Ey + F = 0$$
$$(A = C \neq 0)$$

is the general equation of a circle (possibly degenerate). If we complete the square on x and y, the result is the standard equation of a circle

$$(x - h)^2 + (y - k)^2 = r^2$$

(a) If $r^2 > 0$, the graph is a circle centered at (h, k) with radius r.

(b) If $r^2 = 0$, the graph is a single point (h, k), a point circle.

(c) If $r^2 < 0$, there is no graph (imaginary circle).

2. *Parabola:* A second-degree equation of the form

$$Ax^2 + Cy^2 + Dx + Ey + F = 0$$
$$(A = 0 \quad \text{or} \quad C = 0)$$

is the general equation of a parabola. Completing the square, we obtain the standard equations of a parabola.

(a)
$$(y - k)^2 = 4p(x - h)$$

Vertex: (h, k) $p > 0$: parabola opens to right

Axis: $y = k$ $p < 0$: parabola opens to left

(b)
$$(x - h)^2 = 4p(y - k)$$

Vertex: (h, k) $p > 0$: parabola opens upward

Axis: $x = h$ $p < 0$: parabola opens downward

3. *Ellipse:* A second-degree equation of the form

$$Ax^2 + Cy^2 + Dx + Ey + F = 0$$
$$(AC > 0)$$

$$Ax^2 + Cy^2 + Dx + Ey + F = 0$$
$$AC < 0$$

is the general equation of an ellipse. Completing the square on x and y, we obtain the standard equations of an ellipse.

is the general equation of a hyperbola. Completing the square on x and y results in the standard equations of a hyperbola

(a)
$$\frac{(x-h)^2}{a^2} + \frac{(y-k)^2}{b^2} = 1$$
$$a > b$$

(a)
$$\frac{(x-h)^2}{a^2} - \frac{(y-k)^2}{b^2} = 1$$
where $c^2 = a^2 + b^2$

Center: (h, k) Major axis parallel to x-axis

Foci: $F_1(h - c, k)$ and $F_2(h + c, k)$, where $c^2 = a^2 - b^2$

Center: (h, k)
Foci: $F_1(h + c, k)$ and $F_2(h - c, k)$
Vertices: $V_1(h + a, k)$ and $V_2(h - a, k)$
Asymptotes: $y = \dfrac{b}{a}(x - h) + k$ and

$$y = \frac{-b}{a}(x - h) + k$$

(b)
$$\frac{(x-h)^2}{b^2} + \frac{(y-k)^2}{a^2} = 1$$
$$a > b$$

$$\frac{(y-k)^2}{a^2} - \frac{(x-h)^2}{b^2} = 1$$
where $c^2 = a^2 + b^2$

Center: (h, k) Major axis parallel to y-axis

Foci: $F_1(h, k - c)$ and $F_2(h, k + c)$, where $c^2 = a^2 - b^2$

Center: (h, k)
Foci: $F_1(h, k + c)$ and $F_2(h, k - c)$
Vertices: $V_1(h, k + a)$ and $V_2(h, k - a)$
Asymptotes: $y = \dfrac{a}{b}(x - h) + k$ and

$$y = \frac{-a}{b}(x - h) + k$$

4. *Hyperbola:* A second-degree equation of the form

Chapter 10 EXERCISES

In Exercises 1–5, find the distance between the given points P_1 and P_2.

1. $P_1(2, 5)$; $P_2(2, 11)$
2. $P_1(-4, 7)$; $P_2(5, 7)$
3. $P_1(1, 5)$; $P_2(4, 9)$
4. $P_1(8, -4)$; $P_2(-1, -6)$
5. $P_1(3, -2)$; $P_2(-4, 5)$

In Exercises 6–10, find the slope of the line passing through the given points.

6. $(2, 4)$, $(1, 1)$ **7.** $(2, 1)$, $(1, 2)$
8. $(0, 3)$, $(2, -4)$ **9.** $(-1, -1)$, $(1, 1)$
10. Origin, $(4, 10)$
11. What are the units of the slope of a line if the abscissa and ordinate units are as given?
(a) abscissa, number of calories; ordinate, total cost
(b) abscissa, hours; ordinate, miles

(c) abscissa, time; ordinate, distance

(d) abscissa, cubic meters; ordinate, square meters

12. Given the equations $ax + by = c$, $dx + ey = f$,

    (a) What conditions on the constants must exist if the lines are parallel?

    (b) Perpendicular?

In Exercises 13–17, find an equation of the line passing through the given points.

13. $(-2, 3)$, $(6, 9)$ 14. $(3, 2)$, $(6, 3)$

15. $(2, -4)$, $(-6, 2)$ 16. $(3, 0)$, $(-5, 2)$

17. $(-2, 5)$, $(4, 5)$

In Exercises 18–22, find an equation of the line passing through the given point and having the indicated slope.

18. $(2, 5)$; $m = 3$ 19. $(1, -2)$; $m = -2$

20. $(4, 6)$; $m = \dfrac{1}{2}$ 21. $(2, 7)$; $m = 0$

22. $(4, -3)$; no slope

In Exercises 23–27, find an equation of the line having the given slope and y-intercept.

23. $m = -3$; $b = 5$ 24. $m = -\dfrac{3}{2}$; $b = \dfrac{7}{4}$

25. $m = 0$; $b = \dfrac{3}{2}$ 26. $m = \dfrac{6}{7}$; $b = \dfrac{1}{7}$

27. $m = \dfrac{3}{4}$; $b = -3$

In Exercises 28–30, write an equation of a line having the given x- and y-intercepts.

28. x-intercept -9 29. x-intercept -8

    y-intercept -6 y-intercept 6

30. x-intercept -15

    y-intercept -10

31. Find an equation of the line parallel to $2x - y = 6$ and passing through $(2, 2)$.

32. Find an equation of the line perpendicular to $2x - y = 6$ and passing through $(2, 2)$.

*33. Find an equation for the perpendicular bisector of the line segment connecting the points $(2, 7)$ and $(14, 3)$.

*34. Given the triangle with vertices $A(-3, -4)$, $B(0, 4)$, and $C(4, -2)$, write an equation of the altitude from B to AC.

In Exercises 35–54, identify each of the conic sections.

35. $7x^2 + 16y^2 + 14x - 64y = 41$

36. $y^2 + 2x + 8y + 6 = 0$

37. $16y^2 - x^2 - 6x - 80y = -75$

38. $4x^2 + 4y^2 - 8x + 18y + 12 = 0$

39. $2x^2 - 24x + 3y + 78 = 0$

40. $8x^2 - 9y^2 - 16x + 54y = 1$

41. $2x^2 - 18x + 15y - 21 = 0$

42. $y = x^2 - 4x + 4$

43. $y^2 - 4x - 4y + 16 = 0$

44. $9x^2 - 16y^2 - 108x + 96y + 36 = 0$

45. $8x^2 + 4y^2 - 64x - 8y + 68 = 0$

46. $4x^2 + 12x - 20y + 49 = 0$

47. $8x^2 - 28y^2 - 8x - 28y = 61$

48. $8x^2 + 9y^2 + 16x - 54y - 1 = 0$

49. $3y^2 + 15x - 12y + 20 = 0$

50. $3y^2 - 4x^2 - 16x - 24y - 52 = 0$

51. $x^2 + y^2 - 6x + 4y + 14 = 0$

52. $x^2 + y^2 - 4x + 4 = 0$

53. $x^2 + y^2 - 12y + 36 = 0$

54. $3x^2 + 3y^2 - 4x + 2y + 6 = 0$

In Exercises 55–58, find an equation of the circle that satisfies the given conditions.

55. center $(3, -2)$; radius $\sqrt{13}$

56. center $(-4, 2)$; diameter 8

57. center $(4, -1)$; passes through the point $(-1, 3)$.

*58. A diameter of the circle is the line segment joining the points $(-3, 5)$ and $(7, -3)$.

In Exercises 59–62, find the center and radius of each circle. State whether the circle is a real circle, a point circle, or an imaginary circle.

59. $2x^2 + 2y^2 - 16x + 20y - 24 = 0$

60. $x^2 + y^2 - \dfrac{4}{3}x + \dfrac{2}{3}y + 2 = 0$

61. $4x^2 + 4y^2 = 0$ 62. $6x^2 + 6y^2 - 3x = 0$

In Exercises 63–66, determine the vertex and sketch the graph of each parabola.

63. $y^2 - 6x + 6y + 9 = 0$

64. $x^2 + 2x + 12y - 71 = 0$

65. $3y^2 - 2x + 7y - 11 = 0$

66. $4x = y^2 + 5y - 6$

In Exercises 67–70, determine the vertices and sketch each ellipse.

67. $9x^2 + 25y^2 = 225$

68. $x^2 + 25y^2 - 25 = 0$ 69. $9x^2 + 16y^2 = 25$

70. $4x^2 + y^2 - 8x + 12y = 0$

In Exercises 71–75, find the vertices, foci, and the equations of the asymptotes of each hyperbola. Sketch the graph.

71. $4x^2 - 3y^2 = 12$ 72. $5x^2 - 4y^2 = 20$

73. $4x^2 - 45y^2 = 180$

74. $49y^2 - 16x^2 = 784$

75. $9x^2 - 16y^2 - 36x - 32y = 124$

11 Systems of Linear Equations

In this chapter, we shall study some of the basic methods for solving systems of linear equations: graphically, elimination method, substitution method, and Cramer's rule. While most of our examples will be restricted to linear equations in two variables and three variables, the elimination method provides a basis for solving systems involving a large number of variables. Such large systems can be easily solved with the aid of a computer.

11.1 SYSTEMS OF LINEAR EQUATIONS IN TWO VARIABLES

If we have a single linear equation containing two variables x and y, we cannot determine anything definite about the values of x and y, because whatever value we choose to give one of them, say x, there will be a corresponding value of the other variable, y.

For example, consider the equation

$$x + y = 3$$

Solving for y, we obtain

$$y = -x + 3$$

However, we cannot find the value of y from this equation unless we know the value of x. We can arbitrarily choose a value for x, and there will be one corresponding value of y. Thus, we can find as many pairs of values as we please that satisfy the given equation.

For example,

$$\text{if } x = 1, \quad \text{then} \quad y = -(1) + 3 = 2$$
$$\text{if } x = 2, \quad \text{then} \quad y = -(2) + 3 = 1$$
$$\text{if } x = 4, \quad \text{then} \quad y = -(4) + 3 = -1$$

and so on. Any of the ordered pairs $(1, 2)$, $(2, 1)$, $(4, -1)$, etc. will satisfy the equation $x + y = 3$. Thus, a single equation containing two unknowns is not sufficient to determine the definite value of either unknown.

Now consider a second linear equation in two variables:

$$x - y = 5$$

As before, this equation is satisfied by many pairs of values. For example,

$$\text{if } x = 1, \quad \text{then} \quad y = -4$$
$$\text{if } x = 2, \quad \text{then} \quad y = -3$$
$$\text{if } x = 4, \quad \text{then} \quad y = -1$$

and so on. Thus, any of the ordered pairs $(1, -4)$, $(2, -3)$, $(4, -1)$, etc. will satisfy the equation $x - y = 5$.

Is there a pair of values, one for x and one for y, that satisfies both equations? We see that the pair of values $x = 4$ and $y = -1$ satisfies *both* of the equations. The pair of equations

$$\begin{cases} x + y = 3 \\ x - y = 5 \end{cases}$$

is called a system of linear equations and the pair of values $x = 4$ and $y = -1$ that satisfies both equations is called a solution of the system.

DEFINITION
A solution of the system

$$a_1 x + b_1 y = c_1$$
$$a_2 x + b_2 y = c_2$$

is any ordered pair (x, y) that satisfies *both* equations.

EXAMPLE 1   (a) The ordered pair $(2, 1)$ is a solution of the system

$$\begin{cases} 3x + 4y = 10 \\ 4x + y = 9 \end{cases}$$

(b) The ordered pair (3, 5) is *not* a solution of the system

$$\begin{cases} x + 2y = 13 \\ 3x - y = 14 \end{cases}$$

While (3, 5) satisfies the equation $x + 2y = 13$, it does not satisfy the equation $3x - y = 14$.

In this section we give some graphical and algebraic methods for solving systems of linear equations in two variables.

Graphic Solution

We know from Section 10.2 that the graph of a linear equation is a straight line. Given a system of two linear equations in two variables, we can graph each of the equations on the same set of coordinate axes. For graphs of two linear equations in two variables, there are three possible cases:

1. The lines intersect in exactly one point.
2. The lines are parallel and have no points in common.
3. The lines coincide and have an infinite number of points in common.

Geometrically, the solution of two linear equations in two variables is the set of ordered pairs representing the points of intersection of these lines, if such points of intersection exist.

EXAMPLE 2   Consider the system

$$\begin{cases} x + y = 3 \\ x - y = 5 \end{cases}$$

Solve the system graphically.

Solution   Putting both equations in slope-intercept form we get

$$\begin{cases} y = -x + 3 \\ y = x - 5 \end{cases}$$

Graphing the equations, we obtain Figure 11.1. The point of intersection has coordinates (4, −1). Since this point lies on both lines, it is a solution of both equations. Hence, the ordered pair (4, −1) is a solution of the linear system. Note that the slopes of the two lines are unequal. ■

DEFINITION

A system of linear equations in which the lines intersect in exactly one point is called a consistent system and the lines are said to be independent. Such a system has a unique solution. (Note that the lines have unequal slopes.)

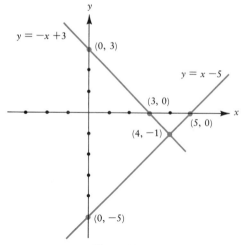

Figure 11.1

EXAMPLE 3   Solve the system of equations

$$\begin{cases} y - x = 1 \\ 2y - 2x = 8 \end{cases}$$

graphically.

Solution   Putting each equation in slope-intercept form gives

$$\begin{cases} y = x + 1 \\ y = x + 4 \end{cases}$$

Graphing each equation, we get Figure 11.2. The system results in parallel lines. Thus, since the lines do not intersect, there is no solution. Note that the lines have the same slope and different y-intercepts. ■

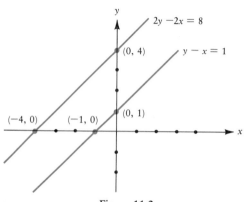

Figure 11.2

DEFINITION

A system of linear equations in which the lines are parallel is called an **inconsistent system** and the lines are said to be **independent**. Such a system has no solution. (Note that the lines have the same slope and different y-intercepts.)

EXAMPLE 4   Solve the following system graphically:

$$\begin{cases} 2x + 3y = -4 \\ -4x - 6y = 8 \end{cases}$$

Solution   Putting each of the equations in slope-intercept form, we have

$$\begin{cases} y = -\dfrac{2}{3}x - \dfrac{4}{3} \\ \\ y = -\dfrac{2}{3}x - \dfrac{4}{3} \end{cases}$$

Both equations have the same slope, $-\frac{2}{3}$, and the same y-intercept, $-\frac{4}{3}$. Thus, the graphs of both equations are represented by the same line (see Figure 11.3). Since the lines coincide, there is an infinite number of points in common. Thus, the system has infinitely many solutions. ■

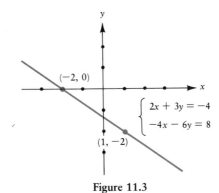

$$\begin{cases} 2x + 3y = -4 \\ -4x - 6y = 8 \end{cases}$$

$(-2, 0)$

$(1, -2)$

Figure 11.3

DEFINITION

A system of linear equations in which the lines coincide is called a **consistent system** and the lines are **dependent**. Such a system has an infinite number of solutions. (Note that the lines have the same slope and same y-intercept.)

Although the graphical method gives useful information, it is not always feasible. For example, the point of intersection may be too far from the origin and thus difficult to graph. If the coefficients are very small fractional numbers, the graphical method may not yield the desired accuracy.

EXERCISES 11.1

In Exercises 1–6, determine which of the systems are consistent and which are inconsistent. Also determine whether the lines are independent or dependent.

1. $\begin{cases} 3x - 2y = 2 \\ 2x + y = 8 \end{cases}$
2. $\begin{cases} 4x - 2y = 7 \\ -2x + y = 3 \end{cases}$

3. $\begin{cases} -x + y = 1 \\ x - y = -1 \end{cases}$

4. $\begin{cases} 4x + 2y = 0 \\ 2x + 6y = 0 \end{cases}$
5. $\begin{cases} -x + y = -1 \\ x + y = 5 \end{cases}$

6. $\begin{cases} x - 3y = 2 \\ -2x + 6y = -8 \end{cases}$

In Exercises 7–20, solve using the graphical method.

7. $\begin{cases} -x + y = 7 \\ 4x + 3y = 0 \end{cases}$
8. $\begin{cases} x + y = 7 \\ -2x + y = 1 \end{cases}$

9. $\begin{cases} 3x + 2y = 0 \\ 2x + 3y = 5 \end{cases}$
10. $\begin{cases} x + 2y = 3 \\ -2x - 4y = -6 \end{cases}$

11. $\begin{cases} 2x - 3y = -14 \\ 3x + 7y = 48 \end{cases}$
12. $\begin{cases} 5x + 2y = 10 \\ x - y = 9 \end{cases}$

13. $\begin{cases} 6x + 2y = -5 \\ 3x = 11 - y \end{cases}$
14. $\begin{cases} 2x = 4 - y \\ 8x + 4y = 16 \end{cases}$

15. $\begin{cases} x = 6y - 10 \\ 2x = 7y - 15 \end{cases}$
16. $\begin{cases} 3x - 2y = -12 \\ -4x - 2y = 2 \end{cases}$

17. $\begin{cases} x - 3y = 6 \\ 2x - 6y = -1 \end{cases}$
18. $\begin{cases} 2x + y = 7 \\ 2x - y = 5 \end{cases}$

19. $\begin{cases} x - 2y = 2 \\ x - y = 5 \end{cases}$
20. $\begin{cases} 3x + 2y = 12 \\ x + 4y = 14 \end{cases}$

11.2 THE SUBSTITUTION METHOD

We now present the first of two algebraic methods used to solve a system of two linear equations in two variables. It is called the substitution method. Consider the following example.

EXAMPLE 1 Solve the system $\begin{cases} x + y = 6 \\ 2x + 3y = 16 \end{cases}$

Solution We first solve one of the equations for one of the variables. From the first equation, we obtain

$$y = 6 - x \qquad (1)$$

This expression is now substituted for y in the second equation and we get a linear equation in one variable, x.

$$2x + 3(6 - x) = 16$$

Solving for x, we get

$$2x + 18 - 3x = 16$$

$$-x = -2$$

$$x = 2$$

Substituting this value of x into expression (1), we get the corresponding y value:

$$y = 6 - 2 \quad \text{or} \quad y = 4$$

Thus, $x = 2$ and $y = 4$ is the solution of the system.   ■

Check $2 + 4 = 6$ and $2(2) + 3(4) = 4 + 12 = 16$

The Substitution Method

1. Select one of the equations and solve for one of the unknowns in terms of the other unknown.
2. Substitute this expression into the equation not used in step 1.

EXAMPLE 2   Solve the system $\begin{cases} 5x - 2y = 4 \\ 2x + 3y = 10 \end{cases}$

Solution   Solving the second equation for x, we get

$$2x + 3y = 10$$

$$2x = -3y + 10$$

$$x = \frac{-3y + 10}{2} \tag{2}$$

We now substitute this expression for x in the first equation:

$$5x - 2y = 4$$

$$5\left(\frac{-3y + 10}{2}\right) - 2y = 4$$

This last equation is a linear equation in one variable, y. Solving for y, we obtain

$$\frac{-15y + 50}{2} - 2y = 4$$

$$-15y + 50 - 4y = 8$$

$$-19y = -42$$

$$y = \frac{42}{19}$$

We can now substitute $y = \dfrac{42}{19}$ in either equation to find x. Using the first equation, we get

$$5x - 2y = 4$$

$$5x - 2\left(\frac{42}{19}\right) = 4$$

$$5x - \frac{84}{19} = 4$$

$$5x = 4 + \frac{84}{19} = \frac{76}{19} + \frac{84}{19} = \frac{160}{19}$$

$$5x = \frac{160}{19}$$

$$x = \frac{32}{19} \quad \blacksquare$$

Does the substitution method warn us about inconsistent systems (no solutions) and consistent systems with dependent lines (infinitely many solutions)?

EXAMPLE 3 Solve, if possible, the system $\begin{cases} x + y = 2 \\ 2x + 2y = 2 \end{cases}$

Solution Solving the first equation for x, we get

$$x = 2 - y$$

and substituting in the second equation,

$$2(2 - y) + 2y = 2$$
$$4 - 2y + 2y = 2$$
$$4 = 2$$

The statement $4 = 2$ is a contradiction. Thus, no values of x and y satisfy the system. Note that the equations represent parallel lines. $\blacksquare$

EXAMPLE 4 Solve, if possible, the system $\begin{cases} x - y = -2 \\ -2x + 2y = 4 \end{cases}$

Solution Solving the first equation for x,

$$x = y - 2$$

and substituting in the second equation,

$$-2(y - 2) + 2y = 4$$
$$-2y + 4 + 2y = 4$$
$$4 = 4$$

The statement $4 = 4$ is always true (an identity). Thus, all solutions of one equation will also satisfy the other equation. Note that the equations both represent the same line. ■

Examples 2, 3, and 4 show us that

---

1. If the system has a unique solution, the substitution method will give that unique solution.
2. If the system has no solution, the substitution method will result in a contradiction.
3. If the system has infinitely many solutions, the substitution method will result in an identity.

The following example shows a method that is sometimes used to solve a system of equations.

EXAMPLE 5   Solve the system $\begin{cases} 3x + 5y = 3 \\ 9x + y = 2 \end{cases}$

Solution   Solving the first equation for y, we get

$$y = \frac{3 - 3x}{5} \qquad (3)$$

Also, solving the second equation for y, we obtain

$$y = 2 - 9x \qquad (4)$$

Since y is to have the same value in the equation for the same value of x, we have equation (3) = equation (4); that is,

$$\frac{3 - 3x}{5} = 2 - 9x$$

Multiplying both sides by 5 gives

$$3 - 3x = 10 - 45x$$

$$42x = 7$$

$$x = \frac{1}{6}$$

Substituting $x = \dfrac{1}{6}$ into either equation, we can solve for y. Using the second equation, we get

$$9\left(\frac{1}{6}\right) + y = 2$$

$$\frac{3}{2} + y = 2$$

$$y = \frac{1}{2}$$

Thus, the solution of the system is $x = \dfrac{1}{6}$ and $y = \dfrac{1}{2}$. ∎

EXERCISES 11.2

Solve using the substitution method.

1. $\begin{cases} x + y = 0 \\ 2x + y = 4 \end{cases}$

2. $\begin{cases} x + y = 3 \\ 2x + y = 4 \end{cases}$

3. $\begin{cases} x + 2y = 5 \\ 2x - y = 5 \end{cases}$

4. $\begin{cases} 2x - 3y = 5 \\ -4x + 6y = -8 \end{cases}$

5. $\begin{cases} 3x - 4y = 8 \\ x - y = 3 \end{cases}$

6. $\begin{cases} 4x + 2y = 7 \\ y = -2x + 3 \end{cases}$

7. $\begin{cases} x - 2y = 1 \\ 2x + y = 12 \end{cases}$

8. $\begin{cases} x + 2y = 4 \\ -x + y = 5 \end{cases}$

9. $\begin{cases} x - 2y = 6 \\ x - y = 5 \end{cases}$

10. $\begin{cases} 4x - 3y = 21 \\ x + 2y = 8 \end{cases}$

11. $\begin{cases} 3x + 6y = 9 \\ 4x + 8y = 12 \end{cases}$

12. $\begin{cases} 3x + 4y = 10 \\ 2x + y = 5 \end{cases}$

13. $\begin{cases} x - 3y = 5 \\ 3x - 5y = 3 \end{cases}$

14. $\begin{cases} 2x - y = 11 \\ 5x - 2y = 27 \end{cases}$

15. $\begin{cases} 3x - y = 14 \\ 5x - 3y = 18 \end{cases}$

16. $\begin{cases} 4x + 5y = -5 \\ x - 4 = -3y \end{cases}$

17. $\begin{cases} 5x + 3y = 29 \\ -2y = -15 - x \end{cases}$

18. $\begin{cases} 4x + 3y = 2 \\ 2x - y = \dfrac{2}{3} \end{cases}$

19. $\begin{cases} \dfrac{3}{8}x + y = \dfrac{1}{4} \\ \dfrac{1}{2}x = \dfrac{5}{7}y + \dfrac{19}{14} \end{cases}$

20. $\begin{cases} \dfrac{1}{2}x = 2 - \dfrac{5}{4}y \\ \dfrac{1}{6}x = \dfrac{3}{2} + \dfrac{5}{3}y \end{cases}$

11.3 THE ELIMINATION METHOD

We now introduce an algebraic procedure for solving systems of linear equations. This method involves replacing a system of equations with a simpler equivalent system. Equivalent systems are systems that have the same solution(s). The following theorem indicates which algebraic operations can be used to replace a given system with an equivalent system.

Operations to Produce an Equivalent System

Given a system of linear equations, each of the following operations can be performed on the equations to produce an equivalent system.

1. Interchange any two equations.
2. Multiply any equation by a nonzero real number.
3. Multiply any equation by a nonzero real number and add the resulting equation to another equation.

EXAMPLE 1   Solve the system $\begin{cases} 3x - y = 14 \\ 2x + y = 1 \end{cases}$

Solution

$$3x - y = 14$$
$$2x + y = 1$$

Adding the equations, we get $5x \quad\quad = 15$
Solving for x, $\qquad\qquad x = 3$

Substituting $x = 3$ into the first equation of the original system gives $3(3) - y = 14$, or $y = -5$. Thus, the solution of the system is $x = 3$ and $y = -5$.   ■

Check $3(3) - (-5) = 14 \quad \text{and} \quad 2(3) + (-5) = 1$

Note that our method involved eliminating an unknown, in this case, the variable y; thus, the name elimination method. The idea is to combine the equations in such a way that we have one equation with one unknown.

Elimination Method

1. Multiply each equation (if necessary) by positive numbers to make the coefficients of one unknown the same in absolute value.
2. Then add or subtract the resulting equations to eliminate that unknown.

EXAMPLE 2   Solve the system $\begin{cases} 2x + y = 9 \\ x - 2y = 7 \end{cases}$

Solution $\begin{cases} 2x + y = 9 \\ x - 2y = 7 \end{cases}$

We shall eliminate the unknown y. Multiply the first equation by 2:

$$\begin{cases} 4x + 2y = 18 \\ x - 2y = 7 \end{cases}$$

Now add the equations:

$$5x = 25$$
$$x = 5$$

Substituting $x = 5$ into the second equation, we obtain

$$5 - 2y = 7$$
$$y = -1$$

Thus, the solution of the system is $x = 5$ and $y = -1$. ■

EXAMPLE 3   Solve the system $\begin{cases} 6x + 2y = -3 \\ -5x + 3y = 6 \end{cases}$

Solution $\begin{cases} 6x + 2y = -3 \\ -5x + 3y = 6 \end{cases}$

We choose to eliminate the unknown x. Multiply the first equation by 5 and the second equation by 6:

$$\begin{cases} 30x + 10y = -15 \\ -30x + 18y = 36 \end{cases}$$

Since the signs of the coefficients of x are different, we add the equations to obtain

$$28y = 21$$
$$y = \frac{3}{4}$$

Substituting $y = \frac{3}{4}$ into the first equation gives

$$6x + 2\left(\frac{3}{4}\right) = -3$$

$$6x + \frac{3}{2} = -3$$

$$6x = -\frac{9}{2}$$

$$x = -\frac{3}{4}$$

Thus, the solution of the system is $x = -\dfrac{3}{4}$ and $y = \dfrac{3}{4}$. ■

Although the elimination method was used in the above examples, the question arises as to whether the method gives information not only about systems that have a unique solution, but also about systems that have an infinite number of solutions or no solutions. Fortunately, the elimination method does yield such information. The following examples illustrate how the elimination method achieves this.

EXAMPLE 4 Solve the system $\begin{cases} -3x + 4y = 5 \\ 6x - 8y = -2 \end{cases}$

Solution We try to eliminate y. Multiplying the first equation by 2, we get

$$\begin{cases} -6x + 8y = 10 \\ 6x - 8y = -2 \end{cases}$$

Now, add the equations to obtain

$$0 = 8$$

This is a contradiction. Since this cannot occur, the system has no solution. This could be seen graphically by noting that the graph of the system consists of two parallel lines. ■

EXAMPLE 5 Solve the system $\begin{cases} x + 2y = 3 \\ 2x + 4y = 6 \end{cases}$

Solution Multiply the first equation by 2:

$$\begin{cases} 2x + 4y = 6 \\ 2x + 4y = 6 \end{cases}$$

Subtracting gives

$$0 = 0$$

which is an identity. Thus, as in the substitution method discussed previously, the elimination method shows us that any point on the coinciding lines is a solution. ■

Note.

1. If the elimination method results in the contradiction

$$0 = c \quad \text{where} \quad c \neq 0$$

then there is no solution of the original system.

2. If the elimination method results in the identity

$$0 = 0$$

then there are infinitely many solutions of the original system.

The elimination method can also be used to solve certain nonlinear systems.

EXAMPLE 6 Solve the nonlinear system $\begin{cases} \dfrac{1}{x} - \dfrac{1}{y} = 1 \\ \dfrac{2}{x} + \dfrac{3}{y} = 7 \end{cases}$

Solution If we let $u = \dfrac{1}{x}$ and $v = \dfrac{1}{y}$, then the system of nonlinear equations in x and y becomes a system of linear equations in u and v. That is,

$$\begin{cases} u - v = 1 \\ 2u + 3v = 7 \end{cases} \tag{1}$$

We now apply the elimination method to this linear system. First, we multiply the first equation by 2 and subtract the result from the second equation:

$$\begin{cases} 2u - 2v = 2 \\ 2u + 3v = 7 \end{cases}$$

Thus, we have

$$-5v = -5$$
$$v = 1$$

Substituting $v = 1$ into the first equation of system (1), we get $u - 1 = 1$, or, $u = 2$. To find x and y, we use the relationships $u = \dfrac{1}{x}$ and $v = \dfrac{1}{y}$. That is,

$$\text{if} \quad u = 2, \quad \text{then} \quad \frac{1}{x} = 2 \quad \text{and} \quad x = \frac{1}{2}$$

$$\text{if} \quad v = 1, \quad \text{then} \quad \frac{1}{y} = 1 \quad \text{and} \quad y = 1$$

Hence, the system has solution $x = \frac{1}{2}$ and $y = 1$. ■

EXAMPLE 7 Solve the system $\begin{cases} 3x + 4y = a \\ 2x - 3y = b \end{cases}$

Solution   Multiply the first equation by 3 and the second equation by 4 to get

$$\begin{cases} 9x + 12y = 3a \\ 8x - 12y = 4b \end{cases}$$

Now add the equations to get

$$17x = 3a + 4b$$

$$x = \frac{3a + 4b}{17}$$

Rather than substituting this expression into any of the equations, we return to the original system and eliminate x and solve for y. Multiply the first equation by 2 and the second equation by 3 to obtain

$$\begin{cases} 6x + 8y = 2a \\ 6x - 9y = 3b \end{cases}$$

Now subtract the equations:

$$17y = 2a - 3b$$

$$y = \frac{2a - 3b}{17}$$

Hence, the system has solution $x = \dfrac{3a + 4b}{17}$ and $y = \dfrac{2a - 3b}{17}$.   ■

EXERCISES   11.3

Solve the systems in Exercises 1–37 using the elimination method.

1. $\begin{cases} x + 2y = -1 \\ 2x + y = 4 \end{cases}$

2. $\begin{cases} x - 2y = -1 \\ 2x + y = -7 \end{cases}$

3. $\begin{cases} 2x - 3y = 5 \\ x - 2y = 1 \end{cases}$

4. $\begin{cases} 3x - 8y = 2 \\ x - 2y = 2 \end{cases}$

5. $\begin{cases} 2x + 3y = -5 \\ 3x + y = 3 \end{cases}$

6. $\begin{cases} x - 3y = 3 \\ 2x - 5y = 4 \end{cases}$

7. $\begin{cases} 6x + 2y = 13 \\ 2x - 3y = 8 \end{cases}$

8. $\begin{cases} 3x - 2y = 6 \\ 4x - 10y = -3 \end{cases}$

9. $\begin{cases} 2\frac{1}{2}x - 3y = 4 \\ 2x - 1\frac{1}{2}y = 5 \end{cases}$

10. $\begin{cases} x + 3y - 1 = 0 \\ x + 5y + 3 = 0 \end{cases}$

11. $\begin{cases} 2x + 5y + 5 = 0 \\ 3x + 2y - 9 = 0 \end{cases}$

12. $\begin{cases} 4x + 3y + 1 = 0 \\ 3x + 2y + 2 = 0 \end{cases}$

13. $\begin{cases} 2x + 1 = 3y \\ 6y + 3 = 3x \end{cases}$

14. $\begin{cases} 3x + 5 = 2y \\ 2y - 11 = x \end{cases}$

15. $\begin{cases} 3x + 9 = -4y \\ 2x - 16 = y \end{cases}$

16. $\begin{cases} 5x - 4y = 23 \\ 11x + 7y = 19 \end{cases}$

17. $\begin{cases} -9x + 13y = -5 \\ 7x + 5y = 19 \end{cases}$

18. $\begin{cases} 13x + 11y = 16 \\ 17x + 12y = 9 \end{cases}$

19. $\begin{cases} 5x + 19y = 1 \\ 11x + 23y = 21 \end{cases}$

20. $\begin{cases} 15x + 19y = 8 \\ 13x + 17y = 0 \end{cases}$

21. $\begin{cases} 23x - 12y = 25 \\ 21x - 13y = -16 \end{cases}$

22. $\begin{cases} 17x + 15y = 15 \\ 25x + 23y = 7 \end{cases}$

23. $\begin{cases} 16x + 25y = 6 \\ 19x + 27y = -9 \end{cases}$

24. $\begin{cases} \dfrac{1}{x} + \dfrac{1}{y} = 4 \\ \dfrac{1}{x} - \dfrac{1}{y} = 2 \end{cases}$

25. $\begin{cases} \dfrac{1}{x} + \dfrac{1}{y} = \dfrac{2}{3} \\ -\dfrac{1}{x} + \dfrac{1}{y} = \dfrac{4}{3} \end{cases}$

26. $\begin{cases} \dfrac{1}{x} + \dfrac{1}{y} = \dfrac{5}{12} \\ \dfrac{4}{x} - \dfrac{3}{y} = \dfrac{1}{2} \end{cases}$

27. $\begin{cases} \dfrac{2}{x} + \dfrac{5}{y} = 1 \\ \dfrac{14}{x} + \dfrac{15}{y} = 5 \end{cases}$

28. $\begin{cases} \dfrac{3}{x} - \dfrac{2}{y} = 6 \\ \dfrac{7}{x} + \dfrac{12}{y} = 4 \end{cases}$

29. $\begin{cases} \dfrac{8}{x} + \dfrac{3}{y} = 8 \\ \dfrac{10}{x} + \dfrac{9}{y} = 3 \end{cases}$

30. $\begin{cases} \dfrac{16}{x} + \dfrac{7}{y} = 5 \\ \dfrac{11}{x} + \dfrac{5}{y} = \dfrac{7}{2} \end{cases}$

31. $\begin{cases} \dfrac{13}{x} + \dfrac{4}{y} = \dfrac{5}{2} \\ \dfrac{12}{x} + \dfrac{7}{y} = -1 \end{cases}$

32. $\begin{cases} cx + 3y = 2d \\ 3cx - 2y = d \end{cases}$

33. $\begin{cases} 5x - 4my = 7n \\ 2x + 3my = 5n \end{cases}$

34. $\begin{cases} ax + by = a \\ ax - by = b \end{cases}$

35. $\begin{cases} 4ax - 3by = 0 \\ 5ax - 2by = 7ab \end{cases}$

36. $\begin{cases} \dfrac{9a}{x} - \dfrac{4a}{y} = 1 \\ \dfrac{6a}{x} + \dfrac{2a}{y} = 3 \end{cases}$

37. $\begin{cases} \dfrac{8a}{x} + \dfrac{2b}{y} = 1 \\ \dfrac{14a}{x} - \dfrac{5b}{y} = 6 \end{cases}$

38. Show that the equations $5x - 2y = 3$ and $4y - 5 = 10x$ are inconsistent.

*39. Solve $\begin{cases} mx + bn = n(a - y) \\ b(mx + an) = an(b - y) \end{cases}$
for x and y.

*40. Solve $\begin{cases} b(x - a) + a(y - b) = 2 \\ ab(x - y) = (a - b)(ab + 1) \end{cases}$
for x and y.

11.4 SYSTEMS OF LINEAR EQUATIONS IN THREE VARIABLES

In Section 11.1 we saw that in order to obtain a unique solution for a system of linear equations in two variables, we must have two independent equations. Similarly, to obtain a unique solution for a system of linear equations in three variables, we must have three independent equations. In this section we show how the elimination method can be used to solve a system of linear equations in three variables.

> **DEFINITION**
>
> A solution of the system
>
> $$\begin{cases} a_1x + b_1y + c_1z = d_1 \\ a_2x + b_2y + c_2z = d_2 \\ a_3x + b_3y + c_3z = d_3 \end{cases}$$
>
> is any ordered triple (x, y, z) that satisfies *all* the equations of the system.

EXAMPLE 1 Given the system $\begin{cases} 2x - y + 3z = -9 & (1) \\ x + 3y - z = 10 & (2) \\ 3x + y - z = 8 & (3) \end{cases}$

Determine whether each of the following ordered triples is a solution of the system: **(a)** $(1, -1, 1)$ **(b)** $(1, 2, -3)$

Solution (a) From the ordered triple $(1, -1, 1)$, we get $x = 1$, $y = -1$, and $z = 1$. Substituting these values into the equations, we obtain

$$2(1) - (-1) + 3(1) = 6 \neq -9 \tag{1}$$

$$(1) + 3(-1) - (1) = -3 \neq 10 \tag{2}$$

$$3(1) + (-1) - (1) = 1 \neq 8 \tag{3}$$

Thus, $(1, -1, 1)$ is *not* a solution of the given system.

(b) for $(1, 2, -3)$, we have $x = 1$, $y = 2$, and $z = -3$.

$$2(1) - (2) + 3(-3) = -9 \checkmark \tag{1}$$

$$(1) + 3(2) - (-3) = 10 \checkmark \tag{2}$$

$$3(1) + (2) - (-3) = 8 \checkmark \tag{3}$$

Hence, $(1, 2, -3)$ is a solution of the given system. ■

To solve a system of linear equations in three variables, we eliminate one variable from one pair of equations, and then eliminate the *same* variable from another pair of equations. This produces a system of two linear equations in the two remaining variables, which can be solved by either of the methods discussed previously.

EXAMPLE 2 Solve the system $\begin{cases} 2x - 3y + 5z = 27 & (1) \\ x + 2y - z = -4 & (2) \\ 5x - y + 4z = 27 & (3) \end{cases}$

Solution Let y be the first variable eliminated. Multiply equation (1) by 2 and equation (2) by 3 and add the results:

$$\begin{array}{r} 4x - 6y + 10z = 54 \\ 3x + 6y - 3z = -12 \\ \hline 7x + 7z = 42 \end{array} \tag{4}$$

Next we eliminate y between equations (2) and (3). Multiply equation (3) by 2 and add the results to equation (2).

$$\begin{array}{r} 10x - 2y + 8z = 54 \\ x + 2y - z = -4 \\ \hline 11x + 7z = 50 \end{array} \tag{5}$$

Thus, we obtain a system of linear equations in the variables x and z:

$$\begin{cases} 7x + 7z = 42 & (4) \\ 11x + 7z = 50 & (5) \end{cases}$$

Subtracting equation (5) from equation (4) gives

$$-4x = -8$$
$$x = 2$$

Substituting $x = 2$ into equation (4) gives

$$14 + 7z = 42 \quad \text{or} \quad z = 4$$

Finally, substituting $x = 2$ and $z = 4$ into (3), we get

$$10 - y + 16 = 27 \quad \text{or} \quad y = -1$$

Hence, the solution of the given system is $x = 2$, $y = -1$, and $z = 4$; that is, the ordered triple $(2, -1, 4)$. ■

EXAMPLE 3 Solve the system $\begin{cases} x + 2y - z = 6 & (1) \\ 2x - y + 3z = -13 & (2) \\ 3x - 2y + 3z = -16 & (3) \end{cases}$

Solution Suppose we choose to eliminate the variable z.

Multiply (1) by 3 and add results to (2): $5x + 5y = 5$ (4)

Subtract (3) from (2): $-x + y = 3$ (5)

Multiply (5) by 5 and add results to (4): $10y = 20$

$$y = 2$$

Substitute $y = 2$ in (5): $x = -1$

Finally, substitute $y = 2$ and $x = -1$ in (1): $z = -3$

Hence the solution of the given system is $x = -1$, $y = 2$, and $z = -3$; that is, the ordered triple $(-1, 2, -3)$. ■

Note. If one of the variables does not appear in one of the given equations, then eliminate this variable from the other two equations. Then solve the resulting system of two linear equations in two variables.

EXAMPLE 4 Solve the system $\begin{cases} 4x + 3y + z = 9 & (1) \\ 5x - 4y - z = 19 & (2) \\ x + 2y = 1 & (3) \end{cases}$

Solution Add (1) to (2): $9x - y = 28$ (4)

Equation (3): $x + 2y = 1$ (5)

Now multiply (4) by 2 and add results to (5):

$$19x = 57$$
$$x = 3$$

Substituting $x = 3$ in (5) gives $y = -1$. Substituting $x = 3$ and $y = -1$ in (1) yields $z = 0$. Thus, the solution is the ordered triple $(3, -1, 0)$. ∎

EXERCISES 11.4

Solve:

1. $\begin{cases} x + y + z = 6 \\ x - y + z = 2 \\ x + y - z = 0 \end{cases}$

2. $\begin{cases} 2x + y + z = 11 \\ x + 2y + z = 2 \\ x + y + 2z = 15 \end{cases}$

3. $\begin{cases} x + 3y - z = -11 \\ 2x + 6y + 3z = -2 \\ 3x - y - 2z = 1 \end{cases}$

4. $\begin{cases} x - 2y + z = 5 \\ 2x + y - z = -1 \\ 3x + 3y - 2z = -4 \end{cases}$

5. $\begin{cases} 3x + y - z = 11 \\ x + 3y - z = 13 \\ x + y - 3z = 11 \end{cases}$

6. $\begin{cases} 2x + 3y - z = -2 \\ 4x - 3y + 2z = 9 \\ 6x - 6y + 3z = 13 \end{cases}$

7. $\begin{cases} x + y + z = 6 \\ x + y - z = 2 \\ 2x + 3y - 5z = 1 \end{cases}$

8. $\begin{cases} x + y + 2z = 0 \\ x + 2y - z = 7 \\ 2x + y + 4z = -1 \end{cases}$

9. $\begin{cases} 3x + 2y = 7 \\ 4x + 3y - z = 4 \\ 3x + 2y + z = 13 \end{cases}$

10. $\begin{cases} 5x + 2z = 17 \\ 8x - y + 6z = 26 \\ 8x + 3y - 12z = 24 \end{cases}$

11. $\begin{cases} 2x - 3y + 2z = 2 \\ x + 2y + z = 1 \\ 3x - 5y + 2z = 6 \end{cases}$

12. $\begin{cases} 14x - 2y + 3z = 12 \\ 10x - 3y + 2z = 2 \\ 14x + y + 3z = 24 \end{cases}$

13. $\begin{cases} 2x - 3y + z = 8 \\ x + 3y + 8z = 1 \\ 3x - y + 2z = -1 \end{cases}$

14. $\begin{cases} -4x + 2y - 9z = 2 \\ 3x + 4y + z = 5 \\ x - 3y + 2z = 8 \end{cases}$

15. $\begin{cases} 10x - 6y - 3z = 11 \\ 5x + 9y + 2z = 3 \\ 15x - 12y - z = 4 \end{cases}$

16. $\begin{cases} x + y = 1 \\ y - z = -3 \\ x - z = 2 \end{cases}$

17. $\begin{cases} x + y = 3 \\ y + z = 5 \\ x + z = 4 \end{cases}$

18. $\begin{cases} x + 2y = 1 \\ 2x + 3z = 3 \\ y + 2z = 8 \end{cases}$

19. $\begin{cases} 2x + 3y = z - 1 \\ 3x = 8z - 1 \\ 5y + 7z = -1 \end{cases}$

20. $\begin{cases} 2x + y = z \\ 4x + z = 4y \\ y = x + 1 \end{cases}$

21. $\begin{cases} 3x + y = z + 2 \\ y = 1 - 2x \\ 3z = -2y \end{cases}$

22. $\begin{cases} \dfrac{2}{x} - \dfrac{1}{y} - \dfrac{3}{z} = 7 \\[2mm] \dfrac{1}{x} + \dfrac{2}{y} - \dfrac{1}{z} = 10 \\[2mm] \dfrac{3}{x} - \dfrac{3}{y} + \dfrac{2}{z} = -7 \end{cases}$

23. $\begin{cases} \dfrac{1}{x} + \dfrac{2}{y} - \dfrac{3}{z} = 1 \\[2mm] \dfrac{5}{x} + \dfrac{4}{y} + \dfrac{6}{z} = 24 \\[2mm] \dfrac{7}{x} - \dfrac{8}{y} + \dfrac{9}{z} = 14 \end{cases}$

24. $\begin{cases} \dfrac{2}{x} + \dfrac{3}{y} - \dfrac{5}{z} = 25 \\[2mm] \dfrac{8}{x} - \dfrac{6}{y} + \dfrac{10}{z} = -26 \\[2mm] -\dfrac{12}{x} - \dfrac{9}{y} + \dfrac{5}{z} = 13 \end{cases}$

25. $\begin{cases} \dfrac{1}{x} + \dfrac{1}{y} = 5 \\[2mm] \dfrac{1}{y} - \dfrac{1}{z} = \dfrac{3}{2} \\[2mm] \dfrac{1}{x} - \dfrac{1}{z} = \dfrac{1}{2} \end{cases}$

26. $\begin{cases} \dfrac{1}{x} - \dfrac{1}{y} = \dfrac{1}{6} \\[2mm] \dfrac{1}{y} + \dfrac{1}{z} = \dfrac{7}{12} \\[2mm] \dfrac{1}{x} - \dfrac{1}{z} = \dfrac{1}{4} \end{cases}$

27. $\begin{cases} \dfrac{2}{x} + \dfrac{3}{y} = -19 \\[2mm] \dfrac{4}{x} - \dfrac{5}{z} = -23 \\[2mm] \dfrac{2}{x} + \dfrac{3}{z} = 5 \end{cases}$

*28. $\begin{cases} 2x + y = a \\ x - z = 2a - b \\ 3z + y = 3(b - a) \end{cases}$

*29. $\begin{cases} 4x - 3y + 2z = \dfrac{3}{2}a \\[2mm] x - 6y + 4z = -\dfrac{1}{2}a \\[2mm] 3x - 2y - z = \dfrac{7}{12}a \end{cases}$

*30. $\begin{cases} ax - by + cz = 2 \\ 3ax - 4cz = -9 \\ 5ax + 2by = 9 \end{cases}$

*31. $\begin{cases} cx + by = r \\ by + az = s \\ az + cx = t \end{cases}$

*32. $\begin{cases} bx + ay = c \\ cy + bz = a \\ cx + az = b \end{cases}$

*33. $\begin{cases} ax + by + cz = 3 \\ a^2x + b^2y + c^2z = a + b + c \\ ax - by + cz = 1 \end{cases}$

*34. $\begin{cases} x + y + z = 0 \\ ax + by + cz = 0 \\ bx + ay - cz = b^2 - a^2 \end{cases}$

*35. $\begin{cases} \dfrac{1}{x} - \dfrac{1}{y} = a \\[2mm] \dfrac{1}{x} - \dfrac{1}{z} = b \\[2mm] \dfrac{1}{y} + \dfrac{1}{z} = c \end{cases}$

*36. $\begin{cases} \dfrac{1}{x} - a = \dfrac{1}{y} - c \\[2mm] \dfrac{1}{y} - b = \dfrac{1}{z} - a \\[2mm] \dfrac{1}{x} = a - \dfrac{1}{z} \end{cases}$

11.5 DETERMINANT SOLUTIONS TO LINEAR SYSTEMS: CRAMER'S RULE

Solution of Systems of Two Linear Equations by Determinants

Consider the general system of two equations in two variables. Such a system can be written

$$\begin{cases} a_1 x + b_1 y = c_1 & (1) \\ a_2 x + b_2 y = c_2 & (2) \end{cases}$$

where a_1, a_2, b_1, b_2, c_1, and c_2 are real numbers.

The solution of this system can be found using the elimination method. In order to eliminate y, multiply equation (1) by b_2 and equation (2) by b_1 to obtain

$$\begin{cases} a_1 b_2 x + b_1 b_2 y = c_1 b_2 & (3) \\ a_2 b_1 x + b_1 b_2 y = c_2 b_1 & (4) \end{cases}$$

Subtracting equation (4) from equation (3), we have

$$(a_1 b_2 - a_2 b_1)x = c_1 b_2 - c_2 b_1$$

Therefore, if $(a_1b_2 - a_2b_1) \neq 0$, we have

$$x = \frac{c_1b_2 - c_2b_1}{a_1b_2 - a_2b_1}$$

Similarly, multiplying equation (1) by a_2 and equation (2) by a_1, and subtracting the resulting equations, gives

$$y = \frac{a_1c_2 - a_2c_1}{a_1b_2 - a_2b_1}$$

Thus, the solution to the general system

$$\begin{cases} a_1x + b_1y = c_1 \\ a_2x + b_2y = c_2 \end{cases}$$

is the ordered pair (x, y) where

$$x = \frac{c_1b_2 - c_2b_1}{a_1b_2 - a_2b_1} \quad \text{and} \quad y = \frac{a_1c_2 - a_2c_1}{a_1b_2 - a_2b_1}$$

Note that the denominators in the formulas for x and y are identical and are given by the expression

$$a_1b_2 - a_2b_1$$

We now introduce a symbol that will help us in computing such expressions.

DEFINITION

The expression $a_1b_2 - a_2b_1$ is called a **determinant of the second order** and it is denoted by the symbol:

$$\begin{vmatrix} a_1 & b_1 \\ a_2 & b_2 \end{vmatrix}$$

That is,

$$\begin{vmatrix} a_1 & b_1 \\ a_2 & b_2 \end{vmatrix} = a_1b_2 - a_2b_1$$

The four numbers a_1, a_2, b_1, and b_2 are arranged in two horizontal lines called **rows** and two vertical lines called **columns**. Each number is called an **element** of the determinant. Thus, the first row of the determinant consists of elements a_1, b_1 and the second row consists of elements a_2, b_2. The first column consists of the elements a_1, a_2 and the second column consists of b_1, b_2. See Figure 11.4.

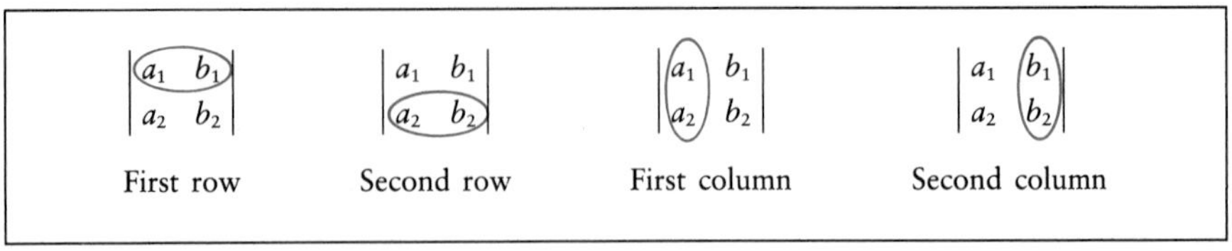

Figure 11.4

Note that the vertical bars used to denote a determinant should not be confused with the absolute value sign. We take the determinant of a **square array** of numbers.

Following is a simple way to remember how to compute a determinant of the second order:

$$\begin{vmatrix} a_1 & b_1 \\ a_2 & b_2 \end{vmatrix} = a_1 b_2 - a_2 b_1$$

The definition of a determinant of the second order and the rule of computing remain the same when the letters are replaced by specific numbers.

EXAMPLE 1 Compute the determinant

$$\begin{vmatrix} 6 & -2 \\ 3 & 5 \end{vmatrix}$$

Solution
$$\begin{vmatrix} 6 & -2 \\ 3 & 5 \end{vmatrix} = (6)(5) - (3)(-2) = 30 + 6 = 36 \quad \blacksquare$$

EXAMPLE 2 Compute the determinants:

(a) $\begin{vmatrix} a_1 & c_1 \\ a_2 & c_2 \end{vmatrix}$

(b) $\begin{vmatrix} c_1 & b_1 \\ c_2 & b_2 \end{vmatrix}$

Solution (a) $\begin{vmatrix} a_1 & c_1 \\ a_2 & c_2 \end{vmatrix} = a_1 c_2 - a_2 c_1$

(b) $\begin{vmatrix} c_1 & b_1 \\ c_2 & b_2 \end{vmatrix} = c_1 b_2 - c_2 b_1 \quad \blacksquare$

As we see from Example 2, the formulas for the solution of a system of two linear equations

$$x = \frac{c_1 b_2 - c_2 b_1}{a_1 b_2 - a_2 b_1} \qquad y = \frac{a_1 c_2 - a_2 c_1}{a_1 b_2 - a_2 b_1}$$

can be written using the determinant notation:

$$x = \frac{\begin{vmatrix} c_1 & b_1 \\ c_2 & b_2 \end{vmatrix}}{\begin{vmatrix} a_1 & b_1 \\ a_2 & b_2 \end{vmatrix}} \qquad y = \frac{\begin{vmatrix} a_1 & c_1 \\ a_2 & c_2 \end{vmatrix}}{\begin{vmatrix} a_1 & b_1 \\ a_2 & b_2 \end{vmatrix}}$$

Cramer's Rule

The values of x and y that satisfy the system

$$\begin{cases} a_1 x + b_1 y = c_1 & (1) \\ a_2 x + b_2 y = c_2 & (2) \end{cases}$$

are given by the formulas

$$x = \frac{D_x}{D} \quad \text{and} \quad y = \frac{D_y}{D}, \text{ provided } D \neq 0$$

and where $D = \begin{vmatrix} a_1 & b_1 \\ a_2 & b_2 \end{vmatrix}$, $D_x = \begin{vmatrix} c_1 & b_1 \\ c_2 & b_2 \end{vmatrix}$, and $D_y = \begin{vmatrix} a_1 & c_1 \\ a_2 & c_2 \end{vmatrix}$.

Note that the entries a_1, a_2, b_1, b_2 of D are just the coefficients of x and y in the equations (1) and (2) and D is called the **coefficient determinant**. The entries for the determinant D_x can be obtained by replacing the x-coefficients a_1, a_2 in D with the constants c_1, c_2. Similarly, the entries for the determinant D_y can be obtained by replacing the y-coefficients b_1, b_2 in D with the constants c_1, c_2.

EXAMPLE 3 Use Cramer's rule to solve the system $\begin{cases} 2x + 8y = 3 \\ 3x - 5y = 4 \end{cases}$

Solution We begin by computing the determinants D, D_x, and D_y.

$$D = \begin{vmatrix} 2 & 8 \\ 3 & -5 \end{vmatrix} = -10 - 24 = -34; \quad D_x = \begin{vmatrix} 3 & 8 \\ 4 & -5 \end{vmatrix} = -15 - 32 = -47;$$

and $D_y = \begin{vmatrix} 2 & 3 \\ 3 & 4 \end{vmatrix} = 8 - 9 = -1.$ Thus $x = \dfrac{D_x}{D} = \dfrac{-47}{-34} = \dfrac{47}{34}$ and $y =$

$\dfrac{D_y}{D} = \dfrac{-1}{-34} = \dfrac{1}{34}.$ Hence, the solution is the ordered pair $\left(\dfrac{47}{34}, \dfrac{1}{34} \right).$ ∎

EXAMPLE 4   Use Cramer's rule to solve the system $\begin{cases} 3x - 2y = 6 \\ 5x + 3y = -9 \end{cases}$

Solution $D = \begin{vmatrix} 3 & -2 \\ 5 & 3 \end{vmatrix} = 19;\ D_x = \begin{vmatrix} 6 & -2 \\ -9 & 3 \end{vmatrix} = 0;\ \text{and}\ D_y = \begin{vmatrix} 3 & 6 \\ 5 & -9 \end{vmatrix} = -57.$

Thus, $x = \dfrac{D_x}{D} = \dfrac{0}{19} = 0$ and $y = \dfrac{D_y}{D} = \dfrac{-57}{19} = -3$. Hence, the solution is the ordered pair $(0, -3)$. ■

EXAMPLE 5   Using Cramer's rule, solve $\begin{cases} 3x - 2y = 4 \\ 6x - 4y = 5 \end{cases}$

Solution $D = \begin{vmatrix} 3 & -2 \\ 6 & -4 \end{vmatrix} = 0;\ \ D_x = \begin{vmatrix} 4 & -2 \\ 5 & -4 \end{vmatrix} = -6;\ \ D_y = \begin{vmatrix} 3 & 4 \\ 6 & 5 \end{vmatrix} = -9.$

Thus, $x = \dfrac{D_x}{D} = \dfrac{-6}{0}$ and $y = \dfrac{D_y}{D} = \dfrac{-9}{0}$, which are undefined fractions. Using the methods of the previous sections, we see that this system is inconsistent. ■

When $D = 0$ and $D_x \neq 0$ or $D_y \neq 0$, the system is inconsistent.

EXAMPLE 6   Use Cramer's rule to solve $\begin{cases} 9x - 6y = 21 \\ 6x - 4y = 14 \end{cases}$

Solution $D = \begin{vmatrix} 9 & -6 \\ 6 & -4 \end{vmatrix} = 0;\ \ D_x = \begin{vmatrix} 21 & -6 \\ 14 & -4 \end{vmatrix} = 0;\ \ D_y = \begin{vmatrix} 9 & 21 \\ 6 & 14 \end{vmatrix} = 0.$

Thus, $x = \dfrac{D_x}{D} = \dfrac{0}{0}$ and $y = \dfrac{D_y}{D} = \dfrac{0}{0}$, which are indeterminate forms.

Using the methods of the previous sections, we see that the equations are dependent. ■

When $D = 0$, $D_x = 0$,   and $D_y = 0$, the equations are dependent.

EXERCISES   11.5

Compute each of the determinants in Exercises 1–16.

1. $\begin{vmatrix} 2 & 3 \\ 3 & 4 \end{vmatrix}$

2. $\begin{vmatrix} -1 & -1 \\ 3 & 0 \end{vmatrix}$

3. $\begin{vmatrix} 3 & 4 \\ 9 & 12 \end{vmatrix}$

4. $\begin{vmatrix} 5 & -3 \\ 10 & 6 \end{vmatrix}$

5. $\begin{vmatrix} -3 & 2 \\ 12 & -8 \end{vmatrix}$

6. $\begin{vmatrix} -\frac{3}{4} & \frac{1}{3} \\ -1 & -\frac{8}{9} \end{vmatrix}$

7. $\begin{vmatrix} 6 & -5 \\ -2 & 3 \end{vmatrix}$

8. $\begin{vmatrix} 9 & -1 \\ 0 & 7 \end{vmatrix}$

9. $\begin{vmatrix} x & 3 \\ 5 & x \end{vmatrix}$

10. $\begin{vmatrix} x-1 & 2 \\ 3 & x \end{vmatrix}$

11. $\begin{vmatrix} -2 & x \\ -8 & y \end{vmatrix}$

12. $\begin{vmatrix} x-3 & 2x-6 \\ 1 & 2 \end{vmatrix}$

13. $\begin{vmatrix} a & b \\ ax & bx \end{vmatrix}$

14. $\begin{vmatrix} a & x \\ b & y \end{vmatrix}$

15. $\begin{vmatrix} 0 & 0 \\ x & y \end{vmatrix}$

16. $\begin{vmatrix} a & b \\ b & a \end{vmatrix}$

In Exercises 17–20, solve each equation for x.

17. $\begin{vmatrix} x & 1 \\ 3 & 2 \end{vmatrix} = 3$

18. $\begin{vmatrix} 3 & x \\ 5 & 2 \end{vmatrix} = -4$

19. $\begin{vmatrix} 1 & -3 \\ x & -4 \end{vmatrix} = 21$

20. $\begin{vmatrix} x & 4 \\ 6 & x \end{vmatrix} = 1$

In Exercises 21–29, solve each of the systems using Cramer's rule.

21. $\begin{cases} 2x - 3y = 13 \\ 3x + 4y = -6 \end{cases}$

22. $\begin{cases} 4x - 3y = -13 \\ 5x + 2y = 1 \end{cases}$

23. $\begin{cases} 8x - 3y = -8 \\ 6x + y = 7 \end{cases}$

24. $\begin{cases} -3x + 6y = 11 \\ x - 3y = -4 \end{cases}$

25. $\begin{cases} -4x - 6y = -12 \\ 2x + 3y = 5 \end{cases}$

26. $\begin{cases} -3x + 2y = -5 \\ 6x - 4y = 10 \end{cases}$

27. $\begin{cases} -3x + 4y = 1 \\ 9x + 8y = 12 \end{cases}$

28. $\begin{cases} 3x + 4y = 1 \\ -2x - 5y = -10 \end{cases}$

29. $\begin{cases} 3x = 2y + 6 \\ y = x - 1 \end{cases}$

In Exercises 30–35, solve for x and y using Cramer's rule.

30. $\begin{cases} 9x - 4y = a \\ -7x + 3y = -a \end{cases}$

31. $\begin{cases} ax + by = c \\ cx + ay = b \end{cases}$

32. $\begin{cases} ax - by = c \\ bx + ay = c \end{cases}$

*33. $\begin{cases} 5b^2x - 4a^2y = 8ab \\ 8b^2x - 7a^2y = 11ab \end{cases}$

*34. $\begin{cases} 2x - 5ay = -13a + 4 \\ 9x + 4ay = 21a + 18 \end{cases}$

*35. $\begin{cases} x + ay = m \\ x + a^2y = n \end{cases}$

*36. For what value(s) of a does the system $\begin{cases} x + ay = 3 \\ x - y = 3 \end{cases}$ have an infinite number of solutions?

*37. For what value(s) of a does the system $\begin{cases} x + ay = 4 \\ x - 2y = 3 \end{cases}$ have no solution?

11.6 SOLUTION OF SYSTEMS OF THREE LINEAR EQUATIONS BY DETERMINANTS

Just as determinants can be used to solve systems of two linear equations, determinants can also be used to solve systems of three linear equations in three variables.

DEFINITION
The symbol

$$\begin{vmatrix} a_1 & b_1 & c_1 \\ a_2 & b_2 & c_2 \\ a_3 & b_3 & c_3 \end{vmatrix}$$

is called a **determinant of the third order** and its value is defined by

$$a_1b_2c_3 + a_2b_3c_1 + a_3b_1c_2 - a_1b_3c_2 - a_2b_1c_3 - a_3b_2c_1$$

The value of a third-order determinant can be evaluated without having to memorize the expression given in the definition.

To evaluate a third-order determinant, copy the 3×3 (read "three-by-three") array and rewrite the first two columns of elements to the right of the array according to the following diagram:

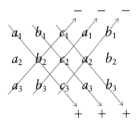

The arrows pointing downward represent the three products having a positive sign, and the arrows pointing upward represent the three products having a negative sign.

EXAMPLE 1 Compute the value of $\begin{vmatrix} 1 & 1 & -2 \\ 4 & -1 & 0 \\ -2 & 0 & 3 \end{vmatrix}$.

Solution
$$\begin{vmatrix} 1 & 1 & -2 \\ 4 & -1 & 0 \\ -2 & 0 & 3 \end{vmatrix} = \begin{array}{ccccc} 1 & 1 & -2 & 1 & 1 \\ 4 & -1 & 0 & 4 & -1 \\ -2 & 0 & 3 & -2 & 0 \end{array}$$

$$= (1)(-1)(3) + (1)(0)(-2) + (-2)(4)(0)$$
$$\quad - (-2)(-1)(-2) - (0)(0)(1) - (3)(4)(1)$$
$$= (-3) + (0) + (0) - (-4) - (0) - (12)$$
$$= -11 \quad \blacksquare$$

EXAMPLE 2 Compute the value of $\begin{vmatrix} 1 & 4 & -5 \\ 2 & 3 & -1 \\ 3 & 7 & 2 \end{vmatrix}$.

Solution
$$\begin{vmatrix} 1 & 4 & -5 \\ 2 & 3 & -1 \\ 3 & 7 & 2 \end{vmatrix} = \begin{array}{ccccc} 1 & 4 & -5 & 1 & 4 \\ 2 & 3 & -1 & 2 & 3 \\ 3 & 7 & 2 & 3 & 7 \end{array}$$

$$= (6) + (-12) + (-70) - (-45) - (-7) - (16)$$
$$= -40 \quad \blacksquare$$

Note. While this diagonal method works for third-order determinants, it does *not* work for higher-order determinants.

Cramer's rule can now be extended to systems of three linear equations in three variables.

Cramer's Rule

The values of x, y, and z that satisfy the system

$$\begin{cases} a_1x + b_1y + c_1z = d_1 \\ a_2x + b_2y + c_2z = d_2 \\ a_3x + b_3y + c_3z = d_3 \end{cases}$$

are given by the formulas

$$x = \frac{D_x}{D}, \; y = \frac{D_y}{D}, \; z = \frac{D_z}{D}, \quad \text{provided } D \neq 0, \quad \text{and where}$$

$$D = \begin{vmatrix} a_1 & b_1 & c_1 \\ a_2 & b_2 & c_2 \\ a_3 & b_3 & c_3 \end{vmatrix} \qquad D_x = \begin{vmatrix} d_1 & b_1 & c_1 \\ d_2 & b_2 & c_2 \\ d_3 & b_3 & c_3 \end{vmatrix}$$

$$D_y = \begin{vmatrix} a_1 & d_1 & c_1 \\ a_2 & d_2 & c_2 \\ a_3 & d_3 & c_3 \end{vmatrix} \qquad D_z = \begin{vmatrix} a_1 & b_1 & d_1 \\ a_2 & b_2 & d_2 \\ a_3 & b_3 & d_3 \end{vmatrix}$$

EXAMPLE 3 Solve using Cramer's rule: $\begin{cases} 3x + y - 2z = -3 \\ 2x + 7y + 3z = 9 \\ 4x - 3y - z = 7 \end{cases}$

Solution $D = \begin{vmatrix} 3 & 1 & -2 \\ 2 & 7 & 3 \\ 4 & -3 & -1 \end{vmatrix} = 2$

$$= (-21) + (12) + (12) - (-56) - (-27) - (-2)$$

$$= 88$$

$$D_x = \begin{vmatrix} -3 & 1 & -2 \\ 9 & 7 & 3 \\ 7 & -3 & -1 \end{vmatrix} = 9$$

$$= (21) + (21) + (54) - (-98) - (27) - (-9)$$

$$= 176$$

$$D_y = \begin{vmatrix} 3 & -3 & -2 \\ 2 & 9 & 3 \\ 4 & 7 & -1 \end{vmatrix} = \begin{matrix} 3 & -3 & -2 & 3 & -3 \\ 2 & 9 & 3 & 2 & 9 \\ 4 & 7 & -1 & 4 & 7 \end{matrix}$$

$$= (-27) + (-36) + (-28) - (-72) - (63) - (6)$$

$$= -88$$

$$D_z = \begin{vmatrix} 3 & 1 & -3 \\ 2 & 7 & 9 \\ 4 & -3 & 7 \end{vmatrix} = \begin{matrix} 3 & 1 & -3 & 3 & 1 \\ 2 & 7 & 9 & 2 & 7 \\ 4 & -3 & 7 & 4 & -3 \end{matrix}$$

$$= (147) + (36) + (18) - (-84) - (-81) - (14)$$

$$= 352$$

Thus, $x = \dfrac{D_x}{D} = \dfrac{176}{88} = 2, \quad y = \dfrac{D_y}{D} = \dfrac{-88}{88} = -1, \quad z = \dfrac{D_z}{D} = \dfrac{352}{88} = 4.$

Hence, the solution is given by the ordered triple $(2, -1, 4)$. ■

EXAMPLE 4 Consider the system $\begin{cases} x - y + z = 1 \\ 2x + 3y - 4z = 6 \\ 4x + 6y - 8z = 5 \end{cases}$

Solution Computation shows that $D = 0$, $D_x = -7$, $D_y = -42$, and $D_z = -35$. Thus we have $x = \dfrac{D_x}{D} = \dfrac{-7}{0}$, $y = \dfrac{D_y}{D} = \dfrac{-42}{0}$, and $z = \dfrac{D_z}{D} = \dfrac{-35}{0}$, which are undefined. Hence, the system is inconsistent and there are no solutions. ■

When $D = 0$ and $D_x \neq 0$ or $D_y \neq 0$ or $D_z \neq 0$, the system is inconsistent.

EXAMPLE 5 Consider the system $\begin{cases} x - y + z = 1 & (1) \\ 2x + 3y - 4z = 6 & (2) \\ 4x + 6y - 8z = 12 & (3) \end{cases}$

Solution Computation shows that $D = 0$, $D_x = 0$, $D_y = 0$, and $D_z = 0$. Thus,

$$x = \frac{D_x}{D} = \frac{0}{0}, \quad y = \frac{D_y}{D} = \frac{0}{0}, \quad z = \frac{D_z}{D} = \frac{0}{0},$$ which are indeterminate forms.

Hence the system contains dependent equations, equations (2) and (3). ■

> When $D = D_x = D_y = D_z = 0$, the system contains dependent equations.

EXERCISES 11.6

Compute the value of the determinants in Exercises 1–12.

1.
$\begin{vmatrix} 5 & 3 & 2 \\ -1 & -1 & 3 \\ 2 & 4 & -1 \end{vmatrix}$

2.
$\begin{vmatrix} 1 & 0 & 5 \\ 3 & -1 & 2 \\ 1 & 3 & -2 \end{vmatrix}$

3.
$\begin{vmatrix} 1 & -2 & 0 \\ 1 & 5 & 3 \\ -2 & 3 & 0 \end{vmatrix}$

4.
$\begin{vmatrix} 2 & -1 & 4 \\ 3 & 0 & 2 \\ 1 & 0 & -1 \end{vmatrix}$

5.
$\begin{vmatrix} 0 & 1 & -2 \\ 4 & 3 & 1 \\ 5 & 2 & -1 \end{vmatrix}$

6.
$\begin{vmatrix} 0 & -2 & 1 \\ -3 & 5 & 0 \\ 4 & 0 & 5 \end{vmatrix}$

7.
$\begin{vmatrix} 3 & -2 & 1 \\ 4 & 5 & 2 \\ -4 & 0 & 3 \end{vmatrix}$

8.
$\begin{vmatrix} 2 & -3 & 5 \\ 1 & 2 & -1 \\ 5 & -1 & 4 \end{vmatrix}$

9.
$\begin{vmatrix} 0 & a & b \\ -a & 0 & c \\ -b & -c & 0 \end{vmatrix}$

10.
$\begin{vmatrix} 1 & c & -b \\ -c & 1 & a \\ b & -a & 1 \end{vmatrix}$

11.
$\begin{vmatrix} 1 & 1 & 1 \\ a & b & c \\ a^2 & b^2 & c^2 \end{vmatrix}$

12.
$\begin{vmatrix} (k-1) & 2 & 3 \\ 1 & (k-1) & 2 \\ -1 & -2 & (k-1) \end{vmatrix}$

In Exercises 13–15, solve for k.

13.
$\begin{vmatrix} 1 & k & 0 \\ 8 & 2 & 1 \\ 3 & -1 & 2 \end{vmatrix} = 18$

14.
$\begin{vmatrix} 0 & 3 & 1 \\ k & 2 & -1 \\ 4 & -2 & 2 \end{vmatrix} = 9$

15.
$\begin{vmatrix} k & 0 & 0 \\ 0 & k & 0 \\ 0 & 0 & k \end{vmatrix} = -27$

In Exercises 16–31, solve each of the systems using Cramer's rule.

16. $\begin{cases} x + 2y - z = 5 \\ -6x + 9y - 3z = -2 \\ 2x - 3y + z = 4 \end{cases}$

17. $\begin{cases} 3x - 2y + 4z = 1 \\ -2x - y + z = -3 \\ 6x - 4y + 8z = 2 \end{cases}$

18. $\begin{cases} -x + 3y = -5 \\ 2x + y + 3z = 5 \\ 2y + z = -1 \end{cases}$

19. $\begin{cases} x + 2y - 3z = 6 \\ 2x + 4y - 6z = 7 \\ 3x - y - 2z = 5 \end{cases}$

20. $\begin{cases} 2x + 3z = 2 \\ 4x + y + 3z = 1 \\ -6x + y + 9z = -2 \end{cases}$

21. $\begin{cases} x - 4y + z = 5 \\ 3x - 12y + 3z = 15 \\ 2x - 3z = 4 \end{cases}$

22. $\begin{cases} 3y - 5z = 0 \\ 5z + 7x = 0 \\ 3y - 7x = 0 \end{cases}$

23. $\begin{cases} x + y - z = 3 \\ y + 4z = 2 \\ 3x + y = 5 \end{cases}$

24. $\begin{cases} x + 3y + 4z = 14 \\ x + 2y + z = 7 \\ 2x + y + 2z = 2 \end{cases}$

25. $\begin{cases} \dfrac{1}{x} + \dfrac{2}{y} - \dfrac{3}{z} = 1 \\[2mm] \dfrac{5}{x} + \dfrac{4}{y} + \dfrac{6}{z} = 24 \\[2mm] \dfrac{7}{x} - \dfrac{8}{y} + \dfrac{9}{z} = 14 \end{cases}$

26. $\begin{cases} -\dfrac{2}{x} + \dfrac{2}{y} + \dfrac{1}{z} = 15 \\[2mm] \dfrac{4}{x} - \dfrac{3}{y} + \dfrac{1}{z} = 4 \\[2mm] -\dfrac{6}{x} + \dfrac{1}{y} + \dfrac{3}{z} = 25 \end{cases}$

27. $\begin{cases} \dfrac{2}{x} + \dfrac{3}{y} - \dfrac{5}{z} = 25 \\[2mm] \dfrac{8}{x} - \dfrac{6}{y} + \dfrac{10}{z} = -26 \\[2mm] -\dfrac{12}{x} - \dfrac{9}{y} + \dfrac{5}{z} = 13 \end{cases}$

*28. $\begin{cases} x + y + z = a \\ x - y + z = b \\ x + y - z = c \end{cases}$

*29. $\begin{cases} ax + by - cz = 1 \\ ax - by + 2cz = 2 \\ -ax + 2by - 3cz = 4 \end{cases}$

*30. $\begin{cases} ax + by - cz = c \\ ax + cz = 2b \\ -ax + 3by + cz = 3c \end{cases}$

*31. Show that $\begin{vmatrix} 1 & a & b + c \\ 1 & b & c + a \\ 1 & c & a + b \end{vmatrix} = 0.$

32. Prove that the determinant $\begin{vmatrix} x & y & 1 \\ x_1 & y_1 & 1 \\ x_2 & y_2 & 1 \end{vmatrix} = 0$ is an equation of the line passing through (x_1, y_1) and (x_2, y_2).

The area K of the triangle whose vertices, in counterclockwise order, are $P_1(x_1, y_1)$, $P_2(x_2, y_2)$, *and* $P_3(x_3, y_3)$ *is given by* $K = \dfrac{1}{2}\begin{vmatrix} x_1 & y_1 & 1 \\ x_2 & y_2 & 1 \\ x_3 & y_3 & 1 \end{vmatrix}$. *(See Figure 11.5.) Find the area of the triangle having the given vertices in Exercises 33–35.*

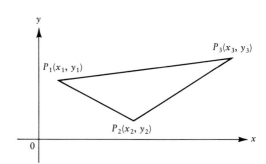

Figure 11.5

33. (2, 3), (4, 1), and (6, 5)
34. (1, 3), (−2, 5), and (7, −1). What can you say about these points?
35. (0, 0), (a, 0), and (0, b), where $a > 0$ and $b > 0$.

11.7 PROBLEM SOLVING INVOLVING SYSTEMS OF LINEAR EQUATIONS AND SYSTEMS OF LINEAR INEQUALITIES

In Chapter 4 we saw how linear equations in one variable were used to solve word problems. In the following sections, we shall see how many problems that involve more than one unknown quantity can be solved by means of a system of linear equations. We shall restrict ourselves to problems where the number of equations must be equal to the number of unknowns.

Number Problems

EXAMPLE 1 The sum of two numbers is 43 and their difference is 11. Find the numbers.

Solution Let x be the large number and y the smaller number. The conditions of the problem yield the equations

$$\begin{cases} x + y = 43 & \text{(1)} \\ x - y = 11 & \text{(2)} \end{cases}$$

Adding equations (1) and (2), we obtain

$$2x = 54$$

$$x = 27$$

Substituting $x = 27$ into equation (1), we get $y = 16$. Hence, the numbers are 27 and 16. ∎

EXAMPLE 2 The sum of three numbers is 180. The third number is 5 less than the second, and the first plus the second is 10 less than the third. What are the numbers?

Solution Let x be the first number, y the second number, and z the third number. Then

$$\begin{cases} x + y + z = 180 \\ \phantom{x + y + {}} y = z + 5 \\ \phantom{x + y + {}} z = x + y + 10 \end{cases}$$

Equivalently,

$$\begin{cases} x + y + z = 180 & \text{(1)} \\ \phantom{x + {}} y - z = 5 & \text{(2)} \\ -x - y + z = 10 & \text{(3)} \end{cases}$$

Adding equations (2) and (3) gives $x = -15$. Substituting $x = -15$ into equations (1) and (3) results in the system

$$\begin{cases} y + z = 195 & \text{(4)} \\ y - z = 5 & \text{(2)} \\ -y + z = 25 & \text{(5)} \end{cases}$$

Equation (4) + equation (2) gives $y = 100$. Substituting $y = 100$ into equation (2) gives $z = 95$. Hence, the three numbers are -15, 100, and 95. ∎

EXAMPLE 3 The sum of the digits of a two-digit number is 13 and the number is 9 more than the number obtained when the digits are reversed. Find the number.

Solution Let $\quad t =$ the tens digit and $u =$ the units digit

$\quad 10t + u =$ the given number,

$\quad 10u + t =$ the number with digits reversed

Using the conditions stated in the problem, we have

$$\begin{cases} t + u = 13 & (1) \\ 10t + u = 10u + t + 9 & (2) \end{cases}$$

Equivalently,

$$\begin{cases} t + u = 13 & (1) \\ t - u = 1 & (2) \end{cases}$$

Adding equations (1) and (2) gives

$$2t = 14$$

$$t = 7$$

Using $t = 7$ and equation (1) or (2), we get $u = 6$. Thus, the required number is 76. ∎

Check $7 + 6 = 13$ and the number 76 is 9 more than 67.

EXAMPLE 4   If a certain two-digit number is divided by the sum of its digits, the quotient is 4 and the remainder is 3. If the digits are reversed, the sum of the resulting number and 23 is twice the given number. Find the number.

Solution   Let $t = $ the tens digit and $u = $ the units digit.

$10t + u = $ the given number

$10u + t = $ the number with digits reversed.

The conditions stated in the problem give the equations

$$\begin{cases} 10t + u = 4(t + u) + 3 & (1) \\ 10u + t + 23 = 2(10t + u) & (2) \end{cases}$$

Equivalently,

$$\begin{cases} 6t - 3u = 3 & (1) \\ -19t + 8u = -23 & (2) \end{cases}$$

or,

$$\begin{cases} 2t - u = 1 & (1) \\ -19t + 8u = -23 & (2) \end{cases}$$

Multiplying equation (1) by 8 and adding the results to equation (2) gives

$$-3t = -15$$

$$t = 5$$

Substituting $t = 5$ in equation (1) yields $u = 9$. Thus, the number is 59. ■

Check The sum of the digits is $5 + 9 = 14$. Dividing 59 by 14 gives a quotient of 4 and a remainder of 3. Also,

$$95 + 23 = 2(59)$$
$$118 = 118$$

EXAMPLE 5 If a certain number is added to both the numerator and denominator of a given fraction, its value is $\frac{1}{2}$. However, if the same number is subtracted from both numerator and denominator, its value is $\frac{1}{7}$. If, however, 1 is added to both numerator and denominator, the fraction is $\frac{2}{5}$. Find the fraction.

Solution Let x denote the numerator of the fraction, and let y denote the denominator of the fraction.

1. Let z be the number added in the first condition. Then

$$\frac{x + z}{y + z} = \frac{1}{2}$$

Clearing the fractions, we get

$$2(x + z) = y + z$$
$$2x + 2z = y + z$$

Thus, $\qquad\qquad\qquad\qquad 2x - y + z = 0 \qquad\qquad\qquad (1)$

2. If we subtract z from numerator and denominator, then

$$\frac{x - z}{y - z} = \frac{1}{7}$$

Then, $\qquad\qquad\qquad\qquad 7(x - z) = y - z$
$$7x - 7z = y - z$$

Thus, $\qquad\qquad\qquad\qquad 7x - y - 6z = 0 \qquad\qquad\qquad (2)$

3. If 1 is added, the fraction equals $\frac{2}{5}$. That is,

$$\frac{x + 1}{y + 1} = \frac{2}{5}$$

Then $\qquad\qquad\qquad\qquad 5(x + 1) = 2(y + 1)$
$$5x + 5 = 2y + 2$$

Thus, $\qquad\qquad\qquad\qquad 5x - 2y = -3 \qquad\qquad\qquad (3)$

We now have to solve the following system of three linear equations in three unknowns:

$$\begin{cases} 2x - y + z = 0 & (1) \\ 7x - y - 6z = 0 & (2) \\ 5x - 2y = -3 & (3) \end{cases}$$

Eliminating z in equations (1) and (2), we obtain

$$6 \times \text{equation (1)} + \text{equation (2):} \qquad \begin{cases} 19x - 7y = 0 & (4) \\ 5x - 2y = -3 & (3) \end{cases}$$

a system of two equations in two unknowns, x and y. Eliminating y in equations (4) and (3), we get

$$2 \times \text{equation (4)} - 7 \times \text{equation (3):} \qquad 3x = 21$$
$$x = 7$$

Substituting $x = 7$ into equation (4), we find that $y = 19$. Thus, the required fraction is $x/y = \frac{7}{19}$. ■

EXERCISES 11.7

Numbers

1. Find two numbers whose sum is 12 and whose difference is 4.

2. Find two numbers whose sum is 25 and whose difference is 11.

3. If one-sixth of the sum of two numbers is 8, and half their difference is 13, what are the numbers?

4. Find two numbers such that one-seventh of their sum is equal to two-fifths of their difference, and twice the smaller is 3 less than the larger.

5. Find two numbers such that one-quarter of the greater exceeds one-sixth of the smaller by 2, and twice the smaller increased by the greater is equal to 56.

*6. Five times the difference of two numbers equals half their product, while five times the sum of their reciprocals is $\frac{3}{2}$. Find the two numbers.

7. Find three numbers whose sum is 75, if the sum of the first and third is twice the second, and the third is $\frac{3}{2}$ the first.

8. The sum of three numbers is 45. One-half the first plus $\frac{2}{3}$ the second plus $\frac{3}{4}$ the third equals 30. The first plus $\frac{1}{3}$ the second plus $\frac{2}{3}$ the third equals 23. What are the numbers?

*9. The sum of the reciprocals of three numbers is 4. The reciprocal of the first number exceeds the sum of the reciprocals of the other two by 2. The reciprocal of the third number is 1 less than twice the reciprocal of the second number. Find the numbers.

*10. The sum of three numbers is 51. If the first number is divided by the second, the quotient is 2 and the remainder is 5. However, if the second number is divided by the third, the quotient is 3 and the remainder is 2. What are the numbers?

Digits

11. A certain two-digit number is equal to five times the sum of its digits. If 9 is added to the number, the result is equal to the digits reversed. Find the number.

12. A certain two-digit number is four times the sum of its digits. If 27 is added to the number, the digits will be reversed. Find the number.

13. The middle digit of a three-digit number is

zero and the sum of the other digits is 11. If the nonzero digits are reversed, the number formed exceeds the original number by 495. Find the number.

*14. In a two-digit number the tens digit is 3 more than the units digit. The number divided by the sum of the digits is 7. What is the number?

15. The sum of the three digits of a number is 14. The hundreds digit is 3 less than the units digit. If 63 is subtracted from the number, the tens digit will become hundreds, hundreds will become units, and units will become tens. What is the number?

16. A certain three-digit number is equal to 48 times the sum of its digits. If 198 is subtracted from the number, the digits will be reversed. Also, the sum of the extreme digits is equal to twice the middle digit. Find the number.

17. A two-digit number is such that four times the reciprocal of the tens digit plus three times the reciprocal of the units digit is 2. If five times the reciprocal of the tens digit minus twice the reciprocal of the units digit is $\frac{7}{12}$, what is the number?

*18. A number consists of two digits. If 3 is added to the number, and the result divided by the sum of the digits, the quotient is 6. If 3 is subtracted from the number, and the result divided by the units digit, the quotient is 12. What is the number?

*19. A given number consists of two digits whose sum is 11. If a certain number is added to the given number, the sum is 44. However, if this number is added to the sum of the digits of the given number, the result is 17. What is the given number?

Fractions

20. If the numerator of a fraction is increased by 2 and the denominator by 1, it becomes equal to $\frac{5}{8}$. If the numerator and denominator are both diminished by 1, it becomes equal to $\frac{1}{2}$. Find the fraction.

21. Find the fraction that is equal to $\frac{1}{2}$ when its numerator is increased by 1, and equal to $\frac{1}{3}$ when its denominator is increased by 1.

22. If a certain number is added to the numerator and denominator of a given fraction, its value becomes $\frac{4}{7}$, but if the same number is subtracted from both numerator and denominator, its value is $\frac{1}{4}$. If, however, 7 is added to both numerator and denominator, the fraction becomes $\frac{2}{3}$. What is the fraction?

23. Find the values of a, b, c, so that the fraction $\dfrac{a + bx}{3 + cx}$ will have the values 2, 5, $\frac{1}{3}$ when $x = 1, -\frac{1}{2}, 3$, respectively.

*24. The numerator of a fraction is 5 less than its denominator. If 17 is added to the numerator and 2 added to the denominator, the result is a fraction that is the reverse of the original fraction. Find the original fraction.

11.8 MOTION PROBLEMS

In Chapter 4 we learned how to solve uniform motion problems that require the use of only one unknown. We now discuss motion problems that are more easily solved using two unknowns.

EXAMPLE 1 If a boat is rowed downstream 10 km in 2 hr and the same distance upstream in $3\frac{1}{3}$ hr, find the rate of rowing in still water and the rate of the current.

Solution Let

$x =$ the rate of rowing in still water measured in km/hr

$y =$ the rate of the current in km/hr

Then

$x + y =$ the rate downstream in km/hr

$x - y =$ the rate upstream in km/hr

Using the relationship time = distance/rate, we obtain:

$$\text{time downstream: } \frac{10 \text{ km}}{(x + y) \text{ km/hr}} = 2 \text{ hr} \tag{1}$$

$$\text{time upstream: } \frac{10 \text{ km}}{(x - y) \text{ km/hr}} = 3\frac{1}{3} \text{ hr} \tag{2}$$

Thus, we have the following system of equations:

$$\begin{cases} x + y = 5 \\ x - y = 3 \end{cases} \begin{matrix} (1). \\ (2) \end{matrix}$$

Adding equations (1) and (2), we get $2x = 8$; that is,

$$x = 4$$

Substituting $x = 4$ into equation (1), we find that $y = 1$. Thus, the rate of rowing in still water is 4 km/hr and the rate of the current is 1 km/hr. ■

Check

$$\frac{10 \text{ km}}{(4 + 1) \text{ km/hr}} = \overset{2}{\cancel{10}} \; \cancel{\text{km}} \cdot \frac{1}{\cancel{5}} \frac{\text{hr}}{\cancel{\text{km}}} = 2 \text{ hr}$$

$$\frac{10 \text{ km}}{(4 - 1) \text{ km/hr}} = 10 \; \cancel{\text{km}} \cdot \frac{1}{3} \frac{\text{hr}}{\cancel{\text{km}}} = \frac{10}{3} \text{ hr} = 3\frac{1}{3} \text{ hr}$$

EXAMPLE 2 Two women are running on a circular race track 300 ft in circumference. Running in opposite directions, they meet every 10 sec. Running in the same direction, the faster runner passes the slower runner every 50 sec. Find their rates.

Solution Let

$$x = \text{rate, in feet per second, of the faster runner}$$

$$y = \text{rate, in feet per second, of the slower runner}$$

When they run in opposite directions, they meet at the end of 10 sec. Thus, after 10 sec, both runners together cover the entire 300 ft. Using the relationship distance = (rate) · (time), we have (see Figure 11.6):

$$\text{distance covered by faster runner} = (10 \text{ sec})\left(x \frac{\text{ft}}{\text{sec}}\right) = (10x) \text{ ft}$$

$$\text{distance covered by slower runner} = (10 \text{ sec})\left(y \frac{\text{ft}}{\text{sec}}\right) = (10y) \text{ ft}$$

Therefore, we have

$$(10x) \text{ ft} + (10y) \text{ ft} = 300 \text{ ft} \tag{1}$$

When they run in the same direction, the faster runner must gain 300 ft each

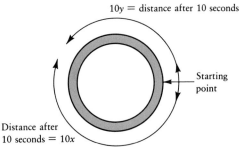

Figure 11.6

time she overtakes the slower runner. Thus, the distance that the faster runner covers in 50 sec is equal to the distance covered by the slower runner plus the circumference of the track.

$$\text{distance covered by faster runner} = (50 \text{ sec})\left(x\frac{\text{ft}}{\text{sec}}\right) = (50x) \text{ ft}$$

$$\text{distance covered by slower runner} = (50 \text{ sec})\left(y\frac{\text{ft}}{\text{sec}}\right) = (50y) \text{ ft}$$

Therefore, we have

$$(\text{distance of faster}) = (\text{distance of slower}) + (\text{circumference})$$

That is,

$$(50x) \text{ ft} = (50y) \text{ ft} + 300 \text{ ft} \qquad (2)$$

After simplifying equations (1) and (2), we have the system

$$\begin{cases} x + y = 30 \\ x - y = 6 \end{cases} \qquad \begin{matrix} (1) \\ (2) \end{matrix}$$

Adding equations (1) and (2) and solving for x, we obtain

$$x = 18 \quad \text{and} \quad y = 12$$

Hence, the faster runner runs at a rate of 18 ft/sec and the slower runner runs at a rate of 12 ft/sec. ■

EXAMPLE 3 A train travelled a certain distance at a uniform rate. If the speed had been 6 mph more, the trip would have taken 4 hr less. If the speed had been 6 mph less, the trip would have taken 6 hr more. Find the distance travelled by the train.

Solution Let

$$x = \text{rate of the train in miles per hour}$$

$$y = \text{the time of trip in hours}$$

Then xy = distance travelled

From the conditions stated in the problem, we have:

$$(x + 6)\frac{\text{mi}}{\text{hr}} \cdot (y - 4) \text{ hr} = (x + 6)(y - 4) \text{ mi} = \text{distance travelled}$$

$$(x - 6)\frac{\text{mi}}{\text{hr}} \cdot (y + 6) \text{ hr} = (x - 6)(y + 6) \text{ mi} = \text{distance travelled}$$

Thus, we have the system

$$\begin{cases} (x + 6)(y - 4) = xy & (1) \\ (x - 6)(y + 6) = xy & (2) \end{cases}$$

Simplifying,

$$\begin{cases} -4x + 6y = 24 & (1) \\ 6x - 6y = 36 & (2) \end{cases}$$

Adding equations (1) and (2) and solving for x, we get

$$x = 30$$

Substituting $x = 30$ into equation (1), we find that $y = 24$. Hence, the distance (xy) miles = 720 mi. ∎

EXERCISES 11.8

1. A crew can row 18 mi downstream in 2 hr, and back again in 3 hr. What is the rate of the crew in still water and the rate of the current?

2. A woman can row downstream 3 mi in 20 min, but it takes her 1 hr to return. What is her rate in still water and what is the rate of the current?

3. On a stream whose current is known to be $2\frac{1}{4}$ mph, a crew rowed downstream for 2 hr. It took them 6 hr to return. How far did the crew row, and what was their rate of rowing in still water?

4. A canoeist rows on a river whose current is 2.5 mph. He finds that it takes him as long to go upstream 2 mi as downstream 7 mi. Find his rate.

5. With a given head wind, a certain plane can travel 1500 mi in 4 hr. But flying in the opposite direction with the same wind blowing,

the plane can fly that distance in 3.5 hr. Find the wind speed and the speed of the plane in still air.

6. City A and City B are 400 mi apart, and City B is due east of City A. A plane flew from City A to City B in 2 hr and then returned to City A in 2.5 hr. If the wind blew with a constant velocity from the west during the entire trip, find the speed of the plane in still air and the speed of the wind.

7. Two points, C and D, are 45 km apart. A car leaves C for D at the same time that a truck leaves D for C, going the same route. The car and truck travel at uniform rates and meet at the end of 2 hr. Upon reaching the destination, each driver starts the return trip without delay, and they meet for the second time 15 km from B. Find the rate of car and the rate of truck.

8. In 8 hr A walks 12 mi more than B does in

7 hr, and in 13 hr B walks 7 mi more than A does in 9 hr. How many miles does each walk per hour?

9. A and B ran a race that lasted 5 min. B had a head start of 20 yd, but A ran 3 yd while B was running 2 yd, and won by 30 yd. Find the length of the course and the rate of each per minute.

*10. Frank and Neelam both run a mile. In the first heat, Frank gives Neelam a head start of 20 yd, and beats her by 30 sec. In the second heat, Frank gives Neelam a head start of 32 sec, and beats her by $9\frac{5}{11}$ yd. Find the rate per hour at which Frank runs.

*11. A and B are two towns situated 24 mi apart, on the same bank of a river. A camper goes from A to B in 7 hr by rowing the first half of the distance and walking the second half. In returning he walks the first half at $\frac{3}{4}$ of his former rate, but the stream being with him he rows at double his rate in going, and he completes the entire distance in 6 hr. Find his rates of walking and rowing upstream.

*12. Two trains, 92 ft long and 84 ft long, respectively, are moving with uniform velocities on parallel tracks. When they move in opposite directions, they are observed to pass each other in $1\frac{1}{2}$ sec. However, when they move in the same direction, the faster train is observed to pass the other in 6 sec. Find the rate at which each train moves.

11.9 WORK PROBLEMS

EXAMPLE 1 A job can be done by A and B working together in 10 days. After working together for 7 days, A stops, and B completes it in 9 days. How long would it take each to do the work alone?

Solution Let

x = the number of days it would take A alone

y = the number of days it would take B alone

Person	Time Alone	Amount in 1 Day	Amount in 10 Days	Amount in 7 Days	Amount in 9 Days
A	x days	$\dfrac{1}{x}$	$\dfrac{10}{x}$	$\dfrac{7}{x}$	$\dfrac{9}{x}$
B	y days	$\dfrac{1}{y}$	$\dfrac{10}{y}$	$\dfrac{7}{y}$	$\dfrac{9}{y}$

In ten days, A and B can complete 1 entire job.

$$\frac{10}{x} + \frac{10}{y} = 1 \tag{1}$$

Working together for 7 days can be expressed by $\dfrac{7}{x} + \dfrac{7}{y}$. Then B completes the job in 9 days; that is, B completes $\dfrac{9}{y}$ of the job. Thus, the second condition gives the equation

$$\frac{7}{x} + \frac{7}{y} + \frac{9}{y} = 1 \tag{2}$$

Therefore, we have the system

$$
\begin{cases}
\dfrac{10}{x} + \dfrac{10}{y} = 1 & \text{(1)} \\[2mm]
\dfrac{7}{x} + \dfrac{7}{y} + \dfrac{9}{y} = 1 & \text{(2)}
\end{cases}
$$

Equivalently,

$$
\begin{cases}
\dfrac{10}{x} + \dfrac{10}{y} = 1 & \text{(1)} \\[2mm]
\dfrac{7}{x} + \dfrac{16}{y} = 1 & \text{(2)}
\end{cases}
$$

Let $u = \dfrac{1}{x}$ and $v = \dfrac{1}{y}$. Then the system becomes

$$
\begin{cases}
10u + 10v = 1 \\
7u + 16v = 1
\end{cases}
$$

Solving for u and v, we get $u = \frac{1}{15}$ and $v = \frac{1}{30}$. Hence, $x = 15$ and $y = 30$. A takes 15 days and B takes 30 days. ∎

EXAMPLE 2 A, B, and C can cut a given quantity of timber in 4 days. A and B can do it in 6 days; B and C, in 7 days. How long would it take each separately?

Solution Let x represent the number of days in which A can cut the timber

y represent the number of days in which B can cut the timber

z represent the number of days in which C can cut the timber

Then $\dfrac{1}{x}$ represents the part of timber A can cut in one day

So $\dfrac{1}{y}$ represents the part of timber B can cut in one day

And $\dfrac{1}{z}$ represents the part of timber C can cut in one day

Conditions in the problem give

$$
\begin{cases}
\dfrac{1}{x} + \dfrac{1}{y} + \dfrac{1}{z} = \dfrac{1}{4} & \text{(1)} \\[3mm]
\dfrac{1}{x} + \dfrac{1}{y} \phantom{{}+ \dfrac{1}{z}} = \dfrac{1}{6} & \text{(2)} \\[3mm]
\phantom{\dfrac{1}{x} + {}}\dfrac{1}{y} + \dfrac{1}{z} = \dfrac{1}{7} & \text{(3)}
\end{cases}
$$

Let $u = \dfrac{1}{x}$, $v = \dfrac{1}{y}$, and $w = \dfrac{1}{z}$. Then our system becomes

$$\begin{cases} u + v + w = \dfrac{1}{4} & (1) \\[2mm] u + v = \dfrac{1}{6} & (2) \\[2mm] v + w = \dfrac{1}{7} & (3) \end{cases}$$

Subtracting equation (2) from equation (1), we get

$$w = \frac{1}{12}$$

Substituting $w = \frac{1}{12}$ into equation (3) gives $v = \frac{5}{84}$. Then equation (2) yields $u = \frac{3}{28}$. Thus,

$$u = \frac{1}{x} = \frac{3}{28} \quad \text{or} \quad x = \frac{28}{3} = 9\frac{1}{3}$$

and

$$v = \frac{1}{y} = \frac{5}{84} \quad \text{or} \quad y = \frac{84}{5} = 16\frac{4}{5}$$

and

$$w = \frac{1}{z} = \frac{1}{12} \quad \text{or} \quad z = 12.$$

Hence, A can cut the timber in $9\frac{1}{3}$ days, B in $16\frac{4}{5}$ days, and C in 12 days. ∎

EXERCISES 11.9

1. A and B, working together, did a job in 6 hr. At another time, they did the same amount when A worked 3 hr and B 10 hr. How many hours would it take for either to do the job alone?

2. One of two machines will burn 10 cans of oil in 24 days; the other, 10 cans in 30 days. How long will 10 cans last if both machines are run at the same time?

3. Tom and Jerry can together do a certain work in 30 days. At the end of 18 days, Jerry is called off and Tom finishes the work in 20 more days. Find the time each could do the work alone.

4. Art and Janet together can paint the house in 15 hr. After they have painted together for 6 hr, Janet can finish the job in 15 hr. How long would it take each to do it alone?

5. A and B can do a job in $9\frac{3}{5}$ days; A and C in $11\frac{3}{7}$ days; and B and C in 15 days. In what time can each do it alone?

6. A, B, and C together do a piece of work in 30 days. A and B can together do it in 32 days; and B and C can together do it in 120 days. Find the time in which each can do it alone.

*7. A and B together can do a job in $5\frac{1}{7}$ hr; A and C, in $4\frac{4}{5}$ hr. All three of them work at it for 2 hr when A drops out and B and C finish it in $1\frac{9}{17}$ hr. How long would it take each man separately to do the job?

11.10 MIXTURE PROBLEMS

EXAMPLE 1 How many gallons of milk containing 18% butterfat and of cream containing 45% butterfat should be combined to give 12 gallons of milk containing 36% butterfat?

Solution Let x be the number of gallons of milk containing 18% butterfat

and let y be the number of gallons of cream containing 45% butterfat

Then x gallons $+ y$ gallons $= 12$ gallons   (1)

and   18%x butterfat $+ 45$%y butterfat $= 36$%(12) butterfat   (2)

Thus, we have the system

$$\begin{cases} x + y = 12 & \text{(1)} \\ .18x + .45y = 4.32 & \text{(2)} \end{cases}$$

or

$$\begin{cases} x + y = 12 & \text{(1)} \\ 18x + 45y = 432 & \text{(2)} \end{cases}$$

Multiplying equation (1) by 18 and subtracting the result from equation (2) gives $y = 8$. Substituting this value into one of the original equations yields $x = 4$. Hence, we combine 4 gallons of milk containing 18% butterfat and 8 gallons of cream containing 45% butterfat. ■

EXAMPLE 2 A chemist has a 3% solution and an 18% solution of a disinfectant. How many ounces of each should be used to make 25 ounces of a 5% solution?

Solution Let $x =$ the number of ounces from the 3% solution

$y =$ the number of ounces from the 18% solution

Then x oz $+ y$ oz $= 25$ oz   (1)

and

3%x(disinfectant) $+ 18$%y(disinfectant) $= 5$%(25)(disinfectant)   (2)

Thus, we have the system

$$\begin{cases} x + y = 25 & \text{(1)} \\ 3x + 18y = 125 & \text{(2)} \end{cases}$$

Multiplying equation (1) by 3 and subtracting the result from equation (2) gives

$y = \frac{10}{3}$. Substituting this value in equation (1) yields $x = \frac{65}{3}$. Hence, the chemist should use $21\frac{2}{3}$ oz of the 3% solution and $3\frac{1}{3}$ oz of the 18% solution. ∎

EXAMPLE 3 A nutritionist wishes to prepare a food mixture that contains 40 g vitamin A and 50 g vitamin B. The two mixtures that are available contain the following percentages of vitamin A and vitamin B:

	Vitamin A	Vitamin B
Mixture 1	10%	4%
Mixture 2	5%	12%

How many grams of each mixture should be used to obtain the desired diet?

Solution Let

x = the number of grams of mixture 1 to be used

y = the number of grams of mixture 2 to be used

The equation describing the amount of vitamin A we must have from the two mixtures is

$$.1x + .05y = 40 \tag{1}$$

Similarly, for vitamin B we have

$$.04x + .12y = 50 \tag{2}$$

Thus, we have the system

$$\begin{cases} .1x + .05y = 40 & (1) \\ .04x + .12y = 50 & (2) \end{cases}$$

or

$$\begin{cases} 10x + 5y = 4000 & (1) \\ 4x + 12y = 5000 & (2) \end{cases}$$

Multiplying equation (1) by $\frac{2}{5}$ and subtracting the result from 10 times equation (2) gives $y = 340$. Substituting this value into equation (1) yields $x = 230$. Thus, the nutritionist needs 230 g mixture 1 and 340 g mixture 2. ∎

EXAMPLE 4 Two test tubes, A and B, contain mixtures of zinc and iron. A mixture of 3 parts from A and 2 parts from B will contain 40% zinc. A mixture of 1 part from A and 2 parts from B will contain 32% zinc. What are the percentages of zinc in A and B, respectively?

Solution Let

$$x = \text{the percentage of zinc in } A$$
$$y = \text{the percentage of zinc in } B$$

First, we note that in the first mixture, we have

$$\left(\begin{array}{c}\text{percentage of zinc} \\ \text{in } A\end{array}\right) + \left(\begin{array}{c}\text{percentage of zinc} \\ \text{in } B\end{array}\right) = 40\% \qquad (1)$$

$$\frac{3}{5}x \qquad + \qquad \frac{2}{5}y \qquad = \frac{40}{100} \qquad (1)$$

Similarly, for the second mixture we have

$$\left(\begin{array}{c}\text{percentage of zinc} \\ \text{in } A\end{array}\right) + \left(\begin{array}{c}\text{percentage of zinc} \\ \text{in } B\end{array}\right) = 32\% \qquad (2)$$

$$\frac{1}{3}x \qquad + \qquad \frac{2}{3}y \qquad = \frac{32}{100} \qquad (2)$$

Thus, we have the system

$$\begin{cases} \dfrac{3x}{5} + \dfrac{2y}{5} = \dfrac{40}{100} & (1) \\[3mm] \dfrac{x}{3} + \dfrac{2y}{3} = \dfrac{32}{100} & (2) \end{cases}$$

Simplifying the system, we get

$$\begin{cases} 60x + 40y = 40 & (1) \\ 100x + 200y = 96 & (2) \end{cases}$$

Multiplying equation (1) by 5 and subtracting the result from equation (2) yields

$$x = \frac{52}{100} = 52\%$$

Solving for y, we get

$$y = \frac{22}{100} = 22\%$$

Hence, A is 52% zinc and B is 22% zinc. ■

EXERCISES 11.10

1. How many quarts of a mixture containing 45% acid should be combined with a mixture containing 70% acid to obtain 6 quarts of a mixture containing 55% acid?

2. One alloy contains 60% iron and 40% copper, and another contains 35% iron and 65% copper. How much of each must be

mixed to produce 200 kg of an alloy that is 45% iron and 55% copper?

3. One tank contains a mixture of 20 gal water and 4 gal chlorine. Another tank has 12 gal water and 4 gal chlorine. How many gallons must be drawn from each to have 6 gallons of a mixture that is 20% chlorine?

4. Find the number of ounces each of silver 80% pure and of silver 64% pure that is required to make 20 ounces of silver 70% pure.

5. A woman wants to obtain 15 quarts of a 24% acid solution by combining a quantity of 20% acid solution, a quantity of 30% acid solution, and 1 quart of pure water. How many quarts of each of the acid solutions must be used?

6. Pure gold is 24 carats fine. How many ounces each of pure gold and of gold 10 carats fine must be combined with 24 ounces of gold 14 carats fine to make 51 ounces of gold 16 carats fine?

*7. Of two kinds of gold, a carats fine and b carats fine, how many ounces each are required to make p ounces c carats fine?

8. A psychologist is conducting an experiment using rats and she must administer to them a blend that contains 46 oz carbohydrate, 32 oz protein, and 12.4 oz of cholesterol. The following compositions are available:

	Carbohydrate	Protein	Cholesterol
Blend A	40%	30%	4%
Blend B	20%	20%	12%
Blend C	30%	10%	10%

How many ounces of each blend should be used to meet the required amounts of carbohydrate, protein, and cholesterol?

*9. A and B are alloys of gold and copper. An alloy that is 5 parts A and 3 parts B is 52% gold. One that is 5 parts A and 11 parts B is 42% gold. What are the percentages of gold in A and B, respectively?

*10. Given three alloys of the following composition: A, 5 parts (by weight) lead, 2 zinc, 1 copper; B, 2 parts lead, 5 zinc, 1 copper; C, 3 parts lead, 1 zinc, 4 copper. To obtain 9 ounces of an alloy containing equal quantities (by weight) of lead, zinc, and copper, how many ounces of A, B, and C must be taken and melted together?

11.11 GEOMETRIC PROBLEMS

EXAMPLE 1 The degree measure of one of two supplementary angles is 5 more than one-fourth that of the other. What are the degree measures of the angles?

Solution Let
$$x = \text{the number of degrees in the smaller angle}$$
$$y = \text{the number of degrees in the larger angle}$$

Since the angles are supplementary, we have

$$x \text{ degrees} + y \text{ degrees} = 180 \text{ degrees} \qquad (1)$$

The other given condition yields

$$x \text{ degrees} = \frac{1}{4}y \text{ degrees} + 5 \text{ degrees} \qquad (2)$$

Thus, we have the following system

$$\begin{cases} x + y = 180 & (1) \\ x = \frac{1}{4}y + 5 & (2) \end{cases}$$

or

$$\begin{cases} x + y = 180 & (1) \\ 4x - y = 20 & (2) \end{cases}$$

Adding the equation, we obtain

$$5x = 200$$

$$x = 40$$

Substituting $x = 40$ into one of the original equations gives $y = 140$. Thus, the angles are 40° and 140°. ∎

EXAMPLE 2 Twice the base of a parallogram is 4 in. more than four times the altitude. If the altitude remains unchanged and the base is decreased by 2 in., the area will be decreased by 6 sq in. Find the altitude and the base of the parallelogram (see Figure 11.7).

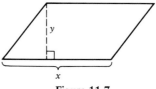

Figure 11.7

Solution Let

x represent the base measured in inches

y represent the altitude measured in inches

The first stated condition requires

$$(2x) \text{ in.} = (4y) \text{ in.} + 4 \text{ in.} \qquad (1)$$

Using the fact that the area of a parallelogram is base times altitude, and the second stated condition, we have

$$(\text{base})(\text{altitude}) - 6 \text{ sq in.} = (\text{new base})(\text{altitude})$$

$$(xy) \text{ sq in.} - 6 \text{ sq in.} = (x - 2)(y) \text{ sq in.} \qquad (2)$$

Thus, we have the system

$$\begin{cases} 2x = 4y + 4 & (1) \\ xy - 6 = (x - 2)(y) & (2) \end{cases}$$

or

$$\begin{cases} x - 2y = 2 & (1) \\ -6 = -2y & (2) \end{cases}$$

Equation (2) yields $y = 3$, and substituting this value into equation (1) gives $x = 8$.

Hence, the base is 8 in. and the altitude is 3 in. ∎

EXAMPLE 3   The general equation of a circle is given by

$$x^2 + y^2 + Dx + Ey + F = 0$$

Find the equation of the circle passing through the three points (5, 3), (6, 2), and (3, −1) by determining the constants D, E, and F.

Solution   Since each point lies on the circle, they must satisfy the given equation.

For (5, 3), we have:	$25 + 9 + 5D + 3E + F = 0$	(1)
For (6, 2), we have:	$36 + 4 + 6D + 2E + F = 0$	(2)
For (3, −1), we have:	$9 + 1 + 3D - E + F = 0$	(3)

Thus, we have the following system in the unknowns D, E, and F:

$$\begin{cases} 5D + 3E + F = -34 & (1) \\ 6D + 2E + F = -40 & (2) \\ 3D - E + F = -10 & (3) \end{cases}$$

Eliminating F from (1) and (2), and then from (2) and (3) gives

$$\begin{cases} -D + E = 6 & (4) \\ 3D + 3E = -30 & (5) \end{cases}$$

Multiplying equation (4) by 3 and adding the result to (5), we obtain

$$6E = -12$$

$$E = -2$$

Substituting $E = -2$ into equation (4) gives $D = -8$. Finally, using these values and equation (3), we have $F = 12$. Hence, the equation of the circle passing through (5, 3), (6, 2), and (3, −1) is

$$x^2 + y^2 - 8x - 2y + 12 = 0 \quad ∎$$

EXAMPLE 4   In Figure 11.8, each circle is tangent to the other two as shown. The points A, B, and C are the respective centers. Find the radius of each circle if $AB = 4$ ft, $AC = 5$ ft, and $BC = 7$ ft.

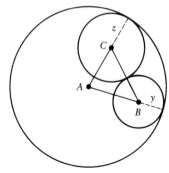

Figure 11.8

Solution   Let

$$x = \text{radius of circle with center at } A$$
$$y = \text{radius of circle with center at } B$$
$$z = \text{radius of circle with center at } C$$

From the figure we have:

$$x \text{ ft} = 5 \text{ ft} + z \text{ ft} \tag{1}$$
$$x \text{ ft} = 4 \text{ ft} + y \text{ ft} \tag{2}$$
$$y \text{ ft} + z \text{ ft} = 7 \text{ ft} \tag{3}$$

Thus, we have the system

$$\begin{cases} x - z = 5 & (1) \\ x - y = 4 & (2) \\ y + z = 7 & (3) \end{cases}$$

Eliminating z, we obtain

$$x - y = 4 \tag{2}$$
$$x + y = 12 \tag{4}$$

Solving for x, we get

$$2x = 16$$
$$x = 8$$

Substituting into the original system, we find that $y = 4$ and $z = 3$.

Hence,   the radius of circle centered at A is 8 ft

the radius of circle centered at B is 4 ft

the radius of circle centered at C is 3 ft   ◾

EXERCISES 11.11

1. Find the two acute angles of a right triangle if one of them is 12° greater than five times the other.

2. One angle of two complementary angles is 30° less than twice that of the other. Find the angles.

3. Three times the width of a rectangle exceeds twice its length by 3 ft, and four times its length is 12 ft more than its perimeter. Find the dimensions of the rectangle.

4. Three times the length of a given rectangle is 10 less than its perimeter and three times the width is 6 more than twice the length. What are the dimensions of the rectangle?

5. If the altitude of a triangle is increased by 2 in. and the base is decreased by 3 in., the area is decreased by 16 sq in. If the altitude is decreased by 3 in. and the base is increased by 2 in., the area is increased by $1\frac{1}{2}$ sq in. Find the base and the altitude.

6. Find the three angles of a triangle if the sum of the first and second is twice the third angle, and if the second angle less the third angle is 10 less than the first angle.

7. Find the three angles of a triangle if the first exceeds the third by 45°, and the sum of twice the second angle and the third angle is 9° more than the first angle.

8. Determine the values of a, b, c so that the graph of $y = ax^2 + bx + c$ will pass through the points $(2, -12)$, $(3, -10)$, and $(5, 0)$.

*9. It is proved in geometry that two distances such as AB and AC in Figure 11.9 are equal. Find the lengths of AC, CD, and BE if $AD = 12$ ft, $DE = 10$ ft, and $AE = 14$ ft.

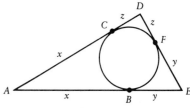

Figure 11.9

*10. Given three circles with centers A, B, and C (see Figure 11.10). Find the radii of the circles if $AB = 23$, $BC = 20$, and $AC = 25$.

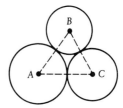

Figure 11.10

11.12 MISCELLANEOUS PROBLEMS

EXAMPLE 1 *Investment Problem* A woman invests a total of $7800 in government bonds yielding an annual return of 4% and in stocks yielding an annual return of 6%. If the income of the 4% investment exceeded the investment at 6% by $92, how much was invested at each rate?

Solution Let

$x =$ amount (in dollars) invested at 4%

$y =$ amount (in dollars) invested at 6%

Then

$$(x) \text{ dollars} + (y) \text{ dollars} = 7800 \text{ dollars} \tag{1}$$

Also,

$$\left(\begin{array}{c}\text{income from 4\%}\\ \text{investment}\end{array}\right) = \left(\begin{array}{c}\text{income from 6\%}\\ \text{investment}\end{array}\right) + \$92$$

that is, $\qquad .04x \qquad = \qquad .06y \qquad + \$92 \qquad$ (2)

Thus, we have the system

$$\begin{cases} x + y = 7800 & \text{(1)} \\ 4x - 6y = 9200 & \text{(2)} \end{cases}$$

Solving the system, we find $x = 5600$ and $y = 2200$. Hence, she invested $5600 at 4% and $2200 at 6%. ∎

EXAMPLE 2 *Business Problem* At the local flea market, a shopper bought 3 belts and 4 towels for $18.70. Another shopper bought 4 belts and 7 towels from the same vendor for $27.75. What was the price per belt and price per towel?

Solution Let $\qquad x =$ cost of one belt, in cents

$\qquad\qquad y =$ cost of one towel, in cents

Then

$$\begin{cases} 3x + 4y = 1870 & \text{(1)} \\ 4x + 7y = 2775 & \text{(2)} \end{cases}$$

Solving the system we find that $x = 398$ and $y = 169$. Thus, the price of the belt is $3.98 and the price of the towel is $1.69. ∎

EXAMPLE 3 *Business Problem* A sum of money was divided equally among a certain number of children. If there had been three more children, each would have received $1 less, and if there had been two less children, each would have received $1 more. Find the number of children, and what each received.

Solution Let $\qquad x =$ the number of children

$\qquad\qquad y =$ the number of dollars that each received

Then $\qquad xy =$ the number of dollars to be divided

From the given conditions we have

$$\begin{cases} (x + 3)(y - 1) = xy & \text{(1)} \\ (x - 2)(y + 1) = xy & \text{(2)} \end{cases}$$

That is,

$$\begin{cases} -x + 3y = 3 & \text{(1)} \\ x - 2y = 2 & \text{(2)} \end{cases}$$

Solving, we obtain $y = 5$ and $x = 12$. Thus, there were 12 children and each received $5. ■

EXAMPLE 4 *Age Problem* Four years ago, Brad was five times as old as Nate was at that time. Eight years from now, Brad will be twice as old as Nate. Find their present ages.

Solution Let

$$x = \text{Brad's present age in years}$$
$$y = \text{Nate's present age in years}$$

Then

$$(x - 4) \text{ years was Brad's age 4 years ago, and}$$
$$(y - 4) \text{ years was Nate's age 4 years ago}$$

Also,

$$(x + 8) \text{ years will be Brad's age 8 years from now, and}$$
$$(y + 8) \text{ years will be Nate's age 8 years from now.}$$

From the given conditions we have

$$\begin{cases} x - 4 = 5(y - 4) & (1) \\ x + 8 = 2(y + 8) & (2) \end{cases}$$

That is,

$$\begin{cases} x - 5y = -16 & (1) \\ x - 2y = 8 & (2) \end{cases}$$

Solving, we obtain $y = 8$ and $x = 24$. Thus, Brad is 24 years old and Nate is 8 years old. ■

EXERCISES 11.12

1. Mr. Beebe sold some of his property and invested the money at 6% and 4% interest, which yield $700 per year. If the amount invested at 6% was $5000 less than the amount at 4%, find the total amount invested.

2. A broker invests a sum of money, part at 3% and part at 5%, and receives a yearly income of $380 from her investments. If she doubled the 5% investment and decreased the 3% investment accordingly, her income would be $80 more. Find the amount invested at each rate.

3. One part of $3000 is invested at 5% and the remainder at 8%. The yearly income from both is $186. How much is invested at each of these rates?

4. A sum of money at simple interest amounted to $16,000 in 16 years, and to $17,000 in 18 years. What was the sum of money and the rate of interest?

5. Three radios and five tape recorders cost $700. Five radios and eight tape recorders cost $1150. Find the cost of each radio and each tape recorder.

6. A painter bought a dozen brushes and six

pens for $10.50. His wife also bought nine brushes and a half dozen pens for $9.15. What is the cost of each?

7. A woman sells her cabin and invests one-half of the price plus $500 at 5%. She takes a 6% mortgage for the balance. From the two investments together, she receives $490 annually in interest. Find the price of the cabin and the sum invested each way.

8. The ages of three students total 49 years. Three times Peggy's age is 9 more than twice Brenda's age, and three times Joe's age is 12 less than four times Peggy's age. What are the ages of the three students?

9. After inheriting $10,000, a man invested the money partly at 5%, partly at 4%, and partly at 3%, yielding an annual interest of $390. The 5% investment brings $10 more interest yearly than the 4% and 3% investments together. How much money is invested at each rate?

*10. A landlord received $1300 in rents on two apartments last year, and one of the apartments brought $20 per month more than the other. Find the monthly rental on each apartment if the more expensive apartment was vacant for 2 months.

11.13 SYSTEMS OF LINEAR INEQUALITIES

In this section, we briefly discuss the graph of an inequality in two variables and graphs of systems of linear inequalities.

Graph of an Inequality in Two Variables

Consider the graph of the line $y = 1$. See Figure 11.11. Note that the line $y = 1$ divides the plane into two regions: the half-plane above the line and the half-plane below the line. The half-plane above the line consists of all points for which $y > 1$, and is called the graph of that inequality. The half-plane below the line is the graph of the inequality $y < 1$. See Figure 11.12. The line $y = 1$ is called the boundary line. The graph of the inequality $y \geq 1$ consists of the half-plane above the line together with the boundary line. Similarly, the graph of $y \leq 1$ is the boundary line and the half-plane below the line. See Figure 11.13. A half-plane without its boundary is called an open half-plane. A half-plane together with its boundary is called a closed half-plane.

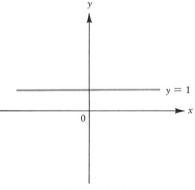

Figure 11.11

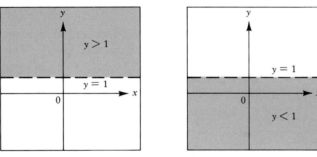

Figure 11.12

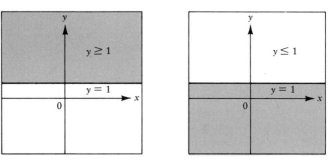

Figure 11.13

EXAMPLE 1 Graph the inequality $y > x$.

Solution We begin by determining the boundary. This can be achieved by graphing the equality $y = x$, shown as the dashed line in Figure 11.14. Thus, $y > x$ is the shaded open half-plane *above* the dashed line. ■

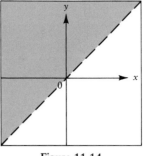

Figure 11.14

EXAMPLE 2 Graph the inequality $x - y \geq -2$.

Solution (a) Solve for y:
$$x - y \geq -2$$
$$-y \geq -x - 2$$
$$y \leq x + 2$$

(b) Determine the boundary by graphing the equation $y = x + 2$.

(c) Since the inequality is "less than or equal to," its graph is the closed half-plane below the boundary line (see Figure 11.15). ■

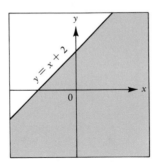

Figure 11.15

In general, the graph of a linear inequality in two variables, such as $Ax + By \geq C$, is either an open or closed half-plane (open for the strict inequalities $<$ and $>$, and closed for the inequalities $\leq$ and $\geq$). Its boundary is the graph of the associated linear equation $Ax + By = C$.

EXAMPLE 3   Graph the inequality $x + y < 3$.

Solution   (a) Solve for y: $y < -x + 3$

(b) The boundary is given by the graph of the equation $y = -x + 3$. See the dashed line in Figure 11.16.

(c) The graph of the inequality $y < -x + 3$ is the shaded open half-plane in Figure 11.16. ■

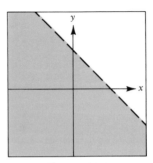

Figure 11.16

Graphs of Systems of Linear Inequalities

Graphs can also be used to solve systems of inequalities. To find the solution set of a system of linear inequalities, graph all the inequalities in the system on the same graph and shade each in a different way. The area where the shading overlaps describes the solution of the system.

EXAMPLE 4 Solve the system $\begin{cases} y \geq -2 \\ x > 1 \end{cases}$

Solution (a) The boundaries of the inequalities are given by the associated linear equations $y = -2$ and $x = 1$. Note that the inequality $y \geq -2$ includes its boundary, while the inequality $x > 1$ does not include its boundary. See Figure 11.17.

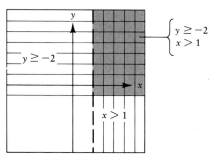

Figure 11.17

(b) The solution is the shaded area on and above the line $y = -2$, and to the right of the line $x = 1$. ■

EXAMPLE 5 Solve the system $\begin{cases} x + y \geq 4 \\ -2x + y \leq -3 \end{cases}$

Solution (a) Solving for y, we obtain $\begin{cases} y \geq -x + 4 \\ y \leq 2x - 3 \end{cases}$

(b) The boundaries are given by the equations $y = -x + 4$ and $y = 2x - 3$. See Figure 11.18.

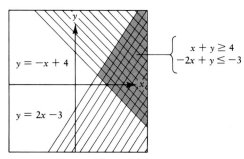

Figure 11.18

(c) The solution is the shaded area on and above the line $y = -x + 4$, and on and to the right of the line $y = 2x - 3$. ■

EXERCISES 11.13

Solve graphically the following systems of inequalities.

1. $\begin{cases} x \geq 0 \\ y \geq 0 \end{cases}$ 2. $\begin{cases} x \geq 0 \\ y \leq 0 \end{cases}$ 3. $\begin{cases} x \leq 0 \\ y \geq 0 \end{cases}$

4. $\begin{cases} x \leq 0 \\ y \leq 0 \end{cases}$ 5. $\begin{cases} x > 1 \\ y > -2 \end{cases}$ 6. $\begin{cases} x \leq 2 \\ y \geq 3 \end{cases}$

7. $\begin{cases} y \geq x \\ x > 2 \end{cases}$ 8. $\begin{cases} y \leq x \\ x \geq -1 \end{cases}$

9. $\begin{cases} -2x + y > -4 \\ -3x + y < 6 \end{cases}$ 10. $\begin{cases} 2x + y \geq 1 \\ x - 2y \geq 2 \end{cases}$

11. $\begin{cases} 2x + y \leq 3 \\ 4x + 2y \leq 6 \end{cases}$

12. $\begin{cases} y - x - 1 < 0 \\ 3x + y - 6 > 0 \end{cases}$

13. $\begin{cases} 2x + y - 2 > 0 \\ x - 2y + 2 < 0 \end{cases}$ 14. $\begin{cases} x + y > 3 \\ x - y < 6 \end{cases}$

15. $\begin{cases} x - y \leq -2 \\ x + y \geq 2 \end{cases}$ 16. $\begin{cases} 3x + 4y \leq 15 \\ x - 2y \leq -5 \end{cases}$

17. $\begin{cases} 2x + 3y \geq 6 \\ 6x + 9y \geq 0 \end{cases}$ 18. $\begin{cases} x + y \leq 1 \\ 2x + y \leq 2 \end{cases}$

19. $\begin{cases} x + 3y \leq 6 \\ 2y \geq 2 \\ x \geq -2 \end{cases}$ 20. $\begin{cases} y \geq x \\ x + y \geq 3 \\ x - 4 \leq 0 \end{cases}$

Chapter 11 SUMMARY

SYSTEMS OF LINEAR EQUATIONS: METHODS FOR FINDING SOLUTIONS

1. Graphical method
2. Substitution method
3. Elimination method

DETERMINANTS

1. Second-order determinant

$$\begin{vmatrix} a_1 & b_1 \\ a_2 & b_2 \end{vmatrix} = a_1 b_2 - a_2 b_1$$

2. Third-order determinant

$$\begin{vmatrix} a_1 & b_1 & c_1 \\ a_2 & b_2 & c_2 \\ a_3 & b_3 & c_3 \end{vmatrix} = \begin{matrix} a_1 & b_1 & c_1 & a_1 & b_1 \\ a_2 & b_2 & c_2 & a_2 & b_2 \\ a_3 & b_3 & c_3 & a_3 & b_3 \end{matrix}$$

CRAMER'S RULE

1. Solutions to the system $\begin{cases} a_1 x + b_1 y = c_1 \\ a_2 x + b_2 y = c_2 \end{cases}$ are

given by $x = \dfrac{D_x}{D}$ and $y = \dfrac{D_y}{D}$, provided $D \neq 0$,

and where $D = \begin{vmatrix} a_1 & b_1 \\ a_2 & b_2 \end{vmatrix}$, $D_x = \begin{vmatrix} c_1 & b_1 \\ c_2 & b_2 \end{vmatrix}$,

$D_y = \begin{vmatrix} a_1 & c_1 \\ a_2 & c_2 \end{vmatrix}$.

2. Solutions to the system

$$\begin{cases} a_1 x + b_1 y + c_1 z = d_1 \\ a_2 x + b_2 y + c_2 z = d_2 \\ a_3 x + b_3 y + c_3 z = d_3 \end{cases}$$

are given by $x = \dfrac{D_x}{D}$, $y = \dfrac{D_y}{D}$, and $z = \dfrac{D_z}{D}$, provided $D \neq 0$, and where

$$D = \begin{vmatrix} a_1 & b_1 & c_1 \\ a_2 & b_2 & c_2 \\ a_3 & b_3 & c_3 \end{vmatrix}, D_x = \begin{vmatrix} d_1 & b_1 & c_1 \\ d_2 & b_2 & c_2 \\ d_3 & b_3 & c_3 \end{vmatrix},$$

$$D_y = \begin{vmatrix} a_1 & d_1 & c_1 \\ a_2 & d_2 & c_2 \\ a_3 & d_3 & c_3 \end{vmatrix}, D_z = \begin{vmatrix} a_1 & b_1 & d_1 \\ a_2 & b_2 & d_2 \\ a_3 & b_3 & d_3 \end{vmatrix}.$$

Chapter 11 EXERCISES

In Exercises 1–6, use the graphical method to determine whether the equations are: (a) a consistent system and independent equations, (b) a consistent system and dependent equations, or (c) an inconsistent system and independent equations.

1. $\begin{cases} x - 2y = 8 \\ 2x + 5y = -11 \end{cases}$

2. $\begin{cases} x - 4y = 3 \\ -3x + 8y = 1 \end{cases}$

3. $\begin{cases} 3x + 2y = 10 \\ 6x + 4y = 7 \end{cases}$

4. $\begin{cases} 2x - y = 12 \\ 6x - 3y = 36 \end{cases}$

5. $\begin{cases} x = y + 2 \\ 3y = -2 - 5x \end{cases}$

6. $\begin{cases} -2x + 5y = 5 \\ 4x - 2y + 8 = 8y - 2 \end{cases}$

In Exercises 7–12, solve using the substitution method.

7. $\begin{cases} 2x + 3y = 8 \\ x - 2y = -10 \end{cases}$

8. $\begin{cases} -5x + y = -7 \\ 2x + 5y = 19 \end{cases}$

9. $\begin{cases} -x + 2y = -8 \\ -2x - 5y = 11 \end{cases}$

10. $\begin{cases} 3x - 8y = 1 \\ -x + 4y = -3 \end{cases}$

11. $\begin{cases} -x + y = -2 \\ 5x + 3y = -2 \end{cases}$

12. $\begin{cases} -3x - 7y = -5 \\ x - y = 1 \end{cases}$

In Exercises 13–27, solve using the elimination method.

13. $\begin{cases} 2x + 3y = 1 \\ -x + y = 1 \end{cases}$

14. $\begin{cases} 3x + 5y = 11 \\ 15x - 15y = 7 \end{cases}$

15. $\begin{cases} 2x - y = 3 \\ 4x + 2y = 50 \end{cases}$

*16. $\begin{cases} 3x + 2ay = 8a^2 \\ 2x + 3by = 27b^2 \end{cases}$

*17. $\begin{cases} ax + by = a^2 - b^2 \\ bx + ay = a^2 - b^2 \end{cases}$

*18. $\begin{cases} \dfrac{x}{a} + \dfrac{y}{b} = 2 \\ bx = ay \end{cases}$

19. $\begin{cases} \dfrac{6}{x} + \dfrac{5}{y} = 16 \\ \dfrac{9}{x} - \dfrac{1}{y} = -10 \end{cases}$

20. $\begin{cases} \dfrac{1}{x} - \dfrac{1}{y} = 8 \\ \dfrac{1}{x} + \dfrac{1}{y} = -4 \end{cases}$

21. $\begin{cases} \dfrac{a}{x} + \dfrac{b}{y} = 1 \\ \dfrac{b}{x} - \dfrac{a}{y} = 0 \end{cases}$

22. $\begin{cases} 4x + y + 3z = 1 \\ -8x + y - z = -5 \\ 2x - y - 2z = 5 \end{cases}$

23. $\begin{cases} 4x - 2y + z = 3 \\ 3x - y - z = -2 \\ 2x + y - z = 1 \end{cases}$

24. $\begin{cases} x + 2y + 4z = 2 \\ 2x + 3y - 2z = -3 \\ 2x - y - 4z = -8 \end{cases}$

25. $\begin{cases} 3x + 2y - 5z = -2 \\ 2x + 3y = -3 \\ -3y + 4z = 8 \end{cases}$

26. $\begin{cases} x - y - 2z = 3 \\ -3x + 3y + 6z = 5 \\ 2x + 3y + z = 1 \end{cases}$

27. $\begin{cases} \dfrac{1}{x} - \dfrac{1}{y} - \dfrac{2}{z} = -3 \\ \dfrac{3}{x} + \dfrac{2}{y} + \dfrac{1}{z} = 4 \\ \dfrac{2}{x} - \dfrac{1}{z} = 0 \end{cases}$

In Exercises 28–33, evaluate the determinants.

28. $\begin{vmatrix} 2 & -1 \\ 5 & -4 \end{vmatrix}$

29. $\begin{vmatrix} 17 & -8 \\ -9 & 16 \end{vmatrix}$

30. $\begin{vmatrix} -19 & 7 \\ 35 & -3 \end{vmatrix}$

31. $\begin{vmatrix} 3 & 0 & 0 \\ 1 & 3 & 2 \\ 2 & 3 & 1 \end{vmatrix}$

32. $\begin{vmatrix} 1 & -1 & -6 \\ -2 & -3 & 0 \\ 2 & -5 & 0 \end{vmatrix}$

33. $\begin{vmatrix} 2 & 6 & -5 \\ 2 & 6 & -5 \\ 2 & 6 & -5 \end{vmatrix}$

In Exercises 34–42, solve by Cramer's rule.

34. $\begin{cases} 4x + 7y = -19 \\ 5x - 3y = 35 \end{cases}$

35. $\begin{cases} 4x - 6y = 8 \\ 2x - 3y = 7 \end{cases}$

36. $\begin{cases} 6x + 9y = 12 \\ 8x + 12y = 16 \end{cases}$

37. $\begin{cases} 10x - 9y = 2c \\ 4x + 15y = 7c \end{cases}$

*38. $\begin{cases} ax + by = (a - b)^2 \\ ax - by = a^2 - b^2 \end{cases}$

*39. $\begin{cases} x - y = a - b \\ ax - by = 2a^2 - 2b^2 \end{cases}$

40. $\begin{cases} x + y + 3z = 3 \\ 2x - y = 0 \\ y - z = 8 \end{cases}$

41. $\begin{cases} 5x + 2y = -17 \\ 3x + 7z = 23 \\ 4y + 6z = 36 \end{cases}$

*42. $\begin{cases} x - 7y + 4z = -16c \\ x + y - 4z = -8b \\ -7x + y + 4z = -16a \end{cases}$

43.

Solve for x: $\begin{vmatrix} x & 3 & x \\ 1 & x & -2 \\ 4 & 2 & 1 \end{vmatrix} = 0.$

12 Logarithms

One of the most effective ways of saving time and mathematical computations when a calculator is not available is to use logarithms. Logarithms greatly simplify the processes of multiplication, division, raising to a power, and extracting a root: multiplication is replaced by addition, division by subtraction, raising to a power by a simple multiplication, and extracting a root by division.

Logarithms were invented in 1614 by John Napier, Baron of Merchiston of Scotland, who lived from 1550 to 1617. Burgi, a Swiss (1552–1632), was an independent inventor of logarithms, though his work was not so elegant as Napier's. Burgi published his work in 1620.

Exponents

We shall see later that a logarithm is an exponent. Thus, we shall first review the definitions and laws of exponents.

Definitions

1. $a^n = \underbrace{a \cdot a \cdot a \cdot \ldots \cdot a}_{n \text{ factors}},$ n a positive integer

   For example, $2^4 = 2 \cdot 2 \cdot 2 \cdot 2$ and $3 \cdot 3 \cdot 3 \cdot 3 \cdot 3 = 3^5$.

2. $a^{-n} = \dfrac{1}{a^n},$ $a \neq 0.$

   For example, $5^{-3} = \dfrac{1}{5^3}$ and $\dfrac{1}{6^9} = 6^{-9}.$

3. $a^0 = 1, \quad a \neq 0$

For example, $14^0 = 1$.

4. $a^{m/n} = \sqrt[n]{a^m} = (\sqrt[n]{a})^m$, a nonnegative

For example, $32^{3/5} = \sqrt[5]{(32)^3} = (\sqrt[5]{32})^3$ and $(\sqrt[3]{7})^2 = \sqrt[3]{7^2} = 7^{2/3}$.

Laws of Exponents

1. $a^m \cdot a^n = a^{m+n}$

For example, $3^7 \cdot 3^8 = 3^{7+8} = 3^{15}$.

2. $\dfrac{a^m}{a^n} = a^{m-n}, \quad a \neq 0$

For example, $\dfrac{8^6}{8^2} = 8^{6-2} = 8^4$.

3. $(a^m)^n = a^{mn}$

For example, $(13^5)^4 = 13^{5 \cdot 4} = 13^{20}$.

12.1 DEFINITION OF LOGARITHM

If 4 is taken as a base and 2 as an exponent, then $4^2 = 16$. The exponent 2 is the logarithm of 16 when the base is 4. Similarly, since $5^{-3} = \frac{1}{125}$, -3 is the logarithm of $\frac{1}{125}$ when the base is 5. Since other numbers may be chosen as bases, we give the following definition.

DEFINITION

The logarithm of a number to a given base is the exponent that must be placed on the base to produce the number.

Thus, the logarithm of 16 base 4, written $\log_4 16$, is 2 because $4^2 = 16$, and $\log_5(\frac{1}{125}) = -3$ because $5^{-3} = \frac{1}{125}$.

EXAMPLE 1 (a) $\log_3 9 = 2$ because $3^2 = 9$

(b) $\log_5 1 = 0$ because $5^0 = 1$

(c) $\log_4\left(\dfrac{1}{4}\right) = -1$ because $4^{-1} = \dfrac{1}{4}$

(d) $\log_{16} 4 = \dfrac{1}{2}$ because $16^{1/2} = 4$

(e) $\log_{27} 9 = \dfrac{2}{3}$ because $27^{2/3} = 9$

(f) $\log_7 7 = 1$ because $7^1 = 7$

Using symbols, the definition may be stated as:

DEFINITION

Given N and b, both positive and $b \neq 1$, $\log_b N = x$ if and only if $b^x = N$.

EXAMPLE 2

Exponential Form	Logarithmic Form
$3^4 = 81$	$\log_3 81 = 4$
$2^5 = 32$	$\log_2 32 = 5$
$7^0 = 1$	$\log_7 1 = 0$
$25^{1/2} = 5$	$\log_{25} 5 = \frac{1}{2}$
$27^{1/3} = 3$	$\log_{27} 3 = \frac{1}{3}$
$8^{2/3} = 4$	$\log_8 4 = \frac{2}{3}$
$10^2 = 100$	$\log_{10} 100 = 2$
$5^{-2} = \frac{1}{25}$	$\log_5(\frac{1}{25}) = -2$
$10^{-1} = .1$	$\log_{10}(.1) = -1$

Since, by definition, $\log_b N$ is the exponent that must be placed on b to produce N, it follows that

$$b^{\log_b N} = N$$

EXAMPLE 3 $7^{\log_7 30} = 30$

EXAMPLE 4   Determine the unknowns N, b, or x.

(a) $\log_b\left(\dfrac{1}{27}\right) = -3$ (b) $\log_5 N = 4$ (c) $\log_{10} 1000 = x$

(d) $12^{\log_{12} 46} = x$ (e) $\log_{11} N = 0$

Solution   (a) If $\log_b\left(\dfrac{1}{27}\right) = -3$, then $b^{-3} = \left(\dfrac{1}{27}\right) = (3)^{-3}$. Thus, $b = 3$.

(b) If $\log_5 N = 4$, then $N = 5^4 = 625$.

(c) If $\log_{10} 1000 = x$, then $10^x = 1000 = 10^3$. Thus, $x = 3$.

(d) If $12^{\log_{12} 46} = x$, then $x = 46$.

(e) If $\log_{11} N = 0$, then $N = 11^0 = 1$. ■

EXERCISES 12.1

In Exercises 1–20, express in logarithmic form.

1. $2^3 = 8$

2. $2^{-6} = \dfrac{1}{64}$

3. $7^{-2} = \dfrac{1}{49}$

4. $5^4 = 625$

5. $3^{-3} = \dfrac{1}{27}$

6. $\left(\dfrac{2}{3}\right)^3 = \dfrac{8}{27}$

7. $3^{-4} = \dfrac{1}{81}$

8. $\left(\dfrac{1}{5}\right)^{-2} = 25$

9. $16^{3/4} = 8$

10. $25^{1/2} = 5$

11. $27^{2/3} = 9$

12. $27^{-1/3} = \dfrac{1}{3}$

13. $16^{-5/4} = \dfrac{1}{32}$

14. $36^{-3/2} = \dfrac{1}{216}$

15. $17^0 = 1$

16. $10^0 = 1$

17. $\sqrt{49} = 7$

18. $\sqrt[3]{125} = 5$

19. $a^b = c$

20. $c^a = b$

In Exercises 21–40, express in exponential form.

21. $\log_6 216 = 3$

22. $\log_9\left(\dfrac{1}{27}\right) = -\dfrac{3}{2}$

23. $\log_{10} 0.01 = -2$

24. $\log_{27} 81 = \dfrac{4}{3}$

25. $\log_\pi 1 = 0$

26. $\log_e 1 = 0$

27. $\log_{15} 225 = 2$

28. $\log_{1/2}\left(\dfrac{1}{16}\right) = 4$

29. $\log_{1/2} 16 = -4$

30. $\log_a 8 = 3$

31. $\log_b 7 = -2$

32. $\log_{10} 10 = 1$

33. $\log_4 4 = 1$

34. $\log_a a^{1/3} = \dfrac{1}{3}$

35. $\log_{10} 72 = y$

36. $\log_{12} 64 = x$

37. $\log_c 1 = 0$

38. $\log_a 1 = 0$

39. $\log_c b = a$

40. $\log_a c = b$

Find the values of the logarithms in Exercises 41–60.

41. $\log_5 25$

42. $\log_2 8$

43. $\log_3 81$

44. $\log_9 9$

45. $\log_9 81$

46. $\log_5 125$

47. $\log_{1/2}\left(\dfrac{1}{16}\right)$

48. $\log_{10} 10,000$

49. $\log_7 343$

50. $\log_{20} 8,000$

51. $\log_{10} 0.1$

52. $\log_{10} 0.01$

53. $\log_{125} 25$

54. $\log_2\left(\dfrac{1}{8}\right)$

55. $\log_{10} 0.001$

56. $\log_4\left(\dfrac{1}{32}\right)$

57. $\log_{16} \sqrt{4}$

58. $\log_8 16$

59. $\log_a a^4$

60. $\log_c c^{-5}$

Determine the value of N in Exercises 61–70.

61. $\log_4 N = 3$

62. $\log_5 N = 4$

63. $\log_4 N = 5$

64. $\log_8 N = \dfrac{2}{3}$

65. $\log_2 N = -3$

66. $\log_3 N = -2$

67. $\log_8 N = -\dfrac{4}{3}$

68. $\log_b N = 1$

69. $\log_{169} N = \dfrac{1}{2}$

70. $\log_{169} N = -\dfrac{1}{2}$

Determine the value of the base b in Exercises 71–80.

71. $\log_b 10 = 1$

72. $\log_b 64 = 6$

73. $\log_b 49 = 2$

74. $\log_b 81 = 4$

75. $\log_b 1000 = \dfrac{3}{2}$

76. $\log_b 2 = \dfrac{1}{8}$

77. $\log_b 11 = \dfrac{1}{2}$

78. $\log_b 32 = \dfrac{5}{4}$

79. $\log_{b+1} 4 = 2$

80. $\log_{b+2} 10,000 = 4$

Determine the values in Exercises 81–85.

81. $2^{\log_2 8}$

82. $3^{\log_3 7.5}$

83. $5^{\log_5 1}$

*84. $3^{-\log_3 3}$

*85. $5^{-\log_5 10}$

Evaluate each of the following in Exercises 86–90.

*86. $\dfrac{\log_8 64 + \log_{64} 8}{\log_9 27 - \log_{27} 9}$

*87. $\dfrac{\log_{25} 125 - \log_{16} \dfrac{1}{8}}{\log_4 \dfrac{1}{32} - \log_{15} 1}$

*88. $\dfrac{\log_{49} 7 - \log_6 36\sqrt{6}}{\log_\pi 1 - \log_{1/9} 81}$

*89. $\dfrac{\log_5 0.04 + \log_{10} 0.1}{\log_{1/8}\left(\dfrac{1}{4}\right) + \log_{1/4} 32}$

*90. Find $-\log_8 \log_4 \log_2 16$.

12.2   PROPERTIES OF LOGARITHMS

Since a logarithm is an exponent, the properties of logarithms are based on the laws of exponents. These properties are useful in computations involving logarithms.

Property 1

The logarithm of a product is equal to the sum of the logarithms of the factors; that is,

$$\log_b MN = \log_b M + \log_b N$$

Proof   Let $x = \log_b M$ and $y = \log_b N$. Then, expressing each in exponential form, we have

$$M = b^x \quad \text{and} \quad N = b^y$$

Multiplying these equations, we obtain

$$MN = b^x b^y = b^{x+y}$$

Changing to logarithmic form, we get

$$\log_b MN = x + y$$
$$\log_b MN = \log_b M + \log_b N$$

The proof is similar for a product having more than two factors. For example,

$$\log_b MNP = \log_b M + \log_b N + \log_b P \quad \blacksquare$$

EXAMPLE 1   (a) $\log_b (25)(62) = \log_b 25 + \log_b 62$

(b) $\log_b 26 = \log_b (13)(2) = \log_b 13 + \log_b 2$

(c) $\log_b 64 = \log_b (4)(4)(4) = \log_b 4 + \log_b 4 + \log_b 4 = 3 \log_b 4$

Note. $64 = 4^3$; and thus, $\log_b 64 = \log_b 4^3 = 3 \log_b 4$.

CAUTION

Be careful to recognize that

$$\log_b (M + N) \neq \log_b M + \log_b N$$

We have shown that $\log_b M + \log_b N = \log_b MN$ and, if the above expression were true, then $\log_b (M + N) = \log_b MN$ and $M + N = MN$ for any positive real numbers, and this is definitely not true.

Property 2

The logarithm of the quotient of two numbers is equal to the logarithm of the numerator minus the logarithm of the denominator; that is,

$$\log_b \frac{M}{N} = \log_b M - \log_b N$$

Proof   Let

$$x = \log_b M \quad \text{and} \quad y = \log_b N$$

In exponential form:

$$M = b^x \quad \text{and} \quad N = b^y$$

Divide the equations:

$$\frac{M}{N} = \frac{b^x}{b^y} = b^{x-y}$$

Change to logarithmic form:

$$\log_b \frac{M}{N} = x - y = \log_b M - \log_b N \quad \blacksquare$$

EXAMPLE 2   (a) $\log_b \dfrac{7}{5} = \log_b 7 - \log_b 5$

(b) $\log_b \dfrac{14}{33} = \log_b \dfrac{7 \cdot 2}{3 \cdot 11} = \log_b (7 \cdot 2) - \log_b (3 \cdot 11)$

$$= \log_b 7 + \log_b 2 - [\log_b 3 + \log_b 11]$$

$$= \log_b 7 + \log_b 2 - \log_b 3 - \log_b 11$$

CAUTION

Again, be careful to recognize that

$$\frac{\log_b M}{\log_b N} \neq \log_b M - \log_b N$$

Property 3

The logarithm of a power of a number is equal to the exponent times the logarithm of the number; that is,

$$\log_b M^p = p \log_b M$$

Proof   Let
$$x = \log_b M$$

Express in exponential form:
$$M = b^x$$

Raise to the p-th power:
$$M^p = (b^x)^p = b^{px}$$

Change to logarithmic form:
$$\log_b M^p = px = p \log_b M \quad \blacksquare$$

EXAMPLE 3   (a) $\log_b 4^3 = 3 \log_b 4$

(b) $\log_b \sqrt{3} = \log_b 3^{1/2} = \dfrac{1}{2} \log_b 3$

(c) $\log_b \dfrac{49}{64} = \log_b \dfrac{7^2}{4^3} = \log_b 7^2 - \log_b 4^3 = 2 \log_b 7 - 3 \log_b 4$

Part **(b)** can be generalized:

Since $\sqrt[n]{M} = M^{1/n}$, $\log_b \sqrt[n]{M} = \dfrac{1}{n} \log_b M$.

EXAMPLE 4 $\log_b \sqrt[5]{106} = \dfrac{1}{5} \log_b 106$

EXAMPLE 5   Use the properties of logarithms to write the following expression as the sums and differences of logarithms:

$$\log_3 \sqrt{\frac{2x(x-y)^5}{7z^4 w^{1/2}}}$$

Solution $\log_3 \sqrt{\dfrac{2x(x-y)^5}{7z^4 w^{1/2}}} = \log_3 \left[\dfrac{2x(x-y)^5}{7z^4 w^{1/2}}\right]^{1/2}$

$= \dfrac{1}{2} \log_3 \left[\dfrac{2x(x-y)^5}{7z^4 w^{1/2}}\right]$ [by Property 3]

$= \dfrac{1}{2} \{\log_3 [2x(x-y)^5] - \log_3 [7z^4 w^{1/2}]\}$ [by Property 2]

$= \dfrac{1}{2} \{\log_3 2 + \log_3 x + \log_3 (x-y)^5 - [\log_3 7 + \log_3 z^4 + \log_3 w^{1/2}]\}$

   [by Property 1]

$= \dfrac{1}{2} \{\log_3 2 + \log_3 x + 5 \log_3 (x-y) - [\log_3 7 + 4 \log_3 z + \dfrac{1}{2} \log_3 w]\}$

   [by Property 3]

$= \dfrac{1}{2} \left\{\log_3 2 + \log_3 x + 5 \log_3 (x-y) - \log_3 7 - 4 \log_3 z - \dfrac{1}{2} \log_3 w\right\}$ ■

EXAMPLE 6   Express the following as a single logarithm:

$$\log_b x - \log_b 3 + \log_b 4$$

Solution   We arrange the expression so that all the positive logarithms are together and all the negative logarithms are together:

$$\log_b x + \log_b 4 - \log_b 3 = \log_b 4x - \log_b 3 = \log_b \left(\frac{4x}{3}\right)$$ ■

EXAMPLE 7   Simplify $\log_4 M + \log_5 N - \log_4 P - \log_4 Q + \log_5 R - \log_5 S$.

Solution   Note that not all of the bases are the same, and that the properties apply only for logarithms with the same base. In order to simplify the expression, we gather those logarithms with like bases before using the properties:

$\log_4 M - \log_4 P - \log_4 Q + \log_5 N + \log_5 R - \log_5 S$

$= [\log_4 M - (\log_4 P + \log_4 Q)] + [(\log_5 N + \log_5 R) - \log_5 S]$

$= [\log_4 M - \log_4 (PQ)] + [\log_5 (NR) - \log_5 S]$

$= \log_4 \left(\dfrac{M}{PQ}\right) + \log_5 \left(\dfrac{NR}{S}\right)$ ■

EXAMPLE 8   Express $\dfrac{1}{2}\left\{\log_b 7 - 3 \log_b x + \dfrac{2}{3} \log_b 4\right\}$ as a single logarithm.

Solution $\dfrac{1}{2}\left\{\log_b 7 - 3 \log_b x + \dfrac{2}{3} \log_b 4\right\}$

$$= \dfrac{1}{2}\{\log_b 7 - \log_b x^3 + \log_b 4^{2/3}\} \qquad\qquad \text{[by Property 3]}$$

$$= \dfrac{1}{2}\left\{\log_b \left(\dfrac{7 \cdot 4^{2/3}}{x^3}\right)\right\} \qquad\qquad \text{[by Properties 1 and 2]}$$

$$= \log_b \left(\dfrac{7 \cdot 4^{2/3}}{x^3}\right)^{1/2} \qquad\qquad \text{[by Property 3]}$$

$$= \log_b \sqrt{\dfrac{7 \cdot 4^{2/3}}{x^3}} \quad \blacksquare$$

EXAMPLE 9   Given $\log_{10} 2 = 0.3010$, $\log_{10} 3 = 0.4771$, and $\log_{10} 5 = 0.6990$, find $\log_{10}\left(\dfrac{48}{\sqrt[3]{5}}\right)$.

Solution   We start by writing the expression in terms of 2, 3, and 5.

$$\log_{10}\left(\dfrac{48}{\sqrt[3]{5}}\right) = \log_{10} 48 - \log_{10} \sqrt[3]{5}$$

$$= \log_{10} (3 \cdot 16) - \log_{10} 5^{1/3}$$

$$= \log_{10} (3 \cdot 2^4) - \log_{10} 5^{1/3}$$

$$= \log_{10} 3 + 4 \log_{10} 2 - \dfrac{1}{3} \log_{10} 5$$

$$= (0.4771) + 4(0.3010) - \dfrac{1}{3}(0.6990)$$

$$= 0.4771 + 1.2040 - 0.2330$$

$$= 1.4481 \quad \blacksquare$$

EXERCISES 12.2

Given $\log_{10} 2 = 0.3010$, $\log_{10} 3 = 0.4771$, $\log_{10} 5 = 0.6990$, and $\log_{10} 7 = 0.8451$, find each of the following in Exercises 1–20.

1. $\log_{10} 63$

2. $\log_{10} 98$

3. $\log_{10} 140$

4. $\log_{10} 300$

5. $\log_{10} 1050$

6. $\log_{10}\left(\dfrac{9}{35}\right)$

7. $\log_{10}\left(\dfrac{21}{10}\right)$

8. $\log_{10} 0.35$

9. $\log_{10} \sqrt{3}$

10. $\log_{10} \sqrt{42}$

11. $\log_{10} \sqrt{210}$

12. $\log_{10} \sqrt[3]{0.12}$

13. $\log_{10} 18^{2/3}$

14. $\log_{10}\left(\dfrac{7}{12}\right)^{3/5}$

15. $\log_{10}\left[\dfrac{48}{\sqrt{5}}\right]$

16. $\log_{10}\sqrt[3]{168}$

*17. $\log_{10}\left[\dfrac{\sqrt{7}}{120}\right]$

*18. $\log_{10}\left[\dfrac{\sqrt[3]{7}}{\sqrt[5]{3}}\right]$

*19. $\log_{10}(126)^3$

*20. $\log_{10}\left[\dfrac{35^3}{\sqrt[4]{15}}\right]$

In Exercises 21–35, using the properties of logarithms, write each as the sum and difference of logarithms. (Assume that all logarithms have the same base.)

21. $\log PQR$

22. $\log\left[\dfrac{AB}{C}\right]$

23. $\log\left[\dfrac{C}{AB}\right]$

24. $\log M^4$

25. $\log(A^3B^5)$

26. $\log\left[\dfrac{1}{A^n}\right]$

27. $\log A^{-n}$

28. $\log\sqrt[n]{B^5}$

29. $\log\left[\dfrac{1}{\sqrt[n]{c}}\right]$

30. $\log\sqrt[3]{P^2Q^4}$

31. $\log\dfrac{M^n\sqrt[m]{R}}{S^5}$

32. $\log\sqrt{\dfrac{AB^n}{C^m}}$

33. $\log\sqrt[5]{\dfrac{MN}{Q}}$

34. $\log(\sqrt[5]{x}\,\sqrt[6]{y})$

35. $\log\sqrt{\dfrac{x^2y^5}{z}}$

Express each of the following in Exercises 36–47 as a single logarithm:

36. $\log_a x + \log_a y$

37. $3\log_c z - 2\log_c x$

38. $3\log_e x + 2\log_e y$

39. $\log_2 5 + \log_2 6 - \log_2 7$

40. $\log_e a + \log_e b - \log_e c$

41. $\dfrac{1}{3}\log_e 3 + \log_e \pi - \log_e x$

42. $\dfrac{3}{4}\log_{10} b - 6\log_{10} c - \dfrac{4}{5}\log_{10} a$

43. $\log_b 2 + \log_b \pi + \dfrac{1}{2}\log_b l - \dfrac{1}{2}\log_b g$

44. $\dfrac{1}{2}\log_{10} 3 + \dfrac{1}{2}\log_{10} 5 - \dfrac{1}{6}\log_{10} 15$

45. $\dfrac{1}{4}\log_5 28 - \dfrac{1}{2}\log_5 7 + \dfrac{1}{6}\log_5 14$

$\quad -\dfrac{1}{3}\log_5 2$

*46. $\dfrac{1}{2}[\log_{10}(s-a) + \log_{10}(s-b)$

$\quad + \log_{10}(s-c) + \log_{10}s]$

*47. $\log_b k + \log_b(k-1) + \log_b$

$\quad (k-2) + \cdots + \log_b 2 + \log_b 1$

In Exercises 48–60, determine whether each equation is true or false. (Assume that the base in each equation is the same for all the logarithms.)

48. $\log\dfrac{1}{x} = -\log x$

49. $(\log x)^2 = 2\log x$

50. $\dfrac{1}{2}\log a = \sqrt{a}$

51. $\log x - \log y = \dfrac{x}{y}$

52. $x^3 = 3\log x$

53. $\log(xy)^4 = (\log x + \log y)^4$

54. $\dfrac{1}{n}\log a^n = \log a$

55. $6^{2\log_6 x + \log_6 y} = x^2y$

56. $10^{4+\log_{10} 5} = 50{,}000$

57. $\dfrac{1}{2}\log A - \dfrac{1}{3}\log B - \dfrac{1}{6}\log C = \dfrac{1}{6}\log\dfrac{A^3}{B^2C}$

58. $\log\dfrac{A}{BC} = \dfrac{\log A}{\log B + \log C}$

59. $3\log(\log z) = \log(\log z)^3$

60. $\log(x^7 - y^3) = 7\log x - 3\log y$

Using the properties of logarithms, prove the identities in Exercises 61–68.

*61. $\log x^2 + \log\dfrac{1}{x} + \log\sqrt{x} = \dfrac{3}{2}\log x$

*62. $\log xy + \log\dfrac{x}{y} + \log\sqrt{xy}$

$\quad = \dfrac{5}{2}\log x + \dfrac{1}{2}\log y$

*63. $\dfrac{1}{2}\log\left[\dfrac{x^2-1}{2}\right] + \dfrac{1}{2}\log\left[\dfrac{x-1}{2x+2}\right]$

$\quad = \log(x-1) - \log 2$

*64. $\dfrac{1}{4}\log(x^2+3x+2) + \dfrac{1}{4}\log\left(\dfrac{x+1}{x+2}\right)$

$\quad = \log\sqrt{x+1}$

*65. $\log\left[x - \dfrac{1}{x}\right]^2 = 2[\log(x+1) + \log(x-1)$

$\quad - \log x]$

*66. $\log\left[\dfrac{x\sqrt{x+1}}{x^2-1}\right] = \log x - \dfrac{1}{2}\log(x+1)$
$- \log(x-1)$

*67. $\log(N \times 10^n) = n + \log N$

*68. $\log(N \div 10^n) = -n + \log N$

12.3 LOGARITHMIC EQUATIONS

An equation that involves logarithms is called a logarithmic equation. In order to solve such equations, we attempt to combine the given logarithmic expressions into a single logarithm using the properties of logarithms. The solution is then obtained from the definition of a logarithm.

EXAMPLE 1 Solve the equation $\log_{10}(3x-5) = 1$.

Solution We begin by writing $\log_{10}(3x-5) = 1$ in exponential form,

$$3x - 5 = 10^1, \quad 3x - 5 = 10, \quad 3x = 15, \quad \text{or} \quad x = 5. \quad \blacksquare$$

EXAMPLE 2 Solve the equation $\log_{10}(x-21) + \log_{10} x = 2$.

Solution Using Property 1, we obtain:

$$\log_{10}(x-21) + \log_{10} x = 2$$
$$\log_{10}[(x-21)x] = 2$$
$$\log_{10}[x^2 - 21x] = 2$$

Now we convert this logarithmic equation into its exponential form. Thus,

$$x^2 - 21x = 10^2$$
$$x^2 - 21x - 100 = 0$$
$$(x-25)(x+4) = 0$$
$$x = 25 \quad \text{or} \quad x = -4$$

Recall that we can take the logarithm of only positive numbers. If we substitute $x = -4$ into the *original* equation, we obtain $\log_{10}(-17) + \log_{10}(-4)$, which are undefined. Thus, $x = -4$ is *not* a solution to the logarithmic equation. However, substituting $x = 25$, we get $\log_{10}(4) + \log_{10}(25) = \log_{10}(25 \cdot 4) = \log_{10} 100 = 2$. Hence, $x = 25$ is a solution. $\blacksquare$

EXAMPLE 3 Solve the equation $2 \log_{10} x - \log_{10}(30 - 2x) = 1$.

Solution Using Properties 3 and 2, we obtain

$$2 \log_{10} x - \log_{10}(30 - 2x) = 1$$

$$\log_{10} x^2 - \log_{10}(30 - 2x) = 1$$

$$\log_{10}\left[\frac{x^2}{30 - 2x}\right] = 1$$

Converting to exponential form, we get

$$\frac{x^2}{30 - 2x} = 10$$

Solving for x, $$x^2 + 20x - 300 = 0$$

$$(x + 30)(x - 10) = 0$$

$$x = -30 \quad \text{or} \quad x = 10$$

When $x = -30$: $2 \log_{10}(-30) - \log_{10}(90) \overset{?}{=} 1$. But $\log_{10}(-30)$ is undefined.
When $x = 10$: $2 \log_{10} 10 - \log_{10} 10 = 2(1) - 1 = 1$.
Hence, $x = 10$ is the solution of the logarithmic equation. ■

EXERCISES 12.3

Solve the following logarithmic equations:

1. $\log_{10}(2x + 3) = 1$

2. $\log_2(2x - 7) = 4$

3. $\log_{10} x + \log_{10} 4 = 2$

4. $\log_{10} x + \log_{10}(x - 3) = 1$

5. $\log_6(x - 9) + \log_6 x = 2$

6. $\log_{10} x - \log_{10}(x - 3) = 1$

7. $-\log_{10}(x - 2) = 0$

8. $-\log_{10}(x - 1) = 2$

9. $\log_{10} x + \log_{10} 2 + \log_{10} 7 = 1$

10. $\log_{10} x + \log_{10} 5 - \log_{10} 3 = 2$

11. $\log_{10}(2x + 1) - \log_{10}(3x - 4) = 1$

12. $\log_2(3x + 7) - \log_2(2x - 5) = 5$

13. $2 + \log_2 x = \log_2(x + 5)$

14. $1 + \log_{10} x = \log_{10}(x + 1)$

15. $\log_{10}\left(x + \dfrac{16}{x}\right) = 1$

16. $\log_6(x + 3) = 1 - \log_6(x + 4)$

17. $\log_6(x + 2) + \log_6(x + 3) = 1$

18. $\log_7(x + 4) + \log_7 2 = 2 \log_7 x$

***19.** $\log_x(2x^2 - 3x) = 1$

***20.** $\log_{x+2}(3x^2 + 4x - 14) = 2$

***21.** $\log_{x+2}(17x^2 - 6x + 8) = 3$

***22.** $\log_x(3x^2 + 10x) = 3$

***23.** $\log_{\sqrt{2x^2 + 1}}(5x^4 - 3x^2 + 7) = 4$

***24.** $\log_{10} \log_{10} \log_{10} x = 0$

***25.** $\log_2 \log_2 \log_2 x = 0$

***26.** $\log_{\sqrt{x-1}}(x^4 - 8x^2 - 2x + 1) = 4$

***27.** $\log_{\sqrt{2}} \log_2 \log_4(x - 15) = 0$

***28.** $\log_{10}(\log_2 \log_3 \sqrt{x} + 1) = 0$

***29.** $\log_x \log_3 \log_x 2x^2 = 0$

***30.** $\log_{10}[3 + 2 \log_{10}(1 + x)] = 0$

12.4 COMMON LOGARITHMS

There are two important systems of logarithms in use today: The natural, or Napierian, system, which has base e, where e is the irrational number approximated by 2.71828; and the common, or Briggs, system, which has base 10. We shall use 10 as the base for our computations with logarithms since 10 is the base of our number system. Logarithms to the base 10 are called common logarithms, and we shall write log x rather than $\log_{10} x$. Logarithms to the base e are called natural logarithms, and are denoted by ln x. Natural logarithms will be discussed in the next section.

Characteristic and Mantissa

One reason base 10 is used so often in computational work is because every positive real number N can be written in the scientific form

$$N = A \times 10^n, \text{ where } 1 \le A < 10 \text{ and } k \text{ an integer}$$

(see Section 7.3).

EXAMPLE 1 (a) $372 = 3.72 \times 10^2$ (b) $.0716 = 7.16 \times 10^{-2}$

(c) $.0000587 = 5.87 \times 10^{-5}$ (d) $4.95 = 4.95 \times 10^0$

If N is any positive real number and we write

$$N = A \times 10^n$$

where $1 \le A < 10$ and n is an integer, then using Property 3 of logarithms we obtain

$$\log N = \log A + \log 10^n$$

Since $\log 10^n = n \log 10 = n$, we have

$$\boxed{\log N = \log A + n = n + \log A}$$

DEFINITION
The form $\log N \doteq \log A + n$ is called the standard form of log N, where the number $\log A$ is called the mantissa and the integer n is called the characteristic of log N:

$$\log N = \underset{\text{characteristic}}{n} + \underset{\text{mantissa}}{\log A}$$

Notice that for $1 \le A < 10$, it follows that $\log 1 \le \log A < \log 10$; that is, $0 \le \log A < 1$. Thus, the mantissa of a logarithm is a number between 0 and 1. Hence, the equation $\log N = \log A + n$ implies that if N is any positive real number, then *log N can be written as the sum of a positive decimal fraction* $\log A$ *(mantissa) and an integer n* (characteristic).

Theorem 1

If N is any positive real number written in its scientific notation, $N = A \cdot 10^n$, where $1 \le A < 10$ and n is an integer, then

$$\log N = \log(A \times 10^n) = n + \log A$$

EXAMPLE 2

$$\log 534 = \log[5.34 \times 10^2]$$
$$= \log 5.34 + \log 10^2$$
$$= \log 5.34 + 2$$

Thus, the mantissa is $\log 5.34$ and the characteristic is 2.

We saw in the previous section that $\log_{10} N = N$, for $N > 0$. That is, $\log N$ is the exponent of 10 such that the result is N. When N is an integral power of 10, $\log N$ is easily determined. Consider the following list of integral powers of 10 and their corresponding logarithmic form:

$$\log 100 = 2 \quad \text{is equivalent to} \quad 10^2 = 100$$
$$\log 10 = 1 \quad \text{is equivalent to} \quad 10^1 = 10$$
$$\log 1 = 0 \quad \text{is equivalent to} \quad 10^0 = 1$$
$$\log 0.1 = -1 \quad \text{is equivalent to} \quad 10^{-1} = 0.1$$
$$\log 0.01 = -2 \quad \text{is equivalent to} \quad 10^{-2} = 0.01$$

Note that $\log 100 = \log 10^2 = 2$ and $\log 0.1 = \log 10^{-1} = -1$. That is, when N is an integral power of 10, the logarithm of N is the exponent of 10. However, most values of N are not integral powers of 10. Therefore, we use a table to compute $\log N$. We shall show that if we can find $\log N$ for $1 \le N < 10$, then $\log N$ can be determined for any $N > 0$. Consider the partial table showing some values of $\log N$ for $1 \le N < 10$ (Figure 12.1). The common logarithm table, Table 2, is given in Appendix I.

N	0	1	2	3	4	5	6	7	8	9
4.8	.6812	.6821	.6830	.6839	.6848	.6857	.6866	.6875	.6884	.6893
4.9	.6902	.6911	.6920	.6928	.6937	.6946	.6955	.6964	.6972	.6981
5.0	.6990	.6998	.7007	.7016	.7024	.7033	.7042	.7050	.7059	.7067
5.1	.7076	.7084	.7093	.7101	.7110	.7118	.7126	.7135	.7143	.7152
5.2	.7160	.7168	.7177	.7185	.7193	.7202	.7210	.7218	.7226	.7235
5.3	.7243	.7251	.7259	.7267	.7275	.7284	.7292	.7300	.7308	.7316

Figure 12.1

EXAMPLE 3   Find log 4.83.

Solution   We locate the row containing 4.8 and then move to the column 3. Thus, log 4.83 = 0.6839. Hence, $10^{0.6839}$ = 4.83.   ∎

EXAMPLE 4   Find log 5.34.

Solution   We locate the intersection of the row containing 5.3 under N and the column containing 4. Therefore,

$$\log 5.34 = 0.7275$$

What this says is: $10^{0.7275}$ = 5.34.   ∎

Note that even though we have used an equals sign, the logarithms we found were only approximations. The reason for this is that most logarithms are irrationals and cannot be represented by a terminating decimal.

Until now, we have limited ourselves to finding log N where $1 \le N < 10$ and integral powers of 10. How do we compute log N when $0 < N < 1$ or $N > 10$? We now show that this can be achieved by representing N in scientific notation.

EXAMPLE 5   Find log 534.

Solution   Writing 534 in scientific notation, we get $534 = 5.34 \times 10^2$. Thus,

$$\log 534 = \log[5.34 \times 10^2]$$
$$= 2 + \log 5.34$$
$$= 2 + 0.7275 \text{ (see Example 4)} \quad ∎$$

Examples 4 and 5 show that 5.34 and 534 have the same mantissa (0.7275), but different characteristics—0 and 2, respectively.

EXAMPLE 6   Find log 0.00485.

Solution   Writing 0.00485 in scientific notation, we get $0.00485 = 4.85 \times 10^{-3}$. Thus,

$$\log 0.00485 = \log[4.85 \times 10^{-3}]$$
$$= -3 + \log 4.85$$
$$= -3 + 0.6857 \quad ∎$$

EXAMPLE 7   Find log 9070.

Solution Writing 9070 in scientific notation, we get $9070 = 9.070 \times 10^3$. Thus,

$$\log 9070 = \log[9.070 \times 10^3]$$
$$= 3 + \log 9.070$$
$$= 3 + 0.9576$$

log 9.070 was found in Table 2 of Appendix I. Note that the characteristic and the mantissa are written separately. This is done for computational reasons that will be explained later. ■

Antilogarithms

Until now, we have been given a positive real number N and found $\log N$. However, suppose we are given $\log N$ and told to find N. We have solved a few such problems in previous sections. For example, if $\log N = 3$, then $N = 10^3 = 1000$. Suppose $\log N = 2 + 0.7340$ and we want to find N. With the help of the logarithm table, the procedure is basically the same as before. If $\log N = 2 + 0.7340$, then $N = 10^{2+0.7340}$. Using the properties of exponents, we get

$$N = 10^{2+0.7340} = 10^2 \cdot 10^{0.7340} = 10^2 \cdot (5.42) = 542$$

where $10^{0.7340}$ was obtained from Table 2 of Appendix I.

DEFINITION

Given a number x, the number N such that $\log N = x$ is called the antilogarithm of x and is written as $N = $ antilog x.

EXAMPLE 8 Find N if $\log N = -1 + 0.6964$.

Solution Using the table in Figure 12.1, we look for the mantissa 0.6964 inside the table. The mantissa appears in row 4.9 and the 7 column. Hence, by Theorem 1,

$$N = 4.97 \times 10^{-1} = .497 \quad ■$$

EXAMPLE 9 Find N if $\log N = 3.8938$.

Solution First we write $\log N = 3.8938$ in standard form; that is, as the sum of an integer and a *nonnegative* decimal fraction. Thus, $\log N = 3.8938 = 3 + 0.8938$. Table 2 of Appendix I gives $0.8938 = \log 7.83$. Using Theorem 1, we get

$$N = 7.83 \times 10^3 = 7830 \quad ■$$

EXAMPLE 10   Find N if $\log N = -2.2874$.

Solution   Recall that the mantissa of any logarithm is a *nonnegative decimal fraction.* The number given above is negative, so we first convert the given number -2.2874 to an *equivalent* number having a positive decimal fraction by adding and subtracting 3. Thus,

$$\log N = \underbrace{-2.2874 + 3}\ - 3 = 0.7126 - 3$$

and we have a logarithm with mantissa 0.7126 and characteristic -3. Table 2 of Appendix I gives $0.7126 = \log 5.16$, and Theorem 1 yields

$$N = 5.16 \times 10^{-3} = .00516 \quad \blacksquare$$

Interpolation

So far we have computed logarithms and antilogarithms that we could find in Table 2 of Appendix I. What happens if we cannot find such values in the table? For example, we used Table 2 to find logarithms of N when N had three digits. However, what occurs if N has four or more significant digits, say, log 5.235? Similarly, what if we are looking for the antilogarithm and there are no values in the table corresponding to the mantissa; say, $\log N = 1.7206$? These problems can be accomplished by using an approximation method called linear interpolation. This process is based on the assumption that small differences between numbers are proportional to the differences between their corresponding logarithms. While this assumption is not strictly true, the approximations obtained are sufficiently accurate. Thus, we obtain good approximations for the logarithms of four-digit numbers and antilogarithms to four significant digits.

EXAMPLE 11   Find log 5.235.

Solution   Writing 5.235 in scientific notation, we have $5.235 = 5.235 \times 10^0$ and the characteristic $n = 0$. We find log 5.230 and log 5.240 in Table 2 (actually log 5.23 and log 5.24) and arrange the larger number first to make computations easier.

$$.010\left[\ .005\left[\begin{array}{c|c} N & \log N \\ \hline 5.230 & 0.7185 \\ 5.235 & \text{unknown} \\ 5.240 & 0.7193 \end{array}\right]d\ \right].0008$$

We assume that the change in the value of the logarithms is proportional to the change in the corresponding numbers. So,

$$\frac{.005}{.010} = \frac{d}{.0008}$$

and, $$d = \left(\frac{.005}{.010}\right)(.0008) = 0.0004$$

Hence, $\log 5.235 = \log 5.230 + d = 0.7185 + 0.0004 = 0.7189.$ ■

EXAMPLE 12   Find $\log 0.04236$.

Solution   (a) Writing 0.04236 in scientific notation, we get $0.04236 = 4.236 \times 10^{-2}$. Thus, $\log 0.04236 = -2 + \log 4.236$.

(b) We now determine the mantissa $= \log 4.236$ using linear interpolation:

N	log N
4.230	0.6263
4.236	unknown
4.240	0.6274

$$.010\left[.006\left[\begin{matrix}4.230 & 0.6263\\4.236 & \text{unknown}\end{matrix}\right]d\right].0011$$

So, $\dfrac{.006}{.010} = \dfrac{d}{.0011}$ and $d \approx .0007$. Hence, mantissa $= \log 4.236 = 0.6263 + 0.0007 = 0.6270$.

(c) Finally, $\log 0.04236 = -2 + \log 4.236 = -2 + 0.6270.$ ■

EXAMPLE 13   If $\log N = 2.7206$, find N.

Solution   (a) $\log N = 2.7206 = 2 + 0.7206$. Therefore, the characteristic is 2 and the mantissa is 0.7206.

(b) We now find the antilogarithm of the mantissa 0.7206. From Table 2 of Appendix I, we find that the given mantissa lies between 0.7202 and 0.7210.

N	log N
5.250	0.7202
unknown	0.7206
5.260	0.7210

$$.010\left[d\left[\begin{matrix}5.250 & 0.7202\\\text{unknown} & 0.7206\end{matrix}\right].0004\right].0008$$

So, $\dfrac{d}{.010} = \dfrac{.0004}{.0008}$, and $d = .005$. Hence, the antilog $0.7206 = 5.250 + .005 = 5.255$.

(c) Finally, antilog $2.7206 = N = 5.255 \times 10^2 = 525.5.$ ■

EXERCISES 12.4

Determine the characteristic of the logarithm of the numbers in Exercises 1–24.

1. 25	2. 625	3. 8023
4. 94	5. 6	6. .7
7. 6542	8. 645.2	9. 64.52
10. 6.452	11. .6452	12. .0645
13. .5421	14. .0758	15. .0075
16. .0003	17. .1009	18. .0304
19. .2022	20. 20.22	21. 2.022
22. .0202	23. .0418	24. 785.6

In Exercises 25–28, how many digits precede the decimal point of the number whose logarithm has the given characteristic?

25. 0	26. 3	27. 1	28. 2

In Exercises 29–60, find the logarithms.

29. log 3.24	30. log 78.2
31. log 495	32. log 362
33. log 2570	34. log 453
35. log 89	36. log 87.9
37. log 3.54	38. log 58.4
39. log 972	40. log 9.21
41. log 0.00672	42. log 0.864
43. log 69.5	44. log 4.23 × 10⁵
45. log 6.43 × 10⁸	46. log 0.06390
47. log 19,800	48. log 0.002540
49. log 7720	50. log 0.3170
51. log 566,000	52. log 0.0006330
53. log 8,000,000	54. log 5.96 × 10⁻³
55. log 0.0001	56. log 2.06000
57. log 0.00000192	58. log 0.000071
59. log 0.000574	60. log 9,830,000

In Exercises 61–80, find the antilogarithm of the logarithms:

61. 2.5198	62. 1.5922
63. 0.2810 − 1	64. 4.6532
65. 2.9053	66. 0.8261 − 2
67. 0.8943 − 3	68. −3.0570
69. 5.5315	70. 0.6021
71. 3.4378	72. 0.5378
73. −0.3526	74. 2.8976
75. 1.9350	76. 1.4048
77. 8.4871 − 10	78. 3.6990
79. 4.9745	80. −2.4976

In Exercises 81–100, use interpolation to find the logarithms:

81. log 7.564	82. log 107.9
83. log 1,487,000	84. log 0.2372
85. log 0.09355	86. log 0.1056
87. log 149.6	88. log 389.4
89. log 64,010	90. log 3.142
91. log 3.572 × 10¹²	92. log 0.007948
93. log(5.321 × 10²)	
94. log(6.829 × 10⁻⁵)	
95. log(8.476 × 10⁻⁸)	
96. log(8.561 × 10⁻²)	
97. log(9.111 × 10³)	
98. log(9.401 × 10⁻³)	
99. log(2.147 × 10)	
100. log(1.143 × 10²)	

Use interpolation to find the antilogarithm of the logarithms in Exercises 101–112.

101. 0.8165	102. 5.8551
103. 0 4774	104. −2 + 0.6086
105. −4 + 0.3040	106. 0.6994 − 4
107. 0.5983	108. 1.0150
109. 4.3002	110. 0.8424 − 2
111. 6.6004 − 10	112. 9.7997 − 10

12.5 NATURAL LOGARITHMS AND CHANGE OF BASES

Natural Logarithms (Logarithms Base *e*)

While most computations use logarithms to the base 10, there are many formulas in biology, electricity, physics, chemistry, and other sciences that use logarithm base *e*, where *e* is the irrational number approximated by 2.71828. Logarithms to the base *e* are called natural logarithms or Napierian logarithms and are written ln *x*. Thus,

$$\ln x = \log_e x$$

Recall that common logarithms (log) are based upon the fact that every positive number can be expressed as a power of 10. Similarly, it is true that every positive number may be expressed as a power of e. For example, $7.16 = e^{1.9685}$ and $10 = e^{2.3026}$. Therefore, we have

$$\ln 7.16 = 1.9685 \quad \text{and} \quad \ln 10 = 2.3026$$

Table 3 in Appendix I lists the values of ln N, for $1 \le N < 10$. Figure 12.2 contains a partial table showing *some* values of ln N.

2.5	0.9163	0.9203	0.9243	0.9282	0.9322	0.9361	0.9400	0.9439	0.9478	0.9517
2.6	0.9555	0.9594	0.9632	0.9670	0.9708	0.9746	0.9783	0.9821	0.9858	0.9895
2.7	0.9933	0.9969	1.0006	1.0043	1.0080	1.0116	1.0152	1.0188	1.0225	1.0260
2.8	1.0296	1.0332	1.0367	1.0403	1.0438	1.0473	1.0508	1.0543	1.0578	1.0613
2.9	1.0647	1.0682	1.0716	1.0750	1.0784	1.0818	1.0852	1.0886	1.0919	1.0953
3.0	1.0986	1.1019	1.1053	1.1086	1.1119	1.1151	1.1184	1.1217	1.1249	1.1282
3.1	1.1314	1.1346	1.1378	1.1410	1.1442	1.1474	1.1506	1.1537	1.1569	1.1600
3.2	1.1632	1.1663	1.1694	1.1725	1.1756	1.1787	1.1817	1.1848	1.1878	1.1909
3.3	1.1939	1.1969	1.2000	1.2030	1.2060	1.2090	1.2119	1.2149	1.2179	1.2208
3.4	1.2238	1.2267	1.2296	1.2326	1.2355	1.2384	1.2413	1.2442	1.2470	1.2499
3.5	1.2528	1.2556	1.2585	1.2613	1.2641	1.2669	1.2698	1.2726	1.2754	1.2782
3.6	1.2809	1.2837	1.2865	1.2892	1.2920	1.2947	1.2975	1.3002	1.3029	1.3056
3.7	1.3083	1.3110	1.3137	1.3164	1.3191	1.3218	1.3244	1.3271	1.3297	1.3324
3.8	1.3350	1.3376	1.3403	1.3429	1.3455	1.3481	1.3507	1.3533	1.3558	1.3584
3.9	1.3610	1.3635	1.3661	1.3686	1.3712	1.3737	1.3762	1.3788	1.3813	1.3838

Figure 12.2

EXAMPLE 1 Find ln 3.8.

Solution Since $3.8 = 3.80$, we look at row 3.8 and column 0 to find that ln 3.8 = 1.3350. ■

As mentioned above, Table 3 gives ln N for $1 \le N < 10$. When computing ln N for $N \ge 10$ or $0 < N < 1$, we must be careful because the natural logarithms of integral powers of 10 are *not* integers. Once again, we use scientific notation to find the natural logarithm for any N.

To find ln N, write N in scientific notation: $N = A \times 10^n$, where $1 \le A < 10$ and n is an integer. Then

$$\ln N = \ln(A \times 10^n) = \ln A + n \ln 10$$

Since $\ln 10 = 2.3026$, we have

$$\ln N = \ln A + n(2.3026)$$

EXAMPLE 2   Use the partial table in Figure 12.2 to find:

(a) $\ln 3.26 = 1.1817$

(b) $\ln 32.6 = \ln(3.26 \times 10) = \ln 3.26 + \ln 10 = 1.1817 + 2.3026$

   $= 3.4843$

(c) $\ln 3260 = \ln(3.26 \times 10^3) = \ln 3.26 + 3 \ln 10$

   $= 1.1817 + 3(2.3026)$

   $= 1.1817 + 6.9078 = 8.0895$

(d) $\ln(.0326) = \ln(3.26 \times 10^{-2}) = \ln 3.26 - 2 \ln 10$

   $= 1.1817 - 2(2.3026) = 1.1817 - 4.6052$

   $= -3.4235$

We now give a method showing how to find the anti-ln of a number.

Theorem 1

To find N, given $\ln N$, we do the following:
1. If $0 \le \ln N < 2.3026$, then we merely look in Table 3.
2. If $\ln N$ does not lie in the above interval, then we first express the given value $\ln N$ in the form $b + n(2.3026)$, where $0 < b < 2.3026$ and n is some integer. Thus, N is 10^n times the anti-ln of b.

EXAMPLE 3   Find N if $\ln N = 2.2576$.

Solution   Note that $0 \le \ln N < 2.3026$; that is, $0 \le 2.2576 < 2.3026$. Using Table 3, we find that $N = 9.56$.   ■

EXAMPLE 4   Find N if $\ln N = 4.0092$.

Solution   We write 4.0092 in the form $b + n(2.3026)$. Thus, $4.0092 = 1.7066 + 1(2.3026)$, and the anti-ln of 1.7066 is 5.51. Hence, $N = 5.51 \times 10^1 = 55.1$.   ■

EXAMPLE 5   Find N if $\ln N = -2.4024$.

Solution $\ln N = -2.4024 = 2.2028 - 2(2.3026)$, and the anti-ln of 2.2028 is 9.05. Thus, $N = 9.05 \times 10^{-2} = .0905$.   ■

Change of Bases

We now state a formula that enables us to go from $\log_a N$ to $\log_b N$. Recall that for $a > 0$, $a \neq 1$, and $N > 0$,

$$N = a^{\log_a N}$$

However, we know that if $x = y$, where $x > 0$ and $y > 0$, then $\log_b x = \log_a y$. Applying this principle, we obtain

$$\log_b N = \log_b a^{\log_a N}$$

and, by Property 3, we get

$$\log_b N = \log_a N \cdot \log_b a$$

or

$$\log_a N = \frac{\log_b N}{\log_b a}$$

We can now state a theorem relating logarithms with different bases.

Theorem 2

Let a and b be positive real numbers and $a \neq 1$ and $b \neq 1$. Then

$$\log_a N = \frac{\log_b N}{\log_b a}$$

Given $\ln 10 = 2.3026$ and $\log e = \log 2.718 = 0.4343$, the relations between common logarithms and natural logarithms can be stated in the following formulas, where N is any positive number.

$$\ln N = (2.3026) \log N$$
$$\log N = (0.4343) \ln N$$

EXAMPLE 6 $\ln 225 = (2.3026) \log 225$

$$= (2.3026)(2.3522)$$
$$= 5.416$$

EXAMPLE 7 Find $\log_3 8.2$ using common logarithms.

Solution $$\log_3 8.2 = \frac{\log 8.2}{\log 3} = \frac{0.9138}{0.4771} \approx 1.9153 \quad ■$$

EXAMPLE 8 Determine $\log_a b$ in terms of $\log_b$.

Solution $$\log_a b = \frac{\log_b b}{\log_b a} = \frac{1}{\log_b a}, \text{ provided } \log_b a \neq 0. \quad ■$$

EXERCISES 12.5

In Exercises 1–12, find the natural logarithms.

1. $\ln 5$
2. $\ln 9.1$
3. $\ln 263$
4. $\ln 0.302$
5. $\ln 41.3$
6. $\ln 0.0631$
7. $\ln \dfrac{1}{3}$
8. $\ln \sqrt{6}$
9. $\ln \sqrt{13}$
10. $\ln \pi$
11. $\ln 0.216$
12. $\ln 0.0117$

In Exercises 13–21, find the antilogarithms of the natural logarithms:

13. $\ln x = 0.6931$
14. $\ln x = 3.2958$
15. $\ln x = 0.9555$
16. $\ln x = 8.9783 - 10$
17. $\ln x = 4.5747$
18. $\ln x = 9.9695 - 10$
19. $\ln x = -0.0619$
20. $\ln x = -1.0498$
21. $\ln x = -3.1129 + 5$

In Exercises 22–30, use the change of bases formula and common logarithms to find the logarithms.

22. $\log_2 5$
23. $\log_{0.3} 0.912$

24. $\log_5 10$
25. $\log_5 e$
26. $\log_7 10$
27. $\log_8 7$
28. $\log_{6.7} 4.27$
29. $\log_{0.048} 72$
30. $\log_\pi e$

Use the properties of logarithms to prove the identities in Exercises 31–35.

31. $\ln e^2 + \ln\left(\dfrac{1}{e}\right) + \ln(e^0) = 1$

32. $\ln e + \ln 1 + \ln\sqrt{e} = \dfrac{3}{2}$

33. $\ln 6 + \ln\left(\dfrac{2}{3}\right) + \ln\left(\dfrac{1}{4}\right) = 0$

34. $\ln 2 + \ln\left(\dfrac{3}{4}\right) - \ln\left(\dfrac{1}{2}\right) = \ln 3$

35. $\ln(2 + \sqrt{3}) + \ln(2 - \sqrt{3}) = 0$

Use the change of bases formula to solve the logarithmic equations in Exercises 36–40.

36. $\log_2 x + \log_3 x = 1$
*37. $\log_2 x + \log_x 2 = 2$
*38. $\log_5 x + \log_x 5 = 2.5$
*39. $\log_{16} x + \log_4 x + \log_2 x = 7$
*40. $\log_{a^2} x + \log_{x^2} a = 1$

12.6 APPLICATIONS OF LOGARITHMS

We begin this section with examples illustrating logarithmic computations. We then show how logarithms can be used to help solve many problems in such diverse areas as biology, business, physics, and chemistry. Despite the present trend toward high-speed calculators and computers, the techniques used in the following examples give more insight into the properties of logarithms.

Computations Using Logarithms

EXAMPLE 1 Compute $\dfrac{(0.0416) \times \sqrt[3]{0.00157}}{(0.315)^2 \times 5.12}$ by logarithms.

Solution Let $N = \dfrac{(0.0416) \times \sqrt[3]{0.00157}}{(0.315)^2 \times 5.12}$.

Then

$$\log N = \log \text{ numerator} - \log \text{ denominator}$$
$$= \log[(0.0416) \times \sqrt[3]{0.00157}] - \log[(0.315)^2 \times 5.12]$$
$$= \log(0.0416) + \frac{1}{3}\log(0.00157) - 2\log(0.315) - \log(5.12)$$

Now

$$\log(0.0416) = 0.6191 - 2$$

$$\log(0.00157) = 0.1959 - 3, \text{ and } \frac{1}{3}\log(0.00157) = 0.0653 - 1$$

$$\log(0.315) = 0.4983 - 1, \text{ and } 2\log(0.315) = 0.9966 - 2$$

$$\log(5.12) = 0.7093$$

Thus,

$$\log N = (0.6191 - 2) + (0.0653 - 1) - (0.9966 - 2) - (0.7093)$$
$$= (0.6191) - 2 + (0.0653) - 1 - (0.9966) + 2 - (0.7093)$$
$$= (-1.0215) - 1$$
$$= -2.0215$$

Note that -2.0215 has a negative fractional decimal part ($-2.0215 = -2 - 0.0215$), and thus cannot represent a mantissa. So we add and subtract 3 and obtain a positive fractional decimal that can represent a mantissa. Hence,

$$\log N = \underbrace{-2.0215 + 3} - 3$$
$$= 0.9785 - 3$$

We find N by computing the anti-log (0.9785):

$$N = 9.517 \times 10^{-3} = .009517 \quad \blacksquare$$

Exponential Equations

An equation in which the unknown occurs in an exponent is called an *exponential equation*. While some exponential equations can be solved by inspection, in general, such equations are solved by taking the logarithm and applying the properties of logarithms.

EXAMPLE 2   Solve $2^{x-3} = 64$.

Solution   Since 64 is a power of 2, we write it in the form 2^6. Thus, our exponential equation becomes

$$2^{x-3} = 2^6$$

This equation is satisfied when

$$x - 3 = 6$$

that is, $x = 9$. ■

EXAMPLE 3   Solve $3^{x+1} = 4^{x-1}$.

Solution   To solve this exponential equation we use the fact that:

If $M = N$ and $M > 0$, $N > 0$, then $\log_b M = \log_b N$.

Thus, if

$$3^{x+1} = 4^{x-1}$$

then

$$\log 3^{x+1} = \log 4^{x-1}$$

$$(x + 1) \log 3 = (x - 1) \log 4$$

$$x \log 3 + \log 3 = x \log 4 - \log 4$$

$$x \log 3 - x \log 4 = -\log 3 - \log 4$$

$$x(\log 3 - \log 4) = -\log 3 - \log 4$$

$$x = \frac{-\log 3 - \log 4}{\log 3 - \log 4}$$

$$= \frac{-0.4771 - 0.6021}{0.4771 - 0.6021}$$

$$= \frac{-1.0792}{-0.1250}$$

$$= 8.6336 \quad ■$$

Exponential Growth and Decay

Many laws of growth and decay that occur in biology and chemistry can be described by an exponential equation.

When a quantity grows exponentially from an initial amount a, the amount y resulting after a period of time t is

$$y = ae^{kt}$$

where k is the growth rate per unit time and e is the base of the natural logarithm.

EXAMPLE 4 The size of a population of bacteria at time t in hours is given by $y = ae^{.5t}$. If the initial population is 10^3, what is the population after 6 hours?

Solution We have $a = 10^3$ and $t = 6$. Thus, the population of the bacteria will be

$$y = 10^3 e^{.5(6)} = 10^3 e^3 = 20,086 \quad \blacksquare$$

EXAMPLE 5 How long, in years, will it take for the amount of a substance to triple when it grows exponentially according to the equation $y = ae^{0.15t}$?

Solution We are looking for the time t when $y = 3a$, (three times the initial amount). Setting $y = 3a$, we get

$$3a = ae^{0.15t}$$
$$3 = e^{0.15t}$$

Taking the natural logarithm of both sides, we obtain

$$\ln 3 = \ln e^{0.15t}$$
$$\ln 3 = 0.15t$$
$$7.32 = \frac{\ln 3}{0.15} = t$$

Thus,
$$t = 7.32 \text{ years.} \quad \blacksquare$$

The form for exponential decay is the same as that for exponential growth except that the exponent is *negative*:

$$y = ae^{-kt}$$

EXAMPLE 6   The half-life of a substance is the time it takes for it to decay exponentially to half its original amount. Given that a radioactive material disintegrates according to the exponential equation

$$y = ae^{-kt}$$

where a is the original amount of active material. Show that the period of half-life is equal to $0.6932/k$.

Solution   We want to find t when $y = \frac{1}{2}a$. Thus, we solve the equation

$$\frac{1}{2}a = ae^{-kt}$$

$$\frac{1}{2} = e^{-kt}$$

Taking the natural logarithm of both sides gives

$$\ln\left(\frac{1}{2}\right) = \ln e^{-kt}$$

$$-0.6931 = -kt$$

$$\frac{0.6931}{k} = t \quad \blacksquare$$

Exponential Equations in Business

Suppose we invest a certain amount of money, P, which is compounded annually at a rate of r percent. Also, suppose that we leave the original principal P and the interest $P \cdot r$ in the account at the end of the year. The amount left on deposit for the second year is

$$P + P \cdot r = P(1 + r)$$

If we continue to leave the interest, then at the end of two years the new principal is

$$P(1 + r) + P(1 + r) \cdot r = P(1 + r)(1 + r) = P(1 + r)^2$$

If the money is left for t years, then the original amount has been compounded to

$$\boxed{A = P(1 + r)^t}$$

EXAMPLE 7   In how many years will $250 amount to $1000 at 4% compounded semiannually?

Solution Let $x =$ the number of years it takes for \$250 to compound to \$1000. In x years, there are $2x$ half-year intervals in each of which the rate is 2%. Thus, using the formula $A = P(1 + r)^t$, with $A = \$1000$, $P = \$250$, $r = .02$, and $t = 2x$, we obtain

$$1000 = 250(1 + .02)^{2x}$$
$$1000 = 250(1.02)^{2x}$$
$$4 = (1.02)^{2x}$$

Taking logarithms of both sides gives

$$\log 4 = 2x \log 1.02$$

and

$$x = \frac{\log 4}{2 \log 1.02} = \frac{.6021}{.0172} = 35$$

Hence, it will take approximately 35 years. ◼

Logarithms in Chemistry and Physics

The pH value of a solution is defined as the negative of the logarithm (base 10) of the hydrogen ion concentration. If $[H^+]$ is the hydrogen ion concentration, then the pH of the solution is

$$pH = -\log[H^+]$$

EXAMPLE 8 If the hydrogen ion concentration of a solution is 3.86×10^{-10}, then

$$pH = -\log[3.86 \times 10^{-10}] = -[-10 + 0.5866] = 9.4134$$

The power used to produce a sound M is called the power level P_M of the sound and is measured in a power unit, usually watts. It is difficult to define a unit for the loudness of a sound. However, we compare the loudness of two sounds and define a unit for the difference in loudness between sounds. This unit is called a **decibel**.

Let M be the initial sound produced by a source, and N be a new sound produced by the same source after some change. If P_M and P_N are the power levels of two sounds M and N, then the difference in loudness between the sounds M and N is measured in decibels (db) by the expression

$$\text{decibel loss or gain} = 10 \log\left(\frac{P_N}{P_M}\right)$$

If the ratio $0 < \dfrac{P_N}{P_M} < 1$, then $\log\left(\dfrac{P_N}{P_M}\right) < 0$ and we have a decibel loss. This occurs when the power has been diminished. In the case of a gain, the later power is greater than the initial power.

EXAMPLE 9 A certain amplifier gives a power output of 40 watts for an input of 100 milliwatts. Determine the decibels gained or lost.

Solution We have $P_M = 100/1000$ watts $= 0.1$ watts and $P_N = 40$ watts. Since the output (P_N) is greater than the input (P_M), there is a decibel gain. Thus,

$$\text{db gain} = 10 \log \frac{40}{0.1} = 10 \log 400 = 26.02 \text{ db} \quad \blacksquare$$

EXERCISES 12.6

Evaluate the expressions in Exercises 1–5 using logarithms.

1. $\sqrt{\dfrac{3500}{(1.06)^5}}$

2. $\sqrt{\dfrac{0.434 \times 96^4}{64 \times 1500}}$

3. $\left[\dfrac{31.4 \times 5.2}{7.8 \times 0.091}\right]^{4/3}$

4. $\left(\dfrac{4400}{69.37}\right)^{2/5}$

5. $\dfrac{\sqrt[5]{0.05287}}{\sqrt[3]{0.374} \times \sqrt[9]{0.07836}}$

In Exercises 6–20, solve the exponential equations:

6. $9^{2x} = 3 \cdot 27^x$

7. $4 \cdot 16^x = 64^{x-1}$

8. $243(\sqrt{3})^x = 27^x$

9. $(2^x)^x = (0.25)8^x$

10. $(5^x)^x = 25^{x+1}$

11. $2^x = 3$

12. $2^x = 30$

13. $10^x = 7$

14. $(0.1)^x = \frac{1}{2}$

15. $100^x = 102$

16. $e^x = 5$

17. $e^{-x} = 0.302$

*18. $a^{(x+1)(x-2)} = 1$

*19. $7^{x2-4x+6} = 343$

*20. $e^{3x+5} = 2e^{x+1}$ (Hint: Divide both sides by e^{x+1}.)

21. Solve $y = ae^{bx}$ for x.

22. Solve $x = ae^{-by}$ for y.

23. Solve $l = ar^{n-1}$ for n.

*24. Solve $s = a\dfrac{(1 - r^n)}{(1 - r)}$ for n.

*25. Solve $A = P(1 + r)^n$ for n.

26. One thousand units of a substance grows exponentially at the rate of 3% per year.

How long will it take for the quantity to reach 1160 units?

27. How long will it take for a quantity that grows exponentially at an annual rate of 15% to triple?

28. Determine the rate of growth necessary to increase a quantity exponentially from 10^4 units to 18,000 units in 90 days.

29. Find the half-life of a material that decays exponentially at the rate of 1.25% per year.

30. If 3% of a certain radioactive material disintegrates in 1 hr, find the period of half-life.

*31. If people begin life as a single biological cell and by cell division become fully grown with about ten billion cells, how many generations of cell division are required? Neglect the effect of cell deaths and assume that all cells require the same time from one division to the next.

32. In how many years will $300 amount to $500 at 4% compounded annually?

33. In how many years will $350 amount to $1000 at 6% compounded semiannually?

34. In how many years will $250 double itself at 4% compounded semiannually?

35. In how many years will $100 double itself at 4% compounded quarterly?

36. In how many years will P dollars double itself at 3% compounded semiannually?

37. At what interest (compounded annually) must a sum of money, say \$1, be invested if it is to be doubled in 15 years?
38. At what interest will money triple itself in 30 years?
39. Given that the hydrogen ion concentration of a solution is 1.65×10^{-12}, find its pH.
40. Given a solution with hydrogen ion concentration of 6.19×10^{-9}, find the pH of the solution.
41. If the pH value of a solution is 8.5, determine the hydrogen ion concentration.

42. Determine the hydrogen ion concentration of a solution with a pH value of 3.2.
43. What is the decibel loss in a circuit if the output is 0.07 watt and the input is 0.10 watt?
44. What is the decibel gain in an amplifier with 0.005-watt input and 0.46-watt output?
45. How much must the output of an 8-watt amplifier be increased if the output is to be raised 3.4 db?

Chapter 12 SUMMARY

EXPONENTS

If $a > 0$, $b > 0$, and x, y are real numbers,

1. $a^x \cdot a^y = a^{x+y}$ 2. $(a^x)^y = a^{xy}$

3. $\dfrac{a^x}{a^y} = a^{x-y}$ 4. $(ab)^x = a^x b^x$

5. $\left(\dfrac{a}{b}\right)^x = \dfrac{a^x}{b^x}, b \neq 0$

LOGARITHM

The logarithm, base b, of a number N is the exponent to which the base b must be raised to give N.

logarithmic form   *exponential form*
$\log_b N = x$ if and only if $N = b^x$

PROPERTIES OF LOGARITHMS

Let b, M, and N be any positive real numbers and p any real number. Then

1. $\log_b (M \cdot N) = \log_b M + \log_b N$

2. $\log_b \left(\dfrac{M}{N}\right) = \log_b M - \log_b N$

3. $\log_b M^p = p \log_b M$
4. If $M = N$, then $\log_b M = \log_b N$.
5. If $\log_b M = \log_b N$, then $M = N$.

LOGARITHMIC EQUATION

A *logarithmic equation* is an equation in which the unknown appears in a logarithm.

LOGARITHM BASE 10: COMMON LOGARITHM (log)

If N is any positive real number written in scientific notation, $N = A \times 10^n$, where $1 \leq A < 10$ and n is an integer, then

$$\log N = n + \log A$$
$$\uparrow \qquad \uparrow$$
$$\text{characteristic} \quad \text{mantissa}$$

ANTILOGARITHM

If $\log_b N = x$, then $N = \text{antilog}(x)$

LOGARITHM BASE e: NATURAL LOGARITHM (ln)

If $N = A \times 10^n$, then

$$\ln N = \ln(A \times 10^n) = \ln A + n \ln 10$$
$$= \ln A + n(2.3026)$$

CHANGE OF BASES

$$\log_a N = \frac{\log_b N}{\log_a b}$$

EXPONENTIAL EQUATION

An exponential equation is an equation in which the unknown appears in one or more exponents.

Chapter 12 EXERCISES

In Exercises 1–10, express the quantities in logarithmic form.

1. $2^2 = 4$
2. $3^3 = 27$
3. $10^{-2} = 0.01$
4. $8^2 = 64$
5. $10^{-3} = 0.001$
6. $11^2 = 121$
7. $15^2 = 225$
8. $10^{-1} = \dfrac{1}{10}$
9. $2^{-3} = \dfrac{1}{8}$
10. $3^{-4} = \dfrac{1}{81}$

In Exercises 11–20, express the quantities in exponential form.

11. $\log_2 32 = 5$
12. $\log_5 125 = 3$
13. $\log_{10} 1000 = 3$
14. $\log_2\left(\dfrac{1}{4}\right) = -2$
15. $\log_{11}\left(\dfrac{1}{121}\right) = -2$
16. $\log_{15}\left(\dfrac{1}{15}\right) = -1$
17. $\log_{10} \sqrt{10} = \dfrac{1}{2}$
18. $\log_{1.8}(1.8) = 1$
19. $\log_2\left(\dfrac{1}{32}\right) = -5$
20. $\log_{16} \sqrt{4} = \dfrac{1}{4}$

Find the values of the logarithms in Exercises 21–30.

21. $\log_5\left(\dfrac{1}{625}\right)$
22. $\log_9 81$
23. $\log_2\left(\dfrac{1}{8}\right)$
24. $\log_3 27$
25. $\log_3\left(\dfrac{1}{9}\right)$
26. $\log_{14} 1$
27. $\log_8 2$
28. $\log_8 4$
29. $\log_6 216$
30. $\log_4\left(\dfrac{1}{32}\right)$

Determine the value of x in Exercises 31–40.

31. $\log_2 x = 3$
32. $\log_5 x = -1$
33. $\log_8 x = 2$
34. $\log_5 x = 3$
35. $\log_{1.5} x = 2$
36. $\log_{10} x = 0$
37. $\log_2 x = 5$
38. $\log_2 x = \dfrac{1}{2}$
39. $\log_{15} x = 1$
40. $\log_{20} x = 0$

Determine the value of b in Exercises 41–50.

41. $\log_b 9 = 2$
42. $\log_b 8 = 3$
43. $\log_b 4 = \dfrac{2}{3}$
44. $\log_b 17 = 1$
45. $\log_b 1 = 0$
46. $\log_b\left(\dfrac{1}{4}\right) = -2$
47. $\log_b 9 = -\dfrac{2}{3}$
48. $\log_b 0.01 = 2$
49. $\log_b\left(\dfrac{1}{16}\right) = -\dfrac{4}{3}$
50. $\log_b 6 = -\dfrac{1}{2}$

In Exercises 51–57, express each quantity as a single logarithm.

51. $\log\left(\dfrac{21}{8}\right) + \log \sqrt{24} + \log \sqrt{\dfrac{8}{27}}$
52. $\log\left(\dfrac{27}{28}\right) + \log \sqrt{98} + \log \dfrac{1}{9}\sqrt{8}$
53. $\ln x + \ln y$
54. $5 \log_2 a + 3 \log_2 b - 2 \log_2 c$
55. $\dfrac{1}{2} \log_a x + \dfrac{1}{3} \log_a y - \dfrac{1}{4} \log_a z$
56. $\log_8 a + \log_8 b - \dfrac{1}{2} \log_8 c$
57. $\dfrac{1}{2} \log(x + y) + \dfrac{1}{2} \log(x - y)$

In Exercises 58–60, express each quantity as the sum and/or difference of logarithms.

58. $\log\left[\sqrt[5]{\dfrac{x^2 y^{-2/3}}{(x + y)}} \right]$
59. $\log \sqrt{s(s - a)(s - b)(s - c)}$
60. $\log\left[\dfrac{P \cdot R^m}{\sqrt[n]{Q}} \right]$

Solve the logarithmic equations for x in Exercises 61–65.

61. $\log_6(x + 3) + \log_6(x + 4) = 1$
62. $\log_2 x + \log_2(x + 6) = 4$
63. $2 + \log_2 x = \log_2(x + 5)$
64. $\log A = \dfrac{1}{n} \log x$
65. $n \log b = \log x - \log a$

Find the common logarithm of each of the numbers in Exercises 66–70.

66. 7329   **67.** .1083   **68.** 5.342

69. 61.24   **70.** 326.5

Find the antilogarithm of each of the numbers in Exercises 71–75.

71. 8.9194   **72.** 7.7978

73. $-3 + 0.2306$ **74.** 1.8817

75. -2.0014

Find the natural logarithm of each of the numbers in Exercises 76–80.

76. 6.14   **77.** 9.08   **78.** 417

79. 0.216   **80.** 0.0117

Use change of bases formula to find the logarithms in Exercises 81–85.

81. $\log_2 7$ **82.** $\log_3 10$

83. $\log_{15} 11$ **84.** $\log_5 9$ **85.** $\log_{17} 52$

In Exercises 86–90, solve the exponential equations.

86. $2^x = 7$ **87.** $6^x = (4.7)^{2x+3}$

88. $(12.4)^{4-3x} = (1.03)^{6x+2}$

89. $125^x = 48(5^x)$ **90.** $(28)^{x^2-4x} = 15$

91. How long will it take for a quantity to increase fourfold when growing exponentially at an annual rate of 13%?

92. The number of bacteria in a culture at time t was given by $y = ae^{5t}$. When will the colony double its initial size a?

93. Radium decomposes according to the formula $y = ae^{-0.038t}$, where a is the initial amount and t is measured in centuries. Determine the half-life of radium.

94. One healing law for a skin wound is $A = A_0e^{-n/10}$, where A (square centimeters) is the unhealed area after n days and A_0 (square centimeters) is the area of the original wound. Find the number of days required to cut the wound down to one-half the area.

95. The pH value of a solution is 3.2. Find the hydrogen ion concentration.

96. A speaker is supplied by 2.5 watts producing a certain sound volume. If the power is increased to 3.5 watts, what is the gain in decibels?

97. How many annual deposits of $1000 each must be made in order to accumulate a fund of $19,600 if interest is at 5% and compounded annually?

*98. If P dollars is invested at r percent annually and is compounded continuously, then at the end of t years it will be worth $Pe^{rt/100}$ dollars. If $100,000 is invested at an interest rate of 6% per year compounded continuously, how many years will it take for the original amount to double?

*99. *Newton's law of cooling:* When á body, initially at a temperature t_0, is placed in cooler surroundings, its temperature will drop exponentially according to the formula $t = t_0e^{-kT}$, where k is a constant and t is the temperature at time T. If $k = 0.81$, how long will it take for a metal at a temperature of 1500°C to cool to 100°C (time is measured in hours)?

*100. The current i in a certain electrical circuit is given by the formula

$$i = \frac{E}{R}(1 - e^{-Rt/L}).$$

Use natural logarithms (ln) to solve for t.

13 Relations and Functions

In this chapter we introduce relations and functions, two of the most important concepts in mathematics. We then discuss special types of functions: polynomial functions, exponential functions, and logarithmic functions.

In this chapter we also study variations. Many of the laws of physics, chemistry, engineering, and other branches of science are stated in terms of variation, an equation that relates one variable to one or more variables using multiplication, division, or both. In this chapter, we classify variations into four types: direct, inverse, joint, and combined.

13.1   RELATIONS AND FUNCTIONS

Often it occurs that certain members of a first set A are related in some manner to certain members of a second set B. For example, let A be the set of natural numbers, $A = \{0, 1, 2, 3, \ldots\}$, and let B be the set of nonnegative even integers, $B = \{0, 2, 4, 6, \ldots\}$. If $x \in A$ and $y \in B$ and if $y = 2x$, we can associate with each x the corresponding y. This correspondence may be expressed as an infinite set of ordered pairs:

$$R = \{(0, 0),(1, 2),(2, 4),(3, 6,), \ldots\}$$

Such a set of ordered pairs is called a relation.

> **DEFINITION**
>
> A relation is a set of ordered pairs. The domain of a relation is the set of all first members of the ordered pairs and the range of a relation is the set of all second members of the ordered pairs.

Hence, for the relation given above, the domain is $A = \{0, 1, 2, 3, \ldots\}$ and the range is $B = \{0, 2, 4, 6, \ldots\}$.

Since we will consider only those relations formed from real numbers, we can use the Cartesian coordinate system to represent relations as points in the plane. The graph of a relation is the graph of all the ordered pairs of the relation.

EXAMPLE 1   The set of ordered pairs $\{(1, 2), (5, 5), (1, 3), (-3, 5), (0, 4), (-1, -1)\}$ defines a relation. The set $\{1, 5, -3, 0, -1\}$ is the domain and the set $\{2, 5, 3, 4, -1\}$ is the range. The graph of the relation is shown in Figure 13.1.

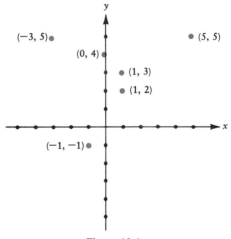

Figure 13.1

EXAMPLE 2   Graph the relation $\{(x, y)\,|\,y = -3x,\ x \text{ a real number}\}$.

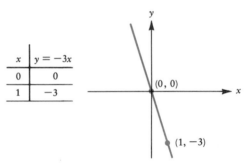

x	$y = -3x$
0	0
1	-3

Figure 13.2

Solution The graph of the equation $y = -3x$ is a straight line. Thus, we need only find two points that lie on the line (see Figure 13.2). ■

EXAMPLE 3 Graph the relation $\{(x, y)| y < x + 1, x$ a real number$\}$.

Solution The graph of $y < x + 1$ consists of all points (x, y) that lie below the graph of the line $y = x + 1$ (see Figure 13.3). ■

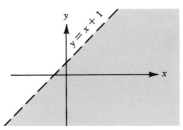

Figure 13.3

Functions

We are now interested in special types of relations called functions.

> **DEFINITION**
>
> A function is a relation in which no two distinct ordered pairs have the same first element and different second elements. The set of all first elements of the ordered pairs is called the domain of the function. The set of all second elements is called the range of the function.

EXAMPLE 4 Let $M = \{(1, 2), (-1, 3), (0, 6), (-7, 5)\}$ and $N = \{(6, 3), (-2, 1), (5, -4), (-2, 8)\}$. Then M is a function, while N is *not* a function. Note that the two distinct ordered pairs $(-2, 1)$ and $(-2, 8)$ in N have the same first element but different second elements.

EXAMPLE 5 Consider the equation $y = x^2$. This represents a function because for any value

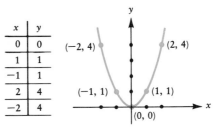

x	y
0	0
1	1
−1	1
2	4
−2	4

Figure 13.4

chosen for x, there is one and only one corresponding value for y. Graphically, the equation $y = x^2$ is called a parabola (see Figure 13.4).

EXAMPLE 6 The equation $y^2 = x$ does *not* represent a function of x. Choosing $x = 9$, we see that $y = 3$ or $y = -3$. Thus, the ordered pairs (9, 3) and (9, -3) satisfy the equation and have the same first elements. Graphing the equation $y^2 = x$, we obtain the results in Figure 13.5.

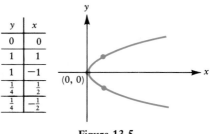

y	x
0	0
1	1
1	-1
$\frac{1}{4}$	$\frac{1}{2}$
$\frac{1}{4}$	$-\frac{1}{2}$

Figure 13.5

Graphically, how can we tell whether the graph of a relation represents a function?

Vertical Line Test

If any vertical line meets the graph of an equation in more than one point, then the equation does not determine a function.

EXAMPLE 7 The graphs in Figure 13.6 represent functions.
The graphs in Figure 13.7 do *not* represent functions.

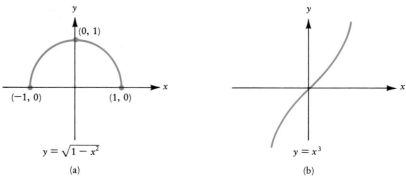

$y = \sqrt{1 - x^2}$

(a)

$y = x^3$

(b)

Figure 13.6

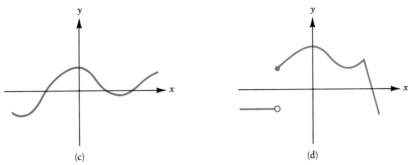

Figure 13.6 (continued)

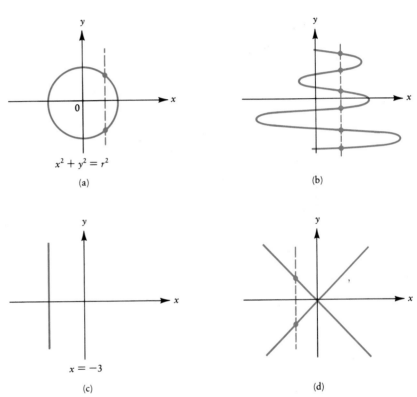

Figure 13.7

Functional Notation

Functions can be thought of as mappings. A function maps members of the domain to corresponding members of the range. In general, if (x, y) is a member of the function f, then f associates x in the domain with y in the range, and we say that f maps x to y. The functional notation used is $y = f(x)$, which is read "y equals f of x." When using this notation, x is called the independent variable and y is called the dependent variable. The graph of the function f is the set of points $(x, f(x))$ for which the function is defined.

EXAMPLE 8 Let f be a function defined by the equation $f(x) = 4x - 3$. Determine the values: $f(0)$, $f(1)$, $f(-1)$, $f(a)$, $f(a + b)$, and $f(x + h)$.

Solution Note that f takes the value of x and multiplies it by 4 and then subtracts 3 from that product.

$$f(0) = 4(0) - 3 = -3 \qquad\qquad f(1) = 4(1) - 3 = 1$$
$$f(-1) = 4(-1) - 3 = -7 \qquad\qquad f(a) = 4a - 3$$
$$f(a + b) = 4(a + b) - 3 \qquad\qquad f(x + h) = 4(x + h) - 3 \quad \blacksquare$$

EXAMPLE 9 Let g be a function defined by the equation $g(x) = x^2 + 5$. Determine $g(0)$, $g(2)$, $g(-2)$, $g(a + b)$, and the expression $\dfrac{[g(x + h) - g(x)]}{h}$, where $h \neq 0$.

Solution The function g squares the value of x and adds 5 to the result.

$$g(0) = (0)^2 + 5 = 5$$
$$g(2) = (2)^2 + 5 = 4 + 5 = 9$$
$$g(-2) = (-2)^2 + 5 = 9$$
$$g(a + b) = (a + b)^2 + 5 = a^2 + 2ab + b^2 + 5$$
$$\frac{g(x + h) - g(x)}{h} = \frac{[(x + h)^2 + 5] - [x^2 + 5]}{h}$$
$$= \frac{x^2 + 2xh + h^2 + \cancel{5} - x^2 - \cancel{5}}{h} = \frac{2xh + h^2}{h} = \frac{\cancel{h}(2x + h)}{\cancel{h}} = 2x + h$$

$$\blacksquare$$

EXAMPLE 10 Let f be a function defined by $f(x) = 7$. Compute $f(0)$, $f(-9)$, $f(a)$, and the expression $\dfrac{f(x + h) - f(x)}{h}$, $h \neq 0$.

Solution The function f maps any value of x to the constant 7. $f(0) = 7$; $f(-9) = 7$; $f(a) = 7$; and $\dfrac{f(x + h) - f(x)}{h} = \dfrac{7 - 7}{h} = \dfrac{0}{h} = 0$. $\blacksquare$

DEFINITION

The domain, D_f, of a function f is the set of values of x for which f is defined. The range, R_f, of f is the set of values $f(x)$. (See Figure 13.8.)

EXAMPLE 11 Determine the domain of the function f defined by the equation $f(x) = -5x + 1$.

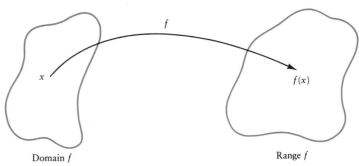

Figure 13.8

Solution Since the equation $f(x) = -5x + 1$ is defined for all real numbers, $D_f = \{x \mid x$ is a real number$\}$. ∎

EXAMPLE 12 Determine the domain of the function g defined by the equation $g(x) = \sqrt{x - 5}$.

Solution Since we can only take the square root of nonnegative numbers over the set of real numbers, the expression under the radical sign, $x - 5$, must be nonnegative; that is,

$$x - 5 \geq 0$$

$$x \geq 5$$

Hence, the $D_g = \{x \mid x$ is a real number and $x \geq 5\}$. ∎

EXAMPLE 13 Determine the domain of the function f defined by $f(x) = \dfrac{(x - 2)(x + 4)}{(x - 5)}$.

Solution The function f is defined for all real numbers except the value $x = 5$. For $x = 5$, the denominator becomes zero and f is undefined. Thus, $D_f = \{x \mid x$ is a real number and $x \neq 5\}$. ∎

EXAMPLE 14 Determine the domain of the function h defined by $h(x) = \dfrac{\sqrt{x^2 - 4}}{x^2 - 3x - 10}$.

Solution The domain of h consists of all real numbers for which the expression $x^2 - 4$ is nonnegative and the denominator is not zero. That is, $x^2 - 4 \geq 0$ and $x^2 - 3x - 10 \neq 0$. Using the methods discussed in Chapter 9, we find that $x^2 - 4 \geq 0$ if x belongs to the set $S = \{x \mid x \leq -2$ or $x \geq 2\}$. The expression $x^2 - 3x - 10$ equals zero at the values $x = 5$ and $x = -2$. These points, 5 and -2, must be eliminated from the set S. Hence,

$$D_h = \{x \mid x < -2 \text{ or } x \geq 2 \text{ and } x \neq 5\} \quad ∎$$

EXERCISES 13.1

In Exercises 1–6, determine which of the relations define a function.

1. $\{(-1, 2), (0, 5), (2, 4), (3, -1)\}$
2. $\{(2, 3), (3, 4), (4, 3), (5, 4)\}$
3. $\{(2, 3), (2, 4), (3, 5), (3, 6)\}$
4. $\{(3, 3), (4, 3), (5, 3), (6, 3)\}$
5. $\{(4, 2), (4, 3), (4, 4), (4, 5)\}$
6. $\left\{\left(\frac{1}{2}, -4\right), \left(-\frac{1}{2}, 4\right), \left(\frac{3}{2}, -4\right)\left(-\frac{3}{2}, 4\right)\right\}$

For Exercises 7–12, determine the domain and range of the relations in Exercises 1–6.
Determine which of the graphs in Exercises 13–21 represent a function.

13.

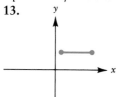

14.

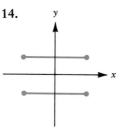

15.

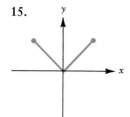

16.

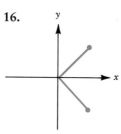

17.

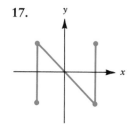

18.

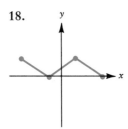

19.

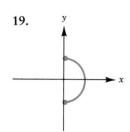

20.

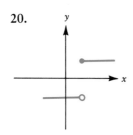

21.

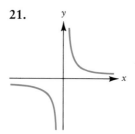

In Exercises 22–30, find the indicated values of the functions defined by the given equations.

22. $f(x) = -3x + 5$
    $f(0); f(-5); f(a); f(a + b)$
23. $g(x) = x^2 + 3x + 1$
    $g(1); g(-1); g(b); g(a - b)$
24. $h(x) = |x - 2|$
    $h(0); h(2); h(4); h(a + b)$
25. $p(x) = \sqrt{x^2 - 9}$
    $p(3); p(-3); p(a); p(x + 1)$
26. $q(x) = 3x - \dfrac{1}{x}$

$q(1); q\left(\dfrac{1}{2}\right); q(y); q(a + b)$

27. $s(x) = |x| - x$
    $s(0); s(4); s(-4); s(x + h)$
28. $v(x) = \left|\dfrac{x - 8}{x + 2}\right|$

$v(0); v(8); v(1); v(a + 8)$
29. $g(x) = x^2(x - 2)$
    $g(0); g(2); g(a); g(a + b)$
30. $f(x) = ax + b$
    $f(0); f(-2); f(a); f(x + h)$

In Exercises 31–35, compute the difference quotient $\dfrac{f(x + h) - f(x)}{h}$, $h \neq 0$, *and simplify the expression for the functions defined by the given equations.*

31. $f(x) = x$ 32. $f(x) = x^2 + 3$

33. $f(x) = 9$ 34. $f(x) = \dfrac{1}{x}$

35. $f(x) = \sqrt{x}$

In Exercises 36–41, find the domain, over the real numbers, of the functions defined by the given equations.

36. $f(x) = |x|$ 37. $g(x) = \dfrac{x}{x - 2}$

38. $p(x) = -9$

39. $h(x) = 4x + \dfrac{3}{x}$

40. $s(x) = \sqrt{x^2 - 3x - 10}$

41. $f(x) = \dfrac{\sqrt{3x - 4}}{x^3 - 9x}$

42. Express the area of a circle as a function of (a) its radius r; (b) its diameter d.

43. Express the circumference of a circle as a function of (a) its radius r; (b) its diameter d.

44. (a) Express the Fahrenheit temperature F of a body as a function of the centigrade temperature C;

(b) the centigrade temperature C as a function of Fahrenheit temperature F.

45. Express the area A and the perimeter P of an isosceles triangle as a function of its base b and its altitude h.

13.2 INVERSE RELATIONS AND FUNCTIONS

Recall that a relation is a set of ordered pairs. Consider the relation R defined by

$$R = \{(4, 6), (-1, 2), (3, 6), (-3, -2)\}$$

The domain of R is $\{4, -1, 3, -3\}$ and the range of R is $\{6, 2, -2\}$. If we interchange the first and second coordinates in each ordered pair in R, we obtain another relation, R^{-1}:

$$R^{-1} = \{(6, 4), (2, -1), (6, 3), (-2, -3)\}$$

Note. The domain of R is the range of R^{-1} and the range of R is the domain of R^{-1}. We call R^{-1} the inverse relation of R and R the inverse relation of R^{-1}.

CAUTION

The notation R^{-1} is read "R inverse" or "the inverse of R." The superscript -1 is not to be interpreted as an exponent. That is,

$$R^{-1} \neq \frac{1}{R}$$

In Figure 13.9, we plot the points R and R^{-1}. Note that each point of the inverse relation R^{-1} is the reflection (mirror image) of the corresponding point of the relation R about the line $y = x$. That is, if we plot the points of a relation R and its inverse relation R^{-1} and sketch the line $y = x$, and if we fold the paper along the line $y = x$, the points of R and R^{-1} will match. This symmetry of R and R^{-1} about the line $y = x$ makes it easy to obtain the graph of R^{-1} if we know the graph of R and vice versa.

The relation R is a function since no two distinct ordered pairs have the same first coordinate. However, R^{-1} is *not* a function because it has two distinct

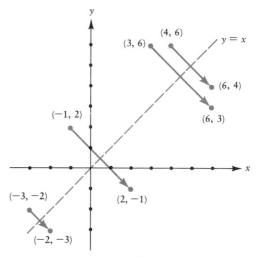

Figure 13.9

ordered pairs, (6, 4) and (6, 3), having the same first coordinate. This is not always true. The inverse of some functions may also be functions.

EXAMPLE 1   Given $F = \{(2, 1), (3, 2), (-3, 3), (0, 4)\}$, find F^{-1}, the domain of F^{-1}, the range of F^{-1}, and graph F and F^{-1}. Is F^{-1} a function? (See Figure 13.10.)

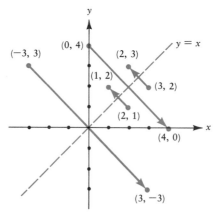

Figure 13.10

Solution   If
$$F = \{(2, 1), (3, 2), (-3, 3), (0, 4)\}$$

Then
$$F^{-1} = \{(1, 2), (2, 3), (3, -3), (4, 0)\}$$

Domain of F^{-1}: $D_{F^{-1}} = \{1, 2, 3, 4\} = R_F$

Range of F^{-1}: $R_{F^{-1}} = \{2, 3, -3, 0\} = D_F$

F^{-1} is a function.   ■

If F is a function and F^{-1} is also a function, then F^{-1} is called the inverse function of F and vice versa.

If a function is given using functional notation $y = f(x)$, then the inverse function, provided it exists, can be found by interchanging x and y in the equation and solving the resulting equation for y.

EXAMPLE 2 Find the inverse function for $y = f(x) = 4x - 5$. See Figure 13.11.

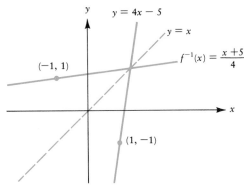

Figure 13.11

Solution

$$y = 4x - 5$$

$$x = 4y - 5 \qquad \text{[interchanged } y \text{ and } x\text{]}$$

$$x + 5 = 4y$$

$$\frac{x + 5}{4} = y \qquad \text{[solved for } y\text{]}$$

Thus, $f^{-1}(x) = \dfrac{x + 5}{4}$. ■

EXAMPLE 3 Given $f(x) = 3x + 1$, find $f^{-1}(x)$. Find $f^{-1}(0)$, $f^{-1}(1)$, and $f^{-1}[f(x)]$.

Solution Let

$$y = 3x + 1.$$

Then

$$x = 3y + 1 \qquad \text{[interchanged } y \text{ and } x\text{]}$$

$$x - 1 = 3y$$

$$\frac{x - 1}{3} = y \qquad \text{[solved for } y\text{]}$$

Thus, $f^{-1}(x) = \dfrac{x - 1}{3}$. $f^{-1}(0) = \dfrac{0 - 1}{3} = \dfrac{-1}{3}$; $f^{-1}(1) = \dfrac{1 - 1}{3} = \dfrac{0}{3} = 0$.

Since $f(x) = 3x + 1$, $f^{-1}[f(x)] = f^{-1}[3x + 1] = \dfrac{(3x + 1) - 1}{3} = x$. ■

EXERCISES 13.2

In Exercises 1–6, find the inverse for each of the relations.

1. $F = \{(1, 2), (2, 3), (3, 6), (4, 8)\}$
2. $G = \{(-1, 2), (1, 4), (3, 3)\}$
3. $H = \{(2, 2), (3, 5), (4, 10), (0, 2), (-1, 5)\}$
4. $R = \{(-2, 0), (1, 2), (4, 3)\}$
5. $S = \{(-1, 5), (0, 2), (1, 1), (2, 2), (3, 5)\}$
6. $T = \{(-4, 1), (-1, 2), (1, 1), (3, 2), (6, 1)\}$

In Exercises 7–12, indicate which are functions.

7. F or F^{-1} in Exercise 1.
8. G or G^{-1} in Exercise 2.
9. H or H^{-1} in Exercise 3.
10. R or R^{-1} in Exercise 4.
11. S or S^{-1} in Exercise 5.
12. T or T^{-1} in Exercise 6.

In Exercises 13–25, each of the equations of the form $y = f(x)$ defines a function f^{-1}. Find the equation $y = f^{-1}(x)$ that defines f^{-1}. State the domain of f^{-1}.

13. $y = 2x + 6$

14. $y = -\dfrac{1}{2}x + 5$

15. $y = 7x + 8$

16. $y = \dfrac{x - 2}{3x - 4}, x \neq \dfrac{4}{3}$

17. $y = \dfrac{2}{x - 1}, x \neq 1$

18. $y = x^3$

19. $y = \sqrt{x}, x \geq 0$

20. $y = \sqrt{x - 3}, x \geq 3$

21. $y = -\sqrt{x - 3}, x \geq 3$

22. $y = 1 + \sqrt{x + 2}, x \geq -2$

23. $y = 1 - \sqrt{x + 2}, x \geq -2$

*24. $y = \dfrac{1}{x}, x \neq 0$

*25. $y = \dfrac{1}{2}\sqrt{1 - x^2}, 0 \leq x \leq 1$

26. Show that the inverse of each of the following functions is the function itself.
   (a) $y = 5 - x$ (b) $y = x$

27. Let $y = mx + b, m \neq 0$. Find $f^{-1}(x)$.

*28. (a) Let $y = |x|, x \geq 0$. Find $f^{-1}(x)$.
   (b) Let $y = |x|, x \leq 0$. Find $f^{-1}(x)$.

*29. Given $y = x^5 - 6$, find $f^{-1}(x)$ and show that $f^{-1}[f(x)] = x$.

*30. Show that the function f defined by
$$y = \frac{mx + b}{ax - m}, x \neq \frac{m}{a},$$ is identical to its inverse, provided $m^2 + ab \neq 0$.

13.3 DIRECT VARIATION

If two variables are related by an equation of the form

$$y = kx$$

where k is a constant, we say that y varies directly as x (or y is directly proportional to x). The constant k is called the constant of variation (or constant of proportionality).

EXAMPLE 1   The formula for the circumference of a circle of radius r is $C = 2\pi r$. Thus, we see that the circumference varies directly as the radius, and that the constant of variation is 2π.

EXAMPLE 2   If y varies directly as x and $y = 35$ when $x = 14$, find y when $x = 16$.

Solution   Since y varies directly as x, we have

$$y = kx$$

We now use the condition that $y = 35$ when $x = 14$ to solve for the constant of variation k. By substitution, we have $35 = 14k$, or $k = \frac{5}{2}$. Thus, the *exact* relation between the variables x and y is

$$y = \frac{5}{2}x$$

Hence, if $x = 16$, then $y = 40$. ∎

CAUTION

Be very careful when determining the constant of variation k in different examples. The value of k depends on the kind of units used in the example.

EXAMPLE 3   (*Hooke's law*). In physics, Hooke's law states that the force applied to stretch an elastic spring varies directly as the amount of elongation. The value of k in Hooke's law is a characteristic of the spring itself and is known as the modulus of elasticity. If a force of 20.5 lb stretches a spring by 3.4 in., find the modulus of elasticity.

Solution   Let F represent the force applied and s represent the amount of elongation. Then $F = ks$ and substituting the given values into the equation, we obtain

$$20.5 \text{ lb} = k(3.4 \text{ in.})$$
$$k = \frac{20.5 \text{ lb}}{3.4 \text{ in.}} = 6.03 \frac{\text{lb}}{\text{in.}} \quad ∎$$

EXAMPLE 4   If an object falls from a high place, the distance (measured in feet) that it falls varies as the square of the time (measured in seconds). If s = the distance in feet and t = time in seconds, write an equation for this relation. Find the value of the constant of variation k if the body falls 100.625 ft in 2.5 sec.

Solution   Since the distance varies directly as the square of the time, we have

$$s = kt^2$$

If $s = 201.25$ ft and $t = 2.5$ sec, then

$$100.625 \text{ ft} = k(2.5 \text{ sec})^2$$

Solving for k,

$$k = 16.1 \text{ ft/sec}^2$$

Thus, the formula to find the distance in feet that an object will fall in t seconds is $s = 16.1t^2$. This is commonly written

$$s = \frac{1}{2}gt^2$$

where $g = 32.2$ ft/sec^2. ■

EXAMPLE 5   The area of a sphere is directly proportional to the square of the radius. If a sphere of radius 5 in. has an area of 100π sq in., express the area of a sphere as a function of its radius.

Solution   Let A represent the area of the sphere, in square inches, and r represent the radius, in inches. Since the area is directly proportional to the square of the radius, we have

$$A = kr^2$$

If $A = 100\pi$ sq in. and $r = 5$ in., then

$$100\pi \text{ sq in.} = k(5 \text{ in.})^2 = 25k \text{ sq in.}$$

Solving for k, we obtain

$$k = 4\pi$$

Thus, $$A = 4\pi r^2$$ ■

EXERCISES   13.3

1. If x varies directly as z and $x = 9$ when $z = \frac{1}{2}$, find x when $z = \frac{2}{3}$.

2. If c varies directly as d and $c = 24$ when $d = 4$, find c when $d = 7\frac{1}{2}$.

3. If d varies directly as t and $d = 6$ when $t = 180$, find d when $t = 120$.

4. If w varies directly as u and $w = 420$ when $u = 7$, find w when $u = 9$.

5. If t varies directly as s and $t = 39$ when $s = 3\frac{1}{4}$, find t when $s = 8$.

6. If y varies directly as x and $y = \frac{3}{8}$ when $x = \frac{3}{4}$, find y when $x = \frac{2}{3}$.

7. If B varies directly as l and $B = 240$ when $l = 30$, find l when $B = 96$.

8. If c varies directly as m and $c = 22$ when $m = 3\frac{1}{2}$, find c when $m = 4\frac{2}{3}$.

9. If s varies directly as t^2 and $s = 64$ when $t = 2$, find s when $t = 3$.

10. If A varies directly as r^2 and $A = 75$ when $r = 5$, find A when $r = 3$.

11. If an object falls freely from a high place, the distance that it falls can be measured by the number of seconds during which it falls. The distance varies as the square of the time. If a body falls 144 ft in 3 sec, how far will it fall in 1 sec?

12. Use the variation determined in Exercise 11 to find the time taken for a stone to drop 788.9 ft.

13. The surface area of a sphere varies as the square of the radius. The area is 16π square units when the radius is 2. Determine a formula expressing the surface area as a function of the radius.

14. The surface area of a cube varies directly as the square of the length of an edge. The sur-

face area is 24 square units when the length of the edge is 2 units. Determine the formula expressing the surface area as a function of the length of an edge.

15. The volume of a pyramid of fixed altitude varies directly as the area of the base. The volume is 35 cubic units when the rectangular base is 5 units wide and 9 units long. Determine a formula expressing the volume as a function of the area of the base.

16. The area of an equilateral triangle varies as the square of its perimeter. If the area is 1.732 sq in. when the perimeter is 6 in., find the area when the perimeter is 11 in. Find the perimeter when the area is 17 sq in.

17. The time of swing of a pendulum varies as the square root of its length. If the pendulum is 1 ft long and the time is 0.225 sec, find the time if the length is 20 in.

18. According to Ohm's law, the voltage drop V across a resistor varies directly as the current I in the resistor. If a current of 2.45 amps causes a voltage drop of 55.3 volts, find the voltage drop caused by 3.16 amps.

13.4 INVERSE VARIATION

If two variables are related by an equation of the form

$$y = \frac{k}{x}$$

where k is a constant, we say that y varies inversely as x.

EXAMPLE 1 If y varies inversely as the cube of x and $y = 7$ when $x = 2$, express y as a function of x.

Solution Since y varies inversely as x, we have

$$y = \frac{k}{x^3}$$

We now use the condition that $y = 7$ when $x = 2$ to solve for the constant of variation k:

$$7 = \frac{k}{(2)^3} \quad \text{or} \quad 7 = \frac{k}{8}$$

and we find that $k = 7 \cdot 8 = 56$. Thus, expressing y as a function of x, we obtain $y = \frac{56}{x^3}$. ■

EXAMPLE 2 (*Newton's law of gravitation*). The force of attraction between two particles of matter varies inversely as the square of the distance between them. If F denotes the force and d the distance, then

$$F = \frac{k}{d^2}$$

EXAMPLE 3 (*Gas law for a perfect gas: Boyle's law*). At constant temperatures, the pressure of a given amount of gas varies inversely as the volume. If P denotes pressure and V volume, then

$$P = \frac{k}{V}$$

EXAMPLE 4 The number of hours required to finish a job varies inversely as the number of people employed. If 6 people can finish the job in 21 hr, how many hours will it take 9 people to do the job?

Solution Let

$$f = \text{the number of people employed}$$

$$h = \text{the number of hours to do the job}$$

Since the number of hours varies inversely as the number of people employed, we have

$$h = \frac{k}{f}$$

Using the condition that $h = 21$ hr when $f = 6$ people, we find that

$$21 \text{ hr} = \frac{k}{6 \text{ people}}$$

or

$$k = 126 \text{ person-hr}$$

Thus,

$$h = \frac{126}{f}$$

and when $f = 9$ people,

$$h = \frac{126 \text{ person-hr}}{9 \text{ people}} = 14 \text{ hr} \quad \blacksquare$$

EXERCISES 13.4

In Exercises 1–9, form an equation to express each of the following.
1. x varies inversely as y
2. a varies inversely as b^2
3. p is inversely proportional as q^4
4. Altitude varies inversely as base
5. Price per article varies inversely as number of articles
6. Speed is inversely proportional to time
7. The density D of a given mass varies inversely as its volume V
8. The intensity I of illumination from a given

source varies inversely as the square of the distance d from the source

9. *Coulomb's law:* The force F of repulsion between two charged particles (with like charges) varies inversely as the square of the distance d between the particles.

10. If x varies inversely as y and $x = 45$ when $y = 16$, find x when $y = 36$.

11. If W varies inversely as l and $W = 6,000$ when $l = 20$, find W when $l = 12$.

12. If l varies inversely as n and $l = 28$ when $n = 2\frac{1}{4}$, find n when $l = 27$.

13. If r varies inversely as d^2 and $r = 30$ when $d = .4$, find r when $d = .2$.

14. If a varies inversely as b^2 and $a = \frac{1}{8}$ when $b = 6$, find a when $b = 3$.

15. If 8 women can finish a job in 90 hours, how many women would be required to finish it in 60 hours?

16. If 12 men can fix a roof in $3\frac{1}{2}$ days, how long will it take 7 men to do the work?

17. (Use gas law: Example 3). If 5 cu ft of gas under pressure of 10 lb/sq in. expands to 20 cu ft, what will the pressure become?

18. (Use Newton's law of gravitation: Example 2). If the force of attraction upon a meteorite 100 mi above the earth's surface is 10 lb, what would the meteorite weigh at the surface?

19. (Use Coulomb's law: Exercise 9). If two charged particles 1 cm apart repel each other with a force of 1 dyne, what would be the force if they were 10 cm apart? 0.1 cm apart?

20. The pressure of a given amount of gas in a spherical balloon varies inversely as the cube of the radius. If the pressure is 15 lb/sq in. when the radius is 20 ft, what will the pressure be when the radius increases to 24 ft?

*21. On a certain journey the time varies inversely as the speed, and the time is 4 hr when the speed is 30 mph. Find the time when the speed is 45 mph and the speed when the time is 5 hr.

*22. The frequency of the vibration of a stretched string varies inversely as the length of the string and the frequency is 320 vibrations per second when the length is 4 ft. How long must the string be to make 7680 vibrations in a minute?

13.5 JOINT AND COMBINED VARIATION

Joint Variation

Sometimes a quantity depends on the variation of two or more other quantities whose values are independent of one another. If a variable w varies directly as the product of x and y; that is,

$$w = kxy$$

where k is a constant, we say that w varies jointly as x and y.

EXAMPLE 1   The area of a triangle varies jointly as its base b and its altitude h, $A = \frac{1}{2}bh$.

EXAMPLE 2   The distance which a train moving with a uniform speed travels varies jointly as the speed v and the time t; that is,

$$d = kvt$$

EXAMPLE 3   The volume of a right circular cylinder varies jointly as the square of its radius r and its height h. Thus, $V = \pi r^2 h$.

EXAMPLE 4   The absolute temperature T of a perfect gas varies jointly as its pressure p and its volume v; that is,

$$T = kpv$$

Combined Variation

Direct, inverse, and joint variations may be combined.

EXAMPLE 5   Determine the equation connecting x, y, and z if z varies directly as x and inversely as the square of y, and $z = 12$ when $x = 2$ and $y = 3$.

Solution   The variation can be expressed as

$$z = \frac{kx}{y^2}$$

Substituting $z = 12$, $x = 2$, and $y = 3$, we obtain

$$12 = \frac{k \cdot 2}{(3)^2}$$

Solving for k, we get $k = 54$. Thus, the required equation is

$$z = \frac{54x}{y^2} \quad \blacksquare$$

EXAMPLE 6   (*Attraction of two masses*). The attraction F of any two masses m_1 and m_2 for each other varies as the product of the masses and inversely as the square of the distance r between the two bodies; that is,

$$F = \frac{km_1 m_2}{r^2}$$

EXAMPLE 7   The brightness of illumination from a lamp varies directly as the strength of the lamp and inversely as the square of the distance from the lamp. At what distance from a 32-candlepower lamp must a page be held in order that it may be illuminated twice as brightly as a page that is held 3 ft away from a 24-candlepower lamp?

Solution   The variation can be expressed as

$$L = \frac{ks}{d^2}$$

where L is the intensity of illumination, s is the strength of the lamp, and d is the distance from the lamp. If we let L_1 represent the intensity of illumination of the page held 3 ft from the 24-candlepower lamp, then $2L_1$ represents the intensity of illumination of the page held at the unknown distance d from the 32-candlepower lamp. Thus, we have the following two equations:

$$L_1 = \frac{24k}{(3)^2} \tag{1}$$

and

$$2L_1 = \frac{32k}{d^2} \tag{2}$$

Multiplying equation (1) by 2 gives

$$2L_1 = \frac{48k}{9} \tag{3}$$

From equations (2) and (3), we obtain

$$\frac{32k}{d^2} = \frac{48k}{9}$$

Therefore,

$$d^2 = \frac{9 \cdot 32k}{48k} = 6$$

and

$$d = \sqrt{6} = 2.45$$

Hence, the page must be held 2.45 ft away from a 32-candlepower lamp to have twice the illumination. ■

EXERCISES   13.5

In Exercises 1–5, write the equation that expresses the relation between the variables, using k as the constant of variation.

1. L varies directly as the fourth power of t and inversely as the square of l.
2. The lateral surface S of a right circular cylinder varies jointly as its base-radius r and its altitude h.
3. The weight w of a body above the surface of the earth varies inversely as the square of the distance d of the body from the center of the earth.
4. The number of units of heat H generated by an electric current of I amperes in a circuit varies jointly as the square of the current I, as the resistance R, and as the time t.
5. In vibrating strings, the vibrating frequency, or pitch, f varies directly as the square root of

the tension t on the string, inversely as the length l of the string, and inversely as the square root of the mass m of the string.

6. If x varies jointly as y and z and $x = 60$ when $y = 8$ and $z = 3$, find x when $y = 10$ and $z = 5$.

7. If r varies jointly as s and t and $r = 72$ when $s = 6$ and $t = 9$, find r when $s = 5$ and $t = 6$.

8. If m varies jointly as n and p and $m = 90$ when $n = 6$ and $p = 16$, find m when $n = 8$ and $p = 10$.

9. If a varies jointly as b and c and $a = 70$ when $b = 5$ and $c = 12$, find a when $b = 6$ and $c = 8$.

10. If $z = kxy$ and $z = 50$ when $x = 3$ and $y = 25$, find z when $x = 8$ and $y = 18$.

11. If $r = kst$ and $r = 144$ when $s = 6$ and $t = 32$, find r when $s = 15$ and $t = 24$.

12. If a varies jointly as b and c and inversely as d, and $a = 10$ when $b = 6$, $c = 5$, and $d = 3$, find a when $b = 8$, $c = 9$, and $d = 6$.

13. If x varies jointly as y and z and inversely as v, and $x = 33$ when $y = 22$, $z = 6$, and $v = 4$, find x when $y = 12$, $z = 14$, and $v = 8$.

14. Y varies jointly as b and c and inversely as d^3, and $Y = 27$ when $b = 3$, $c = 2$, and $d = 2$. Write Y as a function of b, c, and d and then determine the value of Y when $b = 5$, $c = 3$, and $d = 4$.

15. T varies directly as x^2 and inversely as the product y^3z, and $T = 36$ when $x = 3$, $y = 2$, and $z = 5$. Write T as a function of x, y, and z and then determine the value of T when $x = 5$, $y = 4$, and $z = 3$.

*16. The gravitational attraction between two bodies varies jointly as their masses and inversely as the square of the distance between their centers. Two bodies whose centers are 4000 mi apart attract each other with a force of 10 lb. What would be their force of attraction if their masses were doubled and the distance between their centers was 6000 mi?

*17. The current i in amperes in an electric circuit varies inversely as the resistance R in ohms when the electromotive force is constant. If a certain circuit i is 10 amperes when R is 2 ohms, find (a) i when R is 0.02 ohms; (b) R when i is 100 amperes.

*18. The mass of a sphere of given density varies as the cube of its diameter. If a body weighs 10 lb on the earth's surface, what would it weigh on the surface of a planet 16,000 mi in diameter and having the same density as the earth? (See Exercise 16; take the earth's radius as 4000 mi.)

*19. The volume v of a gas varies directly as the absolute temperature T and inversely as the pressure p. If a certain amount of a gas occupies 200 cu ft at a pressure of 12 lb per sq in. and at $T = 240°$, find its volume when the pressure is 20 lb per sq in. and $T = 420°$.

*20. In projecting moving pictures, the brightness required in the lamps varies as the square of the distance from the machine to the screen. If a machine operating 12 ft from the screen requires a 200-candlepower lamp, how close to the screen must the machine be placed in order to operate with 150 candlepower?

13.6 POLYNOMIAL FUNCTIONS

In Chapter 2 we discussed polynomials and in Chapter 3 and Chapter 8 we discussed first- and second-degree equations. In this section, we shall consider polynomial functions and, in particular, linear (first-degree) and quadratic (second-degree) functions.

Functions that can be expressed in polynomial form are called polynomial functions

---

DEFINITION

A function f is a **polynomial function** if

$$f(x) = a_n x^n + a_{n-1} x^{n-1} + \cdots + a_1 x + a_0$$

where the coefficients $a_0, a_1, \ldots, a_n$ are real numbers and the exponents are nonnegative integers.

We say f has **degree** n if $f(x)$ has degree n.

EXAMPLE 1 (a) $f(x) = 10x^4 - 3x^3 + x - 2$ is a polynomial function of degree 4.

(b) $f(x) = -x^2 - 4x + 5$ is a polynomial function of degree 2, often referred to as a **quadratic function**.

(c) $f(x) = 5x - 7$ is a polynomial function of degree 1, often referred to as a **linear function**.

Linear Functions

DEFINITION

A function f, defined by the equation $f(x) = mx + b$, where m and b are constants, $m \neq 0$, is called a **linear function**. The graph of a linear function is a straight line.

EXAMPLE 2 The function f defined by $f(x) = 3x - 1$ is a linear function. The graph of $f(x) = 3x - 1$ (or $y = 3x - 1$) is obtained by plotting two points, since two distinct points uniquely determine a straight line (see Figure 13.12).

If $m = 0$, then $f(x) = b$ defines what is called a **constant function**. The graph of a constant function is the set of all ordered pairs with a second coordinate of b; that is, it is a line parallel to the x-axis passing through the point $(0, b)$.

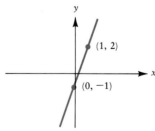

Figure 13.12

EXAMPLE 3 The graph of the constant function $f(x) = 2$ is the line parallel to the x-axis and passing through the point $(0, 2)$; see Figure 13.13.

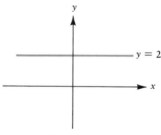

Figure 13.13

EXAMPLE 4 Let f be the linear function defined by $f(x) = mx + b$. Compute $\dfrac{f(x + h) - f(x)}{h}$, $h \neq 0$.

Solution First, note that f multiplies any expression by m and then adds b. Thus,

$$\frac{f(x + h) - f(x)}{h} = \frac{[m(x + h) + b] - [mx + b]}{h}$$

$$= \frac{mx + mh + b - mx - b}{h} = \frac{mh}{h} = m$$

Hence, $\dfrac{f(x + h) - f(x)}{h} = m$, the slope of the line. ■

Quadratic Functions

Now consider polynomials of degree two, $ax^2 + bx + c$. Such polynomials define a class of functions called quadratic functions.

DEFINITION
A function f defined by $f(x) = ax^2 + bx + c$, where a, b, and c are real numbers, $a \neq 0$, is called a quadratic function.

EXAMPLE 5 (a) The expression $f(x) = x^2$ defines a quadratic function. Here we have $a = 1$, $b = 0$, and $c = 0$.

(b) The expression $f(x) = x^2 - 2x - 3$ defines a quadratic function. Here $a = 1$, $b = -2$, and $c = -3$.

(c) The expression $f(x) = -x^2 + x + 2$ defines a quadratic function. Here $a = -1$, $b = 1$, and $c = 2$.

Previously, we saw that two points are all we need to graph a straight line. To graph quadratic functions, we must find more than two points.

EXAMPLE 6 Graph the quadratic functions:

(a) $f(x) = x^2$ (b) $f(x) = x^2 - 2x - 3$ (c) $f(x) = -x^2 + x + 2$

Solution (a) $f(x) = x^2$
If $x = 0$, then $f(0) = (0)^2 = 0$.
If $x = 1$, then $f(1) = (1)^2 = 1$.
If $x = -1$, then $f(-1) = (-1)^2 = 1$.
If $x = 2$, then $f(2) = (2)^2 = 4$.
If $x = -2$, then $f(-2) = (-2)^2 = 4$.

When $x =$	0	1	−1	2	−2
$y = f(x) =$	0	1	1	4	4

Now, we graph these points and draw a smooth curve through them (see Figure 13.14).

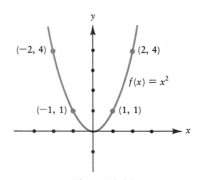

Figure 13.14

(b) $f(x) = x^2 - 2x - 3$
Again, we select different values for x and compute the corresponding $f(x)$. See Figure 13.15.

When $x =$	0	1	−1	2	−2	3
$y = f(x) =$	−3	−4	0	−3	5	0

(c) $f(x) = -x^2 + x + 2$

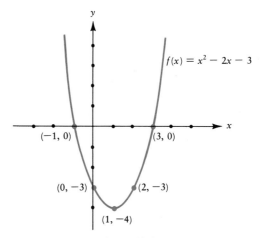

Figure 13.15

When $x =$	0	1	-1	2	-2	3
$y = f(x) =$	2	2	0	0	-4	-4

See Figure 13.16. ■

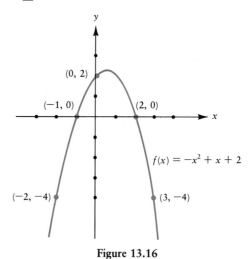

Figure 13.16

As seen by the previous examples, the graph of any quadratic function will have the same general shape. The location of the graph will depend upon the particular values of a, b, and c. The graph of the quadratic function $f(x) = ax^2 + bx + c$ is called a *parabola,* one of the conic sections discussed in Chapter 10.

From Example 6 and Figures 13.14, 13.15, and 13.16, we note that the parabola opens upward when the coefficient of the x^2 term is positive (Figures 13.14 and 13.15) and the parabola opens downward when the coefficient of the x^2 terms is negative (Figure 13.16).

In general,

The graph of the quadratic function $f(x) = ax^2 + bx + c$ opens upward when $a > 0$ and downward when $a < 0$.

Of special interest is the point on the parabola called the vertex. This point represents the *lowest* point of the parabola if the parabola opens upward and the *highest* point on the parabola if the parabola opens downward. If the vertex is the lowest point of the parabola, then the vertex is called the minimum point. If the vertex is the highest point of the parabola, then the vertex is called the maximum point. The vertex of the quadratic function f, defined by $f(x) = ax^2 + bx + c$, is the point (h, k) where $h = \dfrac{-b}{2a}$ and $k = \dfrac{4ac - b^2}{4a}$. Thus we have the following theorem:

Theorem

The quadratic function f, defined by $f(x) = ax^2 + bx + c$, has

$$\text{a maximum at } x = -\frac{b}{2a} \text{ if } a < 0$$

$$\text{a minimum at } x = -\frac{b}{2a} \text{ if } a > 0$$

In either case, the value of the function f is

$$f\left(-\frac{b}{2a}\right) = \frac{4ac - b^2}{4a} = \frac{-(b^2 - 4ac)}{4a}$$

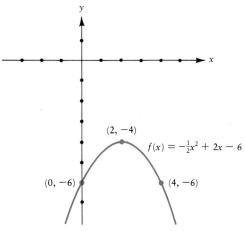

Figure 13.17

EXAMPLE 7 Given $f(x) = -\frac{1}{2}x^2 + 2x - 6$, we have $h = -\frac{b}{2a} = \frac{-2}{2(-\frac{1}{2})} = \frac{-2}{-1} = 2$ and $k = \frac{4ac - b^2}{4a} = \frac{4(-\frac{1}{2})(-6) - (2)^2}{4(-\frac{1}{2})} = \frac{8}{-2} = -4.$. Therefore, the vertex is the point $(2, -4)$. Since $a = -\frac{1}{2} < 0$, the vertex is a maximum. See Figure 13.17 at the bottom of p. 523.

EXAMPLE 8 Given $f(x) = x^2 + 1$, we have $h = -\frac{b}{2a} = \frac{0}{2} = 0$ and $k = \frac{4ac - b^2}{4a} = \frac{4(1)(1) - (0)^2}{4(1)} = \frac{4}{4} = 1$. Thus, the vertex is the point $(0, 1)$ and since $a = 1 > 0$, the vertex is a minimum. See Figure 13.18.

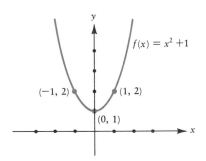

Figure 13.18

The graphs of quadratic functions can be used to illustrate the nature of the roots of the associated quadratic equation $ax^2 + bx + c = 0$, $a \neq 0$ (see Chapter 8). Recall that the nature of the roots of the quadratic equation $ax^2 + bx + c = 0$, $a \neq 0$, is given by the discriminant $b^2 - 4ac$:

1. If $b^2 - 4ac > 0$, the roots are real and unequal and the parabola intersects the x-axis twice.
2. If $b^2 - 4ac = 0$, the roots are real and equal and the parabola touches the x-axis once.
3. If $b^2 - 4ac < 0$, the roots are not real and the parabola does not intersect the x-axis.

We summarize the information concerning quadratic functions using the following graphs (see Figures 13.19 and 13.20).

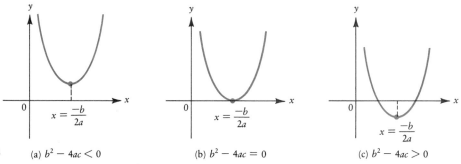

(a) $b^2 - 4ac < 0$ (b) $b^2 - 4ac = 0$ (c) $b^2 - 4ac > 0$

Figure 13.19 *Case 1: $a > 0$; the vertex* $\left(\frac{-b}{2a}, \frac{4ac - b^2}{4a}\right)$ *is a minimum.*

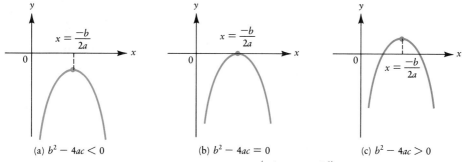

(a) $b^2 - 4ac < 0$ (b) $b^2 - 4ac = 0$ (c) $b^2 - 4ac > 0$

Figure 13.20 *Case* 2: $a < 0$; the vertex $\left(\dfrac{-b}{2a}, \dfrac{4ac - b^2}{4a}\right)$ is a maximum.

Applications

We now consider a few of the many applications of quadratic functions.

EXAMPLE 9 (*Physics*). An object is projected vertically upward with initial velocity of v_0 feet per second. The distance s above the ground at any time t in seconds is described by the quadratic function s defined by

$$s(t) = v_0 t - 16t^2$$

If a ball is thrown vertically upward from the ground with an initial velocity of 64 feet per second, then

(a) How many seconds will it take the ball to reach its maximum height?

(b) What is the maximum height?

(c) How many seconds will it take the ball to reach the ground?

Solution The equation that describes the motion of the ball is $s(t) = 64t - 16t^2$. Since the coefficient of t^2 is -16, the graph is a parabola that opens downward (see Figure 13.21).

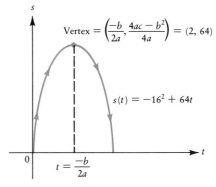

Figure 13.21

(a) The fact that the graph of the equation $s = 64t - 16t^2$ is a parabola that opens downward means that the vertex is a maximum. The time it takes the ball to reach its maximum height is given by the first coordinate of the vertex; that is, $t = \dfrac{-b}{2a} = \dfrac{-64}{-32} = 2$ sec. Thus, it takes the ball 2 sec to reach its maximum height.

(b) It follows that the maximum height is given by the second coordinate of the vertex; that is, $s = \dfrac{4ac - b^2}{4a} = \dfrac{(-16)(0) - (64)^2}{4(-16)} = \dfrac{-(64)^2}{-64} = 64$ ft. The maximum height is 64 ft.

(c) When the ball returns to the ground, the distance will be zero. Therefore we are looking for the value(s) for which $s(t) = 0$; that is,

$$64t - 16t^2 = 0$$

$$16t(4 - t) = 0$$

$$t = 0 \quad \text{and} \quad t = 4$$

There are two times at which the distance is zero: when the ball is thrown and when it reaches the ground after its flight. The former is represented by the time $t = 0$. Hence, $t = 4$ seconds is the time it will take the ball to reach the ground after the ball is thrown. ∎

EXAMPLE 10 The number 24 is written as the sum of two numbers so that the product of the number is maximized. Find the numbers.

Solution Let x be one of the numbers. Then $24 - x$ is the other number. The product p is defined by

$$p(x) = x(24 - x)$$

$$p(x) = 24x - x^2$$

Since p is a quadratic function and is represented by a parabola opening downward, the maximum value of p occurs at the vertex. The coordinates of the vertex (h, k) are

$$h = \frac{-b}{2a} = \frac{-24}{-2} = 12 \quad \text{and} \quad k = \frac{4ac - b^2}{4a} = \frac{-(24)^2}{-4} = (12)^2 = 144$$

and the vertex is $(12, 144)$. Thus, the numbers we were to find are $x = 12$ and $24 - x = 12$. The value of the maximized product is $(12)(12) = (12)^2 = 144$. Note that this is the value of the second coordinate of the vertex. ∎

EXAMPLE 11 *Economics* The total cost of producing x radios per day is $(x^2 + 4x + 5)$ dollars and the price per set at which they may be sold is $(100 - 2x)$ dollars. What should be the daily output to obtain a maximum total profit?

Solution   Profit = (Revenue) − (Cost). If x radios are sold per day at a price of $(100 − 2x)$ dollars each, then the revenue generated is $[x(100 − 2x)]$ dollars. Thus, the profit on the sale of x sets per day is

$$P(x) = [x(100 − 2x)] \text{ dollars} − (x^2 + 4x + 5) \text{ dollars}$$

$$= [x(100 − 2x) − (x^2 + 4x + 5)] \text{ dollars}$$

$$= (−3x^2 + 96x − 5) \text{ dollars}$$

The profit is represented by the quadratic function whose graph is a parabola opening downward. The maximum value of the profit P is given by the first coordinate of the vertex. The first coordinate of the vertex $h = \dfrac{−b}{2a} = \dfrac{−96}{−6} = 16$. Thus, the production required to yield maximum profit is 16 sets per day and the radios sell for $68. ◼

EXERCISES   13.6

Complete the following table on linear functions:

	Linear Function	y-intercept
	$f(x) = 3x − 2$	$y = −2$
1.	$f(x) = 4x + 6$	
2.	$f(x) = 5x − 3/2$	
3.	$2f(x) = x − 7$	
4.	$3f(x) = 4x − 5$	
5.	$−f(x) = 6x + 11$	

Complete the following table on linear functions:

	Linear Function	Slope up or down towards the right
6.	$f(x) = −2x + 7$	
7.	$f(x) = \frac{3}{2}x − 5$	
8.	$5f(x) = 10x − 1$	
9.	$7f(x) = −3x + 6$	
10.	$−f(x) = −x + 1$	

In Exercises 11–16, find the difference quotient
$$\dfrac{f(x + h) − f(x)}{h}, h \neq 0, \text{ and simplify the results.}$$

11. $f(x) = x + 5$

12. $f(x) = −3x + 7$

13. $f(x) = \dfrac{x}{2} − 8$

14. $f(x) = −x$

15. $f(x) = 4 − 5x$

16. $f(x) = ax + b$

17. Given that for $x = 2$ the linear function $f(x) = 3x + k$ takes the value 7.5, find k.

18. The graph of the linear function $f(x) = cx − 4$ is known to pass through the point $(−2, 5)$. Determine c.

*19. Find a linear function such that $3f(x) = f(3x)$ for all x.

*20. Show that if f is a linear function and $f(0) = 0$, then $f(x + y) = f(x) + f(y)$ for all real x and y.

In Exercises 21–24, determine the vertex for each quadratic function.

21. $g(x) = −5x^2 − 14x − 8$

22. $g(x) = 4x^2 − 9$

23. $g(x) = 2x^2 − 6x + 3$

24. $g(x) = \sqrt{7}x^2 − 4x + \sqrt{5}$

In Exercises 25–28, determine whether the graph of each of the quadratic expressions opens upward or downward. Find the maximum or minimum point of each graph.

25. $f(x) = 1 + 3x − 4x^2$

26. $f(x) = 3x^2 − x − 2$

27. $f(x) = x^2$

28. $f(x) = 6x + 1 − 3x^2$

In Exercises 29–32, graph each quadratic function.

29. $f(x) = x^2 − 2x − 3$

30. $f(x) = −2x^2 + x + 3$

31. $f(x) = \dfrac{1}{2}x^2 - x + 2$

32. $f(x) = -(x - 2)^2 - 2$

33. A stone is thrown upward with an initial velocity of 48 feet per second.
  (a) How many seconds will it take the stone to reach its maximum height?
  (b) What is the maximum height of the stone?
  (c) When does the stone hit the ground?

34. Find two positive numbers, the sum of which is 20, and whose product is maximized.

35. A boat manufacturer can sell x boats per week at $\$(2460 - 5x)$ per boat. It costs $\$(2.4 + 1000x)$ to produce x boats per week. How many boats per week should be sold to maximize profits?

36. A gardener has 100 ft of wire fencing to use around a garden. One side of the garden will be formed by the side of the house, so no wire will be needed for that side of the garden. What are the dimensions of the garden if it is to be rectangular in shape and has maximum area?

*37. Find the area of the largest rectangle that can be inscribed in a triangle whose base is 10 in. and altitude is 8 in.

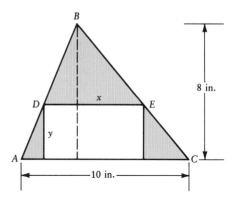

(*Hint*: Use similarity of triangles ABC and BDE to find y. Then maximize area of rectangle: $A = xy$.)

*38. Prove that the maximum area of a rectangle with constant perimeter P units is $\left(\dfrac{1}{16}P^2\right)$ square units.

13.7 EXPONENTIAL AND LOGARITHMIC FUNCTIONS

Exponential Functions

Many applications in mathematics and science can be solved using functional relationships where the exponent is a variable. One such functional relationship is called an exponential function.

> **DEFINITION**
>
> If b is a positive number, then the function f defined by
>
> $$f(x) = b^x, \quad b \neq 1,$$
>
> is called an exponential function with base b.

The domain of the exponential function is the set of real numbers $\mathcal{R}$ and the range is the set of positive real numbers.

EXAMPLE 1   Graph the exponential functions defined by

(a) $f(x) = 2^x$ **(b)** $f(x) = 3^x$ **(c)** $f(x) = \left(\dfrac{1}{2}\right)^x$

Solution

x	-3	-2	-1	0	1	2	3
$f(x) = 2^x$	$\dfrac{1}{8}$	$\dfrac{1}{4}$	$\dfrac{1}{2}$	1	2	4	8
$f(x) = 3^x$	$\dfrac{1}{27}$	$\dfrac{1}{9}$	$\dfrac{1}{3}$	1	3	9	27
$f(x) = \left(\dfrac{1}{2}\right)^x$	8	4	2	1	$\dfrac{1}{2}$	$\dfrac{1}{4}$	$\dfrac{1}{8}$

We now graph all three exponential functions using the same coordinate system (see Figure 13.22). Note that the graphs of $f(x) = 2^x$ and $f(x) = 3^x$ go up to the right as x gets larger. Such graphs represent what is called an increasing function. The graph of $f(x) = (\frac{1}{2})^x$ goes down to the right as x gets larger; thus, we say it is a decreasing function. Finally, notice that each of the graphs passes through the point $(0, 1)$. The properties of exponential functions can be summarized in the following theorem.

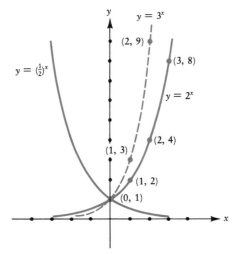

Figure 13.22

Theorem

Given an exponential function defined by $f(x) = b^x$, where $b > 0$ and $b \neq 1$, then

1. If $b > 1$, then $f(x) = b^x$ defines an increasing function.

2. If $b < 1$, then $f(x) = b^x$ defines a decreasing function.

In either case, the graph passes through the point $(0, 1)$.

EXAMPLE 2 Show that if f is an exponential function, and x and y are real numbers, then $f(x + y) = f(x)f(y)$.

Solution If f is an exponential function, then it is defined by $f(x) = b^x, b > 0$ and $b \neq 1$. Thus,

$$f(x + y) = b^{x+y} = b^x \cdot b^y = f(x) \cdot f(y) \quad \blacksquare$$

Logarithmic Functions

In Chapter 12 we saw that if $y = b^x$, $b > 0$ and $b \neq 1$, then the exponent x is called the *logarithm of the number y to the base b*. Therefore, $y = \log_b x$ is equivalent to $x = b^y$. We now use the idea of logarithm to define a function whose domain is the set of positive real numbers.

DEFINITION

The function f defined by

$$f(x) = \log_b x, \quad b > 0 \quad \text{and} \quad b \neq 1,$$

for all positive real numbers x is called the logarithmic function with base b.

In order to graph a logarithmic function $f(x) = \log_b x$, we shall use the equivalent equation $x = b^y$.

EXAMPLE 3 Sketch the graph of $y = \log_2 x$.

Solution The equation $y = \log_2 x$ is equivalent to the equation $x = 2^y$. To find some ordered pairs that lie on the graph, we substitute values for y and find the corresponding values of x. See Figure 13.23.

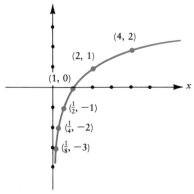

Figure 13.23

y	-3	-2	-1	0	1	2	3
$x = 2^y$	$\frac{1}{8}$	$\frac{1}{4}$	$\frac{1}{2}$	1	2	4	8

Note that $f(x) = \log_2 x$ defines an increasing function. ■

EXAMPLE 4   Sketch the graph of $y = \log_{1/4} x$.

Solution   The equation $y = \log_{1/4} x$ is equivalent to the equation $x = \left(\dfrac{1}{4}\right)^y$. See Figure 13.24.

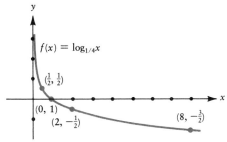

Figure 13.24

y	$-\dfrac{3}{2}$	$-\dfrac{1}{2}$	0	$\dfrac{1}{2}$	$\dfrac{3}{2}$
$x = \left(\dfrac{1}{4}\right)^y$	8	2	1	$\dfrac{1}{2}$	$\dfrac{1}{8}$

Note that $f(x) = \log_{1/4} x$ defines a decreasing function. ■

The properties of logarithmic functions can be summarized in the following theorem.

Theorem

Given a logarithmic function defined by $f(x) = \log_b x$, where $b > 0$ and $b \neq 1$, then

1. If $b > 1$, then $f(x) = \log_b x$ defines an increasing function.
2. If $0 < b < 1$, then $f(x) = \log_b x$ defines a decreasing function.

In either case, the graph passes through the point $(1, 0)$.

EXAMPLE 5   Show that if f is a logarithmic function and x and y are any positive real numbers, then $f(x) + f(y) = f(xy)$.

Solution   If f is a logarithmic function, then it is defined by $f(x) = \log_b x$, $b > 0$ and $b \neq 1$. Thus,

$$f(x) + f(y) = \log_b x + \log_b y = \log_b(xy) = f(xy) \blacksquare$$

EXERCISES 13.7

In Exercises 1–10, determine the base of the exponential function $f(x) = b^x$ whose graph contains the given points.

1. $\left(2, \dfrac{1}{4}\right)$ **2.** $(2, 9)$ **3.** $(3, 64)$

4. $(2, 16)$ **5.** $\left(-2, \dfrac{1}{16}\right)$

6. $(-3, 64)$ **7.** $(-2, 49)$
8. $(3, 125)$ **9.** $(-3, 27)$ **10.** $(0, 1)$

In Exercises 11–20, indicate whether the function is increasing or decreasing. Graph each function.

11. $f(x) = 4^x$ **12.** $f(x) = 2^{x+1}$

13. $f(x) = \left(\dfrac{1}{3}\right)^x$

14. $f(x) = -2^x$ **15.** $f(x) = -\left(\dfrac{1}{3}\right)^x$

16. $f(x) = (\sqrt{9})^x$ **17.** $f(x) = 1^x$

18. $f(x) = \left(\dfrac{1}{5}\right)^{-x}$ **19.** $f(x) = 3^{-x}$

20. $f(x) = 7(3)^x$
*21. How does the graph of $f(x) = b^x$ compare with the graph of $f(x) = -b^x$?
*22. How does the graph of $f(x) = b^x$ compare with the graph of $f(x) = b^{-x}$?
*23. If $f(x) = 2^x$ and $g(x) = x^2$, determine which function, f or g, grows faster for $x \geq 0$ by graphing both functions on the same coordinate system.

*24. Let f be an exponential function. Show that $f(x - y) = \dfrac{f(x)}{f(y)}$ for any real numbers x and y.
*25. Let $f(x) = a \cdot 10^{nx}$ and let $f(0) = 7$ and $f(1) = 700$. Determine the values of a and n.

In Exercises 26–35, determine the base of the logarithmic function $f(x) = \log_b x$ whose graph contains the given points.

26. $(9, 2)$ **27.** $(4, 2)$ **28.** $(125, 3)$

29. $(64, 3)$ **30.** $\left(\dfrac{1}{2}, -1\right)$

31. $\left(\dfrac{1}{16}, -2\right)$ **32.** $\left(8, \dfrac{3}{2}\right)$ **33.** $\left(3, \dfrac{1}{2}\right)$

34. $(9, -2)$ **35.** $(1, 0)$

In Exercises 36–41, indicate whether the function is increasing or decreasing. Graph each function.
36. $f(x) = \log_3 x$ **37.** $f(x) = \log_2 x$
38. $f(x) = \log_{1/3} x$ **39.** $f(x) = \log_{1/2} x$
40. $f(x) = -\log_2 x$ **41.** $f(x) = -\log_3 x$
*42. What is the relationship between the graph of $y = \log_b x$ and $y = b^x$? (Look at the line $y = x$.)

*43. Let f be a logarithmic function. Show that $f(x) - f(y) = f\left(\dfrac{x}{y}\right)$ for any positive real numbers x and y.

Chapter 13 SUMMARY

1. A relation is a set of ordered pairs (x, y). The domain of a relation is the set of all first coordinates of the ordered pairs and the range of a relation is the set of all second coordinates of the ordered pairs.
2. A function is a relation such that no two distinct ordered pairs have the same first coordi-

nates. Any vertical line intersects the graph of a function in at most one point.
3. The inverse relation R^{-1} of a relation R is the set of ordered pairs obtained by interchanging the first and second coordinates in each ordered pair of the relation R.
4. If F is a function and F^{-1} is also a function,

then F^{-1} is called the **inverse function** of F. The inverse function, provided it exists, can be found by interchanging y and x in the equation $y = f(x)$ and then solving for y.

5. If two variables are related by an equation $y = kx$, where k is a constant, we say that y **varies directly** as x. The constant k is called the **constant of variation (proportionality)**.

6. If two variables are related by an equation of the form $y = \dfrac{k}{x}$, we say that y **varies inversely** as x.

7. If a variable w varies directly as the product of x and y, that is, $w = kxy$, we say that w **varies jointly** as x and y.

8. **Linear function** $f(x) = mx + b, m \neq 0$. The graph is a straight line. **Quadratic function** $f(x) = ax^2 + bx + c, a \neq 0$. The graph is a parabola that opens upward if $a > 0$ and opens downward if $a < 0$. The vertex $\left(\dfrac{-b}{2a}, \dfrac{4ac - b^2}{4a} \right)$ is a minimum if $a > 0$ and a maximum if $a < 0$.

9. If b is a positive number, then the function f defined by $f(x) = b^x, b \neq 1$, is called an **exponential function** with base b. If $b > 1$, then $f(x) = b^x$ defines an increasing function. If $b < 1$, then $f(x) = b^x$ defines a decreasing function. The graph of an exponential function passes through the point $(0, 1)$.

10. The function f defined by $f(x) = \log_b x$ for all positive real numbers x is called the **logarithmic function** with base b. If $b > 1$, then $f(x) = \log_b x$ defines an increasing function. If $b < 1$, then $f(x) = \log_b x$ defines a decreasing function. The graph of a logarithmic function passes through the point $(1, 0)$.

Chapter 13 EXERCISES

1. If $f(x) = 6x^2 - 2x + 1$, find $f(0)$, $f(1)$, $f(-1)$, and $f(a + b)$.

2. If $f(x) = x^3 - 5$, find $f(-2)$, $f(-1)$, $f(0)$, and $f(a - b)$.

3. If $f(x) = \sqrt{x - 1} + 3x$, find $f(1)$, $f(5)$, $f(10)$, and $f(a + b)$.

4. If $f(x) = \dfrac{x + 2}{x - 4}$, find $f(0)$, $f(2)$, $f(-4)$, and $f(a - b)$.

In Exercises 5–10, compute the expression $\dfrac{f(x + h) - f(x)}{h}$, provided $h \neq 0$, for each of the following.

5. $f(x) = -3x + 2$
6. $f(x) = 7 - 2x$
7. $f(x) = x^2 - 4x$
8. $f(x) = 8x - 7x^2$
9. $f(x) = 19$
10. $f(x) = -5$

In Exercises 11–22, determine the domain of the function f.

11. $f(x) = x^3 + 2x^2 - 7$
12. $f(x) = -2x^4 + 7$
13. $f(x) = \sqrt{x + 5}$
14. $f(x) = \sqrt{2x - 1}$
15. $f(x) = \sqrt{x^2 - 4}$
16. $f(x) = \sqrt{9 - x^2}$
17. $f(x) = \dfrac{2x + 9}{x^2 - x - 20}$
18. $f(x) = \dfrac{x^2 - 6}{x(x^2 - 25)}$

19. $f(x) = \dfrac{\sqrt{x}}{x - 3}$
20. $f(x) = \dfrac{3x + 5}{\sqrt{x}}$
21. $f(x) = \dfrac{\sqrt{x + 1}}{x^2 + 6}$
22. $f(x) = \dfrac{x^2 + 5}{\sqrt[3]{x + 7}}$

23. Use the vertical line test to determine which of the accompanying graphs represents a function.

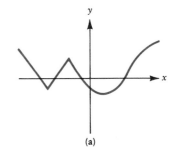

(a)

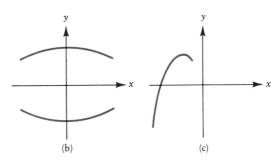

(b)　(c)

*24. A function f is said to be even if, for every number x in the domain of f, $-x$ is also in the domain of f and $f(-x) = f(x)$. A function f is odd if $f(-x) = -f(x)$. Determine whether each of the following equations defines a function that is even, odd, or neither.

(a) $f(x) = x^3$ (b) $f(x) = |x|$
(c) $f(x) = 2x - 3$
(d) $f(x) = \sqrt{x^2 + 3}$ (e) $f(x) = x|x|$
(f) $f(x) = \dfrac{\sqrt{x^2 - 5}}{|x|}$

*25. What are the conditions on a, b, and c such that $f(x) = ax^2 + bx + c$ defines an even function?

26. If s varies directly as t, and s is 8 when $t = 5$, find s when $t = 20$.

27. If u varies directly as v, and u is 27 when v is 24, find u when v is 72.

28. If y varies inversely as x, and $y = 30$ when $x = 84$, find y when $x = 63$.

29. If c varies inversely as d^2, and $c = 10{,}890$ when $d = 2$, find c when $d = 5\frac{1}{2}$.

30. If r varies jointly as s and t, and $r = 6$ when $s = 4$ and $t = 3$, find r when $s = 4$ and $t = 5$.

31. If w varies jointly as u and v, and $w = \frac{3}{2}$ when $u = 8$ and $v = \frac{1}{3}$, find w when $u = 4$ and $v = 12$.

32. If a varies directly as the cube of b and inversely as c, and $a = 4\frac{1}{2}$ when $b = 6$ and $c = \frac{2}{3}$, find b when $a = 32$ and $c = \frac{1}{9}$.

33. If z varies directly as the square of x and inversely as the square root of y, and $z = 1\frac{1}{2}$ when $x = 3$ and $y = 3$, find y when $x = 6$ and $z = 2\frac{2}{3}$.

*34. If x varies inversely as y and y varies directly as z, what is the relationship between x and z?

*35. If x varies inversely as y, y varies directly as z, and z varies directly as u, what is the relationship between x and u?

36. Find a linear function f such that $f(0) = -2$ and $f(-1) = 4$.

37. Find a linear function f such that $f(1) = 3$ and $f(2) = 5$.

*38. For what conditions are the graphs of the linear functions $y = m_1x + b_1$ and $y = m_2x + b_2$ parallel?

In Exercises 39–44, tell whether the graph of the equation will have a minimum or maximum.

39. $f(x) = x^2 - 7x$ 40. $f(x) = -4x^2 - 3$
41. $f(x) = 4 + 7x - x^2$
42. $f(x) = 8x + 5x^2$
43. $f(x) = \dfrac{-3x^2}{2} + 1$
44. $f(x) = 7x^2 - 12x + 1$

In Exercises 45–50, determine the coordinates of the vertex of the graph of each function and sketch the graph.

45. $f(x) = x^2 - 2x - 3$
46. $f(x) = -2x^2 + 8x - 5$
47. $f(x) = -x^2 + 2x$
48. $f(x) = x^2 - 3x + 2$
49. $f(x) = x^2 - 7x + 6$
50. $f(x) = 3 - 5x - 2x^2$

*51. Show that the point $(h + p, q)$ is on the graph of $f(x) = a(x - h)^2 + k$, $a \neq 0$, if and only if the point $(h - p, q)$ is also on the graph.

*52. Show that if the two distinct points $(p_1, 0)$ and $(p_2, 0)$ lie on the graph of $f(x) = a(x - h)^2 + k$, then $\dfrac{p_1 + p_2}{2} = h$.

53. Find the value of k if the vertex of the graph of $f(x) = 2x^2 - kx + 1$ lies on the line $y = x$.

54. Find the maximum value of the product of two numbers if their sum is 4.

55. If an object is shot vertically upward with an initial velocity of 48 feet per second from a height of 160 feet, its distance h above the ground after t seconds is given by the equation $h = 160 + 48t - t^2$. Find the maximum height of the object and at what time it reaches its maximum.

In Exercises 56–60, determine whether each is a decreasing or increasing function.

56. $f(x) = 5^x$ 57. $f(x) = -3x$
58. $f(x) = \left(\dfrac{2}{3}\right)^x$ 59. $f(x) = \left(\dfrac{4}{3}\right)^x$
60. $f(x) = -\left(\dfrac{7}{5}\right)^x$

*61. A light beam, passing through a plastic plate, loses one-fourth of its intensity. What is the intensity of the beam after it has passed x such plates?

In Exercises 62–66, determine whether each is a decreasing or increasing function.

62. $f(x) = \log_2 x$

63. $f(x) = \log_{0.4} x$

64. $f(x) = \log_7 x$

65. $f(x) = \log_\pi x$

66. $f(x) = \log_e x$

67. Find the value of b if the graph of $f(x) = \log_b x$ contains the point $(\frac{1}{3}, -1)$.

Appendix I:
Tables

Table 1. Powers and Roots

No.	Square	Cube	Square root	Cube root	No.	Square	Cube	Square root	Cube root
1	1	1	1.000	1.000	51	2 601	132 651	7.141	3.708
2	4	8	1.414	1.260	52	2 704	140 608	7.211	3.733
3	9	27	1.732	1.442	53	2 809	148 877	7.280	3.756
4	16	64	2.000	1.587	54	2 916	157 464	7.348	3.780
5	25	125	2.236	1.710	55	3 025	166 375	7.416	3.803
6	36	216	2.449	1.817	56	3 136	175 616	7.483	3.826
7	49	343	2.646	1.913	57	3 249	185 193	7.550	3.849
8	64	512	2.828	2.000	58	3 364	195 112	7.616	3.871
9	81	729	3.000	2.080	59	3 481	205 379	7.681	3.893
10	100	1 000	3.162	2.154	60	3 600	216 000	7.746	3.915
11	121	1 331	3.317	2.224	61	3 721	226 981	7.810	3.936
12	144	1 728	3.464	2.289	62	3 844	238 328	7.874	3.958
13	169	2 197	3.606	2.351	63	3 969	250 047	7.937	3.979
14	196	2 744	3.742	2.410	64	4 096	262 144	8.000	4.000
15	225	3 375	3.873	2.466	65	4 225	274 625	8.062	4.021
16	256	4 096	4.000	2.520	66	4 356	287 496	8.124	4.041
17	289	4 913	4.123	2.571	67	4 489	300 763	8.185	4.062
18	324	5 832	4.243	2.621	68	4 624	314 432	8.246	4.082
19	361	6 859	4.359	2.668	69	4 761	328 509	8.307	4.102
20	400	8 000	4.472	2.714	70	4 900	343 000	8.367	4.121
21	441	9 261	4.583	2.759	71	5 041	357 911	8.426	4.141
22	484	10 648	4.690	2.802	72	5 184	373 248	8.485	4.160
23	529	12 167	4.796	2.844	73	5 329	389 017	8.544	4.179
24	576	13 824	4.899	2.884	74	5 476	405 224	8.602	4.198
25	625	15 625	5.000	2.924	75	5 625	421 875	8.660	4.217
26	676	17 576	5.099	2.962	76	5 776	438 976	8.718	4.236
27	729	19 683	5.196	3.000	77	5 929	456 533	8.775	4.254
28	784	21 952	5.292	3.037	78	6 084	474 552	8.832	4.273
29	841	24 389	5.385	3.072	79	6 241	493 039	8.888	4.291
30	900	27 000	5.477	3.107	80	6 400	512 000	8.944	4.309
31	961	29 791	5.568	3.141	81	6 561	531 441	9.000	4.327
32	1 024	32 768	5.657	3.175	82	6 724	551 368	9.055	4.344
33	1 089	35 937	5.745	3.208	83	6 889	571 787	9.110	4.362
34	1 156	39 304	5.831	3.240	84	7 056	592 704	9.165	4.380
35	1 225	42 875	5.916	3.271	85	7 225	614 125	9.220	4.397
36	1 296	46 656	6.000	3.302	86	7 396	636 056	9.274	4.414
37	1 369	50 653	6.083	3.332	87	7 569	658 503	9.327	4.431
38	1 444	54 872	6.164	3.362	88	7 744	681 472	9.381	4.448
39	1 521	59 319	6.245	3.391	89	7 921	704 969	9.434	4.465
40	1 600	64 000	6.325	3.420	90	8 100	729 000	9.487	4.481
41	1 681	68 921	6.403	3.448	91	8 281	753 571	9.539	4.498
42	1 764	74 088	6.481	3.476	92	8 464	778 688	9.592	4.514
43	1 849	79 507	6.557	3.503	93	8 649	804 357	9.644	4.531
44	1 936	85 184	6.633	3.530	94	8 836	830 584	9.695	4.547
45	2 025	91 125	6.708	3.557	95	9 025	857 375	9.747	4.563
46	2 116	97 336	6.782	3.583	96	9 216	884 736	9.798	4.579
47	2 209	103 823	6.856	3.609	97	9 409	912 673	9.849	4.595
48	2 304	110 592	6.928	3.634	98	9 604	941 192	9.899	4.610
49	2 401	117 649	7.000	3.659	99	9 801	970 299	9.950	4.626
50	2 500	125 000	7.071	3.684	100	10 000	1 000 000	10.000	4.642

Table 2. Common Logarithms

	0	1	2	3	4	5	6	7	8	9
1.0	.0000	.0043	.0086	.0128	.0170	.0212	.0253	.0294	.0334	.0374
1.1	.0414	.0453	.0492	.0531	.0569	.0607	.0645	.0682	.0719	.0755
1.2	.0792	.0828	.0864	.0899	.0934	.0969	.1004	.1038	.1072	.1106
1.3	.1139	.1173	.1206	.1239	.1271	.1303	.1335	.1367	.1399	.1430
1.4	.1461	.1492	.1523	.1553	.1584	.1614	.1644	.1673	.1703	.1732
1.5	.1761	.1790	.1818	.1847	.1875	.1903	.1931	.1959	.1987	.2014
1.6	.2041	.2068	.2095	.2122	.2148	.2175	.2201	.2227	.2253	.2279
1.7	.2304	.2330	.2355	.2380	.2405	.2430	.2455	.2480	.2504	.2529
1.8	.2553	.2577	.2601	.2625	.2648	.2672	.2695	.2718	.2742	.2765
1.9	.2788	.2810	.2833	.2856	.2878	.2900	.2923	.2945	.2967	.2989
2.0	.3010	.3032	.3054	.3075	.3096	.3118	.3139	.3160	.3181	.3201
2.1	.3222	.3243	.3263	.3284	.3304	.3324	.3345	.3365	.3385	.3404
2.2	.3424	.3444	.3464	.3483	.3502	.3522	.3541	.3560	.3579	.3598
2.3	.3617	.3636	.3655	.3674	.3692	.3711	.3729	.3747	.3766	.3784
2.4	.3802	.3820	.3838	.3856	.3874	.3892	.3909	.3927	.3945	.3962
2.5	.3979	.3997	.4014	.4031	.4048	.4065	.4082	.4099	.4116	.4133
2.6	.4150	.4166	.4183	.4200	.4216	.4232	.4249	.4265	.4281	.4298
2.7	.4314	.4330	.4346	.4362	.4378	.4393	.4409	.4425	.4440	.4456
2.8	.4472	.4487	.4502	.4518	.4533	.4548	.4564	.4579	.4594	.4609
2.9	.4624	.4639	.4654	.4669	.4683	.4698	.4713	.4728	.4742	.4757
3.0	.4771	.4786	.4800	.4814	.4829	.4843	.4857	.4871	.4886	.4900
3.1	.4914	.4928	.4942	.4955	.4969	.4983	.4997	.5011	.5024	.5038
3.2	.5051	.5065	.5079	.5092	.5105	.5119	.5132	.5145	.5159	.5172
3.3	.5185	.5198	.5211	.5224	.5237	.5250	.5263	.5276	.5289	.5302
3.4	.5315	.5328	.5340	.5353	.5366	.5378	.5391	.5403	.5416	.5428
3.5	.5441	.5453	.5465	.5478	.5490	.5502	.5514	.5527	.5539	.5551
3.6	.5563	.5575	.5587	.5599	.5611	.5623	.5635	.5647	.5658	.5670
3.7	.5682	.5694	.5705	.5717	.5729	.5740	.5752	.5763	.5775	.5786
3.8	.5798	.5809	.5821	.5832	.5843	.5855	.5866	.5877	.5888	.5899
3.9	.5911	.5922	.5933	.5944	.5955	.5966	.5977	.5988	.5999	.6010
4.0	.6021	.6031	.6042	.6053	.6064	.6075	.6085	.6096	.6107	.6117
4.1	.6128	.6138	.6149	.6160	.6170	.6180	.6191	.6201	.6212	.6222
4.2	.6232	.6243	.6253	.6263	.6274	.6284	.6294	.6304	.6314	.6325
4.3	.6335	.6345	.6355	.6365	.6375	.6385	.6395	.6405	.6415	.6425
4.4	.6435	.6444	.6454	.6464	.6474	.6484	.6493	.6503	.6513	.6522
4.5	.6532	.6542	.6551	.6561	.6571	.6580	.6590	.6599	.6609	.6618
4.6	.6628	.6637	.6646	.6656	.6665	.6675	.6684	.6693	.6702	.6712
4.7	.6721	.6730	.6739	.6749	.6758	.6767	.6776	.6785	.6794	.6803
4.8	.6812	.6821	.6830	.6839	.6848	.6857	.6866	.6875	.6884	.6893
4.9	.6902	.6911	.6920	.6928	.6937	.6946	.6955	.6964	.6972	.6981
5.0	.6990	.6998	.7007	.7016	.7024	.7033	.7042	.7050	.7059	.7067
5.1	.7076	.7084	.7093	.7101	.7110	.7118	.7126	.7135	.7143	.7152
5.2	.7160	.7168	.7177	.7185	.7193	.7202	.7210	.7218	.7226	.7235
5.3	.7243	.7251	.7259	.7267	.7275	.7284	.7292	.7300	.7308	.7316
5.4	.7324	.7332	.7340	.7348	.7356	.7364	.7372	.7380	.7388	.7396
5.5	.7404	.7412	.7419	.7427	.7435	.7443	.7451	.7459	.7466	.7474
5.6	.7482	.7490	.7497	.7505	.7513	.7520	.7528	.7536	.7543	.7551
5.7	.7559	.7566	.7574	.7582	.7589	.7597	.7604	.7612	.7619	.7627
5.8	.7634	.7642	.7649	.7657	.7664	.7672	.7679	.7686	.7694	.7701
5.9	.7709	.7716	.7723	.7731	.7738	.7745	.7752	.7760	.7767	.7774
6.0	.7782	.7789	.7796	.7803	.7810	.7818	.7825	.7832	.7839	.7846
6.1	.7853	.7860	.7868	.7875	.7882	.7889	.7896	.7903	.7910	.7917
6.2	.7924	.7931	.7938	.7945	.7952	.7959	.7966	.7973	.7980	.7987
6.3	.7993	.8000	.8007	.8014	.8021	.8028	.8035	.8041	.8048	.8055
6.4	.8062	.8069	.8075	.8082	.8089	.8096	.8102	.8109	.8116	.8122

Table 2. Common Logarithms (*Continued*)

	0	1	2	3	4	5	6	7	8	9
6.5	.8129	.8136	.8142	.8149	.8156	.8162	.8169	.8176	.8182	.8189
6.6	.8195	.8202	.8209	.8215	.8222	.8228	.8235	.8241	.8248	.8254
6.7	.8261	.8267	.8274	.8280	.8287	.8293	.8299	.8306	.8312	.8319
6.8	.8325	.8331	.8338	.8344	.8351	.8357	.8363	.8370	.8376	.8382
6.9	.8388	.8395	.8401	.8407	.8414	.8420	.8426	.8432	.8439	.8445
7.0	.8451	.8457	.8463	.8470	.8476	.8482	.8488	.8494	.8500	.8506
7.1	.8513	.8519	.8525	.8531	.8537	.8543	.8549	.8555	.8561	.8567
7.2	.8573	.8579	.8585	.8591	.8597	.8603	.8609	.8615	.8621	.8627
7.3	.8633	.8639	.8645	.8651	.8657	.8663	.8669	.8675	.8681	.8686
7.4	.8692	.8698	.8704	.8710	.8716	.8722	.8727	.8733	.8739	.8745
7.5	.8751	.8756	.8762	.8768	.8774	.8779	.8785	.8791	.8797	.8802
7.6	.8808	.8814	.8820	.8825	.8831	.8837	.8842	.8848	.8854	.8859
7.7	.8865	.8871	.8876	.8882	.8887	.8893	.8899	.8904	.8910	.8915
7.8	.8921	.8927	.8932	.8938	.8943	.8949	.8954	.8960	.8965	.8971
7.9	.8976	.8982	.8987	.8993	.8998	.9004	.9009	.9015	.9020	.9026
8.0	.9031	.9036	.9042	.9047	.9053	.9058	.9063	.9069	.9074	.9079
8.1	.9085	.9090	.9096	.9101	.9106	.9112	.9117	.9122	.9128	.9133
8.2	.9138	.9143	.9149	.9154	.9159	.9165	.9170	.9175	.9180	.9186
8.3	.9191	.9196	.9201	.9206	.9212	.9217	.9222	.9227	.9232	.9238
8.4	.9243	.9248	.9253	.9258	.9263	.9269	.9274	.9279	.9284	.9289
8.5	.9294	.9299	.9304	.9309	.9315	.9320	.9325	.9330	.9335	.9340
8.6	.9345	.9350	.9355	.9360	.9365	.9370	.9375	.9380	.9385	.9390
8.7	.9395	.9400	.9405	.9410	.9415	.9420	.9425	.9430	.9435	.9440
8.8	.9445	.9450	.9455	.9460	.9465	.9469	.9474	.9479	.9484	.9489
8.9	.9494	.9499	.9504	.9509	.9513	.9518	.9523	.9528	.9533	.9538
9.0	.9542	.9547	.9552	.9557	.9562	.9566	.9571	.9576	.9581	.9586
9.1	.9590	.9595	.9600	.9605	.9609	.9614	.9619	.9624	.9628	.9633
9.2	.9638	.9643	.9647	.9652	.9657	.9661	.9666	.9671	.9675	.9680
9.3	.9685	.9689	.9694	.9699	.9703	.9708	.9713	.9717	.9722	.9727
9.4	.9731	.9736	.9741	.9745	.9750	.9754	.9759	.9763	.9768	.9773
9.5	.9777	.9782	.9786	.9791	.9795	.9800	.9805	.9809	.9814	.9818
9.6	.9823	.9827	.9832	.9836	.9841	.9845	.9850	.9854	.9859	.9863
9.7	.9868	.9872	.9877	.9881	.9886	.9890	.9894	.9899	.9903	.9908
9.8	.9912	.9917	.9921	.9926	.9930	.9934	.9939	.9943	.9948	.9952
9.9	.9956	.9961	.9965	.9969	.9974	.9978	.9983	.9987	.9991	.9996

Table 3. Natural Logarithms

ln x

x	0	0.01	0.02	0.03	0.04	0.05	0.06	0.07	0.08	0.09
1.0	0.0000	0.0100	0.0198	0.0296	0.0392	0.0488	0.0583	0.0677	0.0770	0.0862
1.1	0.0953	0.1044	0.1133	0.1222	0.1310	0.1398	0.1484	0.1570	0.1655	0.1740
1.2	0.1823	0.1906	0.1989	0.2070	0.2151	0.2231	0.2311	0.2390	0.2469	0.2546
1.3	0.2624	0.2700	0.2776	0.2852	0.2927	0.3001	0.3075	0.3148	0.3221	0.3293
1.4	0.3365	0.3436	0.3507	0.3577	0.3646	0.3716	0.3784	0.3853	0.3920	0.3988
1.5	0.4055	0.4121	0.4187	0.4253	0.4318	0.4383	0.4447	0.4511	0.4574	0.4637
1.6	0.4700	0.4762	0.4824	0.4886	0.4947	0.5008	0.5068	0.5128	0.5188	0.5247
1.7	0.5306	0.5365	0.5423	0.5481	0.5539	0.5596	0.5653	0.5710	0.5766	0.5822
1.8	0.5878	0.5933	0.5988	0.6043	0.6098	0.6152	0.6206	0.6259	0.6313	0.6366
1.9	0.6419	0.6471	0.6523	0.6575	0.6627	0.6678	0.6729	0.6780	0.6831	0.6881
2.0	0.6931	0.6981	0.7031	0.7080	0.7129	0.7178	0.7227	0.7275	0.7324	0.7372
2.1	0.7419	0.7467	0.7514	0.7561	0.7608	0.7655	0.7701	0.7747	0.7793	0.7839
2.2	0.7885	0.7930	0.7975	0.8020	0.8065	0.8109	0.8154	0.8198	0.8242	0.8286
2.3	0.8329	0.8372	0.8416	0.8459	0.8502	0.8544	0.8587	0.8629	0.8671	0.8713
2.4	0.8755	0.8796	0.8838	0.8879	0.8920	0.8961	0.9002	0.9042	0.9083	0.9123
2.5	0.9163	0.9203	0.9243	0.9282	0.9322	0.9361	0.9400	0.9439	0.9478	0.9517
2.6	0.9555	0.9594	0.9632	0.9670	0.9708	0.9746	0.9783	0.9821	0.9858	0.9895
2.7	0.9933	0.9969	1.0006	1.0043	1.0080	1.0116	1.0152	1.0188	1.0225	1.0260
2.8	1.0296	1.0332	1.0367	1.0403	1.0438	1.0473	1.0508	1.0543	1.0578	1.0613
2.9	1.0647	1.0682	1.0716	1.0750	1.0784	1.0818	1.0852	1.0886	1.0919	1.0953
3.0	1.0986	1.1019	1.1053	1.1086	1.1119	1.1151	1.1184	1.1217	1.1249	1.1282
3.1	1.1314	1.1346	1.1378	1.1410	1.1442	1.1474	1.1506	1.1537	1.1569	1.1600
3.2	1.1632	1.1663	1.1694	1.1725	1.1756	1.1787	1.1817	1.1848	1.1878	1.1909
3.3	1.1939	1.1969	1.2000	1.2030	1.2060	1.2090	1.2119	1.2149	1.2179	1.2208
3.4	1.2238	1.2267	1.2296	1.2326	1.2355	1.2384	1.2413	1.2442	1.2470	1.2499
3.5	1.2528	1.2556	1.2585	1.2613	1.2641	1.2669	1.2698	1.2726	1.2754	1.2782
3.6	1.2809	1.2837	1.2865	1.2892	1.2920	1.2947	1.2975	1.3002	1.3029	1.3056
3.7	1.3083	1.3110	1.3137	1.3164	1.3191	1.3218	1.3244	1.3271	1.3297	1.3324
3.8	1.3350	1.3376	1.3403	1.3429	1.3455	1.3481	1.3507	1.3533	1.3558	1.3584
3.9	1.3610	1.3635	1.3661	1.3686	1.3712	1.3737	1.3762	1.3788	1.3813	1.3838
4.0	1.3863	1.3888	1.3913	1.3938	1.3962	1.3987	1.4012	1.4036	1.4061	1.4085
4.1	1.4110	1.4134	1.4159	1.4183	1.4207	1.4231	1.4255	1.4279	1.4303	1.4327
4.2	1.4351	1.4375	1.4398	1.4422	1.4446	1.4469	1.4493	1.4516	1.4540	1.4563
4.3	1.4586	1.4609	1.4633	1.4656	1.4679	1.4702	1.4725	1.4748	1.4770	1.4793
4.4	1.4816	1.4839	1.4861	1.4884	1.4907	1.4929	1.4951	1.4974	1.4996	1.5019
4.5	1.5041	1.5063	1.5085	1.5107	1.5129	1.5151	1.5173	1.5195	1.5217	1.5239
4.6	1.5261	1.5282	1.5304	1.5326	1.5347	1.5369	1.5390	1.5412	1.5433	1.5454
4.7	1.5476	1.5497	1.5518	1.5539	1.5560	1.5581	1.5602	1.5623	1.5644	1.5665
4.8	1.5686	1.5707	1.5728	1.5748	1.5769	1.5790	1.5810	1.5831	1.5851	1.5872
4.9	1.5892	1.5913	1.5933	1.5953	1.5974	1.5994	1.6014	1.6034	1.6054	1.6074
5.0	1.6094	1.6114	1.6134	1.6154	1.6174	1.6194	1.6214	1.6233	1.6253	1.6273
5.1	1.6292	1.6312	1.6332	1.6351	1.6371	1.6390	1.6409	1.6429	1.6448	1.6467
5.2	1.6487	1.6506	1.6525	1.6544	1.6563	1.6582	1.6601	1.6620	1.6639	1.6658
5.3	1.6677	1.6696	1.6715	1.6734	1.6752	1.6771	1.6790	1.6808	1.6827	1.6845
5.4	1.6864	1.6882	1.6901	1.6919	1.6938	1.6956	1.6974	1.6993	1.7011	1.7029
x	0	0.01	0.02	0.03	0.04	0.05	0.06	0.07	0.08	0.09

Table 3. Natural Logarithms (*Continued*)

ln x

x	0	0.01	0.02	0.03	0.04	0.05	0.06	0.07	0.08	0.09
5.5	1.7047	1.7066	1.7084	1.7102	1.7120	1.7138	1.7156	1.7174	1.7192	1.7210
5.6	1.7228	1.7246	1.7263	1.7281	1.7299	1.7317	1.7334	1.7352	1.7370	1.7387
5.7	1.7405	1.7422	1.7440	1.7457	1.7475	1.7492	1.7509	1.7527	1.7544	1.7561
5.8	1.7579	1.7596	1.7613	1.7630	1.7647	1.7664	1.7681	1.7699	1.7716	1.7733
5.9	1.7750	1.7766	1.7783	1.7800	1.7817	1.7834	1.7851	1.7867	1.7884	1.7901
6.0	1.7918	1.7934	1.7951	1.7967	1.7984	1.8001	1.8017	1.8034	1.8050	1.8066
6.1	1.8083	1.8099	1.8116	1.8132	1.8148	1.8165	1.8181	1.8197	1.8213	1.8229
6.2	1.8245	1.8262	1.8278	1.8294	1.8310	1.8326	1.8342	1.8358	1.8374	1.8390
6.3	1.8405	1.8421	1.8437	1.8453	1.8469	1.8485	1.8500	1.8516	1.8532	1.8547
6.4	1.8563	1.8579	1.8594	1.8610	1.8625	1.8641	1.8656	1.8672	1.8687	1.8703
6.5	1.8718	1.8733	1.8749	1.8764	1.8779	1.8795	1.8810	1.8825	1.8840	1.8856
6.6	1.8871	1.8886	1.8901	1.8916	1.8931	1.8946	1.8961	1.8976	1.8991	1.9006
6.7	1.9021	1.9036	1.9051	1.9066	1.9081	1.9095	1.9110	1.9125	1.9140	1.9155
6.8	1.9169	1.9184	1.9199	1.9213	1.9228	1.9242	1.9257	1.9272	1.9286	1.9301
6.9	1.9315	1.9330	1.9344	1.9359	1.9373	1.9387	1.9402	1.9416	1.9430	1.9445
7.0	1.9459	1.9473	1.9488	1.9502	1.9516	1.9530	1.9544	1.9559	1.9573	1.9587
7.1	1.9601	1.9615	1.9629	1.9643	1.9657	1.9671	1.9685	1.9699	1.9713	1.9727
7.2	1.9741	1.9755	1.9769	1.9782	1.9796	1.9810	1.9824	1.9838	1.9851	1.9865
7.3	1.9879	1.9892	1.9906	1.9920	1.9933	1.9947	1.9961	1.9974	1.9988	2.0001
7.4	2.0015	2.0028	2.0042	2.0055	2.0069	2.0082	2.0096	2.0109	2.0122	2.0136
7.5	2.0149	2.0162	2.0176	2.0189	2.0202	2.0215	2.0229	2.0242	2.0255	2.0268
7.6	2.0281	2.0295	2.0308	2.0321	2.0334	2.0347	2.0360	2.0373	2.0386	2.0399
7.7	2.0412	2.0425	2.0438	2.0451	2.0464	2.0477	2.0490	2.0503	2.0516	2.0528
7.8	2.0541	2.0554	2.0567	2.0580	2.0592	2.0605	2.0618	2.0631	2.0643	2.0656
7.9	2.0669	2.0681	2.0694	2.0707	2.0719	2.0732	2.0744	2.0757	2.0769	2.0782
8.0	2.0794	2.0807	2.0819	2.0832	2.0844	2.0857	2.0869	2.0882	2.0894	2.0906
8.1	2.0919	2.0931	2.0943	2.0956	2.0968	2.0980	2.0992	2.1005	2.1017	2.1029
8.2	2.1041	2.1054	2.1066	2.1078	2.1090	2.1102	2.1114	2.1126	2.1138	2.1150
8.3	2.1163	2.1175	2.1187	2.1199	2.1211	2.1223	2.1235	2.1247	2.1258	2.1270
8.4	2.1282	2.1294	2.1306	2.1318	2.1330	2.1342	2.1353	2.1365	2.1377	2.1389
8.5	2.1401	2.1412	2.1424	2.1436	2.1448	2.1459	2.1471	2.1483	2.1494	2.1506
8.6	2.1518	2.1529	2.1541	2.1552	2.1564	2.1576	2.1587	2.1599	2.1610	2.1622
8.7	2.1633	2.1645	2.1656	2.1668	2.1679	2.1691	2.1702	2.1713	2.1725	2.1736
8.8	2.1748	2.1759	2.1770	2.1782	2.1793	2.1804	2.1815	2.1827	2.1838	2.1849
8.9	2.1861	2.1872	2.1883	2.1894	2.1905	2.1917	2.1928	2.1939	2.1950	2.1961
9.0	2.1972	2.1983	2.1994	2.2006	2.2017	2.2028	2.2039	2.2050	2.2061	2.2072
9.1	2.2083	2.2094	2.2105	2.2116	2.2127	2.2138	2.2148	2.2159	2.2170	2.2181
9.2	2.2192	2.2203	2.2214	2.2225	2.2235	2.2246	2.2257	2.2268	2.2279	2.2289
9.3	2.2300	2.2311	2.2322	2.2332	2.2343	2.2354	2.2364	2.2375	2.2386	2.2396
9.4	2.2407	2.2418	2.2428	2.2439	2.2450	2.2460	2.2471	2.2481	2.2492	2.2502
9.5	2.2513	2.2523	2.2534	2.2544	2.2555	2.2565	2.2576	2.2586	2.2597	2.2607
9.6	2.2618	2.2628	2.2638	2.2649	2.2659	2.2670	2.2680	2.2690	2.2701	2.2711
9.7	2.2721	2.2732	2.2742	2.2752	2.2762	2.2773	2.2783	2.2793	2.2803	2.2814
9.8	2.2824	2.2834	2.2844	2.2854	2.2865	2.2875	2.2885	2.2895	2.2905	2.2915
9.9	2.2925	2.2935	2.2946	2.2956	2.2966	2.2976	2.2986	2.2996	2.3006	2.3016
x	0	0.01	0.02	0.03	0.04	0.05	0.06	0.07	0.08	0.09

Answers to Selected Odd-Numbered Exercises

CHAPTER 1

Exercises 1.1

1. $\in$ **3.** $\in$ **5.** $\notin$ **7.** $\notin$ **9.** $\in$ **11.** false   **13.** true   **15.** true   **17.** false
19. true   **21.** {Saturday, Sunday}   **23.** {1, 2, 3, 4, 5, 6, 7}   **25.** {. . . , $-8, -7, -6$}
27. {14, 28}   **29.** $E = \{x \mid x$ is an even integer between 1 and 15$\}$
31. $T = \{x \mid x$ is a state of the USA that begins with the letter A$\}$ **33.** $P = \{x \mid x$ is a negative integer less than $-4\}$ or $P = \{x \mid x$ is a negative integer less than or equal to $-5\}$ **35.** finite   **37.** finite
39. infinite   **41.** false   **43.** false   **45.** true   **47.** true   **49.** false   **51.** E is a subset of N
53. F is a subset of A and N **55.** S is a subset of Q and R **57.** true   **59.** false   **61.** false
63. false   **65.** false   **67.** false   **69.** false   **71.** true   **73.** false   **75.** false   **77.** {m, r}
79. $\emptyset$ **81.** {l, n}   **83.** {m, p, r}   **85.** $\emptyset$ **87.** $\emptyset$ **89.** {1, 2, 3, 4, 5}
91. {1, 2, 3, 4, 6, 7, 8, 9}   **93.** {1, 2, 3, 4, 5, 6, 7, 8, 9}
95. {1, 2, 3, 4, 6, 7}   **97.** {1, 2, 3, 4, 6, 7}

99.

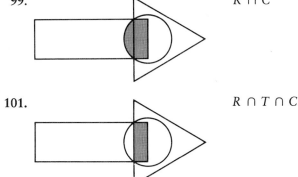

$R \cap C$

101.

$R \cap T \cap C$

103.

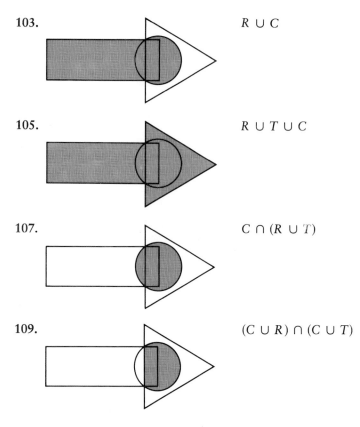

$R \cup C$

105.

$R \cup T \cup C$

107.

$C \cap (R \cup T)$

109.

$(C \cup R) \cap (C \cup T)$

111. Rh negative, type AB   **113.** Rh positive, type B   **115.** Rh negative, type B
117. Rh negative, type O   **119.** $\emptyset$ **121.** A **123.** $\emptyset$ **125.** U **127.** $C = D = \emptyset$
129. $C = D = U$ **131.** $C \subseteq D$ **133.** $C \subseteq U$ **135.** $C = \emptyset$ **137.** C any set

Exercises 1.2

1. (a), (b), (c), (e)   **3.** (c), (e)   **5.** (b), (c), (e)   **7.** (c), (e)   **9.** (c), (e)   **11.** true
13. true   **15.** false   **17.** true   **19.** true   **21.** $0.555\ldots$ **23.** 0.3125 **25.** -6.5
27. $0.777\ldots$ **29.** $-0.3535\ldots$ **31.** $1.8333\ldots$ **33.** $\{\ldots, -3, -2, -1, 0\}$ **35.** $\emptyset$
37. $\emptyset$ **39.** $\{\ldots, -3, -2, -1, 0\}$ **41.** All real numbers which are not integers.
43. $\{-2, -1, 0, -6, 1\}$ **45.** $\{-2, -1, -6\}$ **47.** $\{\sqrt{5}, -\sqrt{2}\}$ **49.** all   **51.** some
53. some   **55.** no   **57.** no

Exercises 1.3

1. commutative property for addition   **3.** commutative property for addition   **5.** additive identity
7. zero-factor property and additive identities   **9.** distributive property   **11.** multiplicative identity
13. associative property for multiplication   **15.** commutative property for multiplication
17. commutative property for addition   **19.** additive inverse   **21.** true   **23.** true   **25.** false
27. (i) multiplicative identity (ii) distributive property (iii) additive inverse (iv) zero-factor property
29. (i) commutative property for addition (ii) associative property for addition (iii) associative
property for addition (iv) additive inverse (v) additive identity (vi) additive inverse

Exercises 1.4

1. < **3.** > **5.** < **7.** < **9.** > **11.** $\sqrt{3} < e$ **13.** $y \geq 0$ **15.** $x \leq -3$
17. (a) $-\sqrt{6}$; $\frac{-2}{3}$; 0; 1.62; e; 3; π; $\frac{22}{7}$ **(b)** -3.13; -2; $-\pi/3$; 0; $\sqrt{11}$; 5 **19.** > or ≥ **21.** > or ≥
23. < **25.** > **33. (a)** $x < 0$ **(b)** $x = 0$ **(c)** $x > 0$
35. (a) $x = 0$ or $x = 1$ **(b)** $x = 0$ or $x = 1$ **(c)** $x > 1$ **(d)** $x < 0$ or $x > 1$ **(e)** $0 < x < 1$

Exercises 1.5

1. 0 **3.** 5 **5.** $\frac{1}{3}$ **7.** e **9.** 1 **11.** 12 **13.** 0 **15.** -7 **17.** true **19.** true
21. false **23.** true **25.** false **27.** false **29.** false **31.** false **33.** 4 **35.** -8
37. 3 **39.** -3 **41.** 16 **43.** 13 **45.** 10 **47.** 22 **49.** -2 **51.** 14 **53.** \$43
55. 6 degrees up

Exercises 1.6

1. -63 **3.** 0 **5.** -36 **7.** 1 **9.** $\frac{-1}{6}$ **11.** -1 **13.** .75 **15.** 1.21 **17.** -6
19. -12 **21.** -27 **23.** 48 **25.** -36 **27.** -60 **29.** 0 **31.** -19 **33.** -24
35. -4 **37.** -70 **39.** -12 **41.** -4 **43.** 4 **45.** 2 **47.** 9 **49.** 6 **51.** -11
53. 0 **55.** not possible **57.** 4 **59.** 20 **61.** -1 **63.** -3 **65.** 2 **67.** -4
69. -7 **71.** 1 **73.** -6 **75.** 15 **77.** 15 **79.** $-\frac{1}{20}$ **81.** \$300 **83.** $\frac{-3}{7}°$C
85. lost 50 cents

Chapter 1 Exercises

1. false **3.** true **5.** true **9.** {spades, hearts, diamonds, clubs}
11. $A = \{x \mid x$ is a multiple of 2 less than $-1\}$ **13.** $C = \{x \mid x$ is a spoken language$\}$
15. $E = \{x \mid x$ is a work by William Shakespeare$\}$ **17.** infinite **19.** finite **21.** equal
23. equal **25.** equal **27.** {0}, {1}, {3}, {0, 1}, {0, 3}, {1, 3}, {0, 1, 3}, Ø **29.** {Ø}; Ø **31.** true
33. true **35.** false **37.** false **39.** false **41.** ⊆ **43.** ⊆ **45.** ⊆ **47.** ⊆ **49.** ⊆

51.

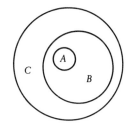

53. {6, 8, 10} **55.** R **57.** {1, 2, 3, 4, 5, 7, 9} **59.** {6, 8, 10} **61.** {2, 4} **63.** U
65. (b) **67.** (c) **71.** B **73.** $A \subseteq B$ **75.** $A = \{1\}$; $B = \{1, 2\}$; $C = \{1, 3\}$ **77.** true
79. true **81.** false **83.** false **85.** true **87.** true
89. commutative property for multiplication **91.** associative property for multiplication
93. additive identity **95.** additive inverse **97.** double negative property
99. commutative property for multiplication **101.** cancellation property for addition
103. transitive property of order **105.** cancellation property for multiplication
107. multiplicative property of order (iii) **109.** 74 **111.** 0 **113.** 0 **115.** -6 **117.** 6
119. 60 **121.** -60 **123.** -5 **125.** 2 **127.** 0 **129.** -3; 1; -2; 3; -4
131. -12; 2; 6; 0; 18 **133.** 5; 0; -7; $\frac{5}{2}$; -10 **135.** \$340 **137.** loss of 19 pounds

CHAPTER 2

Exercises 2.1

1. base = 3; exponent = 5   **3.** base = r; exponent = 4   **5.** base = 10; exponent = 7
7. base = $a + b$; exponent = 6   **9.** base = $4x$; exponent = 2
11. base = x, exponent = 2; base = y, exponent = 3   **13.** 2^3 **15.** z^4 **17.** $(-4y)^3$

19. $-5\left(\dfrac{1}{y}\right)^3$ **21.** πr^2 **23.** $a^2 b^2 - ab^3 c^2$ **25.** $xxxxx$ **27.** $aaabbbb$

29. $(-2x)(-2x)(-2x)(-2x)$ **31.** $-2xxxx$ **33.** $(a - b)(a - b)$ **35.** $(-a)(-a)(-a)(-b)(-b)(-b)(-b)$

37. $xxxy - 4xyy$ **39.** $(2a)(2a)bb - cc$ **41.** 9^2 **43.** $\dfrac{1}{6}b^2$

45. $(8M)^2$ **47.** $(xy)^9$ **49.** $(x + 15)^3$

Exercises 2.2

1. polynomial   **3.** not a polynomial   **5.** polynomial   **7.** polynomial

9. not a polynomial   **11.**

terms:	$15x^2 y$	$16y$
coeff's:	15	16

13.

terms:	x^2	$-7xy$	$3y^2$
coeff's:	1	-7	3

15.

terms:	$\frac{2}{3}x^3 z$	$-\frac{1}{2}xz$	$\frac{4}{7}z^2$
coeff's:	$\frac{2}{3}$	$-\frac{1}{2}$	$\frac{4}{7}$

17.

terms:	$3.1xy^3$	$-10x^2 y^2$	$1.5x^3$
coeff's:	3.1	-10	1.5

19.

terms:	$\sqrt{3}x^2$	πx	15
coeff's:	$\sqrt{3}$	π	15

21. deg. = 2   **23.** deg. = 1   **25.** deg. = 2
27. deg. = 2   **29.** deg. = 0   **31.** deg. = 5   **33.** deg. = 6   **35.** deg. = 4
37. 36   **39.** 15   **41.** 0   **43.** 5   **45.** 3   **47.** 4   **49.** 9
51. perimeter = $2l + 2w$ **53.** (a) area of region $(A + B + C + D)$
    area = lw (b) perimeter of region $(A + D)$
                                          (c) perimeter of region $(B + D)$
                                          (d) perimeter of region C
                                          (e) perimeter of region $(A + B + C + D)$

55. $75 - (11x + 13y)$

Exercises 2.3

1. $7x^2$ **3.** $7x^3$ **5.** $10xy$ **7.** 0   **9.** $3x^2 y$ **11.** $13x + 9y$
13. $3x^2 - x + 4$ **15.** $-10a^2 b^2 - 3c^2 d^2$ **17.** $4a^2 + 13 + 8a$
19. $4x^3 y + 3x^2 y^2 + 2xy^3 + 8a^2 - 9x^2 y$ **21.** $-2x - 5$ **23.** $9x - 12y$
25. $7y^2 + 5y + 2$ **27.** $5x - 5y + 7$ **29.** $-6x + 15$ **31.** $x^2 - 3x - 1$
33. $6x + 8y$ **35.** $5a - 3b$ **37.** $6y - 6x + 5$ **39.** $a + 2b + 1$
41. $13x^2 - (5x - 2)$ **43.** $4 - [(7 + x) - (-3 - x)]$ **45.** $4x^2 - 2(3xy + 2x - 1)$
47. no mistakes   **49.** $-3x^2 y + 6x - 12$ **51.** $2xy - 2$ **53.** $14a - 5b$
55. $x + 5y - 4$ **57.** $-3x^4 + 5x^2 - 7$ **59.** $x^2 y^2 + 3xy - x - 3y - 3$
61. $3x^2 - 6x + 11$ **63.** $16a - 6$ **65.** $7d + 6$ **67.** perimeter = $10x + 6y$
                                                                       area = $9xy$

Exercises 2.4

1. 3^6 **3.** $(-7)^8$ **5.** x^9 **7.** $-a^8$ **9.** 3^{m+4} **11.** 4^{2n+3}
13. b^{2n+4} **15.** y^{n+2} **17.** 3^{6m} **19.** b^{2n+4} **21.** $(y - x)^7$
23. $-2(a + 4b)^7$ **25.** $x^2 y^5$ **27.** $10a^5 b^3$ **29.** $-3a^2 b^3$ **31.** $-15x^4 yz$
33. $-60x^4 y^2$ **35.** $6x^5 y^6$ **37.** $27a^8 b^7$ **39.** $54x^2 y^{10} z^4$ **41.** $-y^4$

43. $-a^7b$ **45.** $-13x^3$ **47.** $4x^3y^2z^2$ **49.** $82a^6b^{10}c$ **51.** 3^6
53. $(-5)^8$ **55.** 2^6 **57.** $-8x^6$ **59.** $-18y^3x^2$ **61.** $64a^4$ **63.** $64a^4b^3$
65. $50a^5c^2$ **67.** $-x^8y^{12}z^4$ **69.** x^8y^9 **71.** $4x^8y^5$ **73.** y^{5n}
75. $27x^{16}y^{11}z^8$ **77.** $(x-y)^{20}$ **79.** $7a^5b^8(a^2-b^2)^7$ **81.** $2x^{10}y$
83. $-616x^6y^6$ **85.** $-730a^{12}b^{12}c^6$

Exercises 2.5

1. $2x - 2y$ **3.** $x^2 - 3xy$ **5.** $7x^2 - 14x$ **7.** $3x^2 - 4x$ **9.** $6x^3 - 3x^2$
11. $5xy^2 - 10xy^3$ **13.** $5x^2 - 5x + 5$ **15.** $16x^2 + 12x - 28$
17. $-2x^4 + 4x^3y + 2x^2y^2$ **19.** $2y^7 + 4y^6 - 6y^4$ **21.** $-3y^4 + 15y^3 - 27y^2$
23. $18a^5b - 15a^3b^2 + 6a^2b^3 - 12a^2b$ **25.** $-4x^7y^3 + 4x^6y^4 - 8x^5y^6 + 4x^3y^8$
27. $144x^2y^2z^2 - 48x^2y^2z^3 + 36x^3y^2z^5 + 12x^2y^3z^2$
29. $-12x^4yz^2 + 18x^3y^2z^2 + 6x^3y^2z^3 - 24x^3y^3z^3 + 30x^3y^2z^4$
31. $x^{n+1}y^{m+1}z^{p+2} - 3x^2y^3z^{n+2} + 2x^{m+1}y^{n+1}z^3$
33. (a) $(6x + 14)$ km   (b) $(30x + 70)$ km   (c) $(3xh + 7h)$ km   (d) $(3x^2 + 7x)$ km
35. $(175h)$ mi $+ [190(h + 3)]$ mi $= (365h + 570)$ mi   **37.** $16x^2$ square inches
39. $(46x^2 + 54x)$ square inches   **41.** $x^2 - x - 6$ **43.** $x^2 - 3x + 2$ **45.** $y^2 - 4y + 3$
47. $x^2 - 7x + 12$ **49.** $y^2 + y - 6$ **51.** $x^2 - 7x + 10$ **53.** $x^4 - 11x^2 + 30$
55. $y^4 + 9y^2 + 18$ **57.** $z^4 - 10z^2 + 24$ **59.** $x^2 + ax - 2a^2$ **61.** $a^2 + 3ax - 18x^2$
63. $x^2 - 5xy + 6y^2$ **65.** $x^4 - 9x^2 + 20$ **67.** $12x^2 - 5x - 2$ **69.** $9y^2 + 18y + 8$
71. $3x^2 - 10x - 15$ **73.** $8x^4 - 18x^2 + 9$ **75.** $3x^2 - 11x + 10$ **77.** $3x^2 + 5x + 2$
79. $3x^2 - 7xy + 2y^2$ **81.** $3x^2 + 8xy - 3y^2$ **83.** $6x^2 + x - 5$ **85.** $4x^2 - 12xy + 9y^2$
87. $9y^4 + 18y^2 + 8$ **89.** $7x^2 + 6xy - y^2$ **91.** $a^3 - 3a^2b + 3ab^2 - b^3$
93. $8x^3 - 12x^2 + 6x - 1$ **95.** $a^4 + a^2x^2 + x^4$ **97.** $12a^3 - 25a^2b + 20ab^2 - 6b^3$
99. $4a^4 - 5a^2 + 1$ **101.** $32x^3y - 52x^2y^2 + 19xy^3 - 5y^4$ **103.** $3a^3 - 20a^2 + 8$
105. $a^2 - 4ab + 4b^2 - c^2$ **107.** $12x^3 - 12x^2 - 17x + 12$ **109.** $8a^3 - 27b^3$
111. $4a^2 + 4ac + c^2 - 9$ **113.** $x^3 + y^3 - x - y$ **115.** $4x^4 + 4x^2y^2 + y^4 - 4z^4$
117. $x^4 + 10x^3 + 35x^2 + 50x + 24$ **119.** $x^6 - y^6$ **121.** $x^{4n+4} - y^{2m}$
123. $4x^2 - 20x + 25$ **125.** $4x^2 - 20xy + 25y^2$ **127.** $4x^4 + 4x^2 + 1$
129. $16 - 40x^3 + 25x^6$ **131.** $4a^2x^2 - 12abxy + 9b^2y^2$ **133.** 25 **135.** 441
137. 9801 **139.** $9x^2 - 16$ **141.** $121x^4 - 49y^2$ **143.** $16x^2 - 25y^4$
145. $x^2 + 4x + 4 - y^2$ **147.** $4a^2 + 4ab + b^2 - c^2$ **149.** $a^4 - (b^4 + 2ab^3 + a^2b^2)$
151. $9 + 6y + y^2 - x^2$ **153.** $9,991$ **155.** $489,996$ **157.** $(1100)^2 - 25$
159. $x(x + c)$ **161.** $\pi(2h - 3)^2 \cdot h$
163. area $= [\frac{1}{4}\pi(4a + 5)^2] - [\frac{1}{4}\pi(3a - 1)^2] = \frac{1}{4}\pi(7a^2 + 46a + 24)$ sq in.

165. distance $=$ rate $\cdot$ time $= (x + 25)\dfrac{\text{mi}}{\text{hr}} \cdot (3x - 1) \,\cancel{\text{hr}} = (3x^2 + 74x - 25)$ mi

167. $x^3 + (2y)^3$ **169.** $125y^3 - 300y^2x^2 + 240yx^4 - 64x^6$ **171.** $(1 + x)^3 + y^3$

Exercises 2.6

1. 7 **3.** 5^9 **5.** 1 **7.** y^2 **9.** x^3 **11.** a^{2n} **13.** a^{n2-2} **15.** $(x + 2)^{2n2+4n}$

17. $\dfrac{81}{y^4}$ **19.** $\dfrac{8x^3}{27}$ **21.** $\dfrac{16x^2}{81y^2}$ **23.** $\dfrac{(x + 21)^3}{27(x - 5)^3}$ **25.** $\dfrac{7^n(x + 3)^n}{(x - 5)^n}$ **27.** $2x^2$

29. $-3x$ **31.** $10x^4y$ **33.** $9x^2y^2$ **35.** $-3x^2y$ **37.** $-11xy^7$ **39.** $-22x^4y^2$
41. $5y^5z^2$ **43.** $-4x^{m-1}y^{4-m}$ **45.** $-4x^{5n-1}y^{5n-5}z$ **47.** $-2ab^2$ **49.** $-8x^{11}y^5z^3$
51. $x^{2m}y^{4m-3}$ **53.** (a) $-3 + 6y - 9y^2$ (b) $y^2 - 2y^3 + 3y^4$
55. $4(-7)^2 - 10(-7) + 13$ **57.** $-4xyz + 5 - 6y$ **59.** $-2xy + xy^2 - 3y$

61. $2x^2 - 3x + 1$ **63.** $12x^2y^3 - xy^2 + 3y$ **65. (a)** $(1 + x)^3 - y(1 + x)^2 + 7(1 + x)$
   (b) $(1 + x)^4 - y(1 + x)^3 + 7(1 + x)^2$ **(c)** $x(1 + x)^3 - xy(1 + x)^2 + 7x(1 + x)$
67. (a) $-1 + 2(x + y)^5$ **(b)** $3(a - b)(x + y)^3 - 6(a - b)(x + y)^8$
69. $1 - (1 - y)y^{-1} + (1 - y)^2y^{-2}$ **71.** length $= (3 + 5x)$ ft
73. (a) $(3x^3 - 9x^2)$ hr   **(b)** $(15x^2 - 45x)$ hr   **(c)** $(15x - 45)$ hr   **(d)** $(x - 3)$ hr

Exercises 2.7

1. $x + 4$ **3.** $x + 7$ **5.** $y^2 - 2y + 1$ **7.** $3x + 4y$ **9.** $5x - 1 - \dfrac{5}{x - 1}$

11. $2x - y$ **13.** $5x^2 + 21x + 4$ **15.** $3x - y$ **17.** $y + x$ **19.** $y^2 + yx + x^2$
21. $4x^2 - y^2$ **23.** $x^2 + 5xy - 4y^2$ **25.** $7x^2 - 2xy + 3y^2$ **27.** $a^3 - 4a^2 - 4a + 16$
29. $12x^3 + 8x^2 + 4x + 1$ **31.** $x - y$ **33.** $a^2 - ab + b^2$ **35.** $a^2 - 3a + 9$
37. $4x^2 - x + 2$ **39.** $a - 3b - 2c$ **41.** $2x + y$ **43.** $2x^2 + 3x + 1$

45. $3a - 2b + \dfrac{4ab^2}{9a^2 + 6ab + 4b^2}$ **47.** $5a^2 - 3a + 1$

49. $4x - 3y$ **51.** $2a^2 - b^2 + \dfrac{2a}{2a^2 - b^2}$

53. $16x^4 + 8x^2 + 4$ **55.** $3a - 2$ **57.** $25a^2 - 10ab + b^2$
59. $y^2 - 4$ **61.** $4a - 5b$ **63.** $x - 3y - 2$ **65.** $3a^2 + 2b$
67. $x + 2$ **69.** $4a^2 + ab - b^2$ **71.** $2b + 3$ **73.** $x - 2y$

75. $x - 3y$ **77.** $a^2 - 3ac - 2c^2$ **79.** $12a^2 - 10ab - 4b^2 + \dfrac{4ab^3 + 8b^4}{a^2 + b^2}$

81. $x + y + a - b$ **83.** $a(x - 2y) + 2$ **85.** $(x + y)^2 + (x + y)(m + n) + (m + n)^2$
87. $5x - 7$ **89.** $k = -6$

Exercises 2.8

1. quotient $= 3x^2 - 7x + 18$; remainder $= -28$
3. quotient $= 2x^2 + x + 3$; remainder $= 3$
5. quotient $= x^2 - 2x + 4$; remainder $= 0$
7. quotient $= x^4 + x^3 - x^2 + 2x + 2$; remainder $= 0$
9. quotient $= 2x^2 + 6x + 37$; remainder $= 217$
11. quotient $= 2x^3 - 6x^2 + 17x - 48$; remainder $= 139$
13. quotient $= x^4 - x^3 + x^2 - x + 1$; remainder $= 0$
15. quotient $= 3x^2 + 4ax + 5a^2$; remainder $= 7a^3$

Exercises 2.9

1. 41   **3.** 1   **5.** 54   **7.** -1 **9.** 5
11. $f(m) = 4$; $m^3 - 3m - 6 = 4$ **13.** yes   **15.** yes   **17.** no
19. no   **21.** yes   **23.** yes   **25.** $m = -7$
27. $P(a) = a^n - a^n = 0$ **29.** $P(-a) = (-a)^n + a^n = a^n + a^n = 2a^n \neq 0$

Chapter 2 Exercises

1. deg. $= 1$ **3.** not a polynomial   **5.** deg. $= 2$ **7.** deg. $= 5$
9. not a polynomial   **11.** deg. $= 2$ **13.** 2400   **15.** 11   **17.** 17,956
19. 205   **21.** 432   **23.** 51   **25.** 343   **27.** 5   **29.** $\frac{1}{144}$

31. $4x$ **33.** $2(x + y) + 4b$ **35.** $2(a + b + d)$ **37.** $\dfrac{x}{2} \cdot (x - 7)$ sq in.

39. $[4(60 + x) + 5(90 + x)]$ mi   **41.** $4(x + 5)$ years

43. $(25x + 19y)$ dollars   **45.** $\dfrac{3x^2 - 2x}{x} = (3x - 2)$ dollars

47. $21x + 22y + 11z$ **49.** $14a + 25x + 14y - 19z$ **51.** $-6ab + 2b^3 - a^2$
53. $x^3 - 5y^3 - xy - 4y$ **55.** $11x^2y - 4xy^2 + 12xy + 4x^3 + 3$
57. $8x^4 - 7a^2 + 11xy$ **59.** $-3x^2 - 15y^2 - 11xy$ **61.** $6x + 7y$
63. $5x + 3y$ **65.** $2x$ **67.** $4x - y$ **69.** $-6y - z$ **71.** $16xy - x^2y - 2xy^2$
73. $36x^3z^4 - 38x^4z^3 + 32a^2b^2x^3z^3$ **75.** $-84abx^4 + 35abx^3y - 42abx^3 + 14ab^2x^2$
77. $80x^6y^4z^2 - 72x^4y^3z^3 + 24x^5y^5z^3 + 16x^5y^3z^4$
79. $18a^3uvz^2w^2 - 24a^4u^2vw^2 + 30a^5u^3v^3$
81. $-20a^2b^3x^2 + 8a^3b^2z^4 - 20ab^3cx^2 - 4abx^4$
83. $-40a^6b^2y + 48ab^2y^6 - 56a^2by^6$
85. $56abe^3x^{19}y^{23}z - 32a^6bd^{19}e^{21}x^{28}y^{23}z^5$
87. $-x^3y^2z^3 + x^3y^2z^4 - x^4y^2z^3 + x^4y^2z^4$
89. $84a^3x^5y^2z - 35a^3x^5y^3z^3 + 28a^6b^2x^4y^5z^3$
91. $x^2 + 2xy + y^2$ **93.** $mp + np + mq + nq$ **95.** $c^2 - d^2$
97. $x^4 + 2x^2y^2 + y^4$ **99.** $c^3 + 3c^2d + 3cd^2 + d^3$ **101.** $m^4 - n^4$
103. $a^8 - 1$ **105.** $p^4 + pq^3 + p^3q + q^4$
107. $21a^5 - 90a^4x + 24a^3x^2 + 14a^2x^3 - 39ax^4 + 10x^5$
109. $154a^3b^2cm - 70a^6my^2 + 28a^3mz + 11b^2cyz - 5a^3y^3z + 2yz^2$
111. $50a^4 + 125a^5b + 14a^6b^2 - 75a^3b - 18a^4b^2 - 21a^5b^3$
113. $21a^4 - 34a^3x + 34a^2x^2 + 2ax^3 - 15x^4$
115. $12x^7 - 3x^6 + 18x^5 - 8x^3 + 22x^2 - 17x + 30$ **117.** $4096b^6 - c^6$
119. $a^9 - 9ay^8 - 8y^9$ **121.** $x^3 + ax^2 + bx^2 + cx^2 + abx + acx + bcx + abc$
123. $a^6 - x^6$ **125.** $xyz + y^2z + xz^2 + yz^2 + xyw + y^2w + xzw + yzw$
127. $x^2 - 10x + 25$ **129.** $81 - 18y + y^2$ **131.** $x^2y^2 - 8xy + 16$
133. $x^4 - 10x^2 + 25$ **135.** $4a^2 + 4a + 1$ **137.** $25 + 30z + 9z^2$

139. $25m^2 + 20mn + 4n^2$ **141.** $x^4 - 10x^2y + 25y^2$ **143.** $\dfrac{x^2y^2}{4} + 6xyz + 36z^2$

145. 361 **147.** $x^2 - 9$ **149.** $x^2 - y^2$ **151.** $x^2y^2 - 25$ **153.** $49 - 9m^2$

155. $\dfrac{a^2}{x^2} - \dfrac{c^2}{y^2}$ **157.** $(a + 2)^2 - c^2$ **159.** $x^4 - (y^2 + xy)^2$

161. $x^6 - 9x^4y + 27x^2y^2 - 27y^3$ **163.** $27x^6 + 108x^4y + 144x^2y^2 + 64y^3$
165. $x^3 + 64y^3$ **167.** $x^3 - 64y^3$ **169.** $a + b + c$
171. $5ab - 3cy^2 + 4x^3y^4$ **173.** $-4ax - 7a^2x^2 + 9a^3x^3$
175. $5y^2 + 3x^3 - 2a^4$ **177.** $-2am + 3am^2 - 4am^3$
179. $5am - 3an + 7n^2y$ **181.** $-2am^3 + 3bc^2 - 5s^2x$ **183.** $a + x$
185. $m^3 + n^3$ **187.** $x^2 - xy + y^2$ **189.** $26x^2y - 20xy + 6xy$

191. $1 + 10x + \dfrac{60x^2}{1 - 6x}$ **193.** $5a^2 + 2ab - 3b^2$ **195.** $y + x$

197. $3x - 3z$ **199.** $a + b$ **201.** $1 + \dfrac{2x^2}{n^2 - x^2}$

CHAPTER 3

Exercises 3.1

1. identity   **3.** conditional   **5.** identity   **7.** conditional   **9.** conditional
11. conditional   **13.** identity   **15.** identity   **17.** conditional   **19.** identity

21. conditional   **23.** conditional   **25.** identity   **27.** conditional   **29.** yes
31. no   **33.** yes   **35.** no   **37.** yes   **39.** no   **41.** yes   **43.** yes   **45.** yes
47. no   **49.** yes

Exercises 3.2

1. $x = 18$ **3.** $x = 4$ **5.** $x = 26$ **7.** $x = 16$ **9.** $x = -2$ **11.** $x = -4$
13. $x = 7$ **15.** $x = -9$ **17.** $x = -14$ **19.** $x = 4$ **21.** $x = -.4$ **23.** $x = -1$
25. $x = -\dfrac{3}{4}$ **27.** $x = -\dfrac{3}{2}$ **29.** -7 **31.** 10 **33.** $\dfrac{1}{4}$ **35.** $\dfrac{17}{6}$ **37.** $x = 3$
39. $x = -3$ **41.** $x = -4$ **43.** $x = 6$ **45.** $x = -5$ **47.** $x = -\dfrac{5}{2}$ **49.** $x = -3$
51. $x = -.7$ **53.** $x = 6$ **55.** $x = -41$ **57.** $x = 13$ **59.** $x = -100$ **61.** $x = -1$
63. $x = -25$ **65.** $x = 32$ **67.** $x = \dfrac{5}{4}$ **69.** $x = -1$ **71.** $x = \dfrac{15}{2}$ **73.** 0 **75.** 8

Exercises 3.3

1. $x = 4$ **3.** $x = \frac{3}{4}$ **5.** $x = \frac{3}{5}$ **7.** $x = -18$ **9.** $x = \frac{1}{2}$ **11.** $x = 3$ **13.** $x = 11$
15. $x = 0$ **17.** $x = 13$ **19.** $x = 10$ **21.** $x = -\frac{65}{4}$ **23.** $x = -1$ **25.** $x = 2$
27. $x = -\frac{1}{4}$ **29.** $x = \frac{1}{3}$ **31.** $x = 3$ **33.** $x = 6$ **35.** $x = \frac{26}{5}$ **37.** $x = 0$
39. $x = \frac{1}{3}$ **41.** $x = -10$ **43.** $x = -4$ **45.** $x = 6$ **47.** $x = 6$ **49.** $x = -\frac{19}{7}$
51. $x = -\frac{14}{5}$ **53.** $x = 6$ **55.** $x = 13$ **57.** $x = \frac{7}{15}$ **59.** $x = 15$ **61.** $x = 9$
63. $x = -\frac{5}{2}$ **65.** $x = \frac{2}{3}$ **67.** $x = -1$ **69.** $x = \frac{6}{5}$ **71.** $x = -\frac{5}{4}$ **73.** $x = -2$
75. $x = \frac{7}{3}$ **77.** $x = 2$ **79.** $x = -\frac{5}{3}$ **81.** $x = -\frac{22}{3}$ **83.** $x = \frac{24}{5}$ **85.** $x = 0$
87. $x = 72$ **89.** $x = -70$ **91.** $x = 2$ **93.** $x = -2$ **95.** $x = 7$ **97.** $x = \frac{5}{3}$
99. $x = 5$

Exercises 3.4

1. $x = \dfrac{5}{m}$ **3.** $x = n + 5$ **5.** $x = ab$ **7.** $x = 3a$ **9.** $x = \dfrac{b - 7}{a}$ **11.** $x = \dfrac{b - 2a}{2}$
13. $x = p - 4m$ **15.** $x = \dfrac{p - mn}{7}$ **17.** $x = \dfrac{a + 2b}{2}$ **19.** $x = \dfrac{5m - 3n}{7}$ **21.** $x = -\dfrac{2np}{m}$
23. $x = 8b$ **25.** $x = \dfrac{2mp + pr}{n}$ **27.** $x = 3m$ **29.** $x = 4m$ **31.** $x = \dfrac{4p + 3n}{m}$
33. $x = \dfrac{7mp + 2r}{2p}$ **35.** $x = m$ **37.** $x = 12a - 8$ **39.** $x = 10c$ **41.** $x = 2a$ **43.** $x = 0$
45. $x = -3m$ **47.** $x = 14 - 3m$ **49.** $x = \dfrac{ac + a}{2}$ **51.** $b = p - a - c$
53. $A = 180° - B - C$ **55.** $\ell = \dfrac{A}{w}$ **57.** $w = \dfrac{V}{\ell h}$ **59.** $n = \dfrac{S + 360}{180}$ **61.** $r = \dfrac{C}{2\pi}$
63. $h = \dfrac{3V}{\pi r^2}$ **65.** $r = \dfrac{S}{2\pi h}$ **67.** $x = \dfrac{y - b}{m}$ **69.** $t = \dfrac{A - p}{rp}$ **71.** $R = \dfrac{W}{I^2}$ **73.** $t = \dfrac{v - v_o}{a}$
75. $n = \dfrac{Vt - ct}{-c}$

Exercises 3.5

1. $<$ **3.** $<$ **5.** $>$ **7.** $>$ **9.** $>$ **11.** $<$ **13.** $<$ **15.** $>$ **17.** $<$
19. $<$ **21.** $>$ **23.** $<$ **25.** $>$ **27.** false: $3 > -4$ but $3^2 < (-4)^2$ **29.** true
31. $m > n$ **33.** $x^n > 0$ **35.** a and b have the same sign: both positive or both negative
37. a and b have the same sign   **39.** b is positive

Exercises 3.6

1. absolute **3.** conditional; let $x = 6$ **5.** absolute **7.** conditional; let $x = 0$ **9.** absolute

11. $x > 7$ **13.** $x \geq 3$ **15.** $x < -1$ **17.** $x < -\dfrac{7}{3}$ **19.** $x \geq -1$ **21.** $x \leq 1$

23. $x < -4$ **25.** $x > -3$ **27.** $x < 2$ **29.** $x \leq 0$ **31.** $x < -4$ **33.** $x \geq 2$
35. $x < -5$ **37.** $x > -\frac{1}{6}$ **39.** $x < -3$ **41.** $x \leq \frac{3}{2}$ **43.** $x \geq \frac{4}{3}$ **45.** $x < 0$
47. $x \leq 1$ **49.** $x < \frac{3}{2}$ **51.** $x \leq 26$ **53.** $x < -\frac{27}{2}$ **55.** $x > 32$ **57.** $x < -35$
59. $x \leq \frac{24}{11}$ **61.** $x \geq -\frac{6}{5}$ **63.** $x \geq -8$ **65.** $x > \frac{37}{11}$ **67.** $x < \frac{9}{4}$ **69.** $x \leq 2$
71. $-3 < x < 4$ **73.** $0 < x < 8$ **75.** $-5 \leq x \leq 2$ **77.** $0 \leq x < 2$ **79.** $-3 \leq x \leq 0$
81. $3 \leq x \leq 10$ **83.** $x \geq -7$ **85.** $x < -1$ or $x > 3$ **87.** $x \leq -5$ or $x > -2$ **89.** $x \geq 0$

Exercises 3.7

1. $-4; 4$ **3.** $-\dfrac{1}{4}; \dfrac{1}{4}$ **5.** 0 **7.** 8; 2 **9.** $\dfrac{25}{4}; \dfrac{31}{4}$ **11.** $\dfrac{7}{3}; -3$ **13.** $-1; \dfrac{1}{2}$

15. $\dfrac{9}{2}$ **17.** $-\dfrac{4}{3}$ **19.** $6; \dfrac{2}{3}$ **21.** $-2 < x < 2$ **23.** $-6 \leq x \leq 6$ **25.** $x > 0$ or $x < 0$

27. $-3 < x < 3$ **29.** $x > 12$ or $x < -12$ **31.** $-8 < x < 2$ **33.** $-7 \leq x \leq 13$
35. 7 **37.** $x \leq -2$ or $x \geq 2$ **39.** $3 \leq x \leq 7$ **41.** $-8 \leq x \leq 6$ **43.** $2 < x < 3$

45. $x \geq \dfrac{2}{5}$ or $x \leq -\dfrac{4}{5}$ **47.** $-5 \leq x \leq -2$ **49.** $x < 1$ or $x > \dfrac{5}{3}$ **51.** $x \geq -\dfrac{3}{5}$ or $x \leq -1$

53. $x \leq 9$ or $x \geq 15$ **55.** $-3 \leq x \leq 0$ **57.** $-\epsilon < y < \epsilon$ **59.** $3 - \delta < x < \delta + 3$
61. $|x| \leq 4$ **63.** $|x - 4| \leq 8$ **65.** $|x - 2.5| \leq 2.5$ **67.** $|x + .55| < .05$ **69.** $|x| > 7$
71. $|4 - x| \geq 6$ **73.** $|x - 5| < \epsilon$

Chapter 3 Exercises

1. conditional **3.** identity **5.** conditional **7.** conditional **9.** identity **11.** 4 **13.** 8
15. 2 **17.** 8 **19.** $-\frac{7}{2}$ **21.** -2 **23.** 3 **25.** 6 **27.** 2 **29.** -3 **31.** $\frac{1}{2}$
33. $\frac{1}{3}$ **35.** 4 **37.** 3 **39.** 3 **41.** $4\frac{1}{2}$ **43.** $-9\frac{1}{3}$ **45.** $-4\frac{4}{5}$ **47.** $-1\frac{1}{15}$ **49.** $\frac{6}{7}$
51. 12 **53.** 15 **55.** -10 **57.** 9 **59.** 6 **61.** -12 **63.** 4 **65.** -1 **67.** 4
69. 5 **71.** 3 **73.** $1\frac{1}{6}$ **75.** 4 **77.** 3 **79.** $-8\frac{3}{4}$ **81.** $1\frac{1}{2}$ **83.** 3 **85.** 3
87. -6 **89.** $-\frac{1}{3}$ **91.** $2\frac{1}{2}$ **93.** $-1\frac{1}{8}$ **95.** 0 **97.** $3\frac{1}{4}$ **99.** $1\frac{1}{2}$ **101.** 2 **103.** -1
105. -2 **107.** 0 **109.** -4 **111.** 5 **113.** 6 **115.** 4 **117.** -5 **119.** 4
121. 4 **123.** 4 **125.** $-4\frac{1}{2}$ **127.** $20 + 5c$ **129.** -7 **131.** $-7b$ **133.** $-4a$

135. $6a$ **137.** $3a + 18b$ **139.** $\dfrac{8a + 15}{11}$ **141.** $x < \dfrac{8}{5}$ **143.** $x < 3$ **145.** $x \geq -3$

147. $x \leq \frac{13}{5}$ **149.** $x > 9$ **151.** $x \geq -\frac{5}{2}$ **153.** $x \leq -\frac{3}{2}$ **155.** $x \leq -2$ **157.** $x < 10$
159. $x < -\frac{1}{2}$ **161.** $x \geq 7$ **163.** $\frac{9}{2} < x < 9$ **165.** $-1 \leq x \leq 4$ **167.** $-4 \leq x \leq 3$
169. $\frac{1}{3} \leq x \leq \frac{10}{3}$ **171.** $4; -10$ **173.** $-4; 12$ **175.** $5; -6$ **177.** $\frac{2}{3}; \frac{5}{3}$ **179.** $21; -9$
181. $-1 < x < 3$ **183.** $\frac{3}{10} \leq x \leq \frac{7}{10}$ **185.** $-\frac{2}{3} < x < 2$ **187.** $-\frac{17}{2} < x < -\frac{13}{2}$
189. $9 \leq x \leq 21$ **191.** $x > 7$ or $x < 1$ **193.** $x \geq \frac{3}{4}$ or $x \leq -\frac{1}{4}$ **195.** $x \geq \frac{11}{3}$ or $x \leq \frac{7}{3}$
197. $x > 6$ or $x < -30$ **199.** all x

CHAPTER 4

Exercises 4.1(A)

1. 18 mi **3.** 10 g **5.** 8000 atoms/cm³ **7.** 21 m **9.** 30 cm³ **11.** 3 min + 45 sec
13. 192 mi/day **15.** 19/6 min **17.** 8 ft/sec² **19.** .42 m **21.** 128 oz **23.** 126,720 in.

25. 113 in.   **27.** 658.17 ft   **29.** 500,000 cm²   **31.** 30 mi/hr   **33.** 88 ft/sec
35. (186,000)(60)(60)(365) mi/day   **37.** 10 ft 4 in.   **39.** 5 mi

Exercises 4.1(B)

1. $4x$ **3.** $\frac{1}{3}a$ **5.** $8x$ **7.** xy **9.** $4 + x$ **11.** $x + 27$ **13.** $5 - x$ **15.** $9 - x$
17. $x - 9$ **19.** $\frac{x}{y}$ or $\frac{y}{x}$ **21.** $2x + 4$ **23.** $3 - 2x$ **25.** $\frac{x - y}{6}$ **27.** $\frac{x}{2} + xy$
29. $(a + b)^2$ **31.** $(17 - x)$ years   **33.** $(x - 2)$ years   **35.** (a) $-13; -11$ (b) $x + 2; x + 4$
(c) $x - 1; x + 1$ (d) $2x + 7; 2x + 9$ **37.** 156 weeks

39. Let x = length of longer piece $x = L - \frac{x}{3}$ **41.** $(100x + 10y)$ cents

43. $2x - 5 = x + 16$ **45.** $x + (x + 1) + (x + 2) = 15$ **47.** $(x + y)^2 = 13$ **49.** $\frac{x}{3} - 7 < 0$

Exercises 4.2

1. 15   **3.** 96   **5.** 7   **7.** 19; 14   **9.** 15; 43   **11.** 3; 12   **13.** 15; 16; 17   **15.** 6; 8; 10
17. 7; 8; 9   **19.** 13; 14   **21.** 68   **23.** 128   **25.** 39

Exercises 4.3

1. 300 km/hr   **3.** 6 hr   **5.** 14 mph; 10 mph   **7.** 5 hr   **9.** 8 mi   **11.** 240 mi

Exercises 4.4

1. Bob 34 yr; Ed 15 yr   **3.** 48 yr; 12 yr   **5.** 14 yr   **7.** 3 yr   **9.** 4 yr; 7 yr; 9 yr   **11.** 7 yr
13. 42 yr; 12 yr; 9 yr   **15.** 56 yr; 17 yr; In 22 yr

Exercises 4.5

1. 7 dimes; 15 quarters   **3.** 12 nickels; 27 dimes; 56 quarters   **5.** 2 adult; 102 children
7. 41 ten dollar bills; 29 twenty dollar bills   **9.** no

Exercises 4.6

1. $2500 at 4%; $6000 at 5%   **3.** $10,000   **5.** $2500 at 8%; $2700 at 6%   **7.** $4\frac{1}{2}\%; 2\frac{1}{4}\%$
9. $1800   **11.** $40   **13.** $2000   **15.** $4225   **17.** $(\frac{1}{3}P)$ dollars at 8%; $(\frac{2}{3}P)$ dollars at 5%

Exercises 4.7

1. $66\frac{2}{3}$ lb of 35¢ candy; $33\frac{1}{3}$ lb of 50¢ candy   **3.** 24 lb of copper   **5.** 2.5 gal   **7.** 15 gal
9. $\frac{25}{3}$ lb   **11.** 287.5 lb

Exercises 4.8

1. 17 ft; 31 ft   **3.** 13 in. by 45 in.   **5.** 13 in.   **7.** 36 sq yd   **9.** 10 yd   **11.** 12 ft
13. 35°; 70°; 210°; 45°   **15.** 35°; 55°   **17.** 30°; 60°; 90°   **19.** 8.02 cm

Exercises 4.9

1. $\frac{12}{7}$ hr   **3.** $\frac{126}{13}$ days   **5.** A $37\frac{1}{2}$ hr; B 25 hr   **7.** same direction at $43\frac{7}{11}$ minutes after 8; opposite
direction at $10\frac{10}{11}$ minutes after 8   **9.** $32\frac{8}{11}$ minutes past 3   **11.** 50 lb
13. Carol 184 lb; Bill 115 lb   **15.** 1 m

Exercises 4.10

1. $14° \le F \le 95°$ **3.** $x \ge 80$ **5.** $x \ge 65$ calories   **7.** no more than 20 quarters
9. at least \$6000

CHAPTER 5

Exercises 5.1

1. $x(x + b)$ **3.** $x(a + 1)$ **5.** $a^2(b - a)$ **7.** $10d(1 - 2c)$ **9.** $ax(a^2 - 2)$
11. $5x^2(a - 2x)$ **13.** $ab(a + 1)$ **15.** $5cx(a + 3)$ **17.** $11a^2c(2ab - 3)$
19. $x^4y^2(3y^2 - 7x)$ **21.** $17p^2q^2(pq - 3)$ **23.** $a(a^2 - ab + b^2)$
25. $3(a + b + c)$ **27.** $19x(3x^5 - m^6x + 2y^6)$ **29.** $2a(3x^2 + 7x - 1)$
31. $7xy(5x^2 - 11x - 9)$ **33.** $m(1 + 23x - 7x^2)$ **35.** $x(yz - zw - yw)$ **37.** 1800
39. 130   **41.** 60   **43.** 9   **45.** $(b - 2c)(a - 2)$ **47.** $(x - 1)(2x^2 - 3)$
49. $(a^2 - 1)(2a + 3)$ **51.** $(m + 2h)(m^2 - 4)$ **53.** $(x - 2y)(4x - 3y)$ **55.** $(c - d)(c^2 - d^2)$
57. $(m - 2n)(3m - 2n)$ **59.** $(x - 1)(x^4 - 8)$ **61.** $(a^2 - b^2)(a + b)$ **63.** $(n - x)(3 - 4x)$
65. $(a - 2)(5x - 1)$ **67.** $(5 + y)(4x - 1)$ **69.** $(m^2 - n^2)(m - n)$ **71.** $(x + 1)(3x^2 - 1)$
73. $(x^2 - 2)(1 - 3x)$ **75.** $(x^2 - y^2)(x - y)$ **77.** $(3c - 4d)(c - 1)$ **79.** $(2x + 3)(3x - 4y)$
81. $2c(3c - d)(3c + 1)$ **83.** $(m + 1)(m^3 + 1)$ **85.** $(4a - c)(3a^2 - c^2)$
87. $b(2b + 1)(a - b)$ **89.** $(m + 2n)(5m - 3)$ **91.** $(a - d)(a + b + c)$
93. $a(c + d)(a - b)$ **95.** $(a + d)(a - b + c)$

Exercises 5.2

1. $(x - 2)(x - 1)$ **3.** $(a - 4)(a - 3)$ **5.** $(t - 5)(t - 4)$ **7.** $(p - 7)(p - 3)$
9. $(b + 5)(b - 3)$ **11.** $(x + 6)(x + 4)$ **13.** $(z - 6)(z - 5)$ **15.** $(r + 7)(r - 3)$
17. $-(m + 4)(m - 3)$ **19.** $-(y - 7)(y - 8)$ **21.** $-(x - 6)(x - 5)$ **23.** $-(a + 4)(a - 9)$

Exercises 5.3

1. $(3x + 2)(x + 1)$ **3.** $(2x + 3)(x + 2)$ **5.** $(2x + 1)(x + 5)$ **7.** $(3x + 2)(x + 3)$
9. $(3x - 2)(x + 1)$ **11.** $(2x - 1)(x + 8)$ **13.** $(3x - 5)(2x - 7)$ **15.** $(4x - 7)(x + 2)$
17. $(7x - 6)(3x + 4)$ **19.** $-(x + 5)(x - 14)$ **21.** $-(h + 5)(h - 4)$ **23.** $-(y - 3)(y - 21)$
25. $2a(a - 2b)(a - 2b)$ **27.** $(4a - 3x)(4a - 3x)$ **29.** $(5x - y)(4x + 3y)$
31. $(3a + 2x)(3a - 4x)$ **33.** $x^2(x + 4)(4x - 1)$ **35.** $(8pq - 21)(2pq + 7)$
37. $(3ab - 2)(19ab + 22)$ **39.** $-(7y^2 + 4)(3y^2 - 1)$ **41.** $(5a^2 + 8)(3a^2 + 1)$
43. $-(4a^2 - 7)(7a^2 + 2)$ **45.** $[(x + y) + 3][(x + y) + 3]$ **47.** $[(x + y) + 3][(x + y) - 5]$
49. $[2(x + y) + 1][(x + y) + 2]$ **51.** $(x - y)[(x - y) - 2][(x - y) + 1]$

Exercises 5.4

1. $(x + a)^2$ **3.** $(5x + 1)^2$ **5.** $(y - 10)^2$ **7.** $(pq - r)^2$
9. $(m + 7a)^2$ **11.** $a(b - c)^2$ **13.** $(2x^2 - 3y^2)^2$ **15.** $(x^2y^2 - c)^2$
17. $(x^2 + 2y^2)^2$ **19.** $(a + 1 - c)^2$ **21.** $(x + y - 1)^2$ **23.** $(3ax + 3bx + 2)^2$
25. $(x + y)(x - y)$ **27.** $(m + 2)(m - 2)$ **29.** $(ab + 6)(ab - 6)$ **31.** $(1 + ab)(1 - ab)$
33. $(a^2 + y)(a^2 - y)$ **35.** $2(5y + 1)(5y - 1)$ **37.** $(3ab + cd)(3ab - cd)$
39. $(m^2 + 2)(m^2 - 2)$ **41.** $(11a^2 + 1)(11a^2 - 1)$ **43.** $(c^3 + 4d)(c^3 - 4d)$
45. $(3a^4 + 7b^2)(3a^4 - 7b^2)$ **47.** $a(b + d)(b - d)$ **49.** $2(1 + 6m^2n^3)(1 - 6m^2n^3)$
51. $a^2(a + 1)(a - 1)$ **53.** $(2x - 2y + 3xy)(2x - 2y - 3xy)$
55. $(x + y + a - b)(x + y - a + b)$ **57.** $ax(x^2 + b)(x^2 - b)$

59. $(a^2 + 1)(a + 1)(a - 1)$ **61.** $(a + b)^2(a^2 + 2ab - b^2)$
63. $(a + b + c - 2)(a + b - c + 2)$ **65.** $(2xy + z + 3)(2xy + z - 3)$
67. $(p^2 + 8pq + q^2)(p + q)^2$ **69.** $(a - x)(y + 5)(y - 5)$
71. 267   **73.** 145,200   **75.** $45 \cdot 95$ **77.** $(a + 1)(a^2 - a + 1)$
79. $(ab + y)(a^2b^2 - aby + y^2)$ **81.** $(m - n)(m^2 + mn + n^2)$
83. $(2a - 1)(4a^2 + 2a + 1)$ **85.** $(a - 5)(a^2 + 5a + 25)$
87. $2(5 - 2m)(25 + 10m + 4m^2)$ **89.** $x(y - 1)(y^2 + y + 1)$
91. $(x + y)(x^2 - xy + y^2)(x - y)(x^2 + xy + y^2)$ **93.** $2b(3a^2 + b^2)$
95. $(c + d)(c^2 - cd + d^2)(c - d)(c^2 + cd + d^2)$ **97.** $(4 + a^2)(16 - 4a^2 + a^4)$
99. $2(4a^2 + 1)(16a^4 - 4a^2 + 1)$

Exercises 5.5

1. $(a - 2b)^3$ **3.** $(1 - b)^3$ **5.** $(y^2 + b^2 + by)(y^2 + b^2 - by)$
7. $(x - 2)(x + 1)(x - 1)$ **9.** $a(3a - 2)(a^2 + 1)$

Chapter 5 Exercises

1. $x(x + y)$ **3.** $5x^3(2 - 5xy)$ **5.** $3a^2(a^2 - ab + 2b^2)$
7. $(a - b)(x - z)$ **9.** $(2x + y)(3x - a)$ **11.** $(3x + 5)(x^2 + 1)$
13. $(y - 1)(y^2 + 1)$ **15.** $(x - 12)(x - 7)$ **17.** $(a - 7b)^2$
19. $(m - 8n)(m - 5n)$ **21.** $(x - 12y)(x - 11y)$ **23.** $(12 - x)(11 - x)$
25. $(5 + xy)(13 - xy)$ **27.** $(x + 2)(x - 13)$ **29.** $(x^2 + 11a^2)(x^2 - 12a^2)$
31. $(3x - 2y)(4x - 5y)$ **33.** $(12x + 5)(x - 3)$ **35.** $(2 + 3x)(3 - 2x)$
37. $(1 + 7x)(5 - 3x)$ **39.** $(5 + 8x)(5 - 8x)$ **41.** $(2500)(1122) = 2,805,000$
43. $47x(x + 2y)$ **45.** $5y(6x - 5y)$ **47.** $(a^2 + 9b)(a^4 - 9a^2b + 81b^2)$
49. $20y(5x + y)(5x - y)$ **51.** $2c(c^2 + 3d^2)$ **53.** $(x^2 + 2xy - y^2)(x^2 - 2xy - y^2)$
55. $(a^2 + bx)(a + x)$ **57.** $(m^2 - n)(m^4 + m^2n + n^2)$
59. $(x^2 + 3xy - 3y^2)(x^2 - 3xy - 3y^2)$

CHAPTER 6

Exercises 6.1

1. not equivalent   **3.** equivalent   **5.** equivalent   **7.** equivalent   **9.** equivalent
11. equivalent   **13.** $\dfrac{51}{57}$ **15.** $\dfrac{4x^2y^6}{34x^3y^4}$ **17.** $\dfrac{7(a - b)^3}{28(a - b)^2}$ **19.** $\dfrac{x^2 + 4x}{x(x^2 - 16)}$
21. $\dfrac{ab(a + b)}{a^3 + b^3}$ **23.** $\dfrac{3n}{5m}$ **25.** $\dfrac{3}{x + y}$ **27.** $\dfrac{3}{x - 4}$ **29.** $\dfrac{4}{x^2 - xy + y^2}$ **31.** $\dfrac{y - 6}{y + 4}$
33. $\dfrac{a(x - 4)}{3x + 4}$ **35.** $\dfrac{x - 2}{3 - x}$ **37.** $\dfrac{-1}{4x + 2}$ **39.** $\dfrac{a - 2x}{a + 3x}$ **41.** $\dfrac{x^2 + 4}{x^2 + 3}$ **43.** $\dfrac{a - x}{b + x}$
45. $\dfrac{-(x + 3)}{(y + z)}$ **47.** $\dfrac{m - 6n}{3m - n}$ **49.** $\dfrac{x^2 + x + 1}{x^2 + 2x + 1}$

Exercises 6.2

1. $\dfrac{15}{8}$ **3.** $\dfrac{12}{7}$ **5.** $\dfrac{5}{4}$ **7.** $\dfrac{1}{2}$ **9.** $\dfrac{a}{y}$ **11.** $\dfrac{q^2}{mn^2p}$ **13.** $\dfrac{2c}{nd}$ **15.** $\dfrac{a^2}{by^2}$ **17.** $\dfrac{-a}{x}$
19. $\dfrac{-5y^3}{4a^4x}$ **21.** $\dfrac{4y(x - 2y)}{3x^4}$ **23.** $\dfrac{8y}{9z(x - 4)}$ **25.** $\dfrac{4x}{15y^2}$ **27.** $\dfrac{27a}{b^2(a - 2b)}$

29. $\dfrac{5(y+3)}{8(x-2)}$ **31.** $\dfrac{(x+3)(x+5)}{3(x-3)(x-5)}$ **33.** -1 **35.** $\dfrac{1}{2x+1}$ **37.** $\dfrac{1}{x-y}$ **39.** $\dfrac{b^2-d^2}{b(a+1)}$

Exercises 6.3

1. $\dfrac{2}{3}$ **3.** $\dfrac{3}{25}$ **5.** $\dfrac{-3}{2}$ **7.** $\dfrac{1}{b}$ **9.** $\dfrac{2}{5b^4}$ **11.** $\dfrac{4b^4x}{3}$ **13.** $\dfrac{8a^3b^4}{5c^2d^3}$ **15.** $\dfrac{3}{2}$

17. $\dfrac{3}{4}$ **19.** $\dfrac{x+2y}{x-2y}$ **21.** $\dfrac{y^2-9}{y^2+12y+36}$ **23.** $\dfrac{(x-7)(x+2)}{(x+4)(x+5)}$ **25.** $\dfrac{(3y+1)(3y-1)}{(4y+1)(4y-1)}$

27. $\dfrac{3a^2-7ab-6b^2}{2a^2-3ab-2b^2}$ **29.** $\dfrac{2+3x}{2x}$ **31.** $4x+y$ **33.** $\dfrac{x}{3(x-y)}$ **35.** $4a$

37. $\dfrac{x+2y}{3}$ **39.** $\dfrac{6(a-b-4)}{a-2b}$

Exercises 6.4

1. $\dfrac{11}{17}$ **3.** $\dfrac{1}{3x}$ **5.** 4 **7.** $\dfrac{3}{b}$ **9.** $\dfrac{k-1}{k-3}$ **11.** $\dfrac{7x^2+2x+4y-2y^2}{x^2-4y^2}$

13. $\dfrac{3x+5y}{3x-2y}$ **15.** $\dfrac{-9a}{x+2a}$ **17.** $\dfrac{2a+7b}{5a+3b}$ **19.** $\dfrac{x^2+x+7a+4a^2}{x+5a}$ **21.** $2\dfrac{4}{5}$

23. $\dfrac{36a^2-25b^2}{30ab}$ **25.** $\dfrac{5a-2b}{a^2-4b^2}$ **27.** $\dfrac{-2}{x^2-16}$ **29.** $\dfrac{x-5}{7(x+2)}$ **31.** $\dfrac{8x-10}{x-2}$

33. $\dfrac{-14}{2x+7}$ **35.** $\dfrac{x^3}{x-1}$ **37.** $\dfrac{2x(x+5y)}{4x^2-9y^2}$ **39.** 0

Exercises 6.5

1. $\dfrac{5}{4}$ **3.** $\dfrac{7x}{2y}$ **5.** $\dfrac{ay}{bx}$ **7.** $\dfrac{9a}{40b}$ **9.** $\dfrac{20x}{3}$ **11.** $\dfrac{9}{8z}$ **13.** 6 **15.** $\dfrac{45}{64}$ **17.** $\dfrac{3a}{a-3}$

19. $\dfrac{x^2}{6}$ **21.** $\dfrac{2b^3}{1-b}$ **23.** $\dfrac{a-b}{b}$ **25.** 4 **27.** $\dfrac{3(x-2y)}{2}$ **29.** $\dfrac{4}{9x+6y}$

31. $\dfrac{3a+b}{3a}$ **33.** $\dfrac{4x}{2x+1}$ **35.** $\dfrac{1}{x}$ **37.** $\dfrac{4x^2-1}{2x}$ **39.** $\dfrac{2(a-b)}{a+b}$ **41.** $\dfrac{1}{ab}$ **43.** $\dfrac{1}{x+y}$

45. $\dfrac{-4}{x^2}$ **47.** $\dfrac{x^2+x+1}{x^3+x^2+2x+1}$ **49.** 0

Exercises 6.6

1. 6 **3.** $\dfrac{-4}{5}$ **5.** $\dfrac{7}{3}$ **7.** -1 **9.** -5 **11.** 1 **13.** 2 **15.** -1 **17.** $\dfrac{3}{5}$ **19.** -2

21. no solution **23.** $\dfrac{1}{a-b}$ **25.** $\dfrac{2a^2}{2a-b}$ **27.** $\dfrac{2ab}{a+b}$ **29.** $\dfrac{a^2b^2-b}{a^2-b^2}$ **31.** $\dfrac{3a^2}{4b^2c^2}$

33. $\dfrac{3b^2d^2}{b^2+2d^2}$ **35.** $\dfrac{a^2-b^2}{4a^2b^2}$ **37.** $r=\dfrac{E-CR}{C}$ **39.** $D=\dfrac{dF}{d-F}$ **41.** $p=\dfrac{VP-GF}{V-G}$

43. $r_1=\dfrac{Rr}{r-R}$ **45.** $r=\dfrac{3V+\pi h^3}{3\pi h^2}$

Exercises 6.7

1. 5 **3.** 5 **5.** $\frac{11}{15}$ **7.** 91; 5 **9.** $\frac{29}{41}$ **11.** $2\frac{2}{9}$ hr **13.** 2 hr **15.** $22\frac{1}{2}$ hr
17. 2 hr; 3 hr **19.** $2\frac{1}{2}$ mph **21.** 40 mph; 50 mph **23.** 1 gal **25.** bond, 6%; stocks, 8%
27. 4 ohms

Chapter 6 Exercises

1. $5x^2y^4$ **3.** $26r^5s^6$ **5.** $rx - ry$ **7.** $(2c - d)^2$ **9.** $(c + d)(f + g)$ **11.** $\dfrac{2ac}{3b}$

13. $\dfrac{4ac^2}{3a - 2x}$ **15.** $\dfrac{2a}{3b}$ **17.** $\dfrac{3}{2b}$ **19.** $4(x - y)$ **21.** $\dfrac{-(a + b)}{4}$ **23.** 1 **25.** $\dfrac{3(a - c)}{5(c + d)}$

27. $\dfrac{8y}{3}$ **29.** $\dfrac{1}{abxy}$ **31.** $\dfrac{x^2 + y^2}{a}$ **33.** 1 **35.** $\dfrac{x^2y^2}{a^2b^2}$ **37.** $\dfrac{x^2 + y^2}{x}$ **39.** $\dfrac{y - 3}{y - 2}$ **41.** $\dfrac{8c}{b}$

43. $\dfrac{t + 2}{t + 6}$ **45.** $\left(\dfrac{x - 1}{x + 1}\right)^2$ **47.** $180x^3$ **49.** $(x + 3)(x - 3)(x + 7)$

51. $12(x + 3)(x - 3)(x - 2)$ **53.** $abc(x + y)(x - y)$ **55.** $(x - 5)(x - 4)(x + 2)$

57. $\dfrac{12x^2 + 28x - 27}{8x^2}$ **59.** $\dfrac{4ab}{4a^2 - b^2}$ **61.** $\dfrac{2(13x + 7)}{3(x - 2)(x + 2)}$ **63.** 0 **65.** $\dfrac{x^2 + x + 2}{(x + 1)(x - 1)^2}$

67. $\dfrac{x}{(x - 2)(x + 8)(x + 9)}$ **69.** $\dfrac{11x^2 + x - 4}{(3x + 1)(3x - 1)}$ **71.** $\dfrac{1}{x + 1}$ **73.** $x + 1$ **75.** 1

77. $\dfrac{1}{8}$ **79.** $-12\frac{1}{2}$ **81.** $\dfrac{2}{15}$ **83.** $\dfrac{-2}{3}$ **85.** 8 **87.** $-2a$ **89.** abc **91.** 28 and 107

93. $\dfrac{xy}{x + y}$ **95.** $1\frac{2}{3}$ mph **97.** 1 qt **99.** Marc $9\frac{3}{4}$ sec; John 10 sec

CHAPTER 7

Exercises 7.1

1. 2^8 **3.** $(-5)^6$ **5.** m^{15} **7.** x^8 **9.** 2^4x^3 **11.** $5^3x^3y^4$ **13.** $4x^4y^6z^6$ **15.** x^{10}

17. $-x^{12}$ **19.** x^6y^6 **21.** -1 **23.** $a^4b^9c^{16}$ **25.** x^3 **27.** 1 **29.** x^2y^2 **31.** $\dfrac{1}{a^2b}$

33. 1 **35.** $-\dfrac{z^2}{y}$ **37.** $8x^9$ **39.** $-x^8y^{12}$ **41.** $x^8y^{12}z^4$ **43.** $16a^4x^{12}y^8$ **45.** $\dfrac{-x^3}{64}$

47. $\dfrac{25x^6}{36y^4}$ **49.** $\dfrac{x^{10}y^{20}z^{10}}{a^{20}b^{30}c^{10}}$ **51.** $25a^4b^9x^{12}y^6$ **53.** x **55.** $\dfrac{y^{10}}{x^3}$ **57.** $a^4b^3x^4y^3$ **59.** $\dfrac{6z^3}{x^2}$

Exercises 7.2

1. 1 **3.** 1 **5.** $\dfrac{1}{a^3}$ **7.** $\dfrac{1}{4}$ **9.** 3 **11.** x^5 **13.** 1 **15.** $\dfrac{2}{x}$ **17.** $\dfrac{y^3}{x^2}$ **19.** $\dfrac{1}{81a^2}$

21. $\dfrac{1}{a^5b^5}$ **23.** $\dfrac{-5}{x^3}$ **25.** -1 **27.** $\dfrac{1}{z^2}$ **29.** $\dfrac{1}{x^3y}$ **31.** $\dfrac{1}{a^3b}$ **33.** $\dfrac{b^2}{x^2}$ **35.** $\dfrac{4}{125}$

37. $\dfrac{1}{y}$ **39.** 1 **41.** 1 **43.** $\dfrac{3c^2}{2}$ **45.** $\dfrac{4d^2x^2}{3}$ **47.** $\dfrac{1}{(x + y)^2}$ **49.** $\dfrac{1}{a^2b}$ **51.** 1

53. x^2 **55.** a^7 **57.** x^2 **59.** x^5 **61.** 125 **63.** 8 **65.** 4^{m+n} **67.** 16 **69.** $x + 1$

71. $a + \dfrac{1}{a^2}$ **73.** $x^3 - \dfrac{3}{x^3}$ **75.** $c^2 - 5$ **77.** $\dfrac{y^6}{(2x^2y^2 - 1)^3}$ **79.** $2a - 1$ **81.** $\dfrac{a^2}{5}$

83. $\dfrac{3y^3}{5x^2}$ **85.** $\dfrac{1}{xy^4}$ **87.** $\dfrac{x}{y^3z}$ **89.** $\dfrac{x^6}{45y^8}$ **91.** $\dfrac{a}{b^3} - \dfrac{b}{a^2}$ **93.** $4xy^2$ **95.** $\dfrac{5x^2y^2}{x^2 + y^2}$ **97.** $\dfrac{1}{4}$

99. -1.5

Exercises 7.3

1. 5.64×10^4 **3.** 7.5×10^{-4} **5.** 4.693×10^3 **7.** 4×10^0 **9.** 5.6037×10^3
11. 3.6×10^5 **13.** 7.930×10^3 **15.** 2.4×10^5 **17.** 1.49×10^8 **19.** 3.15×10^9
21. $6{,}600{,}000{,}000{,}000{,}000{,}000{,}000$ **23.** $.000\,000\,000\,000\,4$
25. $.000\,000\,000\,000\,000\,000\,000\,000\,00166$ **27.** $.000\,08$ **29.** $.03$

Exercises 7.4

1. ± 6 **3.** ± 1 **5.** $\pm x$ **7.** $\pm a^2$ **9.** $\pm 6x$ **11.** -3 **13.** -5 **15.** b **17.** $2ab$

19. $\dfrac{z}{2}$ **21.** 12 **23.** -4 **25.** -2 **27.** 15 **29.** $\dfrac{1}{4}$ **31.** $\dfrac{2}{5}$ **33.** y^2 **35.** $-2a^2$

37. -5 **39.** 3 **41.** $-6x^4$ **43.** 0

Exercises 7.5

1. $\sqrt[3]{a^2}$ **3.** $\sqrt[4]{x^3}$ **5.** $\sqrt[3]{c^{-2}}$ **7.** $\sqrt[3]{x^{-5}}$ **9.** $\sqrt[3]{b^a}$ **11.** $\sqrt[3a]{x^4}$ **13.** $x^{1/2}$ **15.** $b^{3/2}$
17. $a^{-2/3}$ **19.** $b^{-3/4}$ **21.** $x^{-n/m}$ **23.** $x^{-4/5}$ **25.** $x^{-q/p}$ **27.** 5 **29.** 2 **31.** 2

33. 2 **35.** -4 **37.** 2 **39.** 8 **41.** 4 **43.** $\dfrac{1}{9}$ **45.** 8 **47.** 27 **49.** -1024

51. $a^{3/2}$ **53.** $x^{-10/3}$ **55.** $x^{5/4}$ **57.** y^4 **59.** $x^{-8/3}$ **61.** $p^{-1/2}$ **63.** x^{-1} **65.** x^{-5}

67. $x^{-2/9}$ **69.** $\dfrac{2}{5}$ **71.** $\dfrac{-2}{5}$ **73.** $\dfrac{-243}{3125}$ **75.** $\dfrac{4}{49}$

Exercises 7.6

1. $2\sqrt{11}$ **3.** $3\sqrt{37}$ **5.** 20 **7.** $2\sqrt[4]{7}$ **9.** $3x\sqrt{5}$ **11.** $2\sqrt[3]{x^2}$ **13.** $2x\sqrt{3x}$
15. $\dfrac{1}{3}\sqrt[3]{m^2}$ **17.** $3b\sqrt[3]{a^2}$ **19.** $7s\sqrt{2rs}$ **21.** $3y\sqrt[3]{2y}$ **23.** $\dfrac{y}{3}\sqrt[4]{x^2y^3}$ **25.** $25\sqrt[6]{xy}$

27. 2 **29.** $\sqrt[3]{12}$ **31.** $4m\sqrt[3]{2m}$ **33.** $\dfrac{x^2}{3}\sqrt[4]{9a}$ **35.** $3\sqrt[5]{x^2}$ **37.** $2x\sqrt[3]{2m^2}$

39. $2a\sqrt[6]{2a}$ **41.** $\dfrac{\sqrt{2}}{2}$ **43.** $\dfrac{\sqrt{x}}{x}$ **45.** $\dfrac{4\sqrt{11}}{11}$ **47.** $\dfrac{x\sqrt{3}}{3}$ **49.** $\dfrac{\sqrt{15}}{5}$ **51.** $\dfrac{\sqrt{xy}}{y}$

53. $\dfrac{3\sqrt{6}}{5}$ **55.** $3\sqrt[3]{4}$ **57.** $3\sqrt{2y}$ **59.** $\dfrac{\sqrt{ax}}{x}$ **61.** $3r\sqrt{5r}$ **63.** $x\sqrt{x}$ **65.** $\sqrt[3]{9x^2}$

67. $\dfrac{5}{2}\sqrt[3]{2z}$ **69.** $\dfrac{\sqrt{x^2 - 1}}{x + 1}$

Exercises 7.7

1. 0 **3.** 0 **5.** $26 + 4\sqrt{13}$ **7.** $12\sqrt{2} + 5\sqrt{11}$ **9.** 0 **11.** 0 **13.** $a\sqrt{2a}$
15. $7\sqrt{3}$ **17.** $2\sqrt{2} + 3\sqrt{3}$ **19.** $\sqrt{2}$ **21.** $2(a + x)\sqrt{a - x}$ **23.** $(a - b + 1)\sqrt[3]{(a - b)^2}$
25. $(a - a^2b - b^2)\sqrt{b}$

Exercises 7.8

1. $2\sqrt{3}$ 3. $2\sqrt{6}$ 5. $8m$ 7. 44 9. $18x$ 11. $6(a + b)$ 13. $2(a - b)$ 15. $\sqrt[4]{539}$

17. $a\sqrt[10]{a^3b^7}$ 19. $ab\sqrt[30]{a^9b^{13}}$ 21. $\sqrt{6} + \dfrac{1}{2}$ 23. $27 - 12\sqrt{2}$ 25. 49

27. $4a - 4\sqrt{ab} + b$ 29. $21 - 26\sqrt{10}$ 31. $x - 6$ 33. $9x - 6\sqrt{xy} + y$ 35. $63x - y$

37. $4x + 5 - 12\sqrt{x - 1}$ 39. $79 - 20\sqrt{3}$

Exercises 7.9

1. $\dfrac{9 - 3\sqrt{5}}{2}$ 3. $6 + 2\sqrt{7}$ 5. $2\sqrt{2} + 1$ 7. $\dfrac{9(5 + 2\sqrt{3})}{13}$ 9. $3 + 2\sqrt{2}$ 11. $\dfrac{11 - 6\sqrt{2}}{7}$

13. $8 + 7\sqrt{2}$ 15. $\dfrac{25 - 7\sqrt{5}}{4}$ 17. $\dfrac{-(33 + 10\sqrt{8})}{17}$ 19. $\dfrac{9 - 4\sqrt{2}}{7}$ 21. $-\sqrt{2}$

23. $-(2 + \sqrt{6})$ 25. $\dfrac{\sqrt{a} + \sqrt{3a + 3}}{-2a - 3}$ 27. $\dfrac{a + b + 2\sqrt{ab}}{a - b}$ 29. $\dfrac{b - \sqrt{b^2 - a^2}}{a^2}$

Exercises 7.10

1. $x = 4$ 3. $x = 27$ 5. $x = 29$ 7. $x = -5$ 9. $x = 78$ 11. $x = 9$ 13. $x = 4$

15. $x = 9$ 17. $x = 9$ 19. $x = 4$ 21. $x = 3$ 23. $x = 9$ 25. $x = 30$ 27. $x = \dfrac{81}{16}$

29. $x = \dfrac{25}{4}$ 31. $x = b^2 - 2ab + a^2$ 33. $x = a^2 - 6a + 9$ 35. $x = c^2 - 4c + 4$

37. $x = a^2 + 6a + 9$ 39. $x = \dfrac{a}{49}$ 41. $x = 25c$ 43. $x = a + 2\sqrt{ab} + b$

45. $x = \dfrac{a + 2\sqrt{ab} + b}{4}$ 47. $x = a$ 49. $x = 2\sqrt{cd} - d$ 51. $x = \dfrac{-1}{32}$ 53. $x = \dfrac{1}{9}$

55. no solution   57. $x = 8$ 59. $x = \dfrac{1}{9}$

Exercises 7.11

1. $6i$ 3. $7i$ 5. $\sqrt{7}i$ 7. $3\sqrt{2}i$ 9. $2\sqrt{6}\,i$ 11. $6\sqrt{2}\,i$ 13. $\dfrac{1}{3}i$ 15. $\dfrac{1}{4}i$

17. $\dfrac{2}{7}i$ 19. $\sqrt{\dfrac{7}{10}}\,i$ 21. $x\sqrt{y}i$ 23. x^2yi 25. $7 + 3\sqrt{5}i$ 27. $-1 + 4\sqrt{6}i$

29. i 31. $-i$ 33. $-i$ 35. -1 37. $x = 14; y = 8$

Exercises 7.12

1. $6 - 7i$ 3. $7 - i$ 5. $-1 - 4i$ 7. $\dfrac{-1}{2} - \dfrac{5}{6}i$ 9. $0.3 + 0.8i$ 11. $2 - i$

13. $11 - 3\sqrt{3}i$ 15. $5 - 3\sqrt{2}i$ 17. $b = -d$ 19. $b = d$ 21. $7 + 26i$ 23. 13

25. $-3 + 54i$ 27. $15 - 23i$ 29. $12 + 16i$ 31. $2i$ 33. $12 - 16i$ 35. $16 + 30i$

37. $\dfrac{4}{5} - \dfrac{8}{5}i$ 39. $\dfrac{-12}{13} + \dfrac{8}{13}i$ 41. $-i$ 43. $\dfrac{2}{3} - \dfrac{5\sqrt{2}}{3}i$ 45. $\dfrac{-1}{2} - \dfrac{3}{2}i$

47. $\left(\dfrac{3\sqrt{2} + \sqrt{3}}{4}\right) + \left(\dfrac{3 - \sqrt{6}}{4}\right)i$ 49. $\left(\dfrac{a^2 - b}{a^2 + b}\right) + \left(\dfrac{2a\sqrt{b}}{a^2 + b}\right)i$ 51. $ad = -bc$

Chapter 7 Exercises

1. x^6 **3.** $3x^4$ **5.** 1 **7.** x^{10} **9.** 1 **11.** a **13.** $(ab)^{1-x}$ **15.** x^{x^2} **17.** 6

19. b^{3a+12} **21.** $\sqrt{x^3}$ **23.** $\sqrt[3]{\dfrac{(1+x^2)^2}{x^2}}$ **25.** $b^{1/4}(c+d)^{-3/4}$ **27.** $2\sqrt{3}$ **29.** $5\sqrt{3}$

31. $5\sqrt{5}$ **33.** $2\sqrt[3]{3}$ **35.** $3\sqrt[4]{2}$ **37.** $abc\sqrt[5]{a^2b}$ **39.** $2x^4\sqrt[n]{2x^5}$ **41.** $\dfrac{\sqrt{3}}{3}$ **43.** $\dfrac{\sqrt{6}}{2}$

45. $\dfrac{\sqrt{10}}{8}$ **47.** $\dfrac{a\sqrt{bc}}{bc}$ **49.** $\dfrac{b^n(\sqrt[4]{c^n})}{c^n}$ **51.** $\dfrac{\sqrt[3]{9}}{3}$ **53.** $\dfrac{\sqrt{x^2-y^2}}{x+y}$ **55.** $\dfrac{(x+y)\sqrt{x(x+y)}}{x}$

57. $(a-b+c)\sqrt[n]{y}$ **59.** $-\sqrt{3}$ **61.** $11\sqrt{2}$ **63.** $7\sqrt{3}$ **65.** $7\sqrt[3]{4}$ **67.** $(x+1+y)\sqrt{xy}$

69. $\dfrac{(x+y-2)\sqrt{x+y}}{x+y}$ **71.** $x+\sqrt{2x}$ **73.** $\sqrt{3x}-6x\sqrt{2}$ **75.** $\dfrac{5}{4}\sqrt{x}+2x^2\sqrt{2}$

77. $\sqrt{3xy}-\sqrt{xy}$ **79.** $20\sqrt{3}-3\sqrt{2}$ **81.** $2\sqrt[3]{6}$ **83.** $60\sqrt{3}$ **85.** $10\sqrt{7}$
87. $6(\sqrt{3}+\sqrt{2}+1)$ **89.** 6 **91.** $3\sqrt[3]{18}-4\sqrt[3]{12}-6$ **93.** $x-2a\sqrt{x-a^2}$
95. $x-1+\sqrt{x^2-2x}$ **97.** $x+\sqrt{x^2-b^2}$ **99.** $a-\sqrt{a^2-b}$ **101.** $\sqrt{5}-2$

103. $3\sqrt{3}-3\sqrt{2}$ **105.** $\dfrac{3\sqrt{15}-11}{7}$ **107.** 8 **109.** no solution   **111.** $11+i$

113. $-2-13i$ **115.** $2-5i$ **117.** $4+3i$ **119.** $58-30i$ **121.** 89 **123.** $-18+6i$

125. $-i$ **127.** $\dfrac{-3+46i}{25}$ **129.** $\dfrac{21+12i}{65}$ **131.** $x=\dfrac{-23}{29}; y=\dfrac{15}{29}$ **133.** $x=0; y=-2$

CHAPTER 8

Exercises 8.1

1. ± 4 **3.** ± 3 **5.** $\pm\dfrac{1}{2}$ **7.** $\pm a$ **9.** ± 4 **11.** ± 3 **13.** ± 7 **15.** ± 3 **17.** $\pm 2i$

19. $\pm\sqrt{6}\,i$ **21.** $\dfrac{\pm\sqrt{35}\,i}{5}$ **23.** $5; -9$ **25.** $3; 1$ **27.** $b-3, -b-3$ **29.** $-2\pm\dfrac{\sqrt{3}}{2}i$

31. ± 4 **33.** $\pm 3\dfrac{1}{2}$ **35.** $\pm\dfrac{1}{2}$ **37.** $\pm 3\sqrt{3}$ **39.** $\pm a$ **41.** ± 5 **43.** ± 3

45. $r=\pm\sqrt{\dfrac{A}{\pi}}$ **47.** $t=\pm\sqrt{\dfrac{2s}{g}}$ **49.** $i=\pm\sqrt{\dfrac{P}{r}}$

Exercises 8.2

1. $3; 2$ **3.** $13; 1$ **5.** $11; 8$ **7.** $-7; -3$ **9.** $-3; -3$ **11.** $5; -2$ **13.** $2; -14$

15. $-5; 3$ **17.** $4; 0$ **19.** $4; -3$ **21.** $11; 13$ **23.** $7a; 0$ **25.** $0; -3$ **27.** $\dfrac{1}{2}; \dfrac{4}{3}$

29. $-3; \dfrac{1}{2}$ **31.** $2; \dfrac{-5}{9}$ **33.** $\dfrac{-2}{11}; -2$ **35.** $\dfrac{-7}{11}; 1$ **37.** $b; b$ **39.** $a; \dfrac{7a}{5}$

41. $\dfrac{-3a}{2}; \dfrac{a}{2}$ **43.** $\dfrac{-2}{3}; \dfrac{5}{2}$ **45.** $\dfrac{1}{3}; \dfrac{5}{4}$ **47.** $3a; \dfrac{-5}{2a}$ **49.** $2a; 2b$ **51.** $1; -1$

53. $2; a$ **55.** $-5; 6$ **57.** $\dfrac{3}{8}; \dfrac{-3}{2}$ **59.** $1; 1$ **61.** $\dfrac{-1}{2}; 2$ **63.** $2; \dfrac{-1}{2}$ **65.** $\dfrac{-a}{3}; \dfrac{4a}{3}$

67. $2a; \dfrac{-a}{2}$

Exercises 8.3

1. $-3; -5$ **3.** $7; -5$ **5.** $5; -1$ **7.** $9; 1$ **9.** $3; 2$ **11.** $1; -4$ **13.** $1; -2$
15. $4; -3$ **17.** $1; -10$ **19.** $-1 \pm \sqrt{5}$ **21.** $-2 \pm \sqrt{7}$ **23.** $-3 \pm 2\sqrt{3}$
25. $\dfrac{-3 \pm \sqrt{13}}{2}$ **27.** $\dfrac{-1 \pm \sqrt{21}}{2}$ **29.** $\dfrac{5 \pm \sqrt{53}}{2}$ **31.** $2; \dfrac{1}{2}$ **33.** $\dfrac{1}{2}; \dfrac{-2}{3}$
35. $1; \dfrac{-7}{5}$ **37.** $\dfrac{1 \pm \sqrt{10}}{3}$ **39.** $\dfrac{3 \pm \sqrt{41}}{8}$ **41.** $\dfrac{5 \pm \sqrt{43}}{2}$ **43.** $4 \pm 3i$ **45.** $\dfrac{3 \pm 5i}{2}$
47. $\dfrac{5 \pm 3\sqrt{7}i}{2}$ **49.** $\dfrac{5a}{3}; -a$ **51.** $-3a; \dfrac{-5a}{4}$

Exercises 8.4

1. $\dfrac{5}{3}; -1$ **3.** $\dfrac{4}{3}; \dfrac{-1}{2}$ **5.** $1; -5$ **7.** $\dfrac{1}{3}; 3$ **9.** $\dfrac{1}{6}; \dfrac{2}{5}$ **11.** $\dfrac{2}{3}; \dfrac{3}{2}$ **13.** $0; \dfrac{2}{9}$
15. $2; \dfrac{-3}{5}$ **17.** $3 \pm 2\sqrt{3}$ **19.** $\dfrac{-1 \pm \sqrt{41}}{4}$ **21.** $\dfrac{1 \pm \sqrt{34}i}{7}$ **23.** $\dfrac{1 \pm \sqrt{3}i}{2}$
25. $\dfrac{-5 \pm \sqrt{5}i}{2}$ **27.** $\dfrac{a}{2}; \dfrac{-2a}{3}$ **29.** $\sqrt{3} \pm 1$ **31.** $m; \dfrac{1}{m}$ **33.** $\dfrac{b \pm \sqrt{b^2 - 4ac + 4c}}{2(a-1)}$
35. $\dfrac{(-1 \pm \sqrt{5})(a+b)}{2}$ **37.** $ac; \dfrac{b}{a}$ **39.** $\dfrac{a}{a+1}; \dfrac{a}{a-1}$ **41.** $n = 14$
43. $y = \dfrac{(1 - 5x) \pm \sqrt{17x^2 - 26x + 29}}{2}$ **45.** $y = \dfrac{-(3x - 1) \pm \sqrt{5x^2 - 14x + 17}}{2}$

Exercises 8.5

1. real, unequal, irrational   **3.** real, equal, rational   **5.** complex conjugates
7. real, unequal, rational   **9.** real, equal, rational   **11.** real, equal, rational
13. real, unequal, rational   **15.** real, equal, rational   **17.** $-1; 0$ **19.** $1; 7$ **21.** 9
23. $7; \dfrac{-128}{15}$ **25.** $b^2 - 4ac = 4b^2 + 4c^2 \geq 0$, therefore real roots   **27.** $k = -4$
29. $h = 2; k = 4$ **31.** $5; \dfrac{4}{3}$ **33.** $\dfrac{3}{2}; \dfrac{7}{2}$ **35.** $-3; -4$ **37.** $0; \dfrac{-49}{16}$ **39.** $\dfrac{7}{3}; \dfrac{-20}{3}$
41. $5; -3$ **43.** $\dfrac{-1}{3}; \dfrac{-4}{3}$ **45.** $0; \dfrac{-7}{8}$ **47.** $12x^2 - 13x + 3 = 0$ **49.** $24x^2 + 17x + 3 = 0$
51. $x^2 + 7x = 0$ **53.** $x^2 + x - 6 = 0$ **55.** $x^2 - 5 = 0$ **57.** $x^2 + 9 = 0$
59. $x^2 - 2x + 2 = 0$ **61.** $x^2 + 3x + 1 = 0$ **63.** $4x^2 - 24x + 99 = 0$
65. $bx^2 - 2ax + a = 0$ **67.** 4 **69.** -3 **71.** 7 **73.** -1 **75.** $\dfrac{1}{3}$

Exercises 8.6

1. $\pm\sqrt{3}; \pm 2$ **3.** $\pm\sqrt{5}i; \pm\sqrt{2}i$ **5.** $\dfrac{\pm\sqrt{6}}{2}; \pm i$ **7.** $\pm 3; \pm 2$ **9.** $\pm 3; \pm i$
11. $\pm 3; \pm 3i$ **13.** $1; 2; \dfrac{-1 \pm \sqrt{3}i}{2}; -1 \pm \sqrt{3}i$ **15.** $\dfrac{1}{2}; 1; \dfrac{-1 \pm \sqrt{3}i}{4}; -1 \pm \sqrt{3}i$
17. $-1; -2; \dfrac{1 \pm \sqrt{3}i}{2}; 1 \pm \sqrt{3}i$ **19.** $\pm 1; \pm i; \pm 2; \pm 2i$ **21.** $-64; 8$ **23.** $-216; -27$

25. $32; -243$ **27.** $\pm 3; \pm\dfrac{1}{2}$ **29.** $9; 16$ **31.** $-10, 4$ **33.** $-1; 1$ **35.** $2; 1$

37. $1; -2; \dfrac{-1 \pm \sqrt{13}}{2}$ **39.** $-1; \dfrac{3}{2}; \dfrac{-1}{2}; 1$ **41.** $\dfrac{-2}{3}; 2; \dfrac{1}{3}; 1$ **43.** $-\dfrac{1}{2}; 5; \dfrac{-5}{2}; 1$

45. $-1; \dfrac{1}{2}; 2; \dfrac{-1}{4}$ **47.** $\dfrac{1}{2}; \dfrac{1}{3}; \dfrac{-1}{3}; \dfrac{-1}{4}$ **49.** $4; 9$ **51.** $9; 5$ **53.** 10 **55.** $3; -5$

57. $\dfrac{1}{3}$ **59.** $4; -8$

Exercises 8.7

1. $-7 < x < 7$ **3.** $x < 2$ or $x > 4$ **5.** no real numbers   **7.** $x < -2$ or $x > \dfrac{-2}{3}$

9. $\dfrac{-3}{4} < x < 0$ **11.** $x < 1$ or $x > 4$ **13.** $x \le \dfrac{-3}{2}$ or $x \ge 5$ **15.** $x < 0$ or $x > 4$

17. no real numbers   **19.** $-9 \le x \le -4$ **21.** all real numbers   **23.** $x \ne \dfrac{3}{2}$

25. $x < -1$ or $x > 2$ **27.** $x \le \dfrac{-9 - \sqrt{41}}{2}$ and $x \ge \dfrac{-9 + \sqrt{41}}{2}$ **29.** all real numbers

31. $x < 2$ or $x > 5$ **33.** $\dfrac{-3}{2} < x < \dfrac{1}{2}$ **35.** $x \ne -1$ **37.** $-2 < x < 1$ or $x > \dfrac{3}{2}$

39. $x < -6$ or $-4 < x < 0$

Chapter 8 Exercises

1. $\pm\sqrt{3}$ **3.** $\pm\sqrt{5}$ **5.** $\pm\sqrt{6}$ **7.** $\pm\sqrt{b - a}$ **9.** $\pm\dfrac{1}{a}\sqrt{ab(d - c)}$

11. $\pm\dfrac{1}{c}\sqrt{c(a + b)}$ **13.** $9; 5$ **15.** $6; -9$ **17.** $-1; b$ **19.** $\dfrac{-3}{2}; 1$ **21.** $1; \dfrac{-5}{7}$

23. $-7; \dfrac{23}{2}$ **25.** $-1 \pm \sqrt{7}$ **27.** $-4 \pm \sqrt{14}$ **29.** $1 \pm i$ **31.** $\dfrac{5 \pm \sqrt{7}i}{4}$

33. $\dfrac{-3 \pm i}{2}$ **35.** $1 \pm \sqrt{1 + m}$ **37.** $\dfrac{1}{2}; -2$ **39.** $1; \dfrac{-2}{5}$ **41.** $2; \dfrac{-6}{5}$ **43.** $10; -\dfrac{5}{3}$

45. $\dfrac{3 \pm 2\sqrt{2}}{2}$ **47.** $3 \pm \sqrt{6}$ **49.** $\dfrac{1 \pm \sqrt{5}}{6}$ **51.** $\dfrac{1 \pm \sqrt{3}}{4}$ **53.** $\dfrac{1 \pm \sqrt{2}i}{3}$

55. $\dfrac{3 \pm i}{2}$ **57.** $\sqrt{2}; \dfrac{-\sqrt{2}}{2}$ **59.** $\dfrac{-3\sqrt{2} \pm \sqrt{66}}{6}$ **61.** $\dfrac{-5\sqrt{7} \pm \sqrt{21}i}{14}$

63. $a(b \pm \sqrt{b^2 - c})$ **65.** $1; \dfrac{2}{b}$ **67.** real, unequal, rational   **69.** real, unequal, irrational

71. complex conjugates   **73.** real, unequal, irrational   **75.** real, unequal, irrational

77. $\pm 2\sqrt{3}$ **79.** $k < -2\sqrt{10}; k > 2\sqrt{10}$ **81.** $-2\sqrt{10} < k < 2\sqrt{10}$ **83.** $\dfrac{-1}{3}; \dfrac{10}{9}$

85. $x^2 - 7x + 10 = 0$ **87.** $x^2 - 5 = 0$ **89.** $4x^2 - 4x + 5 = 0$ **91.** $x = \pm 2; x = \pm\sqrt{7}$

93. $x = 1$ **95.** $6 - \sqrt{57}$ **97.** $4; 20$ **99.** $1; -3 \pm 2\sqrt{2}$ **101.** $1 < x < 7$

103. all x except $x = 3$ **105.** $-3 < x < \dfrac{5}{6}$ **107.** $\dfrac{(-\sqrt{17} - 1)}{4} < x < \dfrac{(\sqrt{17} - 1)}{4}$

109. $-3 < x \le \dfrac{5}{3}$

CHAPTER 9

Exercises 9.1

1. 5; 3   **3.** 5 or −2   **5.** 15 and 5   **7.** 7 or 9   **9.** 15 and 17   **11.** 17; 18   **13.** 2 or $\frac{1}{2}$
15. 8; 12   **17.** 6; 24   **19.** $\frac{4}{9}$ or $\frac{-9}{4}$ **21.** 49 and 25   **23.** 86   **25.** 574

Exercises 9.2

1. base 18 ft; altitude 24 ft   **3.** 9 in.; 12 in.   **5.** base 27 ft; altitude 15 ft   **7.** length 26 yd;
width 12 yd   **9.** 3 in.   **11.** 5 ft   **13.** 1.6 in.   **15.** 12 in.; 2 in.   **17.** $\frac{b}{2}$ **19.** $4(\sqrt{2} - 1)$

Exercises 9.3

1. 40.5 mph   **3.** 12 mph   **5.** 40 mph   **7.** 6 hr; 5 hr 36 min   **9.** 36 mph or 14 mph
11.(a) 1 sec   **(b)** 2 sec   **13.** 5 sec; 260 ft/sec   **15.** 156 ft; 100 ft/sec   **17.** 4 hr
19. 4 mph

Exercises 9.4

1. John 10 hr; Mark 15 hr   **3.** 15 hr   **5.** 3 hr; 2 hr   **7.** 24 days   **9.** 12 min
11. 24 min; 14.4 min   **13.** 12 wk

Exercises 9.5

1. 40 ft   **3.** 42   **5.** 50 boxes   **7.** 11   **9.** 60 and 20   **11.** woman 24 yr; son 4 yr
13. $\frac{\sqrt{2}}{2}r$ **15.** 16 ft; 11 ft

Exercises 9.6

1. (a) $0 < t < 8$ **(b)** $2 < t < 6$ **3.** (a) at least 50   **(b)** $0 < x < 50$

CHAPTER 10

Exercises 10.1

1. QI   **3.** QII   **5.** QIII   **7.** QIV   **9.** x-axis   **11.** y-axis   **13.** 4   **15.** 7
17. $3\sqrt{2}$ **19.** 8   **21.** 7   **23.** $\sqrt{145}$ **43.** $x = 3; -2$ **47.** (7, 1)   **49.** (0, 1)

Exercises 10.2

13.

15.

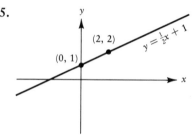

17.

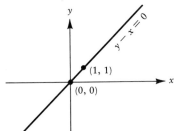

19.

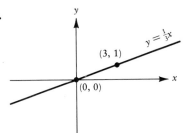

21.

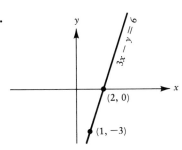

23.

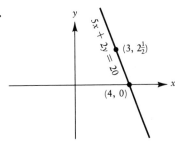

25.

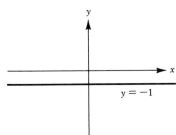

27.

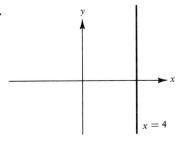

29. y-axis

Exercises 10.3

1. $\dfrac{3}{2}$ **3.** $-\dfrac{5}{3}$ **5.** $\dfrac{4}{3}$ **7.** -2 **9.** 0 **11.** -1 **17.** 6 **19.** -3

Exercises 10.4

1. $x - 1 = 0$ **3.** $x + 11y = 28$ **5.** $3x - 4y = -1$ **7.** $2x + 3y = 5$ **9.** $4x + 3y = -7$
11. $9x - 13y = -84;\ 3x - 7y = -44;\ 3x - 5y = -32$ **13.** $6y - x = 5$ **15.** $x + 2y = 13$

17. $x + 3y = 13$ **19.** $x + y = -1$ **21.** $y = 4x$ **23.** $y = 6x + \dfrac{1}{2}$ **25.** $y = -\tfrac{1}{5}x + 3$

27. $m = \tfrac{5}{2};\ b = -5$ **29.** $m = \tfrac{2}{3};\ b = -2$ **31.** $m = \tfrac{3}{4};\ b = -3$
33. $x - 2y = -8;\ 2x + y = 14$ **35.** $x - 5y = -34;\ 5x + y = 38$ **37.** $x = 0$ **39.** $x = 4$
41. $y = 5$ **45.** $k = \tfrac{-3}{4}$ **47.** $x + y = 4$

Exercises 10.5

1. x-axis; y-axis; origin   **3.** y-axis   **5.** none   **7.** y-axis   **9.** none   **11.** y-axis   **13.** y-axis
15. none   **17.** origin   **19.** y-axis   **21.** x-axis; y-axis; origin   **23.** x-axis; y-axis; origin
25. origin   **27.** origin   **29.** y-axis

Exercises 10.7

1. $(x - 4)^2 + (y - 6)^2 = 9$ **3.** $(x + 2)^2 + (y + 3)^2 = 36$ **5.** $(x + a)^2 + (y + a)^2 = a^2$
7. $(x - 6)^2 + y^2 = 36$ **9.** $(x - 7)^2 + (y - 3)^2 = 2$ **11.** center $(1, -6)$; radius 6 **13.** center
$(0, \frac{1}{3})$; radius $\frac{1}{2}$ **15.** center $(-\pi, 2)$; radius 4 **17.** center $(3, -5)$; radius 6 **19.** imaginary circle
21. point circle **23.** imaginary circle **25.** center $(2, 0)$; radius 4 **27.** $x^2 + y^2 - 3x - 2y = 0$

Exercises 10.8

1. $V(0, 0)$ **3.** $V(0, 0)$ **5.** $V(0, 0)$ **7.** $V(0, 0)$ **9.** $V(0, 0)$ **11.** $V(0, 0)$ **13.** $V(2, 1)$
15. $V(-3, -1)$ **17.** $(y - 1)^2 = 8(x - 3)$ **19.** $(x - 3)^2 = \frac{1}{2}(y + 4)$ **21.** $(y - 3)^2 = 4(x + 2)$
23. $y^2 = 12(x - 1)$

Exercises 10.9

1. $V_1(8, 0)$; $V_2(-8, 0)$ **3.** $V_1(0, 5)$; $V_2(0, -5)$ **5.** $V_1(0, 6)$; $V_2(0, -6)$ **7.** $V_1(4, 0)$; $V_2(-4, 0)$
9. $V_1(0, \frac{\sqrt{115}}{5})$; $V_2(0, -\frac{\sqrt{115}}{5})$ **11.** $V_1(10, 0)$; $V_2(-10, 0)$ **13.** center $(-2, 1)$; $V_1(1, 1)$;
$V_2(-5, 1)$ **15.** center $(2, -2)$; $V_1(6, -2)$; $V_2(-2, -2)$ **17.** center $(1, -2)$; $V_1(5, -2)$; $V_2(-3, -2)$
19. center $(-2, 1)$; $V_1(-6, 1)$; $V_2(2, 1)$ **21.** center $(1, -2)$; $V_1(1, 1)$; $V_2(1, -5)$ **23.** center $(-3, 0)$;
$V_1(-3 + \sqrt{2}, 0)$, $V_2(-3 - \sqrt{2}, 0)$

Exercises 10.10

1. $V(\pm 5, 0)$; $F(\pm 5\sqrt{2}, 0)$; $y = \pm x$ **3.** $V(0, \pm 3)$; $F(0, \pm 3\sqrt{2})$; $y = \pm x$ **5.** $V(\pm 4, 0)$;
$F(\pm 4\sqrt{2}, 0)$; $y = \pm x$ **7.** $V(0, \pm 2)$; $F(0, \pm 2\sqrt{2})$; $y = \pm x$ **9.** $V(0, \pm 5)$; $F(0, \pm\sqrt{34})$; $y = \pm\frac{5}{3}x$
11. $V(\pm 2, 0)$; $F(\pm\sqrt{13}, 0)$; $y = \pm\frac{3}{2}x$ **13.** $V(0, \pm\frac{5}{2})$; $F(0, \pm\frac{5\sqrt{13}}{6})$; $y = \pm\frac{3}{2}x$ **15.** $V(\pm 2, 0)$;
$F(\pm 2\sqrt{26}, 0)$; $y = \pm 5x$ **17.** $16x^2 - 9y^2 = 144$ **19.** $144y^2 - 25x^2 = 3600$
21. center $(1, -1)$; $V_1(1, 1)$; $V_2(1, -3)$; $F_1(1, -1 + \sqrt{5})$; $F_2(1, -1 - \sqrt{5})$; $y = \pm 2(x - 1) - 1$
23. center $(3, 1)$; $V_1(6, 1)$; $V_2(0, 1)$; $F_1(3 + \sqrt{13})$; $F_2(3 - \sqrt{13})$; $y = \frac{2}{3}x - 1$; $y = -\frac{2}{3}x + 3$
25. center $(0, -1)$; $V_1(0, 0)$; $V_2(0, -2)$, $F_1(0, -1 + \sqrt{17})$; $F_2(0, -1 - \sqrt{17})$; $y = \frac{1}{4}x - 1$; $y = -\frac{1}{4}x - 1$
27. center $(3, -7)$; $V_1(3, -3)$; $V_2(3, -11)$; $F_1(3, -7 + 4\sqrt{2})$; $F_2(3, -7 - 4\sqrt{2})$; $y = x - 10$;
$y = -x - 4$ **29.** center $(-2, 3)$; $V_1(-2, 6)$; $V_2(-2, 0)$; $F_1(-2, 3 + \sqrt{13})$; $F_2(-2, 3 - \sqrt{13})$;
$x = \frac{2}{3}(y - 3) - 2$; $x = -\frac{2}{3}(y - 3) - 2$ **31.** center $(1, 0)$; $V_1(3, 0)$; $V_2(-1, 0)$; $F_1(1 + 2\sqrt{5}, 0)$;
$F_2(1 - 2\sqrt{5}, 0)$; $y = 2(x - 1)$; $y = -2(x - 1)$ **33.** $\dfrac{5y^2}{162} - \dfrac{5x^2}{18} = 1$

Chapter 10 Exercises

1. 6 **3.** 5 **5.** $7\sqrt{2}$ **7.** -1 **9.** 1 **11.** (a) cost per calorie (b) miles per hour (c) rate
(d) $\dfrac{1}{\text{meters}}$ **13.** $4y - 3x = 18$ **15.** $3x + 4y = -10$ **17.** $y = 5$ **19.** $2x + y = 0$
21. $y = 7$ **23.** $y = -3x + 5$ **25.** $2y = 3$ **27.** $y = \frac{3}{4}x - 3$ **29.** $3x - 4y = -24$
31. $2x - y = 2$ **33.** $y = 3x - 19$ **35.** ellipse **37.** hyperbola **39.** parabola
41. parabola **43.** parabola **45.** ellipse **47.** hyperbola **49.** parabola
51. no graph (imaginary circle) **53.** point **55.** $x^2 + y^2 - 6x + 4y = 0$
57. $x^2 + y^2 - 8x + 2y - 24 = 0$ **59.** center $(4, -5)$; radius $\sqrt{53}$; real **61.** center $(0, 0)$;
radius 0; point **63.** $V(0, -3)$ **65.** $V(-\frac{181}{24}, -\frac{7}{6})$ **67.** $V_1(5, 0)$; $V_2(-5, 0)$ **69.** $V_1(\frac{5}{3}, 0)$;
$V_2(-\frac{5}{3}, 0)$ **71.** $V(\pm\sqrt{3}, 0)$; $F(\pm\sqrt{7}, 0)$; $y = \pm\frac{2}{\sqrt{3}}x$ **73.** $V(\pm 3\sqrt{5}, 0)$; $F(\pm 7, 0)$; $y = \pm\frac{2\sqrt{5}}{15}x$
75. $V_1(6, -1)$; $V_2(-2, -1)$; $F_1(7, -1)$; $F_2(-3, -1)$; $y = \pm\frac{3}{4}(x - 2) - 1$

CHAPTER 11

Exercises 11.1

1. consistent, independent **3.** consistent, dependent **5.** consistent, independent
7. $x = -3; y = 4$ **9.** $x = -2; y = 3$ **11.** $x = 2; y = 6$ **13.** no solution
15. $x = -4; y = 1$ **17.** no solution **19.** $x = 8; y = 3$

Exercises 11.2

1. $x = 4; y = -4$ **3.** $x = 3; y = 1$ **5.** $x = 4; y = 1$ **7.** $x = 5; y = 2$ **9.** $x = 4; y = -1$
11. infinitely many solutions **13.** $x = -4; y = -3$ **15.** $x = 6, y = 4$ **17.** $x = 1, y = 8$
19. $x = 2; y = -\frac{1}{2}$

Exercises 11.3

1. $x = 3; y = -2$ **3.** $x = 7; y = 3$ **5.** $x = 2; y = -3$ **7.** $x = \frac{5}{2}; y = -1$
9. $x = 4; y = 2$ **11.** $x = 5; y = -3$ **13.** $x = -5; y = -3$ **15.** $x = 5; y = -6$
17. $x = 2; y = 1$ **19.** $x = 4; y = -1$ **21.** $x = 11; y = 19$ **23.** $x = -9; y = 6$
25. $x = -3; y = 1$ **27.** $x = 4; y = 10$ **29.** $x = \frac{2}{3}; y = \frac{-3}{4}$ **31.** $x = 2; y = -1$
33. $x = 41n/23; y = 11n/23m$ **35.** $x = 3b; y = 4a$ **37.** $x = 4a; y = -2b$
39. $x = an/m; y = -b$

Exercises 11.4

1. $x = 1; y = 2; z = 3$ **3.** $x = 2; y = -3; z = 4$ **5.** $x = 2; y = 3; z = -2$
7. $x = 1; y = 3; z = 2$ **9.** $x = 1; y = 2; z = 6$ **11.** $x = 4; y = 0; z = -3$
13. $x = -3; y = -4; z = 2$ **15.** $x = \frac{3}{5}; y = \frac{2}{3}; z = -3$ **17.** $x = 1; y = 2; z = 3$
19. $x = 5; y = -3; z = 2$ **21.** $x = -1; y = 3; z = -2$ **23.** $x = \frac{1}{2}; y = \frac{2}{3}; z = \frac{3}{4}$
25. $x = \frac{1}{2}; y = \frac{1}{3}; z = \frac{2}{3}$ **27.** $x = \frac{-1}{2}; y = \frac{-1}{5}; z = \frac{1}{3}$ **29.** $x = \frac{1}{2}a; y = \frac{1}{3}a; z = \frac{1}{4}a$
31. $x = \dfrac{r - s + t}{2c}; y = \dfrac{r + s - t}{2b}; z = \dfrac{-r + s + t}{2a}$ **33.** $x = 1/a; y = 1/b; z = 1/c$
35. $x = \dfrac{2}{a + b + c}; y = \dfrac{2}{b + c - a}; z = \dfrac{2}{a + c - b}$

Exercises 11.5

1. -1 **3.** 0 **5.** 0 **7.** 8 **9.** $x^2 - 15$ **11.** $-2y + 8x$ **13.** 0 **15.** 0 **17.** $x = 3$
19. $x = \frac{25}{3}$ **21.** $x = 2; y = -3$ **23.** $x = \frac{1}{2}; y = 4$ **25.** no solution **27.** $x = \frac{2}{3}; y = \frac{3}{4}$
29. $x = 4; y = 3$ **31.** $x = \dfrac{ac - b^2}{a^2 - bc}; y = \dfrac{ab - c^2}{a^2 - bc}$ **33.** $x = \dfrac{4a}{b}; y = \dfrac{3b}{a}$
35. $x = \dfrac{n - am}{1 - a}; y = \dfrac{m - n}{a - a^2}$ **37.** $a = -2$

Exercises 11.6

1. -44 **3.** 3 **5.** 23 **7.** 105 **9.** 0 **11.** $-(a - b)(a - c)(b - c)$ **13.** $k = -1$
15. $k = -3$ **17.** dependent equations **19.** no solution **21.** dependent equations
23. $x = 1; y = 2; z = 0$ **25.** $x = \frac{1}{2}; y = \frac{2}{3}; z = \frac{3}{4}$ **27.** $x = \frac{1}{2}; y = \frac{-3}{29}; z = \frac{-1}{10}$
29. $x = -5/a; y = 19/b; z = 13/c$ **33.** $K = 6$ **35.** $K = \dfrac{ab}{2}$

Exercises 11.7

1. 8; 4   **3.** 37; 11   **5.** 20; 18   **7.** 20; 25; 30   **9.** $\frac{1}{3}$; $\frac{3}{2}$; 3   **11.** 45   **13.** 308
15. 437   **17.** 43   **19.** 38   **21.** $\frac{3}{8}$ **23.** $a = 11$; $b = -3$; $c = 1$

Exercises 11.8

1. crew 7.5 mph; current 1.5 mph   **3.** distance 13.5 miles; rate 4.5 mph
5. wind 27 mph; plane 402 mph   **7.** car 10 kph; truck 12.5 kph
9. 150 yards; 30 yards per minute; 20 yards per minute   **11.** 4 mph walking; 3 mph rowing

Exercises 11.9

1. A $10\frac{1}{2}$ hr; B 14 hr   **3.** Tom 50 days; Jerry 75 days   **5.** A 16 days; B 24 days; C 40 days
7. A 12 hr; B 9 hr; C 8 hr

Exercises 11.10

1. 3.6 qt 45% acid; 2.4 qt 70% acid   **3.** $3\frac{3}{5}$ gal from 1st tank; $2\frac{2}{3}$ gal from 2nd tank

5. 6 qt 20% solution; 8 qt 30% solution   **7.** $\dfrac{p(c - b)}{a - b}$ oz a carat fine; $\dfrac{p(a - c)}{a - b}$ oz b carat fine
9. in A, 64%; in B, 32%

Exercises 11.11

1. 13°; 77°   **3.** 15 ft; 21 ft   **5.** base = 5 in.; altitude = 12 in.   **7.** 99°; 27°; 54°
9. $AC = 8$, $CD = 4$, $BE = 6$

Exercises 11.12

1. \$15,000   **3.** \$1800 at 5%; \$1200 at 8%   **5.** radio, \$150; tape recorder, \$50
7. cabin, \$9000; \$5000 at 5%; \$4000 at 6%   **9.** \$5000 at 3%; \$1000 at 4%; \$4000 at 5%

Exercises 11.13

1.

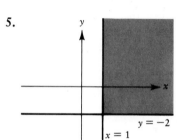

3.

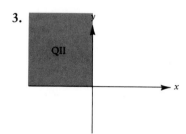

5.

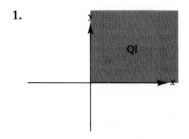

7.

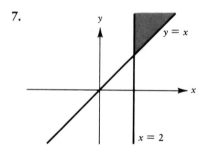

9.

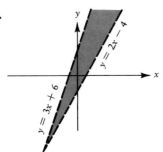

11.

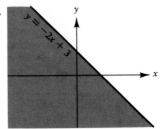

13.

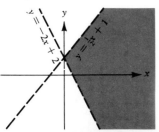

15.

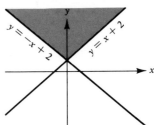

17.

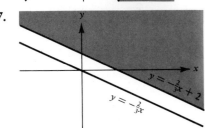

19.

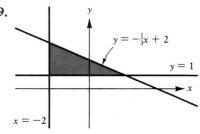

Chapter 11 Exercises

1. consistent; independent   **3.** inconsistent; independent   **5.** consistent; independent
7. $x = -2;\ y = 4$ **9.** $x = 2;\ y = -3$ **11.** $x = \frac{1}{2};\ y = \frac{-3}{2}$ **13.** $x = \frac{-2}{5};\ y = \frac{3}{5}$
15. $x = 7;\ y = 11$ **17.** $x = a - b;\ y = a - b$ **19.** $x = \frac{-3}{2};\ y = \frac{1}{4}$
21. $x = \dfrac{a^2 + b^2}{a};\ y = \dfrac{a^2 + b^2}{b}$ **23.** $x = 1;\ y = 2;\ z = 3$ **25.** $x = \frac{3}{2};\ y = -2;\ z = \frac{1}{2}$
27. $x = \frac{1}{2};\ y = \frac{-1}{3};\ z = \frac{1}{4}$ **29.** 200   **31.** -9 **33.** 0   **35.** no solution
37. $x = c/2;\ y = c/3$ **39.** $x = 2a + b;\ y = a + 2b$ **41.** $x = -4;\ y = \frac{3}{2};\ z = 5$
43. $x = 1 \pm 2\sqrt{2}\,i$

CHAPTER 12

Exercises 12.1

1. $\log_2 8 = 3$ **3.** $\log_7\left(\dfrac{1}{49}\right) = -2$ **5.** $\log_3\left(\dfrac{1}{27}\right) = -3$ **7.** $\log_3\left(\dfrac{1}{81}\right) = -4$ **9.** $\log_{16} 8 = \dfrac{3}{4}$

11. $\log_{27} 9 = \dfrac{2}{3}$ **13.** $\log_{16}\left(\dfrac{1}{32}\right) = \dfrac{-5}{4}$ **15.** $\log_{17} 1 = 0$ **17.** $\log_{49} 7 = \dfrac{1}{2}$ **19.** $\log_a c = b$

21. $3^6 = 216$ **23.** $10^{-2} = 0.01$ **25.** $\pi^0 = 1$ **27.** $15^2 = 225$ **29.** $\left(\dfrac{1}{2}\right)^{-4} = 16$

31. $b^{-2} = 7$ **33.** $4^1 = 4$ **35.** $10^y = 72$ **37.** $c^0 = 1$ **39.** $c^a = b$ **41.** 2   **43.** 4

45. 2   **47.** 4   **49.** 3   **51.** -1 **53.** $\dfrac{2}{3}$ **55.** -3 **57.** $\dfrac{1}{4}$ **59.** 4   **61.** $N = 64$

63. $N = 1024$ **65.** $N = \dfrac{1}{8}$ **67.** $N = \dfrac{1}{16}$ **69.** $N = 13$ **71.** $b = 10$ **73.** $b = 7$

75. $b = 100$ **77.** $b = 121$ **79.** $b = 1$ **81.** 8   **83.** 1   **85.** $\dfrac{1}{10}$ **87.** $\dfrac{-9}{10}$ **89.** $\dfrac{18}{11}$

Exercises 12.2

1. 1.7993   **3.** 2.1461   **5.** 3.0212   **7.** 0.3222   **9.** 0.2386   **11.** 1.1611   **13.** 0.8368
15. 1.3316   **17.** -1.6565 **19.** 6.3009   **21.** $\log P + \log Q + \log R$

23. $\log C - \log A - \log B$ **25.** $3 \log A + 5 \log B$ **27.** $-n \log A$ **29.** $\dfrac{-1}{n} \log c$

31. $n \log M + \dfrac{1}{m} \log R - 5 \log S$ **33.** $\dfrac{1}{5}[\log M + \log N - \log Q]$

35. $\dfrac{1}{2}[2 \log x + 5 \log y - \log z]$ **37.** $\log_c \left(\dfrac{z^3}{x^2}\right)$ **39.** $\log_2 \left(\dfrac{30}{7}\right)$ **41.** $\log_e \left(\dfrac{\pi \sqrt[3]{3}}{x}\right)$

43. $\log_b \left(\dfrac{2\pi \sqrt{l}}{\sqrt{g}}\right) = \log_b \left(2\pi \sqrt{\dfrac{l}{g}}\right)$ **45.** $\log_5 \sqrt[12]{\dfrac{28^3 \cdot 14^2}{7^6 \cdot 2^4}}$ **47.** $\log_b (k\,!)$

49. false: $2 \log x = \log x^2$ **51.** false: $\log x - \log y = \log \left(\dfrac{x}{y}\right)$

53. false: $\log (xy)^4 = 4[\log x + \log y]$ **55.** true   **57.** true   **59.** true

Exercises 12.3

1. $x = \dfrac{7}{2}$ **3.** $x = 25$ **5.** $x = 12$ **7.** $x = 3$ **9.** $x = \dfrac{5}{7}$ **11.** $x = \dfrac{41}{28}$ **13.** $x = \dfrac{5}{3}$
15. $x = 2;\ 8$ **17.** $x = 0$ **19.** $x = 2$ **21.** $x = 0;\ 2;\ 9$ **23.** $x = \pm 1;\ \pm\sqrt{6}$ **25.** $x = 4$
27. $x = 31$ **29.** $x = 2$

Exercises 12.4

1. 1   **3.** 3   **5.** 0   **7.** 3   **9.** 1   **11.** -1 **13.** -1 **15.** -3 **17.** -1 **19.** -1
21. 0   **23.** -2 **25.** 1   **27.** 2   **29.** 0.5105   **31.** 2.6946   **33.** 3.4099   **35.** 1.9494
37. 0.5490   **39.** 2.9877   **41.** $-3 + 0.8274$ **43.** 1.8420   **45.** 8.8082   **47.** 4.2967
49. 3.8876   **51.** 5.7528   **53.** 6.9031   **55.** -4 **57.** $-6 + 0.2833$ **59.** $-4 + 0.7589$
61. 331   **63.** .191   **65.** 804   **67.** .00784   **69.** 340,000   **71.** 2,740   **73.** .4440
75. 86.1   **77.** 0.03070   **79.** 94,300   **81.** 0.8787   **83.** 6.1723   **85.** $-2 + 0.9710$
87. 2.1749   **89.** 4.8063   **91.** 12.5529   **93.** 2.7260   **95.** $-8 + 0.9282$ **97.** 3.9595
99. 1.3318   **101.** 6.554   **103.** 3.002   **105.** 0.0002014   **107.** 3.965   **109.** 19,960
111. 0.0003985

Exercises 12.5

1. 1.6094   **3.** 5.5722   **5.** 3.7209   **7.** -1.0986 **9.** 1.2825   **11.** -1.5325
13. 2   **15.** 2.6   **17.** 97   **19.** 0.94   **21.** 6.6   **23.** 0.0765   **25.** 0.6213   **27.** 0.9358
29. -1.4083 **37.** 2   **39.** 16

Exercises 12.6

1. 51.14   **3.** 1409   **5.** 1.023   **7.** 4   **9.** 1; 2   **11.** 1.585   **13.** 0.8451   **15.** 1.0043

17. 1.197   **19.** 3; 1   **21.** $x = \dfrac{1}{b} \ln\left(\dfrac{y}{a}\right)$ **23.** $n = 1 + \dfrac{\log(l/a)}{\log r}$ **25.** $n = \dfrac{\log(A/P)}{\log(1 + r)}$

27. 7.32 years   **29.** 55.45 years   **31.** 34   **33.** approx. 17.8 years   **35.** 17.4 years
37. 4.7%   **39.** 11.8   **41.** 3.16×10^{-9} **43.** 1.55 db loss   **45.** 3.47 watts

Chapter 12 Exercises

1. $\log_2 4 = 2$ **3.** $\log_{10} 0.01 = -2$ **5.** $\log_{10}(0.001) = -3$ **7.** $\log_{15} 225 = 2$

9. $\log_2\left(\dfrac{1}{8}\right) = -3$ **11.** $2^5 = 32$ **13.** $10^3 = 1000$ **15.** $11^{-2} = \dfrac{1}{121}$ **17.** $10^{1/2} = \sqrt{10}$

19. $2^{-5} = \dfrac{1}{32}$ **21.** -4 **23.** -3 **25.** -2 **27.** $\dfrac{1}{3}$ **29.** 3   **31.** $x = 8$ **33.** $x = 64$

35. $x = (1.5)^2$ **37.** $x = 32$ **39.** $x = 15$ **41.** $b = 3$ **43.** $b = 8$ **45.** $b > 0, b \neq 1$

47. $b = \dfrac{1}{27}$ **49.** $b = 8$ **51.** $\log 7$ **53.** $\ln(xy)$ **55.** $\log_a\left(\dfrac{\sqrt{x}\sqrt[3]{y}}{\sqrt[4]{z}}\right)$ **57.** $\log \sqrt{x^2 - y^2}$

59. $\dfrac{1}{2}\{\log s + \log(s - a) + \log(s - b) + \log(s - c)\}$ **61.** $x = -1$ **63.** $x = \dfrac{5}{3}$ **65.** $x = ab^n$

67. -0.9654 **69.** 1.7870   **71.** 8.306×10^8 **73.** 0.0017   **75.** 0.0100   **77.** 2.2061
79. -1.5325 **81.** 2.807   **83.** 0.8855   **85.** 1.395   **87.** -3.562 **89.** 1.203
91. 10.7 years   **93.** 18.24 centuries   **95.** 6×10^{-4} **97.** 14   **99.** 3.3 hr

CHAPTER 13

Exercises 13.1

1. function   **3.** not a function   **5.** not a function
7. domain = $\{-1, 0, 2, 3\}$; range = $\{2, 5, 4, -1\}$ **9.** domain = $\{2, 3\}$; range = $\{3, 4, 5, 6\}$
11. domain = $\{4\}$; range = $\{2, 3, 4, 5\}$ **13.** function   **15.** function
17. not a function   **19.** not a function   **21.** function
23. $g(1) = 5; g(-1) = -1; g(b) = b^2 + 3b + 1; g(a - b) = (a - b)^2 + 3(a - b) + 1$
25. $p(3) = 0; p(-3) = 0; p(a) = \sqrt{a^2 - 9}; p(x + 1) = \sqrt{(x + 1)^2 - 9}$
27. $s(0) = 0; s(4) = 0; s(-4) = 8; s(x + h) = |x + h| - (x + h)$
29. $g(0) = 0; g(2) = 0; g(a) = a^2(a - 2); g(a + b) = (a + b)^2(a + b - 2)$

31. 1   **33.** 0   **35.** $\dfrac{1}{\sqrt{x + h} + \sqrt{x}}$ **37.** $D_g = \{x \mid x \text{ is real and } x \neq 2\}$

39. $D_h = \{x \mid x \text{ is real and } x \neq 0\}$ **41.** $D_f = \left\{x \mid x \text{ is real, } x \geq \dfrac{4}{3}, x \neq 3\right\}$

43. $C = 2\pi r; C = \pi d$ **45.** $A = \dfrac{1}{2}bh; p = b + \sqrt{b^2 + 4h^2}$

Exercises 13.2

1. $F^{-1} = \{(2, 1), (3, 2), (6, 3), (8, 4)\}$ **3.** $H^{-1} = \{(2, 2), (5, 3), (10, 4), (2, 0), (5, -1)\}$
5. $S^{-1} = \{(5, -1), (2, 0), (1, 1), (2, 2), (5, 3)\}$ **7.** F is a function; F^{-1} is a function
9. H is a function; H^{-1} is not a function   **11.** S is a function; S^{-1} is not a function

13. $y = f^{-1}(x) = \dfrac{x - 6}{2}$; domain = $\{x \mid x \text{ is real}\}$ **15.** $y = f^{-1}(x) = \dfrac{x - 8}{7}$; domain = $\{x \mid x \text{ is real}\}$

17. $y = f^{-1}(x) = \dfrac{x + 2}{x}$; domain = $\{x \mid x \text{ is real and } x \neq 0\}$

19. $y = f^{-1}(x) = x^2$; domain = $\{x \mid x \geq 0\}$ **21.** $y = f^{-1}(x) = x^2 + 3$; domain = $\{x \mid x \leq 0\}$
23. $y = f^{-1}(x) = x^2 - 2x - 1$; domain = $\{x \mid x \leq 1\}$

25. $y = f^{-1}(x) = \sqrt{1 - 4x^2}$; domain $= \left\{x \mid x \text{ is real and } 0 \le x \le \dfrac{1}{2}\right\}$

27. $y = f^{-1}(x) = \dfrac{x - b}{m}$, $m \ne 0$ **29.** $y = f^{-1}(x) = \sqrt[5]{x + 6}$

Exercises 13.3

1. $x = 12$ **3.** $d = 4$ **5.** $t = 96$ **7.** $l = 12$ **9.** $s = 144$ **11.** 16 ft   **13.** $S = 4\pi r^2$

15. $V = \dfrac{7A}{9}$ **17.** 0.29 sec

Exercises 13.4

1. $x = \dfrac{k}{y}$ **3.** $p = \dfrac{k}{q^4}$ **5.** $p = \dfrac{k}{n}$ **7.** $D = \dfrac{k}{V}$ **9.** $F = \dfrac{k}{d^2}$ **11.** $W = 10{,}000$ **13.** $r = 120$

15. 12 women   **17.** 2.5 lb/sq in.   **19.** 0.01 dynes; 100 dynes   **21.** $2\frac{2}{3}$ hr; 24 mph

Exercises 13.5

1. $L = \dfrac{kt^4}{l^2}$ **3.** $w = \dfrac{k}{d^2}$ **5.** $f = \dfrac{k\sqrt{t}}{l\sqrt{m}}$ **7.** $r = 40$ **9.** $a = 56$ **11.** $r = 270$

13. $x = 21$ **15.** $T = \dfrac{160x^2}{y^3z}$; $T = \dfrac{125}{6}$ **17.** (a) 1000 amperes; (b) 0.2 ohm   **19.** 210 ft^3

Exercises 13.6

1. $y = 6$ **3.** $y = -\dfrac{7}{2}$ **5.** $y = -1$ **7.** up   **9.** down   **11.** 1   **13.** $\dfrac{1}{2}$ **15.** -5

17. 1.5   **19.** $f(x) = x$ **21.** $\left(-\dfrac{7}{5}, \dfrac{9}{5}\right)$ **23.** $\left(\dfrac{3}{2}, -\dfrac{3}{2}\right)$ **25.** downward; maximum at $\left(\dfrac{3}{8}, \dfrac{25}{16}\right)$

27. upward; minimum at $(0, 0)$ **33.** $s = 48t - 16t^2$; (a) $t = 1.5$ sec; (b) 36 ft; (c) $t = 3$ sec

35. 146 boats   **37.** $x = 5$ in., $y = 4$ in.; Area $= 20$ in.^{2}

Exercises 13.7

1. $f(x) = \left(\dfrac{1}{2}\right)^x$ **3.** $f(x) = \left(\dfrac{1}{4}\right)^x$ **5.** $f(x) = 4^x$ **7.** $f(x) = \left(\dfrac{1}{7}\right)^x$ **9.** $f(x) = \left(\dfrac{1}{3}\right)^x$

11. increasing   **13.** decreasing   **15.** increasing   **17.** neither   **19.** decreasing

21. reflection in x-axis   **23.** f eventually grows faster than g **25.** $a = 7$, $n = 2$ **27.** $b = 2$

29. $b = 4$ **31.** $b = 4$ **33.** $b = 9$ **35.** $b > 0$, $b \ne 1$ **37.** increasing   **39.** decreasing

41. decreasing

Chapter 13 Exercises

1. $f(0) = 1$; $f(1) = 5$; $f(-1) = 9$; $f(a + b) = 6(a + b)^2 - 2(a + b) + 1$

3. $f(1) = 3$; $f(5) = 17$; $f(10) = 33$; $f(a + b) = \sqrt{a + b - 1} + 3(a + b)$ **5.** -3

7. $2x + h - 4$ **9.** 0   **11.** $D_f = \{x \mid x \text{ is real}\}$ **13.** $D_f = \{x \mid x \ge -5\}$

15. $D_f = \{x \mid x \le -2 \text{ or } x \ge 2\}$ **17.** $D_f = \{x \mid x \text{ is real}, x \ne 5, -4\}$

19. $D_f = \{x \mid x \ge 0, x \ne 3\}$ **21.** $D_f = \{x \mid x \ge -1\}$ **23.** (a) function; (b) not a function;

(c) function   **25.** $b = 0$ **27.** $u = 81$ **29.** $c = 1440$ **31.** $w = 27$

33. $y = 20\frac{1}{4}$ **35.** x varies inversely as u **37.** $y = 2x + 1$ **39.** minimum

41. maximum   **43.** maximum   **45.** $V = (1, -4)$ **47.** $V = (1, 1)$ **49.** $V = \left(\frac{7}{2}, -\frac{25}{4}\right)$

53. $k = -4$ or 2 **55.** $t = 24$ sec; $h = 736$ ft   **57.** decreasing   **59.** increasing

61. $I = \left(\frac{1}{4}\right)^x$ **63.** decreasing   **65.** increasing   **67.** $b = 3$

Index

ENGLISH–METRIC CONVERSIONS

LENGTH

1 inch = 2.540 centimeters
1 foot = 30.48 centimeters
1 yard = 0.9144 meter
1 mile = 1.609 kilometers

VOLUME

1 pint = 0.4732 liter
1 quart = 0.9464 liter
1 gallon = 3.785 liters

WEIGHT

1 ounce = 28.35 grams
1 pound = 453.6 grams
1 pound = 0.4536 kilogram

LENGTH

1 centimeter = 0.3937 inch
1 centimeter = 0.0328 foot
1 meter = 1.0936 yards
1 kilometer = 0.6215 mile

VOLUME

1 liter = 2.1133 pints
1 liter = 1.0567 quarts
1 liter = 0.2642 gallon

WEIGHT

1 gram = 0.0353 ounce
1 gram = 0.002205 pound
1 kilogram = 2.205 pounds